Aquatic Biodiversity II

Developments in Hydrobiology 180

Series editor

K. Martens

Aquatic Biodiversity II

Edited by

H. Segers & K. Martens

Royal Belgian Institute of
Natural Sciences, Belgium

Reprinted from Hydrobiologia, volume 542 (2005)

Springer

Library of Congress Cataloging-in-Publication Data

A C.I.P. Catalogue record for this book is available from the Library of Congress.

ISBN 978-90-481-6951-1 e-ISBN 978-1-4020-4111-2

Published by Springer,
P.O. Box 17, 3300 AA Dordrecht, The Netherlands

Printed on acid-free paper

TABLE OF CONTENTS

Hydrobiologia (2005) 542:ix–x
H. Segers & K. Martens (eds), Aquatic Biodiversity II
DOI 10.1007/s10750-005-0886-x

The future of freshwater biodiversity research: an introduction to the target review

Hendrik Segers
Freshwater Biology, Royal Belgian Institute of Natural Sciences, Vautierstraat 29, 1000 Brussels, Belgium
E-mail: hendrik.segers@naturalsciences.be

During August 2003, an international meeting was organized to mark the handing over of editorial responsibilities over *Hydrobiologia* from Henri Dumont to Koen Martens. Such a milestone in the history of this important scientific journal with aquatic research as scope is a perfect opportunity to reflect on the future directions that this research should, or probably shall, take. Indeed, science and scientific interests not only evolve as a result of inherent drivers like the development of novel paradigms and the application of new techniques, but also from shifts in societal demands towards research: as the human impact, and reliance on natural resources increases, so does the need for a scientifically sound and sustainable management and conservation of these resources increase.

The Belgian Biodiversity Platform, an initiative by the Belgian Science Policy Office, offers scientists and other stakeholders in research a forum to exchange views on biodiversity issues, amongst other services (see http://www.biodiversity.be). Taking advantage of the August 2003 'Aquatic Biodiversity: Past, Present, Future' (Antwerp, Belgium) meeting, the Platform's forum on freshwater biodiversity initiated a discussion on the future direction of freshwater biodiversity research by inviting a target paper, and organizing a workshop on the topic. The following contributions in this section reflect the lively debate generated by this initiative.

In their target paper, De Meester & Declerck (2005) highlight several priorities in freshwater biodiversity research, in a way that takes into account both short-and long-term scales. Their plea for scientific rigor in research, whether fundamental or applied, is supported by all contributors, and reflects the conviction that only high-standard research combined with research management (van der Werf, 2005) can lead to reliable insights.

Interestingly, several of the contributions (e.g. Franklin, 2005; Gopal, 2005a, b) illustrate the need for improved communication at all scales, and understanding between scientists, policy makers and other actors in biodiversity issues. This was exactly the motivation of the Belgian Science Policy to initiate the Biodiversity Platform (see van der Werf, loc. cit.). However, finding an optimal balance between fundamental, bottom-up and science-based vs. applied, top–down or policy-oriented research remains difficult. Whereas De Meester & Declerck (loc. cit.) stress the importance and scientific merit of fundamental ecological research, Franklin (loc. cit.), Gopal (loc. cit. a), Lévêque & Balian (2005) and van der Werf (loc. cit.) highlight the high relevance of more applied research, with a strong focus on conservation needs, not only for society but also for science. Martens & Segers (2005), on the other hand, argue that a renewed effort towards fundamental and applied taxonomic research, as reflected by the Global Taxonomy Initiative (GTI) of the Convention of Biological Diversity (CBD), is called for. Clearly, this debate is by no means exhausted and will undoubtedly remain vigorous. However, it is reassuring to note that, whatever the position of the contributors, all defend the idea that the importance of both approaches calls for a well-guarded balance between them.

Lévêque & Balian (loc. cit.) argue that many ecosystems, especially in temperate regions, are historically relatively young hence immature, being inhabited by almost haphazardous assemblages of organisms. Ecologists should be aware of this, and accordingly should set realistic goals for research on, and the conservation of, such systems.

Further interesting thoughts on the valuation and economic value of biodiversity are developed by Dumont (2005).

While some contributions contain explicit recommendations for research priorities, some also reflect on the science policy and, more specifically, on the structure of research funding opportunities. De Meester & Decerck (loc. cit.) identify a bias in EU programs towards large networks of big institutes and large-scale projects. Whether this policy is justified or not is largely left in limbo, but they argue that national or regional funding agencies should counter the potential risks of this approach (loss of expertise in smaller consortia, few funding opportunities for truly innovative research) by developing an approach that is complementary to that of the EU. Sturmbauer (2005) largely agrees to this, but goes further in stressing the importance of a science policy that is independent from, and in some aspects even opposite to, the strategy of the EU.

In all, it was a gratifying experience to witness the wide range of views expressed during the debate. It would be naïve to think that a single workshop could resolve the problems raised, as a continued and adaptable effort is required to answer many of the needs identified. However, I sincerely hope all contributors will agree that the workshop discussions, and this target review, can be instrumental to the development of a science policy that answers appropriately to the challenges posed by the present-day biodiversity crisis.

Acknowledgements

I gratefully acknowledge the support of the Belgian Science Policy Office to the organization of the workshop and target review, and Koen Martens and Springer publishers for the opportunity to publish this target review in *Hydrobiologia.*

References

De Meester, L. & S. Declerck, 2005. The study of biodiversity in freshwater habitats: societal relevance and suggestions for priorities in science policy. Hydrobiologia 542: 1–9.

Dumont, H. J., 2005. Biodiversity: a resource with a monetary value? Hydrobiologia 542: 11–14.

Franklin, A., 2005. Linking science and policy for biodiversity. Hydrobiologia 542: 15–17.

Gopal, B., 2005a. Relevance and policy dimensions of research on biodiversity in freshwater ecosystems: a developing country perspective. Hydrobiologia 542: 19–21.

Gopal, B., 2005b. Does inland aquatic biodiversity have a future in Asian developing countries? Hydrobiologia 542: 69–75.

Lévêque, C. & E. V. Balian, 2005. Conservation of freshwater biodiversity: does the real world meet scientific dreams? Hydrobiologia 542: 25–26.

Martens, K. & H. Segers, 2005. Taxonomy and systematics in biodiversity research. Hydrobiologia 542: 27–31.

Sturmbauer, C., 2005. Comment on 'The study of biodiversity in freshwater habitats: societal relevance and suggestions for priorities in science policy' by Luc De Meester & Steven Declerck. Hydrobiologia 542: 33–34.

van der Werf, A., 2005. Towards a coherent and high-quality science policy on biodiversity. Hydrobiologia 542: 35–37.

Hydrobiologia (2005) 542:1–9
H. Segers & K. Martens (eds), Aquatic Biodiversity II
DOI 10.1007/s10750-005-0887-9

Opinion paper (target review)

The study of biodiversity in freshwater habitats: societal relevance and suggestions for priorities in science policy

Luc De Meester* & Steven Declerck
Lab. Aquatic Ecology, KULeuven, Ch. De Beriotstraat 32, 3000 Leuven, Belgium
*(*Author for correspondence: Tel.: +32-16-323708, Fax: +32-16-324575, E-mail: Luc.DeMeester@bio.kuleuven.ac.be)*

What should be the priorities in biodiversity research in freshwater habitats? The present target review builds upon a discussion text on this issue that was written upon invitation by Hendrik Segers. The text served as a preparation for a discussion session during the workshop 'the future of biodiversity studies in freshwater habitats', held in honour of Henri Dumont in Beveren, August 2003. The present text is influenced by some of the more striking remarks and ideas that have been formulated during this discussion session, which inspired us to consider different viewpoints. We gratefully acknowledge Hendrik Segers for his invitation to write the discussion paper, the attendants of the workshop for the interesting and lively discussion, and Dr Gopal for giving an impressive and realistic outline of the problems faced by biodiversity research in developing countries. We also want to acknowledge financial support from EU-project BIOMAN EVK2-CT-1999-00046, which got us fully involved in the field of biodiversity research.

In the following text, we first set the scene by positioning biodiversity research within the broader framework of measures and strategies to deal with the biodiversity crisis and sustainable development. We subsequently give an outline of the aspects we think are important from a purely scientific point of view, but with attention for the generation of relevant scientific information needed for the safeguarding of biodiversity. Finally, we discuss a few aspects related to science policy, with special attention to the tension between academic and applied science. Even though some statements are inspired by the discussion session during the workshop, we like to stress that this text mainly reflects our personal opinion. The first version of this opinion paper did not contain any references; during revision, we were asked to include approximately 10 key references.

The relevance of biodiversity research for the management of the biodiversity crisis

How do we cope with the biodiversity crisis, and how can science contribute to its management? One of the basic problems we are facing is that, on a global scale, political decisions are most often inspired by economic motivation (e.g. export), irrespective of improved laws and scientific knowledge. Such decisions have dramatic consequences for biodiversity. During the workshop, Dr Gopal in his lecture 'nicely' illustrated this overwhelmingly sad situation, and illustrated with many examples that scientific knowledge on biodiversity often does not have any power against the laws of economy, where natural resources and the richness of biota are often the victim of the trade-off between protection and the argument of 'feeding the poor'. So we can ask the question: is science on biodiversity necessary and of highest priority?

No doubt, the argument of 'feeding the poor' is a very strong one. It is sometimes abused in the sense that it is used as a synonym for economic development in terms of export which does not always feed back to the benefits of the majority of the poor, local people. If feeding the poor is done in a non-sustainable way, it is clear that it will only help for a short time, and that it may worsen the situation on a longer term, precisely because natural resources have been destroyed. Nevertheless, in practice, the potential long-term benefits of biodiversity conservation are most often ignored, because of the trade-off with the immediate need for the exploitation of natural resources. The sad thing, so often exemplified, is that natural resources are too heavily exploited and that their actual value becomes clear only after they have been destroyed. One then starts to realize that their destruction only temporally supported needs, needs that, however, most often remain unresolved...

No doubt, science on biodiversity can be useful in this context. There is both a need for a good biodiversity science and for good biodiversity scientists (see further). It should, however, also be stressed that many of the problems we are faced with can actually be solved without any additional science on biodiversity itself. Solving the most crucial of problems involves political decisions, which implies a strong motivation to imply true sustainable development combined with political courage. Second, more than data on biodiversity itself, there is a need for technical solutions to the problems. As Henri Dumont formulated it in his final comment during the discussion, we need ideas and solutions to reduce the ecological footprint of people. As the global size of the human population is growing, there is a reduced potential maximal footprint that can be allocated to each person. Reducing population growth combined with technical solutions to live a decent life without having too high an impact on the environment is the only way out. In essence, this often requires technical, engineering solutions. So political courage and technical solutions for the dilemmas we are faced with are the highest priority. However, good scientists on biodiversity, as good environmental scientists in general, are badly needed as they provide data that keep politicians and engineers focused on these issues.

It should be very clearly stated that it is not a good idea to wait to implement proper political decisions and the search for technical solutions because the 'data on biodiversity are incomplete'. Biodiversity researchers should not promote the need for their data to the extent that the lack of data can be used as an excuse not to act now. Courageous political decisions and solutions to deep-cutting dilemmas between economical development and the use of natural resources (e.g. in agriculture, forestry, fisheries) should be implemented as efficiently as possible. Neither 'applied' nor 'academic' science in biodiversity is going to solve the problem. In the meantime, however, science on biodiversity can continue to contribute significantly by providing data that allow to fine-tune and improve the efficiency of the implementation of policies and technical solutions. So the lack of better data should not prevent action, but the collection of better data should be a high priority in order to improve the efficiency of action as soon as possible. We think it is important that this distinction is made.

Scientists have a bad track record in the sense that they often promote their own scientific endeavour, in an effort to attract money for their research group. Within their field, they are the obvious experts, and if they claim that their field is important, who can argue against that? Admittedly, it is realistic to say that scientists sometimes over-emphasize the importance of their own field, inspired by the way funding agencies operate. This is a bad situation, as this may feed the growing perception among non-scientists that scientists only want to promote their own position. Consequently, doubt arises about whether their claims need to be taken seriously. It is clear that many people actually believe that major environmental crises (e.g. global climate change, the biodiversity decline, pollution problems) are perhaps not as bad as depicted by scientists. This is inspired by the combination of the existence of controversies among scientists – a normal scenario for scientists but somewhat difficult to grasp for many non-scientists ('if global climate change is a fact and so overwhelmingly important, how can there be so much discussion and uncertainty about it?') – and the 'suspicion' that scientists overemphasize the problems. Many policy-makers also have this feeling, which is actually one of the reasons why they so strongly emphasize policy-driven research. Unfortunately, this emphasis on policy-driven research may just make things worse, as it stimulates scientists to emphasize the importance of their field of expertise in solving societal problems, and actually selects for scientists that are good in public relations. By itself, it is good for scientists to be good in public relations, but it should be realized that people that are good scientists and good communicators ànd good managers and good politicians at the same time are exceedingly rare...

Science: biodiversity in freshwater habitats

Let us first state that, purely from an ecosystem services to society point of view, the study of freshwater habitats is of utmost importance. Freshwater habitats are among the most threatened and valuable ecosystems we have. Although there is a continued need for describing patterns of biodiversity in freshwater systems, many descriptive studies have already been carried out, so we feel that there is a need for more encompassing and novel approaches. Important and innovative science can be conducted by, for instance, focusing on the issues listed below. In general, we would strongly advocate research that either develops one of the following issues in great detail or combines several of them for a specific model habitat type. The reader will notice that the structure is so that several issues are stressed repeatedly but viewed from different angles – we see this as an illustration of the fact that these issues are indeed important, and that there are different aspects to it. The order in which we deal with the different issues is not reflecting their importance. Rather, there is an underlying logic from more descriptive issues to more mechanistic approaches to the study of biodiversity. Obviously, the list is not exhaustive – after all, it is a discussion text. Our text no doubt is influenced by the fact that we see biodiversity in an ecological context. As such, we did not focus on taxonomic research, even though it is clear that proper taxonomic knowledge is important in collecting the data needed for the questions below.

- An *ecosystem, multigroup or multitrophic* approach: instead of focusing on just one group of model organisms, it can be highly revealing to study patterns of biodiversity at different trophic levels simultaneously: are the patterns related, what are the consequences for quantifying biodiversity at a macro-level and in terms of nature conservation (Declerck et al., in press). Obviously, the study of bacterial diversity may here shed a novel and strongly different light on patterns generally reported.
- *Attention for often-ignored habitat types*: Most studies in the world are being carried out on typical ecosystems such as deep lakes, or streams and rivers. There is a need of attention for specific types of habitats that are often ignored in classical monitoring studies. Typical and understudied habitats are small ponds and pools, ephemeral waters, ephemeral swamps, ponds and lakes in floodplain areas, ... These often smaller systems can also be profitably used as model systems to study specific aspects (e.g. connectivity, metapopulation dynamics, time stress, organism specific dispersal capacities (e.g. Rundle et al., 2002).
- *Attention for often-ignored components of the biota*: By including taxonomic or (better) functional groups that are often ignored, one can contribute to a better understanding of patterns of biodiversity and its consequences. In freshwater systems, such neglected components could be the microbial community in sediments, in the periphyton, and in the plankton. Furthermore, the resting stages of various groups that are present in sediments are an important component to be studied. Resting propagule banks tend to integrate population or community variability in time and space, and may contribute to the improvement of biodiversity assessment techniques (Vandekerkhove, 2004). In addition, resting egg banks are of immediate interest to nature conservation because they may be a prerequisite for the re-establishment of rare or 'extinct' populations following restoration measures. Note that these aspects have also direct relevance to the framework of the Water Directive, in that they may provide better indicator systems to monitor water quality than the ones suggested at this stage. Especially for standing waters, this may be an important contribution. A special and, in our opinion, important issue is the potentially evil side of diversity: it is quite important to engage in studies on the biodiversity of pathogens in relation to the diversity of the habitat, other biota, etc.
- *Genetic diversity*: There is a general misconception that genetic diversity is less important to nature conservation than species diversity. Often, studies on genetic diversity are considered to be of scientific interest only. In contrast, one should realize that genetic diversity and capacity for evolution are central to species survival. Studies on genetic diversity may reveal the underlying processes that determine taxon diversity in natural habitats. It should not be

concluded that this is only interesting from an evolutionary point of view: micro-evolution is only possible when sufficient genetic diversity is present, and given the rapid human-induced changes in our environment, the capacity to evolve in a short term will be crucial to the survival of many populations and species (Frankham et al., 2002). Without insight in the genetic aspects of diversity, we will never be able to understand the loss of taxon diversity or at least not be able to exclude purely genetic effects. It has also direct bearing on conservation practices: if habitat fragmentation leads to population extinction, it is very important to know whether this is due to smaller population sizes per se or by a reduced dispersal and gene flow, because the measures to be taken to mitigate the problem are different. Studies focusing on genetic diversity in addition to taxon diversity are therefore highly wanted. Importantly, recent studies have shown that the use of neutral genetic markers to infer evolutionary potential can be highly misleading. Neutral genetic markers can be very important to study population structure and infer historical processes. For the study of evolutionary potential, however, it is essential that studies are being conducted on ecologically relevant traits. Such quantitative genetic studies are laborious and difficult, but highly needed. The development of methods (e.g. QTL studies, the development of genetic markers for specific traits) can also be highly rewarding in this context.

- *Auto-ecology of specific focal organisms*: For specific organisms that are either rare or functionally important, it may be worthwhile to conduct an encompassing study on its auto-ecology, including studies on habitat requirements, action radius and dispersal capacity, competitive strength, population genetic structure, etc. Such studies can be important in a context of strongly fragmented landscapes. A similar endeavour may also be very useful for more common species that may be considered good model systems for a broad category of freshwater organisms.
- *Determinants of diversity*: A good understanding of the factors and mechanisms that determine diversity is a major scientific challenge and is essential for the development of efficient conservation strategies. Extant patterns of diversity in ecosystem types should be well documented. These patterns can be related to the abiotic environment, community architecture and the landscape context, and be used for the generation of hypotheses. Descriptive studies, however, seldom allow inferring causality. The next step should, therefore, consist of hypothesis testing. This requires well-designed and replicated large-scale experiments in the field (e.g. enclosures) or in the laboratory (mesocosms).
- *The landscape context*: Recently, the relative importance of local and regional factors in determining genetic and taxon diversity is receiving increasing attention (Shurin & Allen, 2001; Cottenie et al., 2003). This issue has strong bearing on nature conservation, since restoration of local conditions may or may not be enough to restore biodiversity, depending on the degree to which regional factors are limiting. Freshwater habitats are among the most attractive model systems to study the effects of landscape context, connectivity, habitat size, etc., because in general, the limits of the habitat are very well delineated in the terrestrial landscape. Corridors, dispersal highways, dispersal bottlenecks and dispersal barriers are easily identified in the field, at least when dealing with obligatory aquatic organisms. An important aspect of the regional setting is the potentially negative aspect of connectedness. It is generally assumed that corridors and landscape connectivity are good for nature conservation. Yet, there are negative aspects to it, such as the spread of exotic species, diseases, etc. Studies encompassing those aspects too would be very welcome.
- *Functional diversity*: It is not enough to study patterns of biodiversity in isolation. What does biodiversity mean in terms of system resilience, ecosystem functioning and ecosystem services to society? It is of the highest priority that well-designed experiments are carried out to determine functional consequences of biodiversity (Giller et al., 2004). What is the importance of taxon or functional diversity for ecosystem functions like productivity, litter decomposition, nutrient cycling, or trophic interactions? Such

experiments have most often been carried out in terrestrial systems (Tilman et al., 1997), and require a very strong design to be interpreted in a straightforward way. The information generated by such experiments, however, is essential to understanding the functional meaning of biodiversity and its added value to society (Chapin et al., 2000), in addition to its intrinsic ethical value. In addition to an experimental approach, one can also measure ecosystem characteristics in a broad survey of habitats that differ in taxon or genetic diversity. However, the patterns generated by these field studies should be complemented with well-designed experiments to allow a straightforward interpretation.

- *Experimental research and modelling*: We think it is imperative that research is not limited to monitoring patterns of diversity, but also involves well-designed experiments oriented at elucidating causal relationships and processes underlying the patterns, as well as investigating the consequences of biodiversity for ecosystem functioning and ecosystem services to society (Giller et al., 2004). Mathematical modelling can also be very instructive, at least to the extent that it is oriented towards the generation of hypotheses rather than to merely describe systems.
- *Aquatic habitats as model systems*: As mentioned already, aquatic habitats are very good model systems to tackle a number of questions, especially related to regional factors: metapopulation dynamics, metacommunity structure, aspects of landscape connectivity and habitat density, etc.
- *Conceptual issues*: Because of the latter point, aquatic habitats are also very good model systems to deal with conceptual issues, such as the discussion on what is a good estimator of biodiversity in relation to nature conservation. Is bacterial biodiversity, presumably directly related to functioning of the system, equally, less or more important than diversity in typical focal taxa of nature conservation such as odonates, birds and amphibians? How to incorporate rareness (Benayas & De La Montana, 2003) and at what scale (local, regional, continental or global) does rareness matter (Hartley & Kunin, 2003)?

Science policy in biodiversity research

We have now outlined a number of issues and approaches that, in our opinion, should allow major breakthroughs in our knowledge of aquatic biodiversity and its consequences for ecosystem services and society. We now want to merely list a few issues of which we think they are important when implementing a science policy oriented at generating the largest benefit to society.

Applied and academic science in biodiversity

Probably, good political decisions can generally be made without detailed biodiversity research. For the implementation of most of these decisions, the need of technical know how may be higher than of scientific data. Why then is science on biodiversity important for the development of an effective biodiversity conservation strategy? In several ways: (1) to provide data that may make the implementation of political decisions more efficient, (2) to provide data that may lead to a fine-tuning of the political decisions themselves, (3) to provide solid theories and data on the causes and consequences of biodiversity that provide real insight and understanding and that allow making reliable predictions about the consequences of human impact (e.g. land use, conservation practices, pollution...), and (4) to guarantee the availability of scientists that have the expertise to discuss biodiversity issues with policy-makers, engineers and the public at large. Reasons 1 and 2 imply data of technically good quality, but not necessarily inspired by strong and innovative scientific hypotheses. It implies the collection of massive amounts of data, just like one needs massive amounts of data to make weather forecasts. Designing the way in which these data can be collected and analyzed is an exercise that should be based on solid, innovative science, but the collection of the data themselves, once the protocols are set, merely requires well-trained and skilled man-power. Reasons 3 and 4 require inspired scientists of the highest possible standards, and reason 4 further requires that these scientists should be open to think and discuss about conservation, environmental and developmental issues.

The seemingly ever-lasting controversy of whether research on environmental issues should be science- or policy-driven stems from a reciprocal distrust. On the one hand, people advocating science-driven research are afraid that all resources will be drained towards the mere collection of descriptive data and monitoring, and that progress will be stopped because of the lack of funds for innovative science and insufficient attention for inspired and well-designed, often experimental research getting to the core of the environmental questions. On the other hand, people advocating policy-driven science are suspicious that scientists tend to just perform 'science for science', and at the same time would depict their results as being relevant while they are not. Whereas it is obvious that policy-makers will often be in need of monitoring data, they should also realize that the quality and relevance of these monitoring programmes depends on the funding of innovative research, going beyond the mere collection of data. Thus, we believe that a two-track approach is needed: funds must be provided for the monitoring and documentation of biodiversity and the assessment of the biodiversity state of sites and ecosystems, but at the same time sufficient funds must be set aside to engage in scientific research of the highest possible quality, aimed at the improvement of our insight into the drivers and consequences of biodiversity. These are both applied and fundamental scientific questions, that need to be addressed according to the highest international standards, and will in many cases require experimental research that at first sight may look quite academic, but has very high relevance in the longer term. There is, for instance, no way to really assess the functional importance of biodiversity unless one invests in very large and well-designed experiments, requiring substantial resources. This research is highly needed in order to be able to address the claims or disclaims that biodiversity is important for ecosystem services, a highly relevant issue.

We suggest to be prudent with too strong emphasis on policy-driven research because it may lead to the culturing of a kind of researcher that accommodates to the needs of policy-makers in a very flexible way, but does not care too much for high scientific standards, because resources for his/her research group are guaranteed by the mere fact that they accommodate so smoothly to the needs of policy-makers. The success of these research groups then reflects their quality in public relations rather than in science. This is a real risk, which then results in a dichotomy among research groups, some of them publishing in high-ranking journals and other ones having a lot of policy-driven science projects and generating a lot of data but do not care about publications in high-ranking journals. As these two kinds of research groups tend to communicate through different channels, there is a risk that standards weaken. Of course, there are many research groups that try to combine relevant research with high standards so that their research can be published in high-ranking journals. Obviously, in our opinion, it is the latter type of research groups that should be promoted, also when allocating resources for applied science. We should design strategies that prevent that policy-driven, so-called relevant research is done according to lower standards than the more academic research. As a matter of fact, the more relevant the work for society, the more important that it is done according to the highest possible scientific standards.

Irrespective of the above, it is also important to acknowledge that not all questions can be tackled in an experimentally equally clean way as the small-scale and well-delimited questions that many scientists like to focus on. Emphasizing on high quality research entails the risk that scientists focus mainly on scientifically clean but less relevant questions. This risk is indeed a realistic one, and is also one of the reasons why policy-makers promote policy-driven research. However, we have good hopes on this issue, in the sense that quite a number of recent papers in high-ranking ecological journals pay attention to this issue, and propose ways to attain high quality standards in research while not avoiding to study the large-scale and complex issues that societal problems are. Complex problems can be tackled by large multidisciplinary projects (e.g. by doing parallel research on a large number of sites, and applying meta-analysis) Given the size of the current EU-IP research projects, for example, very thorough experimental analyses on the functional consequences of biodiversity can be done, even though these complex experiments imply very large designs. In these complex questions too, however, scientific quality and soundness of the design needs to be the prime criterion for project selection, because inadequate designs have a devastating impact on the research of complex issues, and result in a waste of much research money.

In short, scientists should not use complexity of the questions as an excuse not to tackle problems that are relevant to society. At the same time, complexity should be recognized as a problem, and ideally scientists should be very clear about what they think can be tackled at which standard, and for what questions certain criteria (such as statistical independence) may need to be relaxed. Too many research proposals promise to solve problems using designs that are not fully adequate. By overemphasizing societal relevance compared to scientific standards, funding agencies may inadvertently select for project proposals that are successful in polishing away uncertainties. While they aim to select for relevant science, they may sometimes select for science that has been remoulded to fit the question, even though it was intrinsically very difficult to tackle the question in the first place. This is not the type of projects that lead to good science, nor do they lead to good solutions to policy-related problems. The result is that policy-makers mistrust scientists, and that they even more strongly focus on policy-driven process of project selection, leading to a vicious circle.

Three types of research in biodiversity

We suggest distinguishing three kinds of research in biodiversity:

- *Academic research*, following the highest standards, and studying mechanisms that drive diversity or that determine ecosystem functioning *via* diversity. Although these studies tackle fundamental questions in the first place, their results may be of high relevance to society, given the link between the functional consequences of diversity and ecosystem services.
- *Applied research*, including the development of a wide array of techniques, such as the development of optimal cost-effective methods for the large scale assessment of biodiversity and the monitoring of its changes (e.g. indices for ecological integrity), the generation of methods for biodiversity management and conservation, methods that alleviate trade-offs between conservation and exploitation, and that help in conflict resolution between the conservation objective and interests of economic sectors, etc. This research should be done according to the highest scientific standards, should be hypothesis-driven, and should lead to publications in high-ranking journals
- *Application of expertise on biodiversity*. This research involves the implementation of monitoring and assessment programmes, of which the designs are based on the insights obtained by solid, applied research. The measurement of temperature in beer tanks of breweries will not necessarily be called ‘science’ – irrespective of its importance. Similarly, it is perhaps not adequate to call the monitoring of species diversity in a patch of rainforest ‘science’. Such data collection is of utmost importance and it heavily relies on solid scientific expertise, but it should not necessarily be hypothesis-driven. It simply provides highly needed data that can be used by policy-makers to adjust policies. Of course, datasets that have accumulated through time can also yield important insights if appropriately analyzed (e.g. investigation of long term trends *via* time series analysis; incorporation in meta-analyses). Monitoring and survey research must be done according to technically high standards and well-described standardized protocols (important both for the assessment of their value for policy as for later meta-analyses). So quality needs to be guaranteed from a technical, expertise-wise point of view, but this type of research does not necessarily need to lead to publications in high-ranking journals. Noteworthy, the technical, expertise-wise quality implies that the researchers engaged in these monitoring and assessment programs be very well aware of the latest developments in the applied science field.

We feel that distinguishing these three kinds of research on biodiversity is very important. First, it is often claimed that the distinction between academic and applied science should not be overemphasized. Indeed, both academic and applied research adhere to the same scientific standards, lead to novel insights and the development of new methods. Although applied research may be more policy-driven than academic research, the relevancy of both for policy can be very high, and the distinction is often vague. It is clear from the discussion at the workshop, for instance, that the consequences of biodiversity at a functional level and in terms of ecosystem services is a very important question with respect to policy, but that it is of equally high priority from a purely academic point of view. The main difference is then between academic/applied research and the implementation of standard monitoring routines, such as for instance needed for the Water Framework Directive. The difference can also be made in the field of taxonomy, where the innovative science comes from phylogenetic and taxonomic research, whereas the 'monitoring' aspect is equivalent to compiling species lists, faunistic/floristic data, etc.

Towards a strategy in biodiversity research

Policy-makers are often in need of monitoring data, data on biodiversity hotspots, etc., whereas innovative applied or academic science may offer new methods to obtain these data. We feel that it is important that policy-makers and funding agencies make a clear distinction between studies that mainly focus on monitoring and those that go beyond that, and make sure that resources are allocated to both types of projects. Thinking along these lines will automatically solve some of the problems outlined above, because policy-makers show an obvious tendency to consider monitoring studies as very relevant to their policy-making. If funds for monitoring and innovative science are not separated, the bias of policy-makers towards considering monitoring studies as being very relevant may impede investment of sufficient resources into scientific research that improves our understanding in the long run. If organized properly, a certain amount of funds can be allocated to monitoring and faunistic research aimed at the documentation and assessment of biodiversity, the search for diversity hot-spots and the mapping of all these data. Another set of resources can then be allocated to innovative and high quality applied and academic research aimed at the improvement of diversity conservation in the somewhat longer run, by supplying new methods, by investigating the drivers and functional importance of diversity, by assessing potential trade-offs between exploitation and conservation, and by developing methods that maximally alleviate pressures on diversity. Often monitoring and survey programmes give the false feeling that they provide insight into these issues and this is dangerous, because their design is intrinsically inadequate to derive causal relationships. As a first suggestion, a 50:50 ratio of allocating funds to monitoring vs. innovative science seems reasonable. However, this ratio should not necessarily be applied by all funding agencies – rather, it should result at the level of the whole research field on biodiversity.

A note on biodiversity studies in developing countries

As repeatedly mentioned in the discussion during the workshop and so well illustrated in the lecture by Dr Gopal during the meeting in Beveren (see Gopal, 2005), there is a strong need for research on biodiversity in developing countries. However, the conflicts outlined above are even more strongly felt in this case, because of the obvious lack of money for research in general. Our personal feeling is that it should be possible to work according to the strategy outlined above, with funds largely made available in the more industrialized world. It is of crucial importance to also study the drivers and consequences of diversity in the tropics, but it is also important to have faunistic and monitoring data. We sincerely hope that funding agencies come up with inspired solutions to solve the issue of also funding solid research in developing countries.

Science policy on biodiversity: the interplay between Europe and national initiatives

The European Commission has, in its Framework Programme 6, taken a clear-cut position on how it wants to boost scientific research in Europe. Mainly, much attention will go to large networks of big institutes and to large-scale projects. Whether or not this strategy will pay off will be clear in a number of years, but there is some logic to this decision. It should be clear, however, that such a strategy bears some risks, which are, as a matter of fact, recognized by the EU. As only a limited number of large-scale projects will be funded, there is a risk of a loss of a lot of expertise residing in laboratories that are not part of such big networks or institutes. Also, there is a risk that novel and promising approaches that may eventually be developing in small research groups do not get a proper chance to grow and mature. Therefore, it is very important that federal and regional funding agencies follow a science policy that is truly complementary to that taken by the EU. Rather than having the same priorities and funding the same type of research, the best science policy in our opinion is to screen for high quality projects in priority areas that are either completely different from those subsidized by the EU or are nicely complementary to those. These could be actually part of two different types of calls: (1) calls for original projects on biodiversity and (2) calls specifically addressing research groups that would like to carry out projects that are complementary to an existing EU project – these too should also be open to groups that are not yet receiving EU funding, provided that they can prove with a letter of a partner in such an EU project (who may be a foreigner) that she/he is willing to act as a user in the project and is interested in the results of it.

References

Benayas, J. M. R. & E. De La Montana, 2003. Identifying areas of high-value vertebrate diversity for strengthening conservation. Biological Conservation 114: 357–370.

Chapin, F. S., E. S. Zavaleta, V. T. Eviner, R. L. Naylor, P. M. Vitousek, H. L. Reynolds, D. U. Hooper, S. Lavorel, O. E. Sala, S. E. Hobbie, M. C. Mack & S. Diaz, 2000. Consequences of changing biodiversity. Nature 405: 234–242.

Cottenie, K., E. Michels, N. Nuytten & L. De Meester, 2003. Zooplankton metacommunity structure: regional vs. local processes in highly interconnected ponds. Ecology 84: 991–1000.

Declerck, S., J. Vandekerkhove, L. Johansson, K. Muylaert, J. M. Conde-Porcuna, K. Van der Gucht, C. P. Martinez, T. Lauridsen, K. Schwenk, G. Zwart, W. Rommens, J. Lopez-Ramos, E. Jeppesen, W. Vyverman, L. Brendonck & L. De Meester, in press. Multi-group biodiversity in shallow lakes along gradients of phosphorus and water plant cover. Ecology.

Frankham, R., J. D. Ballou & D. A. Briscoe, 2002. Introduction to Conservation Genetics. Cambridge University Press, 615 pp.

Giller, P. S., H. Hillebrand, U. G. Berninger, M. O. Gessner, S. Hawkins, P. Inchausti, C. Inglis, H. Leslie, B. Malmqvist, M. T. Monaghan, P. J. Morin & G. O'mullan, 2004. Biodiversity effects on ecosystem functioning: emerging issues and their experimental test in aquatic environments. Oikos 104: 423–436.

Gopal, B., 2005. Relevance and policy dimensions of research on biodiversity in freshwater ecosystems: a developing country perspective. Hydrobiologia 542: 19–21.

Hartley, S. & W. E. Kunin, 2003. Scale dependency of rarity, extinction risk, and conservation priority. Conservation Biology 17: 1559–1570.

Rundle, S. D., A. Foggo, V. Choiseul & D. T. Bilton, 2002. Are distribution patterns linked to dispersal mechanism? an investigation using pond invertebrate assemblages. Freshwater Biology 47: 1571–1581.

Shurin, J. B. & E. G. Allen, 2001. Effects of competition, predation, and dispersal on species richness at local and regional scales. American Naturalist 158: 624–637.

Tilman, D., J. Knops, D. Wedin, P. Reich, M. Ritchie & E. Siemann, 1997. The Influence of functional diversity and composition on ecosystem processes. Science 277: 1300–1302.

Vandekerkhove, J., 2004. Uncovering hidden biodiversity: the potentials of using zooplankton egg banks. Ph.D. thesis, KULeuven, 169 pp.

Hydrobiologia (2005) 542:11–14
H. Segers & K. Martens (eds), Aquatic Biodiversity II
DOI 10.1007/s10750-005-0888-8

Comment on De Meester and Declerck, 2005 (target review)

Biodiversity: a resource with a monetary value?

Henri J. Dumont
Animal ecology, Ghent University, Ledeganckstraat, 35, B-9000 Ghent
Tel.: +32-9-2645253, Fax: +32-9-2645343, E-mail: Henri.Dumont@ugent.be

Current methods for assigning monetary values to non-domesticated biodiversity, derived from economic theory, fail to capture the true value of ecosystem components and their services. Fundamental research is useful, in that it creates new knowledge, and new knowledge increases the perceived value of natural assets. But how to express that value to things other than domesticates? Two research avenues that might lead to profitable results are the *risk of extinction* concept, which relates value of individuals to population size, and the *international security* avenue. The latter treats biodiversity on equal footing with military security, i.e. conservation expenses need not be justified by things that happen (extinction, or war), but by things that do not happen (no extinction, no war).

The text by De Meester & Declerck (2005) provides an interesting discussion of the three levels at which biodiversity is currently being studied. At the same time, their logic is typical of how contemporary academic biologists tackle a research subject. Nowhere is there an attempt to put an explicit value on any element of biodiversity, be it a gene, a species, or an ecosystem (for a discussion of types of value attached to biodiversity, see Ehrlich & Ehrlich, 1992). There is nothing wrong with this attitude, but the question is whether such research will suffice to stem the tide of the current biodiversity crisis. Further on, I will call what De Meester & Declerck (2005) pursue *exploration value*.

The subject is controversial and difficult, but ultimately unavoidable: if we wish to move away from the position that the gifts of nature are free for all to take, and that nature is as well a source and a sink for the products of human metabolism, we have no alternative but to tag natural values with the measure that pervades modern economics: money. Once nature's free gifts become natural capital, we need a consistent system of property rights and command-and-control measures to deal with their management. Environmental Economists assume that, without a valuation process that attaches a finite monetary value to natural assets, no sustainable use is possible. Or is it?

Internalization of an 'externality', as damage to nature is called in economic jargon, requires, as a rule, that an estimate of the damage is made in monetary terms, and that that amount is paid with an aim at repairing the damage done. A variety of techniques (Garrod & Willis, 2000 provide a useful overview) is available for this, with the contingent valuation technique (CVM) taking an ever more prominent position. CVM requires a demand function for the natural commodity to be preserved or compensated, and this is often derived from probing the willingness to pay (WTP) or willingness to accept (WTA) of the population concerned.

I often discuss this technique with last-year students in biology, and they all accept CVM when it comes to valuing damages (the Exxon Valdez example, a classic in its kind, is one my favorite examples), or to estimate the value of a nature reserve, botanical garden, or zoo, to society (but see further). Estimating the value of ecosystem services relative to that of economic output worldwide, as attempted by Costanza et al. (1997a), is also appreciated, even if found to be clouded in uncertainty. But there is violent disagreement and opposition (with girls distinctly more outspoken than boys) when it comes to valuating genes, species, and assemblages of species, thus biodiversity.

Focusing on the species level, one argument is that WTP invariably leads to a hierarchy of species, with those that are esthetically pleasing to the human eye, or large and charismatic, ending up to be the expensive ones, while the small and ugly (or venomous) would be worth nothing or even fetch negatively prices (people would pay to exterminate them). Everybody would be prepared to pay for ensuring the survival of whales and primates, but very few would favor the black rat. Ethically, this is unacceptable, but the *risk of extinction* approach (see further), if properly applied, might offer a way out of this problem.

Another argument is that tagging a species in monetary units might create the illusion that the numeral thus obtained is the market (or sometimes black market) value of that species, i.e. the amount of money for which it could be bought and sold. Clearly, any values determined by CVM or related methods should be non-market prices, indicative of what society at a given time understood of the nature of a species, but not what that species was actually worth. Whatever the price obtained, it should be clearly understood there is always a shadow price that is higher than the latest estimate.

Here, I argue that natural capital exists, but is largely unknown to us, and will, to a certain degree, forever remain so. What De Meester & DeClerk (2005) advocate, amounts to gather new knowledge by exploration using the scientific method as a tool. Like entropy, knowledge can only increase, and in so doing, it increases the value of natural capital. Knowledge (information) breeds value, and thus the net result of studying nature is an increase in value of the natural capital by *exploration added value*. Because we will never know everything there is to know about nature, its value will keep increasing, and like that of entropy, approach infinity. The increase in knowledge approach sounds elegant, but may not in itself lead to salvation. Ehrlich et al. (1999), for example, are skeptical as to the expectation that more knowledge will automatically alleviate or eliminate future environmental problems in a society with continuing population growth and per capita consumption.

Except in applied science, where intellectual property rights prevail, the *cost of exploration* is paid for by society. A typical example of exploration value is the cost to answer the question "is there life on Mars?" No WTP would ever generate the sums required to send a rocket to Mars; clearly, other motives are at play here. The decision to allocate budgets to such questions (others might be: Is there life in the deepest caves? What lives in the lakes under the ice of Antarctica? How many species of beetles live in the Amazon forest? How many species of zooplankton are there?) is not obtained by WTP questionnaires but through a process of peer review by a mix of political bodies and scientific committees. This procedure is what my female students favor over CVM by WTP, even when I tell them that this process is not necessarily (in fact, it rarely is) objective and transparent, and the outcome depends as much on the bargaining skills of lobbyists as on the merit of the question(s) raised. The financial effort society is prepared to sustain biodiversity studies defines the exploration value that I advocated earlier; it does not capture the entire value of the subject, but at least provides an estimate of the degree to which society attaches relevance to a research subject.

When we think of the biodiversity crisis, we immediately think of the conservation of endangered species, and of command-and-control measures to regulate their trade. We think of pandas, gorillas, over-collected seashells and birdwing butterflies, of mangroves and of the Amazon forest. And when we think of monetarization, we again think of these same subjects. However, biological objects (species or races) are among the oldest items exchanged between people. They include the products of agriculture (preceded by gathering), and of cattle-breeding (preceded by hunting) and fishing (the only domain in which man is still largely a hunter-gatherer, that urgently needs to be protected from overexploitation and replaced by worldwide aquaculture). Their prices are widely agreed upon and are regulated by extensive legal systems and international agreements, and so is the perceived value of the (agro-eco) systems that produce them. It should be noted that these prices apply to samples of these species, expressed in units (individuals, kilograms, cubic meters...) produced in such a way that the survival of the species is not jeopardized. Man's economy needs such systems: he is simply not capable to survive without them. They constitute the economic subsystem of the global natural system, and they will be with us forever.

Essentially, what man has done over the last millennia is to simplify natural ecosystems, in order to maximize his extraction of a desirable biological product (wheat, meat, fish). In that sense, I agree with Rees (2003) that man's economy does not produce anything: it simply extracts, and sometimes modifies products (biological, but also oil, gas, and other fossil energy) from nature, and this severely limits – if not destroys – the idea of substitutability between man-made and natural capital, advocated by many environmental economists (see e.g. Pearce & Turner, 1989). As well, there is only a limited number of plant and animal species that qualify for domestication, and their number has grown insignificantly over time. What has increased over time is the total volume appropriated by the man-made economic subsystem.

Until the mid-20th century, the economic subsystem was growing, but small compared to nature's total production (Goodland et al., 1992; Costanza et al., 1997b, Rees, 2003). Thereafter, man's appropriation of the 'global natural product' soon became a dominant factor to the non-growing natural system (Daly, 1992; Rees, 1995). Few species were added to man's 'survival kit', but many others started disappearing. Thus, when we talk about a biodiversity crisis, we talk about non-domesticated species, and when we talk about valuation, the question is not to value cattle or vegetables, but the surviving 'wild' biodiversity, little of which will ever be domesticated. The choice is therefore between domestication and extinction (see further), and between values derived from WTP or from the insight of panels of peers. Or can we think of other, as yet undeveloped methods?

WTP may seem appropriate to evaluate the value to society of zoos, botanical gardens, *ex situ* wildlife parks and museums, but hardly of natural systems. The reason for this is straightforward: our level of knowledge of the contents (the species) and the metabolism (the functions) of natural systems is simply too low. Man, as a consumer of ecosystems, is uninformed about what these systems are and represent. Even with the man-made systems above, which can be evaluated by their *user value*, problems may arise. Recently, I was told of an incident in a German natural history and arts museum, which was audited for its performance by an outside agency. The value of the museum's library was estimated by the number of users (measured by outside loans) of any given book or journal. The auditors concluded that only recent publications were valuable, and decided to destroy (!) many old books, because there were no records of loans for them. This Fahrenheit 451 story happened because of lack of information: the auditors somehow did not know that old books were never loaned out, because **too** valuable! If such a fiasco can arise by miscommunication in such a simple system, one can easily imagine what could go wrong in the valuation of an ecosystem.

From such pitfalls, I conclude that the research field of valuation lies wide open, and that biologists and ecologists should step in to help economists devise proper measures of value of wild biodiversity.

At this point in time, I see two avenues worthy of further exploration. One addresses the value of individual species in a way that could be termed *survival value*. It is based upon what Dore & Webb (2003) call the probability of extinction of a species. The more resilient a species (or a population) to disturbance, the lower the probability that it will go extinct. The lower the probability of extinction, the lower the value of its individuals. The elegance of this idea is that it couples value of individuals to population size, and that it restricts (monetary) value to only one aspect, *viz.* social value (value to human society). Mathematical tools are available to explore this avenue in full detail, and this would be one of my priorities for biodiversity research.

Of course, there will be problems, one being how to handle invasive species (do we need to assign a negative monetary value to such species in the invaded space?)

My second priority derives from ideas put forward by economists such as Vatn & Bromley (1995), and Kant (2003). Both papers argue, in my opinion correctly, that monetary values obtained by current valuation methods are as arbitrary as values assigned by non-economic methods, and that ecosystem capital is non-reducible to an economic commodity. A single monetary value that captures total ecosystem value simply does not exist. Earlier, I advocated *exploration value* as a measure by which natural capital increases in the eye of an observer, and that this increase is monotonous, and, in principle, unlimited. Kant (2003), on the other hand, reasons that choices of ecosystem capital should move away from the fiction of market-based methodologies, and turn to security, an aspect related to human survival. *International security* used to be more or less synonym to military security, but has been extended, since a decade or so, to include ecological and environmental security. Military expenditure is only justified by the desire to protect humans from the arms created by other humans, and few will object to the fact that all these costly arms are hardly ever used. In fact, it is better to spend money on arms that remain unused than on arms that are used in battle. But humans are not the only danger facing other humans: nature can be as dangerous, if not more dangerous, especially if mankind continues destroying vital components of its machinery. It follows that the policy choices involving the management of natural capital should be done without assigning prices to species or ecosystem processes (and also without the typically economic process of discounting). Military expenditure in 1999 amounted to 2.6% of Gross World Product, an enormous amount of money. I find it perfectly justifiable to spend a commensurate amount on the maintenance of climate, the ozone layer, the ocean level, and the remaining wild biodiversity. Exactly how much money should thus be spent could form the object of a fascinating field of study!

References

Costanza, R., R. d'Arge, R. De Groot, S. Farber, M. Grasso, B. Hannon, S. Naeem, K. Limburg, J. Paruelo, R. V. O'Neill, R. Raskin, P. Sutton & M. Van den Belt, 1997a. The value of the world's ecosystem services and natural capital. Nature 387: 253–260.

Costanza, R., J. Sutherland, H. Daly, R. Goodland & R. Norgaard, 1997b. An Introduction to Ecological Economics. St Lucie Press, Boca Raton.

Daly, H. E., 1992. Steady-state economics: concepts, questions, policies. Gaia 6: 333–338.

De Meester, L. & S. Declerck, 2005. The study of biodiversity in freshwater habitats: societal relevance and suggestions for priorities in science policy. Hydrobiologia 542: 1–9.

Dore, M. H. & D. Webb, 2003. Valuing biodiversity: reality or mirage? Environmental monitoring and Assessment 86: 91–104.

Ehrlich, P. R. & A. H. Ehrlich, 1992. The value of biodiversity. Ambio 21: 219–226.

Ehrlich, P. R., G. Wolff, G. C. Daily, J. B. Hughes, S. Daily, M. Dalton & L. Goulder, 1999. Knowledge and the environment. Ecological Economics 30: 267–284.

Garrod, G. & K. G. Willis, 2000. Economic Valuation of the Environment. Methods and Case Studies. Edward Elgar, Cheltenham.

Goodland, R., H. E. Daly & S. El Serafy, 1992. Population, Technology and Lifestyle. Island Press, Washington DC.

Kant, S., 2003. Choices of ecosystem capital without discounting and prices. Environmental Monitoring and Assessment 86: 105–127.

Pearce, D. W. & R. K. Turner, 1989. Economics of Natural Resources and the Environment. Wheatsheaf, Brighton.

Rees, W. E., 1995. Achieving sustainability: reform or transformation? Journal of Planning Literature 9: 343–361.

Rees, W. E., 2003. Economic development and environmental protection: an ecological economics perspective. Environmental Monitoring and Assessment 86: 29–45.

Vatn, A. & D. W. Bromley, 1995. Choices Without Prices Without Apologies. Handbook of Environmental Economics. Blackwell, Oxford, pp. 3–25.

Hydrobiologia (2005) 542:15–17
H. Segers & K. Martens (eds), Aquatic Biodiversity II
DOI 10.1007/s10750-005-0889-7

Comment on De Meester & Declerck, 2005 (target review)

Linking science and policy for biodiversity

Anne Franklin
Belgian National Focal Point to the Convention on Biological Diversity, Royal Belgian Institute of Natural Sciences, Rue Vautier 29, 1000 Brussels, Belgium
E-mail: anne.franklin@naturalsciences.be

De Meester & Declerck (2005) present several research pathways for the study of biodiversity in freshwater habitats and discuss the relevance of biodiversity research for solving societal problems. In particular, they underline the difficulty of combining economic development and biodiversity conservation. They also highlight that, in addition to scientific knowledge, both political will and technical solutions are needed to progress towards sustainable development.

On all of these points, I agree with the authors. There is an urgent need for solid fundamental science, complemented by applied research and regular monitoring of biodiversity. My main concern relates to their polarization of scientists and policy makers, as belonging to conflicting worlds with incompatible interests. This oversimplification may prove counterproductive when making suggestions for science policy, especially in a field such as biodiversity where environmental, economic and societal aspects are closely interlinked.

Since the Rio Earth Summit and the signature of the Convention on Biological Diversity (CBD), in 1992, there have been major changes in the way biodiversity is conceptualised and studied, with the development of a more dynamic and integrated approach. As summarised by the International Council for Science (2002a), we have 'moved from a static and descriptive approach of biodiversity, with the design of enclosed parks as a conservation goal, to the emergence of a strong awareness of the functional role of biodiversity in maintaining life on earth, and of the dependence of our economic and social development on biological diversity'. Scientific findings have led to the acceptance of new concepts at the policy level, while new ideas and methods have emerged over the past decade under the influence of a new political awareness. For example, increased knowledge of the complex processes in ecosystem functioning helped to formulate and to acknowledge the 'ecosystem approach' in a number of international fora, the first of which the CBD.

As endorsed by the CBD, the ecosystem approach is a 'strategy for the integrated management of land, water and living resources that promotes conservation and sustainable use in an equitable way'. It should be noted that this definition goes far beyond the study of ecosystems, as could be interpreted from a scientific perspective. It is good illustration of how an international policy process builds on improved knowledge to define its framework for action. Because the understanding of ecosystem functioning is still incomplete, the ecosystem approach calls for adaptative management and does not preclude other approaches, such as protected areas or single-species conservation programmes. It rather aims to integrate all of them (Secretariat of the Convention on Biological Diversity, 2001).

The adoption of the ecosystem approach is definitely a prominent achievement of the Convention. However, it is true that more than 10 years after signing the CBD, there are still weaknesses in the linkage between knowledge on biodiversity and the development of policy (whether scientific, environmental or economic). De Meester & Declerck (2005) point out some of the reasons, but do not attempt to explain why and suggest solutions. I would like to highlight two of their comments. Firstly, they state that policy makers, who are often suspicious that environmental problems are overemphasised in order to attract public attention, are largely ignoring scientists. I see several reasons for this lack of dialogue, among others: many policy makers lack time and resources to read the scientific literature, scientists often make little effort to disseminate or explain their work to a non-specialist audience, and scientists do not always ask the questions politicians want answered. This last remark directly links to the authors' second comment – and concern – that policy-driven research will lead to biased research rather than to innovative and 'sound science'.

There is vital need for a strong partnership between science and society in order to solve 'real-world problems' such as feeding a growing population or halting the loss of biodiversity. The ability to carry out elegant fundamental research purely for the sake of scientific curiosity is an essential cradle of new ideas. However, this should remain the privilege of only a fraction of the scientific community. The magnitude of ecological issues facing society makes it essential for researchers to contribute appropriately to those concerns. This does not mean that policy-relevant research should be limited to applied research or monitoring of biodiversity, as suggested by De Meester & Declerck (2005). Good fundamental research programmes can deliver usable knowledge to policy makers, managers, stakeholders and 'simple' citizens, especially in areas where knowledge to underpin management is still inadequate (e.g. effects of biodiversity changes on overall ecosystem functioning, as underlined by Gessner et al. (2004) and other contributors in the same proceedings issue).

Research undertaken with these questions in mind does not undermine the ability of researchers of achieving 'sound science', as feared by the authors. No one would dispute that policy should be based on sound science. The disagreement arises probably over what the expression implies in practice and how it is used by various stakeholders, including policy makers. It is not the aim of this note to define what sound science is, but minimal requirements include science that should be reliable and valid (use adequate reasoning and methodology, be peer-reviewed, be repeatable) and transparent (be readily available for review). In the US, the expression has been used by recent American presidents to describe the basis of their administrations' regulatory decisions; sound science has come to be associated with absolute certainty regarding a particular problem. As biological systems are complex and will always present unexpected behaviour, this is unrealistic and inappropriate. As stated by Carroll (1994) 'scientists must learn how to better communicate to policy makers the important but arcane fact that processes in the natural world are best described as a series of conditional probabilities. We do not live in the "balance of nature", we live in a complex world of uncertainty, risk and environmental change'. Nonetheless, it should remembered that, even with the best scientific information available, policy makers can still face difficult decisions when having to take societal aspects into consideration.

When De Meester & Declerck (2005) write that 'policy-driven research [...] may lead to the culturing of a kind of researcher that accommodates to the needs of policy makers in a very flexible way, but does not care too much for high scientific standards' they tend to confuse two issues. Achieving high scientific quality is a question of appropriate methodology, as highlighted in the previous paragraph. This is true not only for fundamental research, but also for applied research and for the monitoring of biodiversity. The issue of relevance (i.e. asking the right questions), albeit extremely important, comes in addition to scientific excellence. How to determine what is relevant and on what basis this decision will be made is in itself a complex process that is influenced by the underlying socio-economic-political context.

One way to improve relevance for society, without loosing in scientific quality, is to move beyond the traditional 'three-pillar' assessment framework – i.e. economic, social and environment – and to have a better integration between scientific disciplines (International Council for Science, 2002b). Much of the ecological and conservation research is rooted in particular natural sciences disciplines: zoologists, botanists or geneticists each have their own research domains. However, as real-world issues are not bounded by such disciplines, collaboration with economic, political, and social scientists is essential. A closer integration of these disciplines will also help to find better ways of communicating scientific results, including scientific uncertainty, to a non-specialist audience. In this regard, the publication of research results in high-ranking journals as suggested by De Meester & Declerck (2005), while essential from a scientific perspective, is of little direct use to society. It requires additional ways of communication, in order to translate the scientific results into information more readily usable by the wider society. The approach will depend on the issue and the audience, as well as on the context of which they are part.

Exciting new developments are happening in biodiversity research. Initiatives at the European and international level are now in place to build on this new progress, and to bridge the gap between the scientific community and society at large. By encouraging communication, they will help to bring scientific knowledge on biodiversity to the forefront of policy discussions. To list but only a few examples, DG Research of the European Commission addresses science and society issues in general,[1] the US-based Union of Concerned Scientists has launched a Sound Science Initiative,[2] the International Council for Science has published a series of reports on science for sustainable development,[3] while DIVERSITAS is an international research programme promoting, among others, integrative biodiversity science.[4] The Secretariat of the Convention on Biological Diversity provides information on many international programmes linking science and policy for biodiversity on its website. More specifically, freshwater ecosystems are dealt with under the thematic programme on 'inland water biodiversity'.[5] Making these initiatives more visible to the scientific community will help to enhance their coherence and effectiveness. It will also help the ultimate goal of biodiversity conservation, by transforming knowledge into action.

[1] http://europa.eu.int/comm/research/science-society/index_en.html.

[2] http://www.ucsusa.org/global_environment/global_warming/page.cfm?pageID = 939.

[3] http://www.icsu.org/2_resourcecentre/Resource.php4?rub = 8&id = 29.

[4] http://www.diversitas-international.org/.

[5] http://www.biodiv.org/programmes/areas/water/links.aspx.

References

Carroll, C. R., 1994. Improving the link between policy and science. In Meffe, G. K. & C. R Carroll (eds), Principles of Conservation Biology. Sinauer Associates, Sunderland: 481–482.

De Meester, L. & S. Declerck, 2005. The study of biodiversity in freshwater habitats: societal relevance and suggestions for priorities in science policy. Hydrobiologia, present volume.

Gessner, M. O., P. Inchausti, L. Persson, D. G. Raffaelli & P. S. Giller, 2004. Biodiversity effects on ecosystem functioning: insights from aquatic systems. Oikos 104: 419–422.

International Council for Science, 2002a. ICSU Series on Science for Sustainable Development No. 10: Biodiversity, Science and Sustainable Development. 20 pp.

International Council for Science, 2002b. ICSU Series on Science for Sustainable Development No. 8: Making Science for Sustainable Development More Policy Relevant: New Tools for Analysis. 28 pp.

Secretariat of the Convention on Biological Diversity, 2001. Global Biodiversity Outlook. Secretariat of the Convention on Biological Diversity, Montreal. 282 pp.

Hydrobiologia (2005) 542:19–21
H. Segers & K. Martens (eds), Aquatic Biodiversity II
DOI 10.1007/s10750-005-0890-1

Comment on De Meester and Declerck, 2005 (target review)

Relevance and policy dimensions of research on biodiversity in freshwater ecosystems: a developing country perspective

Brij Gopal
School of Environmental Sciences, Jawaharlal Nehru University, New Delhi 110067, India
E-mail: brij@nieindia.org

In their discussion on the future of science and science policy regarding inland aquatic (mostly freshwater) biodiversity, De Meester & Declerck (2005) examine several questions in some detail: Is current biodiversity research necessary and relevant to the management of the biodiversity crisis? Do we need more biodiversity research? Should the scientific research be driven by policy or the science should drive the policy? They refer to some of my observations made at the workshop in Beveren. Recently, I have discussed some of these points in the broader context of tropical wetlands (Gopal et al., 2005) and therefore, I will restrict my response to a few major points only.

The policy and decision makers of practically all nations who are signatories to the Convention on Biological Diversity acknowledge the importance of biodiversity and the need for its conservation. Having signed the Convention, the governments are obliged, in their own interest, to take all appropriate measures to document it and conserve it. However, the relatively low priority accorded to the inland aquatic biodiversity is reflected in the Proceedings of the Convention, which took up the subject only at the COP4 (1998) in Bratislava. The year 2003 was declared the International Year of Freshwater but the issues of biodiversity in freshwaters remained ignored. Unfortunately, freshwater ecosystems are not readily valued for the goods and services provided by them to the human society but only for the water they hold. The functions of biodiversity and its role in the maintenance of ecological integrity of freshwater ecosystems have not been adequately researched and highlighted.

The relevance of current research to solving environmental problems or to policy formulation in the developing countries has several dimensions. A major question is, what kind of biodiversity research is relevant to the policy maker? On one hand, we do not know even what biodiversity exists in our inland aquatic ecosystems. The planktonic and benthic invertebrates, which constitute the greatest proportion of freshwater biodiversity and form the food base of most fishes, are the most poorly known. The taxonomists are themselves a 'threatened species'. On the other hand, endless enumeration of biodiversity in different habitats and their relationships with pollution or other environmental attributes are of no interest to the policy maker though this basic information is required for monitoring and management of our aquatic ecosystems, and also to decide where and which biota need to be conserved. However, the questions of direct interest to policy makers are, for example: What does biodiversity do in a system? What will happen if a few species decline or disappear? Why should one conserve a species which is not well known or of no direct importance? How will fish yield change if a few plankton species change or disappear? In a very general manner, the top down or bottom up food chain effects can be explained but the consequences for humans in terms of water quality or economic benefits/losses, cannot be foretold with the current understanding of aquatic ecosystems in the developing countries. We know that the introduction of an exotic species can have adverse consequences for the whole system and its biodiversity but no one can dare predict the sequence of events. Another question currently being discussed in many countries is related to environmental flows in rivers and streams. Despite considerable research on the subject (Tharme, 2003), there is not enough basic information in the developing countries for the policy makers to decide, assuming that the political will exists, upon the flows in different rivers. There are numerous questions related to the functions of biodiversity and its responses to water resource management practices that need long-term intensive research in all developing countries.

Another dimension is related to what drives research. Besides the most important constraint of funds and facilities, the majority of researchers in the developing countries has very little opportunity to interact with others in their own and related fields and hence, are often unable to conceptualize the issues. Further, most research is oriented towards career promotion – where the number of publications counts. Meaningless repetition of trivial topics is widespread. A few measurements of a few parameters on one or more occasions are considered sufficient for a publication (least publishable unit). There has been little effort devoted to synthesizing the available information and formulate research questions. Short-term degree-oriented research can neither provide solutions to the questions of management nor help the policy maker formulate appropriate policies.

The question of policy-driven science is generally raised by vested interests and policy makers only because the scientific knowledge is often inadequate to answer policy questions. Also, the funding of research by the governments determines the nature and direction of research in all countries. In India, the policy towards cleaning of River Ganga from domestic and industrial wastes promoted research on water quality and its impact on aquatic biota since 1980s but neither the government's remedial steps nor research could help clean the river completely. However, in South Africa studies made for more than two decades on environmental flows in rivers contributed most to the enactment of the Water Act that provides for reserving an amount of water of defined quality (the Environmental Reserve) to sustain the goods and services obtained from the aquatic systems. Even among developed countries research is driven by the policy. For example, the 'no net loss' policy for wetlands in the United States fueled most of the research directed at wetland delineation, functions and mitigation but both the policy and research could not stop wetland loss completely (see Lewis, 2001; Gopal, 2003). Do we not know that the constructed wetlands, particularly the treatment wetlands, cannot be a substitute for natural wetlands.

There are many instances where even good science has not been able to influence the policy because of extraneous considerations by the policy makers. For example, barriers continue to be developed in rivers despite enough sound scientific knowledge about fish migrations and the impacts of migration barriers on fishery resources. The impacts of nutrients, organic wastes and toxic effluent discharges on various organisms, and the pathways of bioaccumulation of toxic substances along the food chain have been well known for decades but appropriate policy for controlling their discharge into freshwaters is generally lacking. Similarly, biodiversity issues are conveniently forgotten in name of the urgency to meet the needs of the poorer sections of the society in the developing countries. Intensive aquaculture to increase fish and prawn yields has taken a heavy toll of the natural fish biodiversity alone notwithstanding the decline in the diversity of other biota. Lake Kolleru on the eastern coast of India is a glaring example where a shallow lake with enormous plant and fish diversity has been converted to a series of large permanent fish tanks for a nearly monoculture of three major Indian carps using organic feeds.

Another problem lies with our brethren: the engineers and technologists. They have made us believe that there is a technological solution to every problem created by another technological intervention, and the policy makers have got used to them for so long. While the ecologists seek refuge under the diversity and complexity of ecosystems, the technocrats are always willing to offer an immediate quick fix solution to every problem, without pondering over its short- or long-term consequences. Such attitudes are a strong barrier to the influence of science on the policy.

The policy maker has to follow two caveats: first, they should always apply the Precautionary Principle, where the lack of scientific information should not be an obstacle to decision making. The Precautionary Principle has been enshrined in the Rio and Johannesburg Declarations, and lies at the foundation of environmental laws in the

European Union. Second the policy has to be necessarily flexible and dynamic, ready to adapt to the growing/developing scientific information and the societal needs.

Finally, the scientific community has no option but to meet simultaneously the dual challenge of conducting both basic and applied research that contributes to the understanding of the functions of biodiversity and addresses the questions of direct concern to the policy maker. The future of inland aquatic ecosystems and their biodiversity lies in the ability of the researchers in the developing countries to formulate socially relevant questions, find answers expeditiously and present them convincingly to the policy makers for their adoption.

References

De Meester, L. & S. Declerck, 2005. The study of biodiversity in freshwater habitats: societal relevance and suggestions for priorities in science policy. Hydrobiologia, present volume.

Gopal, B., 2003. Perspectives on wetland science, application and policy. Hydrobiologia 490: 1–10.

Gopal, B., W. Junk, J. Finlayson, C. Max & Breen, C. M. 2005. Present state and the future of tropical wetlands. In: Polunin, N. (ed), Environmental Future of Aquatic Ecosystems. Cambridge University press, Cambridge, in press.

Lewis, W. M., 2001. Wetlands Explained: Wetland Science, Policy and Politics in America. Oxford University Press, New York. 147 p.

Tharme, R. E., 2003. A global perspective on environmental flow assessment: emerging trends in the development and application of environmental flow methodologies for rivers. River Research and Application 19: 1–45.

Hydrobiologia (2005) 542:23–26
H. Segers & K. Martens (eds), Aquatic Biodiversity II
DOI 10.1007/s10750-005-0891-0

Comment on De Meester and Declerck, 2005 (target review)

Conservation of freshwater biodiversity: does the real world meet scientific dreams?

C. Lévêque[1,]* & E.V. Balian[2]
[1]*Institut de Recherches pour le Développement, 1 rue de Marnes, 92410 Ville d'Avray, France*
[2]*14 rue des Laitières, 94300 Vincennes, France.*
*(*Author for correspondence: E-mail: christian.leveque@cnrs-bellevue.fr)*

The paper from De Meester & Declerck (2005) raises current and future challenges faced by biodiversity research in freshwater habitats. The most relevant question they ask is 'is science on biodiversity necessary?'. This question may open a debate about what kind of information is necessary for stakeholders. Meanwhile if we are talking about setting research priorities, we should more clearly identify the goals we want to reach. This debate is often hidden in scientific forums. We would like here to challenge some of the ideas developed by De Meester & Declerck (2005), partly as a devil's advocate, partly because we believe that scientists are not well prepared conceptually to face the real world which is not based on theories, but on facts.

The accumulation of scientific information is not a solution in itself to our environmental problems such as the erosion of biodiversity (Novacek & Cleland, 2001). As stated in the paper, it is not a good idea to wait for more information to implement proper management. Clearly, we do need strong political decisions to reduce the ecological footprint of people, and such decisions are usually unpopular. Moreover, the recognition that the planet is embraced by human-dominated ecosystems undercuts any assumption that we can restore the biota back to some state recognized as ideally pristine or 'uncontaminated' by human activities (Novacek & Cleland, 2001).

The discussion about policy-driven research vs. academic science is less pertinent to us. For long, limnologists (academic science) did not consider fish populations as part of the ecosystem because fish were studied by fisheries biologists (policy driven?). As a result, most research carried in the IBP focused on plankton, not on fish. On the other hand, the role of, e.g., fish (and other animal species) in nutrient cycling was equally ignored by chemists, who focus on chemical analysis, or the role of sediment, etc. This attitude was not very productive and the recognition of the top–down view (Northcote, 1988) strongly modified our perception of the ecosystem's functioning. Another example of policy-driven research has been the Onchocerciasis control programme in West African rivers (Lévêque et al., 2003). Most of our existing knowledge on freshwater biodiversity in West Africa has been obtained by an aquatic monitoring programme implemented in 1974 to evaluate the impact on the aquatic fauna of the insecticides used to control blackflies.

A first conclusion may be drawn: there is no clear boundary between academic and applied research, the latter one being often a source of new ideas for academic research and, let's be realistic, a major source of funding.

However, our major concern is that the paper enumerates a long list of topics to be covered by academic research that are certainly relevant for ecologists, but not for the purpose of biodiversity conservation. We would suggest, instead, to concentrate on the time scale, that is the past, present and future of freshwater biological diversity, taking the example of the European area. A priority not stated by De Meester & Declerck (2005) is to set up truly relevant study areas based on past history. Actually, it is important for conservation purposes to know what is the legacy and how it could explain the present-day distribution of biodiversity (geographic range, endemic species, hot spots, etc.).

First, it should be stressed that the history of freshwater biodiversity in temperate systems is somewhat different from that in tropical regions. In northern temperate areas, an ice age occurred every 100 000 years since at least 2 million years. Such major events caused extinction and repeated changes in the ranges of taxa that survived (Hewitt, 2004). During glaciations all the freshwater systems disappeared, or were strongly modified, due to climate change. The Danube in Europe and the Mississippi River in North America are assumed to be the refuge zones during these periods (Oberdorff et al., 1996). Recolonization took place progressively when ecological conditions recovered sufficiently to allow life in freshwater systems. Such a recolonization was a matter of chance for many species, which took advantage of opportunities presented by connections between watersheds to migrate. The present biological diversity of freshwater systems in Europe is the legacy of successive glaciation–deglaciation cycles. The last glaciation occurred some 20 000 years ago so that the lakes and river systems are relatively very 'young' and quite impoverished in comparison with tropical systems, which are characterized by their much longer life span (Lévêque, 2003). It is therefore not surprising to find few endemic species in Northern Europe.

However, we can also assume that the freshwater fauna in most European countries include species that have been introduced by humans, either directly or indirectly through physical connections established between river systems. The channels built during the XIXth and XXth centuries directly breached the natural barriers between adjacent watersheds. Even recently (1992), a new canal was opened between the Main-Danube Basin and Rhine River, allowing migration of species. In the absence of written information, we ignore what was the role of humans during historical times, except for the case of carp, which has been introduced everywhere for fish culture. If carp travelled all over Europe (and Asia?), probably other fish and aquatic species also did.

So, the two main components of the present freshwater biodiversity in Europe are the remains of catastrophic climatic events that occurred repeatedly in the last 2 million of years, and the pool of species introduced by humans probably since a long time, but at an increasing rate during the last century. In other words, freshwater biodiversity in Europe has been shaped by chance and by human activities. The fish fauna of lake of Geneva, for instance, is a melting pot of species compared to the fish fauna of Lake Tanganyika which is the product of a long co-evolution over million of years between species and their environment (Lévêque, 2003).

A brief conclusion here is that Europe (and other temperate regions as well) does not offer the best opportunity to develop ecological theoretical models. In Lake Geneva, for instance, the biological diversity is not at all a fine product of evolution, just an assemblage (a melting pot) of species that did not evolve in sympatry and occur, by chance, in the same biota. What is the ecological meaning of such artificial assemblages and what can be learnt from their study? In other words, is it realistic to develop ecological theories (that are in some ways deterministic) based on the study of artificial species assemblages?

Another issue not highlighted by the authors relates to concepts and conservation goals. Today, most of the freshwater systems in Europe are heavily impacted by human activities: damming, canalisation, eutrophication, various types of pollutions as well as exploitation and species introduction. We are far from a mythic pristine system that likely never existed anyway. It would be misleading to claim to decision makers and citizens that we can restore the biodiversity of freshwaters. It should be stressed that there is no way to go back to any past situation that could be presented as a reference point. Of course we can reduce pollution, rehabilitate in some ways the 'natural' functioning of rivers, build new aquatic landscapes that are more pleasant to people than canalized ones. But the main changes cannot be reversed, just because many new living species exist and cannot be eradicated; and because we are also experiencing a period of climate change.

There is a critical need to clarify the current concepts used by ecologists and to identify what they can really propose as operational tools for managers. Quite a number of ecological concepts have been using terms such as ecological equilibrium, stability, steady state, climax, ecosystem integrity, etc. (Naselli-Flores et al., 2003). They are actually misleading for decision makers. The real world of biodiversity is in terms of change, variability, death and recovery.

The real world is therefore not the world of academic ecologists. Only increasing existing knowledge will not solve the questions raised by the loss of biodiversity. In crude terms, the issue should be: what does the society want? What kind of 'nature' do we want? What are our priorities? For example: we surely do not want the reappearance of malaria through protection of mosquitoes habitats; we probably want to increase the production of clean energy such as hydropower (that implies more damming) or to rebuild fine aquatic landscapes attractive for recreation of urban peoples and tourists?

Simply stated: what kind of realistic goals should scientists propose to decision-makers for managing biodiversity?

- Should they propose to manage ecosystems in order to restore 'good ecological condition' that expectedly prevailed in the past? As stated above, there is no way to go back because of species introductions, pollution (including eutrophication) and anthropisation of ecosystems.
- Should they propose to 'freeze' biodiversity at its present state? Why and what for conserving this state that is fairly artificial in temperate areas?
- Should they emphasize the conservation of economically important species, including sport fish? Would it be a more realistic approach considering that managing the ecosystems for the conservation of these species will also protect many other aquatic species? Actually the protection of sport fishing species is one of the major determinant for improving the quality of FW ecosystems in many European countries.
- Should they aim to 'feed the poor' without taking much care of the aquatic environment?
- Should they suggest to protect species of unique heritage value? Which would be these species in temperate zones? How would such an approach cope with global change (and particularly climate change)? Moreover it is not possible to protect such species without controlling the land and water use in the watershed. So the only way to manage freshwater biodiversity would be to develop an integrated ecosystem management (as stated by De Meester & Declerck, loc. cit.), including the control of land use (agriculture, urbanization, transport networks, etc.) in the watershed responsible for change in nutrient inputs, siltation, change in hydrology, etc. Let's be realistic: to what extent can we really control the land use and fluxes to water systems?

The debate is not exhausted. The loss of biodiversity cannot be attributed, on the whole, to ignorance (Orr, 2003). Considerable effort has been made to document the decline of biodiversity and the causes of decline, but it is difficult to say how much information reaches the public or any particular decision maker. Besides, we are ignorant of many reasons why diversity should be preserved.

If changes in freshwater systems are irreversible, what is the meaning of conservation and what strategy should we develop for research? What kind of biodiversity do we want? Here is the split between protectionists and interventionists in the controversial field of conservation. Either we believe that 'mother' nature did very well and that we have to protect the legacy, or we believe that the heritage is only a melting pot of species resulting from hazardous survival and human introductions.

References

De Meester, L. & S. Declerck, 2005. The study of biodiversity ion freshwater habitats: societal relevance and suggestions for priorities in science policy. Hydrobiologia 542: 1–9.

Hewitt G. M., 2004. Genetic consequences of climatic oscillations in the Quaternary. Philosophical Transactions of the Royal Society, London. B 359: 183–195.

Lévêque, C. 2003. Des lacs en équilibre instable La Recherche hors série no. 11: 46–47

Lévêque, C., J. M. Hougard, V. Resh, B. Statzner & L. Yaméogo, 2003. Freshwater ecology and biodiversity in the tropics: what did we learn from 30 years of onchocerciasis control and the associated biomonitoring of West African rivers? Hydrobiologia 500: 23–49.

Naselli-Flores L., J. Padisak, M. Dokulil & I. Chorus, 2003. Equilibrium/steady state concept in phytoplankton ecology. Hydrobiologia 502: 395–403.

Northcote, T. G., 1988. Fish in the structure and function of freshwater ecosystems: a 'top down' view. Canadian Journal of Fisheries and Aquatic Sciences 45: 361–379.

Novacek, M. J. & E. E. Cleland, 2001. The current biodiversity extinction event: scenarios for mitigation and recovery.

Proceedings of the National Academy of Sciences 98: 5466–5470.
Oberdorff, T., B. Hugueny & J. F. Guégan, 1996. Is there an influence of historical events on contemporary fish species richness in rivers? Comparisons between Western Europe and North America. Journal of Biogeography 24: 461–467.
Orr, D., 2003. Diversity. Conservation Biology 17(4): 948–951.

Hydrobiologia (2005) 542:27–31
H. Segers & K. Martens (eds), Aquatic Biodiversity II
DOI 10.1007/s10750-005-0892-z

Comment on De Meester and Declerck, 2005 (target review)

Taxonomy and systematics in biodiversity research

Koen Martens* & Hendrik Segers
Freshwater Biology, Royal Belgian Institute of Natural Sciences, Vautierstraat 29, 1000 Brussels, Belgium
*(*Author for correspondence: E-mail: martens@naturalsciences.be)*

Water is life, and millions of people, mainly children, die each year because they use fresh water of bad quality (www.who.com). Water quality depends on, and can be determined by, extant diversity of organisms within the water body. The fall and decline of biodiversity levels in freshwaters is therefore an issue of more than academic interest, as has been pointed out correctly by De Meester & Declerck (2005). Their opinion paper analyses the ways in which scientists can respond to the needs of society, by formulating priorities in research on aquatic (freshwater) biodiversity. It is pleasing to see that their views take several time scales into account, i.e. they make suggestions for a short-term science policy, but also take into account the needs of future generations, a view all too often neglected.

Hypotheses and databases

The Meester & Declerck (loc.cit.) confirm the need of both fundamental and applied research, and point out that both approaches need to have quality as their first criterion. However, they continue by saying that good quality fundamental research must *an sich* mostly (read always) be hypothesis driven; supporting purely descriptive research must be avoided as much as possible. Such views are pervasive in limnological and ecological (ecosystem) research, but they fall short of being adequate for this and several other research disciplines as well. Moreover, there is a small problem with hypothesis-testing research: where do these hypotheses come from? Mostly this issue is dismissed off-handedly: it doesn't matter very much where they come from, as long as we formulate them so that they can be tested. We argue that inductive research, gathering descriptive information arranged in large databases, will show generalised patterns that allow the formulation of testable hypotheses. As such, inductive (descriptive) science is an essential part of hypothesis driven research. Descriptive research provides the fuel on which the hypothesis-testing engine must run.

Unfortunately, it is increasingly difficult to obtain funding for inductive research, unless we are talking about huge and flashy projects that appeal to the imagination, such as the Hubble telescope in space and the human genome project. Also, it is becoming increasingly difficult to publish the results of descriptive research projects, as international journals, including *Hydrobiologia*, increasingly shy away from it because it is thought that such papers are bad for impact factors. Several initiatives resulting from the Convention on Biological Diversity, such as the Global Biodiversity Initiative (GBIF) set up international networks that will have as main goal to manage, integrate and analyse large databases. This type of work must continue to receive support, even though it will for many years fall short of the requirement to do 'hypothesis-based' research. By datamining existing literature, such projects will validate past research investment and will make existing data available for future meta-analyses. All of that is important and essential for research on biodiversity (Schram, 2004).

However, very few initiatives allow for the collection of new data of this type, and most descriptive surveys must be done 'on the side', or be sneaked into more applied projects. It is vital that such research receives support in its own standing.

Taxonomy and phylogeny

A case-in-point, where reduced recognition of inductive sciences becomes a real problem, is in systematic research, which includes both

Box 1. Taxonomy, Systematics and Classification

Definitions of these fundamental concepts differ according to different authors. Most commonly accepted are those of Mayr & Ashlock (1991), slightly modified here.

Systematics is the scientific study of the kinds and diversity of organisms and of any relationships among them, applying the taxonomic concepts and hypothesis.

- *Taxonomy* is the theory and practice of classifying organisms. ***Classification*** is the resulting ordering of entities in groups, whose members ideally share a common descent. A special tool used in taxonomy, systematics and classification is ***nomenclature***, which sets down rules and recommendations for the vehicular language used in scientific communication on (groups of) organisms.
- *Phylogeny* studies the evolutionary relationships between organisms.

taxonomy (naming and classifying organisms) and phylogeny (analysing extant data to infer past genealogical relationships) (Box 1). Whereas phylogeny is generally perceived as real science, taxonomy has a much harder time to attract external interest and funding. Why would this be so? Let us start by looking for part of the blame in our own ranks and by admitting that this is partly due to taxonomists themselves. Many taxonomists have up to now refused to leave their ivory tower, in which they continue to draw the legs and wings of shrimps and flies, regardless of why this activity would be necessary. They often see the description of their organisms as the endpoint of their research and reply to every question on the 'why' of their choice of model group of organisms with 'because this has not yet been done'. When one realises that 250 years of taxonomic work has lead to the formal description of approximately 5% of the extant biodiversity, such a reply looses a lot of its rationale to conduct research financed by society.

Another part of the problem is caused by the continued image of taxonomy and systematics as dull and dusty disciplines, basically forms of stamp-collecting (i.e. describing as many new species as possible), which contribute very little to assist other scientific disciplines in their efforts to solve problems, both of fundamental and of applied nature. The persistence of this image is in part understandable, again because many taxonomists do very little to reverse it, but is far from congruent with reality. In fact, the past decades have been very exiting times for taxonomy and systematics. A variety of new and powerful tools have been developed to analyse taxonomic data, while new theory and drastically novel initiatives provoke discussion and require systematists to continuously re-evaluate their research discipline.

Already for several decades, systematists have introduced formal and objective techniques to analyse taxonomic (descriptive) data in a statistical and repeatable way. Both phenetic (with distance criterion) and cladistic (parsimony criterion) approaches to phylogeny have yielded testable phylogenies with confidence limits (using bootstrap, jackknife or Bremer support measures). Each new taxon or character added to the matrix constitutes a further test to the hypothesized phylogeny. The availability of genetic data, mostly, but not exclusively, consisting of parts of directly sequenced DNA, lead to a real explosion of novel phylogenetic hypotheses, and present-day analyses use both distance and parsimony criteria, together with maximum likelihood models. Baysean logic, already known for a long time in ecology, found its way into phylogeny, for better or for worse. Increasingly large phylogenetic databases, with date originating from different sources, are being analysed by supertree and supermatrix techniques. The field is most dynamic and will continue to evolve. These developments in phylogeny also have consequences for taxonomy.

Phylocode, barcodes and cryptic species

One aspect of taxonomy which has drawn a lot of critique is the application of Linnean nomenclature, i.e. the use of a set of rules to name levels in a strictly hierarchical organisation of the tree of life into kingdoms, phyla, classes, orders and so on, all the way down to species and infraspecific categories. Also, the traditional binomen to name organisms, a combination of a generic and specific name, such as *Homo sapiens*, has fallen into disgrace. Much of the discomfort by users of scientific nomenclature follows from the perception that it fails to deliver a stable nomenclature for organisms (Donoghue & Gauthier, 2004). This lament is unjustified, as nomenclature has to adapt to new insights in the developing science of taxonomy. Much of the distress also results from a lack of knowledge of this communication tool, especially of the application of the 'principle of priority', stating that the valid name for a taxon is the oldest of any alternative ones. Cladist phylogeneticists have suggested to introduce a new set of rules, the so-called 'phylocode', in which (single) names are used for all categories, reflecting clades in trees, regardless of their hierarchy. Since the Linnean nomenclature is already in use for more than 200 years, it appears quite reasonable to question its continued applicability in view of all the recent developments in taxonomy and nomenclature. Unfortunately, the phylocoders are not very patient people and in stead of waiting until the discussion on advantages and disadvantages of both the old and the new system have been conducted in full, they intend to put the system into function as from early 2006 (Laurin & Cantino, 2004). From that moment onwards, there would be two parallel systems of nomenclature. Needless to say, this risks to create utter nomenclatural chaos and will further confuse users of scientific nomenclature. But to introduce discussion in the traditional field of nomenclature is a worthwhile initiative (Sluys et al., 2004)

Various authors, mostly molecular biologists, have argued that the classical way of identifying units of biodiversity (species) is a slow, cumbersome and unreliable method. They suggest to introduce a system of barcoding by using a single gene sequence as an identifier for species (Tautz et al., 2003). Only in this way, they argue, can real progress be made with the inventory of biodiversity. Greg Venter, the leader of the Human Genome Project, has announced on one of the barcoding meetings that he can sequence 100 000 species per week! Again, this is a worthwhile topic for discussion: should species be identified by their genotype or by their phenotype? But this initiative runs into unsolvable problems and will always fall short of its goals: to provide one universal gene sequence that will identify all organisms. The drawbacks have extensively been discussed by various authors (e.g. Lipscomb et al., 2003; Seberg et al., 2003) and there is no need to elaborate on them here, other than maybe ask which army of taxonomists will identify the 100 000 species that Venter will sequence every week.

But it is interesting to note that the old antagonists, pheneticists (exemplified by the bar coding initiative) and cladists (phylocode), both propose exponents of themselves which reach from phylogenetics into taxonomy and, consequently, nomenclature. May their battles remain dignified and constructive...

Finally, there is mounting evidence that in most invertebrate groups, and possibly in other groups of organisms such as Vertebrates, higher plants, etc. as well, a large number of cryptic species occur that have thus far remained undetected and that have obscured biogeographical patterns. Bar coding will fail to detect them, as this system has no foolproof cut-off point to define species. To identify and characterize cryptic species is a future challenge that has direct relevance to the inventory of extant biodiversity and it is clear that the success of this will rely on more regular taxonomic methods, although an interdisciplinary approach including the use of molecular tools might prove to be invaluable in this exercise.

Recommendations

Systematics and taxonomy are going through interesting, or should we say challenging, times (Schram, 2004). Both fields are increasingly demonstrating that they are valuable scientific disciplines in their own rights, which provide necessary frameworks (such as classifications and phylogenies) which are vital for the future development of biodiversity research.

The following recommendations can be made to further integrate taxonomy in the wider field of biological research, more in particular in biodiversity studies.

1. Alpha-taxonomy, the basic and classical description of new taxa, remains a vital part of the inventory of past and present biodiversity. Few biologists will doubt the value of thorough

taxonomic revisions, and such results should continue to have their place in meanstream international journals (e.g. Short, 2004). However, it would be advisable to consider alpha-taxonomy as a basic methodology, were the description of taxa would be equivalated to a molecular biologist obtaining DNA-sequences in the lab. Such sequences are only a means to conduct further analyses, and so should alpha-taxonomy be seen as a direct means to be able to conduct higher-level analyses, such as phylogenetic reconstructions, or the analyses of larger and integrated databases on distribution patterns, effects of global change etc.

2. However, unlike sequence data, which are stored and made accessible in the web-based database GENBANK, alpha-taxonomy should still have its classical outlets for publication, simply because alpha-taxonomy requires much discussion. Research fields such as ecology, limnology and evolution, can ignore old and obsolete papers, but taxonomy cannot do this because of the principle of priority mentioned above, and have to take into account every paper in which new taxa have been proposed, no matter how bad or old the paper might be. This leads to lengthy treatises on seemingly small details, but this rule is essential to maintain nomenclatorial stability.
3. The above two points illustrate the relevance of museums for this part of the taxonomic and systematic work: museums harbour the necessary collections for comparative purposes, as basic descriptions always require reference to other type material, especially when new characters are tested against traditional data sets. Museums also have in-house journals and monograph-series, which have been excellent vehicles for the publication of alpha-taxonomic results for centuries. International biodiversity funding should give high priority to the maintenance and further development of museums and their research and publication activities.
4. If taxonomists use their alpha-results in ways described above, their research projects will incorporate all levels of taxonomic and systematic work. This can only be achieved if the choice of model groups, on which taxonomic work is conducted, are selected in a broader framework, so that these groups can show wider relevance and application, be that for the monitoring of environmental (water) quality, monitoring biodiversity dynamics (both natural and human induced), and other applied needs of society, but also for fundamental purposes such as the reconstruction of phylogenies of selected higher taxa.
5. It will be easier for funding agencies to accept the necessity of a descriptive component in such projects, even if it requires a large chunk of the budget. It is vital, however, that these same funding agencies also accept the relevance of systematists' efforts to reconstruct the tree of life as an integral part of biodiversity research. Such an effort to deduce processes by the rigorous analysis of patterns is highly relevant to the Convention of Biodiversity, even if no immediate applied spin-offs are apparent.
6. Taxonomists and funding agencies alike should recognize that identification aids for use in applied studies, whether published classically on paper or as web-based resources, present an essential product of taxonomic research. Users (ecologists) should be aware that these aids are not the essence of taxonomic research but are produced specifically to answer to their needs and, if they want to make use of these products, should be willing to reward taxonomists for their work both scientifically – by citing them appropriately – and financially.
7. Systematics requires highly specialized and intense training. Opportunities for this have become increasingly scarce, as recognized by the establishment of initiatives like the Global Taxonomy Initiative (GTI – see http://www.biodiv.org/programmes/cross-cutting/taxonomy/) and the US National Science Foundation's Partnerships for Enhancing Expertise in Taxonomy (PEET) program (see http://web.nhm.ku.edu/peet/). A European or national equivalent of the latter initiative is urgently called for.

A truly new taxonomy?

Taxonomists, their colleague ecological and evolutionary biologists, as well as policy makers must all work towards a new perception of taxonomy and systematics in biodiversity research. There is no doubt that both classical and modern approaches in taxonomy and systematics can make significant contributions to biodiversity research as well as to other disciplines, which study the Earth's biosphere. It will require an effort from all parties involved to make this visible. Only then can these truly exciting times for taxonomy and taxonomists have a positive effect on other research disciplines.

References

De Meester, L. & S. Declerck, 2005. The study of biodiversity ion freshwater habitats: societal relevance and suggestions for priorities in science policy. Hydrobiologia 542: 1–9.

Donoghue, M. J. & J. A. Gauthier, 2004. Implementing the phylocode. Trends in Ecology and Evolution 19: 281–282.

Laurin, M. & P. D. Cantino, 2004. First International Phylogenetic Nomenclature Meeting: a report. Zoologica Scripta 33: 475–479.

Lipscomb D., N. Platnick & Q. Wheeler, 2003. The intellectual content of taxonomy: a comment on DNA taxonomy. Trends in Ecology and Evolution 18: 65–66.

Mayr, E. & P. D. Ashlock, 1991. Principles of Systematic Zoology, McGraw-Hill.

Short, J. W., 2004. A rivision of Australian river prawns, *Macrobrachium* (Crustacea, Decapoda, Palaemonidae). Hydrobiologia 525: 1–100.

Schram, F. R., 2004. The truly new systematics – megascience in the information age. Hydrobiologia 519: 1–7.

Seberg, O. et al., 2003. Shortcuts in systematics? A commentary on DNA-based taxonomy. Trends in Ecology and Evolution 18: 63–65.

Sluys, R., K. Martens & F.R. Schram, 2004. The Phylocode: naming of Biodiversity at a crossroads. Trends in Ecology and Evolution 19: 280–281.

Tautz, D., P. Arctander, A. Minelli, R. H. Thomas & A. P. Vogler, 2003. A plea for DNA taxonomy. Trends in Ecology and Evolution 18: 70–74.

Hydrobiologia (2005) 542:33–34
H. Segers & K. Martens (eds), Aquatic Biodiversity II
DOI 10.1007/s10750-005-0893-y

Future priorities in science policy for biodiversity studies: a comment on the target review by Luc De Meester and Steven Declerck

Christian Sturmbauer
Department of Zoology, University of Graz, Universitätsplatz 2, 8010 Graz, Austria
E-mail: christian.sturmbauer@uni-graz.at

Many scientists are frustrated by the fact that the critical status of many biota and species does not seem to be appreciated by political decision-makers throughout the world. Many decisions remain economy-driven and do not take into account science-based insights. It is often neglected that the necessary trade-off between protection and the need of 'feeding the poor' is biased by short-term goals, which in the long run have an adverse effect. It is the persistence of setting the priority to narrow short-term economic interests vs. more holistic sustainable goals (both economic and biodiversity-oriented) that impedes the change to the better. Clearly, the value of biodiversity *per se* and of natural resources needs to be incorporated in our political action. Such cost–benefit calculations are not easy to undertake to evaluate protective actions in economic terms. However, it is of fundamental importance to find a viable compromise to implement the societal and economic value of nature in economy-driven political processes, to develop a new attitude for approaching the dramatic loss of biodiversity.

The target review defines the structure of biodiversity research necessary to bridge the need of both applied and basic science. It suggests four aspects of science to substantiate political actions for effective environmental conservation: Science should (1) provide data that make the implementation of political decisions more efficient, (2) provide data that may lead to a fine-tuning of the political decisions themselves, (3) provide solid theories and data on the causes and consequences of biodiversity that provide real insight and understanding and that allow making of reliable predictions about the consequences of human impact (e.g. land use, conservation practice, pollution ...), and (4) to guarantee the availability of scientists that have the expertise to discuss biodiversity issues with policy-makers, engineers and the public at large. From these goals the authors derive three types of biodiversity research, academic research, applied research and application of expertise on biodiversity by implementation of monitoring and assessment programs.

An important section of the review suggests a re-structuring of national and European funding landscapes in order to make science more efficient concerning the generation of new insights and their translation into political action. It stresses the need of complementarity among EU programs, biased towards large networks of big institutes and to large-scale projects, and national agencies taking over the role of funding smaller and more strictly focused projects, typically carried out by smaller research groups. Nationally funded research could be part of two types of calls, (1) calls for original projects on biodiversity, and (2) calls specifically addressing research groups that would like to carry out projects complementary to an existing EU project. The authors also point to some risks arising from the funding strategy of the EU supporting large groups and consortia only, to lose the expertise of smaller groups that are not willing to seek funding or fall outside the narrow scope of EC programs. De Meester & Declerck (2005) suggest that national agencies should take over the funding of such groups, but in close coordination to the EU actions. I feel, that this strategy also bears some risks, due to the dictation of the research topics by the EU-calls. In contrast, I feel that some future EU calls should be more flexible in that they (1) explicitly incorporate academic science topics in applied aspects, and (2) that smaller consortia get an equal chance to compete for funding. At present, priorities change from call to call, and often academic research is not covered at all by specific EU programs. Moreover, there is a lot of strategic pressure to form huge groups by complying to the subtle political influence about how the participants of a group should be selected, in order to get funded. It often appears that such large groups are difficult to coordinate and that major efforts must be devoted to communication and management, whereas relatively few researchers have the time, and few institutions have the administrative capacity to cope with this. Thus, I argue to create some opportunities to small consortia, and to acknowledge the need for incorporation of academic science in primarily applied topics.

Reference

De Meester, L. & S. Declerck, 2005. The study of biodiversity in freshwater habitats: societal relevance and suggestions for priorities in science policy. Hydrobiologia 542: 1–9.

Hydrobiologia (2005) 542:35–37
H. Segers & K. Martens (eds), Aquatic Biodiversity II
DOI 10.1007/s10750-005-0894-x

Comment on De Meester and Declerck, 2005 (target review)

Towards a coherent and high-quality science policy on biodiversity

Aline van der Werf
Belgian Science Policy, Rue de la Science 8, B-1000 Brussels, Belgium
E-mail: Aline.vanderwerf@belspo.be

According to De Meester & Declerck (2005), monitoring and survey programmes are likely to become the preferred scientific tools supporting a policy on biodiversity conservation. Nevertheless, without the input of a science carried out in accordance with the highest quality standards, these tools will not be able to guarantee long-term, effective biodiversity management. Scientific principles must indeed guide a systematic and objective documentation, analysis and assessment of biodiversity as well as trends in the state of population, species, habitats and the ecological services that these species and systems provide. Furthermore, one expects science to establish the causes of the loss of biodiversity, particularly those generated by man, to identify priority solutions and to provide the tools to assess the effectiveness of these solutions.

The question at present is how to develop and implement a science capable of covering all these dimensions, both local and global, and that can thus support the conservation and sustainable use of biological diversity. We believe that this must evenly develop along four lines of action: enriching knowledge, improving analysis, furthering the synthesis and integration of information and developing a new way to carry out research in support to biodiversity conservation.

Four lines of action for a coherent science policy on biodiversity

Enriching knowledge

De Meester & Declerck (2005) identify several issues where knowledge about freshwater biodiversity is lacking. To address these, specific research efforts are required, notably to increase our knowledge on structural and functional aspects of biodiversity and its components, and to explore poorly known habitats and species. Developing expertise across a wide range of disciplines: evolutionary biology, ecology, taxonomy, applied biology, conservation biology including a better use of new technologies such as molecular biology (e.g. DNA taxonomy), bio-informatics or satellite imagery ought to improve the base-line knowledge that is a necessary prerequisite to any conservation policy.

Improving analysis

Reliable tools for scientific analysis must be developed with a view to devising more effective and widespread conservation measures. De Meester & Declerck (2005) highlight the need for new models, experimental research at the appropriate scale, new concepts, harmonised methodological protocols and improved sampling methods. Any analysis also implies clearly identifying the hypotheses on which it is based and defining the limits thereof.

Furthering the synthesis and integration of information

At the same time as collecting new data, it is important to make already existing, but patchy data more readily available and more useful. Identifying priority actions and research shortcomings with regard to the preservation of biodiversity will be possible through structured access to information and data collated from different sources and layers, for different purposes and covering numerous research areas, through managing and maintaining access to the data and through the analysis and the interpretation of the gathered information.

Developing a new way to carry out research in support to biodiversity conservation

A science in support to biodiversity conservation should better harness the practical application of certain disciplines such as taxonomy.

Cross-disciplinary research networks, carried out jointly by biodiversity researchers, conservation practitioners, economists, lawyers and social scientists provide a means to develop common approaches to biodiversity conservation problems.

As other sciences in support to policy making, a science devoted to biodiversity conservation must absolutely be implemented in such a manner as to further communication and the exchange of expertise between scientists and the potential users of results: nature conservation associations, the authorities concerned from different sectors relevant to conservation, the private sector, the educational sector and citizens.

A research *continuum* underpinned by high-quality scientific and management standards

The devising and implementing of effective measures for biodiversity conservation is based on both short- and long-term studies, carried out on both small and large scale, at a sectoral, cross-sector or integrated level. Every research type and stage helps to assess the progress made towards the target set by the World Summit on Sustainable Development: 'achieving a significant reduction in the current rate of biodiversity loss by 2010'.

Three categories of research characterise this scientific *continuum*:

An academic, basic research carried out over several years in a context of international research, aimed at anticipating needs, particularly political ones and playing a forward-looking, early-warning role as well as clearing up a number of uncertainties,

A strategic, more applied research that creates reliable and appropriate analytical tools to develop, implement and monitor political decisions in order to assess their effectiveness,

The development of an expertise in biodiversity that trains cross-disciplinary scientists who are able of tap into and transform research experience into useful recommendations for decision-makers or of providing answers to political questions in a short time span.

In concurrence with De Meester & Declerck (2005), we believe that this research, whether it is guided by a 'bottom-up' or a 'top-down' approach, should be carried out in accordance with high standards of scientific but also of management quality.

For this three conditions need to be fulfilled:

- The application of rigorous assessment procedures on submitted proposals as a result of a call, by a panel of international experts who have the requisite cross-disciplinary skills. Such procedures based on precise criteria in terms of relevance and scientific, technical methodological and strategic quality can ensure that the proposal takes account of recent research developments, that it is well designed and that it does not repeat what has been done elsewhere. Furthermore, they enable co-operation with foreign research teams. These assessment procedures are already implemented at European (Sixth Framework Programme, European Science Foundation) and national level (e.g. Belgian Scientific Plan for a Sustainable Development policy (SPSD)).
- The creation of appropriate consultation mechanisms between bodies which fund research both at national (e.g. the SPSD guidance committee of the Belgian science policy), and European level (e.g. the ERA-net scheme). These consultation mechanisms aim at promoting research synergies, preventing duplication, making optimal use of budgets, exchanging good practices for research management and encouraging the running of key projects at the appropriate scale.
- The creation of interface structures at all levels of power (e.g. the European Platform for Biodiversity Research Strategy (EPBRS)), the National Biodiversity Platforms: (e.g. http://www.biodiversity.be/bbpf), users committees of research projects (e.g. SPSD of the Belgian Science Policy),... between research departments and sectoral policy departments: environment,

nature conservation, spatial planning, agriculture, fisheries, development cooperation, the economy, etc..., in order to identify research priorities, to assess and promote the transfer, the harnessing and the utilisation of experience gained through research, notably into monitoring and survey programmes.

The exchange of good practices for programme management between bodies funding research, the optimal use of consultation structures at ministerial and administrative level, the increased development of a science policy interface at all levels of power are a guarantee for the development of a high-quality science in support of an effective, long-term management of biological diversity.

Reference

De Meester, L. & S. Declerck, 2005. The study of biodiversity in freshwater habitats: societal relevance and suggestions for priorities in science policy. Hydrobiologia 542: 1–9.

Hydrobiologia (2005) 542:39–67
H. Segers & K. Martens (eds), Aquatic Biodiversity II
DOI 10.1007/s10750-004-5522-7

Review Paper

An assessment of animal species diversity in continental waters

C. Lévêque[1], E.V. Balian[2] & K. Martens[3,4,*]
[1]*Institut de Recherches pour le Développement, 1 rue de Marnes, 92410 Ville d'Avray, France*
[2]*DIVERSITAS, 51, bd Montmorency 75016 Paris, France*
[3]*Royal Belgian Institute of Natural Sciences, Vautierstraat 29, B1000 Brussels, Belgium*
[4]*Department Biology, University of Ghent, K.L. Ledeganckstraat 35. 9000 Gent, Belgium*
(**Author for correspondence: E-mail: martens@naturalsciences.be*)

Key words: freshwater, global assessment, animal species, biodiversity, species richness

Abstract

There is a need for monitoring the status and trends of freshwater biodiversity in order to quantify the impacts of human actions on freshwater systems and to improve freshwater biodiversity conservation. Current projects carrying assessment of freshwater biodiversity focus mainly on leading-better-known groups such as fish, or identify keystone species and/or endemic freshwater systems for conservation purposes. Our purpose is to complete these existing projects by providing quantitative estimates of species number for all freshwater groups on each continent and/or major eco-regions. This article present the results of the first implementation phase carried out from September 2002 to June 2003 and which addressed only freshwater animal species. The project consisted of: (1) compiling existing data from literature, web sites and museum collections; (2) contacting scientific experts of each group to provide a 'to the best of their knowledge, estimates of species numbers. In this study, we consider true freshwater species to be, those that complete part or all of their life cycle in freshwater, and water-dependent species those that need freshwater for food or that permanently use freshwater habitats. The current order of magnitude for known freshwater animal species world wide is 100 000, of which half are insects. Among other groups, there are some 20 000 vertebrate species; 10 000 crustacean species and 5000 mollusc species that are either true freshwater or water-dependant species. The study highlighted gaps in the basic knowledge of species richness at continental and global scales:

(1) Some groups such as Protozoa, nematodes or annelids have been less studied and data on their diversity and distribution is scarce. Because current richness estimates for these groups are greatly biased by knowledge availability, we can expect that real species numbers might be much higher.
(2) Continents are not equal in the face of scientific studies: South America and Asia are especially lacking global estimates of species richness for many groups, even for some usually well-known ones such as molluscs or insects.

The second phase of the project will address freshwater plants and algae. The present status should be considered as a first sketch of the global picture of freshwater biodiversity. We hope that this project will initiate interactive exchange of data to complete and update this first assessment.

Introduction

It is widely recognised that freshwater biodiversity and habitats are increasingly being affected by human activities (Stiassny, 1999). Available data suggest that around 30% of fish species are threatened (IUCN Red List of Threatened Species, Khlphake et al., 2001). The extinction rate for freshwater animals in North America is estimated to be five times higher than for terrestrial fauna, with 123 freshwater animal species recorded as extinct since 1900 (Ricciardi & Rasmussen, 1999). All estimates of biodiversity loss are conservative considering the lack of knowledge for many groups of species and for extended areas in the tropics and high latitudes (Lévêque & Mounolou, 2003). Consequently, the number of threatened species is expected to be much higher. In addition to species loss, also habitat degradation, eutrophication and species introduction affect freshwater bodies world wide. Despite the signs of rapid and destructive changes in freshwater ecosystems, freshwater biodiversity remains of low priority in global conservation initiatives carried by governmental and intergovernmental organization, (Stiassny, 1999; Balmford et al., 2002; Shumway, 1999). For example, lakes and their watersheds are dramatically under-represented both in protected areas and in Conservation measures although they are experiencing serious degradation (Duker & Borre, 2001). The lack of conservation and management strategies to avert damaging changes in freshwater habitats occurring around the globe increases the potential for permanent loss and ignorance of much of our planet's rich aquatic biota (Lundberg et al., 2000)

Scientists are faced with the challenge to provide baseline studies on freshwater biodiversity at regional and global scales to international policy makers (Postel, 2001). Indeed, monitoring the status and trends of freshwater biodiversity is essential to quantify impacts of human activities on freshwater systems and to improve freshwater biodiversity conservation. Estimating global trends for freshwater biodiversity will also be a critical asset for management strategies considering that studies on climate change and water resources and data management systems are currently being developed at a global scale (GWSP).[1]

Current projects on the assessment of freshwater biodiversity focus mainly on a few better-known groups, such as fish or molluscs, or identify keystone species and/or endemic freshwater systems for conservation purposes (Groombridge & Jenkins, 1998; Lévêque, 2002; Revenga & Kura, 2002). Our aim is to complete these existing projects by providing quantitative estimates of species numbers for all freshwater groups on each continent and/or major eco-regions. The results should highlight gaps and help prioritising the groups and geographical areas where additional investigations are needed.

The first phase of the project has been implemented from September 2002 to June 2003 and has addressed only freshwater animal species. The project consisted of: (1) compiling existing data from literature, web sites and museum collections and reporting the most recent number of known (described) species for each group; (2) contacting scientific experts of each group to provide a 'to-the-best-of-their-knowledge' estimate of species number. The results of this first phase are reported here and were also presented at the conference on 'Aquatic biodiversity: Past-Present-Future, an International Scientific Meeting' held in Antwerp, Belgium, in August 2003. A second project phase should address the plant, fungi and algal species.

What is a freshwater species?

The first problem encountered in assessing freshwater biodiversity is the actual definition of a freshwater species.

Indeed, many animal species depend upon freshwater habitats in many different ways:

- Some species depend on freshwater for all stages of their life cycle, r, i.e. many freshwater fishes (except diadromous species that migrate between marine and freshwater environments), freshwater crustaceans or rotifers;
- Some species need freshwater to complete part of their life cycle, such as many amphibians and insects;
- Some species need humid habitats such as some Collembola;
- Some species are closely dependant on freshwater for food or habitat such as most of the so-called aquatic birds and some mammals, but also parasites whose 'food and habitat' is then a freshwater animal host.

In the present paper, we consider as freshwater species both the 'true freshwater species' and the 'freshwater dependant species'. The latter include for example many aquatic birds, and several mammals or reptiles. In the case of parasites, if

[1]Global Water System Programme, web site http://www.gwsp.org/

they have a freshwater stage, they are considered as true freshwater species; if they have only internal parasitic stages, they would fall into the freshwater dependent species.

An assessment of freshwater invertebrate species diversity

PHYLUM PORIFERA

Habitat and Diversity: The entire phylum Porifera consists of more than 5000 species world wide. Only 197 of these are known to occur in freshwater; a total of less than 300 sponge species are expected to be non-marine (Ruppert & Barnes, 1995). The *Spongillidae* is a major family of freshwater sponges, represented by about 170 species world wide (Smith, 2001), with 27 species being reported from North America (Thorp & Covich, 2001).
Distribution: Of the currently known freshwater sponges, 77 species occur in Eurasia 49 in Africa 27 in North America, 51 in South America and 31 in Oceania (Manconi & Pronzato, 2002; Pronzato & Manconi, 2002). There are likely to be various undescribed species, particularly in tropical regions that have not been thoroughly investigated.
Endemicity: The diversity of sponges in freshwater is low in comparison with those of marine habitats from which they have evolved but there are some valuable hot-spots of endemicity because freshwater sponges have undergone adaptive radiation in ancient lakes such as lakes Baikal and Tanganyika, and in South America (Dumont, 1994b). One sponge family, the Lubomirskiidae, is endemic to Lake Baikal and comprises 14 species (Timoshkin, 1997).

PHYLUM CNIDARIA

Habitat: Also the Cnidaria are a primarily marine phylum with a few freshwater representatives that belong to the class Hydrozoa. There are sessile polyp forms, which are usually colonial and free-swimming medusae, which usually represent the sexually reproducing, dispersal phase.
Diversity and endemicity: The total number of freshwater cnidarians is low, about 30–45 species are estimated to occur world wide (Vervoort, pers. com.). Few comprehensive data could be reported on endemic freshwater cnidarian. In Lake Baikal, one of the two described *Hydra* species is endemic.
Distribution: Fifteen cnidarian species are reported from Europe (Fauna Europaea, 2002) and 22 species are report from freshwaters of North America (Smith, 2001). There are altogether about 10–14 freshwater species of medusa. No freshwater medusa appears to have evolved in Europe and the Americas, but one is currently present there: *Crespadacusta sowerbii* has apparently managed to colonize all continents except Antarctica, in the course of the 20th century (Dumont, 1994b). *Limnocnida* and *Craspedacusta* are the two main genera of cnidarians from continental waters, with a free-swimming medusa in their life cycle. They originated in tropical-subtropical areas, *Limnocnida* is also found in Africa (Williams et al., 1991); *L. tanganicae* is endemic to Lake Tanganyika (Coulter, 1991).

PHYLUM PLATYHELMINTH

There are four classes in this phylum: the Cestoda (tapeworms), Trematoda (flukes), Monogenea and Turbellaria (planarians).
Habitat: The turbellarians are mostly marine with several freshwater orders. Most freshwater turbellarians are free-living and occur in various aquatic systems such as ponds, lakes, streams, hyporheic water, ditches, and temporary puddles. A few freshwater turbellarian species are ectoparasitic on crustaceans. The cestodes, the trematodes and the monogenes are all internal parasites.
Diversity and Endemicity: There are about 20 000 platyhelminth species world wide (Pennak, 1989). The Microturbellaria include 400 species in freshwaters and about 100 species of Macroturbellaria (Triclads or planarians) occur world wide (Giller & Malmqvist, 1998). Free-living platyhelminthes (Turbellaria) comprise 150 described species in Lake Baikal of which about 130 are endemic; several dozens of Baikalian turbellarian species are still awaiting description (Timoshkin, 1997).
Distribution: The total number of non-marine turbellarians reported from Europe is 623 species of which 577 are affiliated to freshwater and 46 species are strictly terrestrial (Faubel, pers. com.; Fauna Europaea, 2002). About 350 microturbel-

larian species have been recorded from Europe, and approximately 150 from North America; 40 macroturbellarian species have been recorded from North America (Smith, 2001).

PHYLUM ROTIFERA

Habitat: The phylum Rotifera is mainly represented by freshwater species, widely distributed in all types of inland waters, and by a few marine species.

Diversity: The phylum currently comprises 1817 valid species, but the total number is doubtlessly higher. There are only 50 exclusively marine species, and a few terrestrial and parasitic species. The Monogononta is the largest group (1441 sp.), followed by Bdelloida (374 sp., Segers, 2002). The latter group is highly enigmatic, as it constitutes the highest taxonomic level of exclusively asexually reproducing multicellular eukaryotes.

Distribution: Rotifers are generally cosmopolitan, with a preponderance of thermophilic or tropicopolitan representatives. Nevertheless, some taxa like *Synchaeta* and *Notholca* are more diverse on higher latitudes. The fauna of tropical South America and Asia, and that inhabiting the freshwaters of Northeast North America are particularly diverse and rich in endemics; the same may hold for Southwest Australia. The fragmentary nature of our knowledge on the rotifer fauna of tropical Africa and Madagascar precludes any reliable assessment (Segers, 1996, 2001, 2003). The number of known rotifer species from Australia is close to 600 spp. (Shiel & Koste, 1986). In Asia, 477 species are recorded in China in 84 genera and 28 families (Zhuge et al., 1998). Only 3% of the Chinese rotifer fauna is restricted to the area, with 15 endemic taxa. Rotifer diversity is also high in Thailand with 310 taxa on record, although some of them may be questionable (Segers, 2001). The majority of rotifers recorded from Southeast Asia are widely distributed and occur in the tropical regions of the old world, Australia and the Austro-Malayan region (Segers, 2001). Remarkably, although the total number of valid species recognised in Rotifera stands at a relatively low 1817 (Monogononta: 1441), local diversity may be fairly high: Dumont & Segers (1996) estimate that some 250 species of Monogononta may co-occur in a single lake.

Endemicity: For a long time, the prevailing idea was that rotifers had a cosmopolitan distribution. However, recent work has demonstrated that the degrees of cosmopolitanism and endemism vary between groups. The same is true for long distance, passive dispersal, and vicariance events, which all play key roles in structuring distribution patterns (Segers, 1996, 2003). Lake Baikal, holds 26 endemic species on a total of 186 rotifer species (Timoshkin, 1997).

PHYLUM NEMERTEA

Habitat: Most nemerteans are marine organisms, and a small minority is terrestrial or lives in freshwater (Moore & Gibson, 1985). They are benthic forms, hermaphroditic and often proteandric.

Diversity and endemicity: Amongst the 900 known species, 20 nemerteans inhabit freshwaters (Gibson, pers. com.). No comprehensive data could be reported about possible nemertean endemicity.

Distribution: Three nemertean species. are found, in North America (Gibson, 1995; Thorp & Covich, 2001), and 15 species are reported from Europe (Fauna Europaea, 2002).

PHYLUM NEMATODA

Habitat: Many nematods are free-living but they also include parasitic species, a number of which affect humans directly or indirectly through their domestic animals (Malakhov, 1994). It is difficult to identify the exact diversity of freshwater nematode species, because some terrestrial nematods live in the water-film surrounding plants or soil particles. In addition, some species are generalist, occurring across wide areas and in many habitats while others are much more specialised. Many of them are terrestrial, but some genera contain obligate freshwater species (or at least, part of their life cycles occurs in freshwater habitats).

Diversity: There are currently about 14 000 known nematode species, about half of which is considered to be non-marine (Bongers, pers com.), but the true number may be closer to 100 000 as suggested by Coomans (2000). About 3000 nematode species have been recorded in freshwater, but some of these might also be found under terrestrial conditions (Bongers, pers. com.).

Distribution: Andrássy (1978) listed 605 species of freshwater nematodes in a checklist for European inland water, Jacobs (1984) listed 327 species of freshwater nematodes in Africa and Thorp & Covich (2001) estimated the number of freshwater nematodes in North American waters at about 300–500, but again the real numbers are expected to be much higher.
Endemicity: Shoshin (1999) registered about 300 (mostly undescribed) species at only six near shore sampling localities in Southern Baikal. Several hundreds of species are awaiting description.

PHYLUM NEMATOMORPHA

Habitat: Most nematomorph species exist in freshwater habitats, except for the genus *Metonema* which comprises pelagic marine species. Nematomorph life cycles involve a short-lived non-feeding adult stage and a more persistent larval stage that commonly parasites arthropods.
Diversity and endemicity: There are currently around 32 described species (Ruppert & Barnes, 1995), which might be a considerable underestimation. For example, Dudgeon (1999) estimated that there are over 100 species of Gordioidea, an order of Nematomorpha confined to freshwaters. No comprehensive data are available on distribution and possible endemicity of Nematomorpha.

PHYLUM GASTROTRICHA

Habitat: Gastrotrichs are free-living in aquatic environments, either marine or freshwater (Boaden, 1985). The freshwater gastrotrich species mainly inhabit the sediment surface and the submerged vegetation in eutrophic waters (Ricci & Balsamo, 2000). Gastrotrichs are amongst the few animal groups commonly found in (near-) anaerobic environments.
Diversity and endemicity: There are 250 known freshwater gastrotrich species; the vast majority of these belong to the order Chaetonotida. (Smith, 2001). This total number is likely to be a considerable underestimation, due to lack of investigations in many geographical areas. We could not find comprehensive data on possible gastrotrich endemicity.
Distribution: Fewer than 100 species of gastrotrichs are reported from North America (Thorp & Covich, 2001) and around 200 species from Europe (Fauna Europaea, 2002).

PHYLUM BRYOZOA

Habitat: Bryozoans are aquatic organisms, living mostly in colonies of interconnected individuals, also called zooids. Most bryozoans are marine, although brackish- and freshwater forms are moderately common. These freshwater forms are generally restricted to warm water, both in lakes and rivers.
Diversity and endemicity: The phylum *Bryozoa* remains largely unknown, although it is a very diverse group with about 5000 living species. The number of freshwater bryozoans world wide is around 70–75 species (Massard, pers. comm.). We could not find comprehensive data on possible gastrotrich endemicity.
Distribution: There are about 15–16 bryozoan species in Europe (Fauna Europaea, 2002) and 24 bryozoan species are found in North America (Thorp & Covich, 2001).

PHYLUM TARDIGRADA

Habitat: Tardigrades, water bears, are present in freshwater, marine and terrestrial habitats (Nelson & McInnes, 2002). All tardigrades are dependent on water to feed, breathe, reproduce and move. In terrestrial habitats, limno-terrestrial tardigrades live in the water film that surrounds litter, moss and lichens where they share a micro-world with other organisms (Collembola, mites, rotifers, and nematodes) and endure extreme environmental conditions from flood to drought. Hydrophilous tardigrades are aquatic species that live only in permanent freshwater habitats, mostly in the littoral zone.
Diversity and distribution: Tardigrades include approximately 800 species world wide of which limno-terrestrial species account for the largest part. In North America, five species are strictly freshwater and 13–14 typically freshwater (Thorp & Covich, 2001), but the actual number is probably much higher. In Europe, there are 1000 species of limno-terrestrial and freshwater tardigrades (Fauna

Europaea, 2002). We could not find comprehensive data on possible tardigrade endemicity.

PHYLUM ANNELIDA

Three classes of annelids occur in freshwaters: Polychaeta or tube worms, Oligochaeta or aquatic earthworms, and Hirudinea including the leeches and blood suckers.

Class Polychaeta

Habitat: Polychaetes are typically marine and only a few of the more than 80 families have freshwater representatives.
Diversity and endemicity: There are 9000 polychaete species currently recognised (Rouse and Pleijel, 2001), but no estimates are available for freshwater species. Four freshwater polychaete species have been described from Lake Baikal (Timoshkin, 1997), but no additional data have been reported on freshwater polychaete endemicity or distribution.

Class Oligochaeta

Habitat and Diversity: The oligochaetes are well-represented in both marine and freshwater environments. Oligochaetes display the greatest diversity of the freshwater annelids with about 700 species world wide (Tim, 1985; Dudgeon, 1999) and are considered to be cosmopolitan (Brinkhurst & Wetzel, 1984).
Distribution: The total number of freshwater oligochaetes is distributed over the following regions: 419 Holarctic species, 170 Sino Indian, .83 Ethiopian, 133 Neotropical, 85 in Antarctica (Tim, 1985; Dudgeon, 1999). Regional estimations gave figures of 50 species in Southern Africa (Day & De Moor, 2002), 73 in North America (Brinkhurst & Gelder, 1991) and 85 species in Europe (Minelli, pers. com.).
Endemicity: Lake Baikal is a centre of diversity and endemicity for oligochaetes, with 164 endemic species out of a total of 194 described species (Timoshkin, 1997).

Class Hirudinea

Habitat: Leeches and blood-suckers are found in fresh and marine waters, but many terrestrial species occur in tropical regions. Leeches form an important component of the benthos of most lakes and ponds (Sawyer, 1986). They are either predators of macroinvertebrates or temporary ectoparasites of freshwater fishes, turtles, amphibians, water birds and occasional mammals (aquatic or terrestrial).
Diversity and endemicity: The total number of Hirudinea world wide might be a little less than 500 species (Minelli, pers. com.). A total of 69 species is recorded in North America (Thorpe & Covich, 1991).
Eleven endemic species of Hirudinea have been described from Lake Baikal out of a total of 13 described species (Timoshkin, 1997). In Lake Tanganyika, 20 Hirudinea species have been described, of which 12 are endemic (Coulter, 1991).

PHYLUM MOLLUSCA

Habitat and Diversity: Molluscs are well distributed in all marine, freshwater and terrestrial habitats. The total number of molluscs from marine, freshwater and terrestrial habitats is estimated to range between 80 000 and 135 000 (Seddon, 2000). In fresh and brackish waters some 5000–6000 species have been identified in 55–60 families of gastropods (snails and slugs) and bivalves (clams and mussels; Taylor, 1988) (Table 1).
Distribution: Molluscs occur throughout, the world except Antarctica. The present distribution is primarily controlled by habitat and life history, and means of dispersion, with a strong influence of geologic history. Most families are widespread, but a few are restricted to single lakes. Widely distributed species (cosmopolitan and circum-boreal) are those that can live in an extreme range of habitats.

The highest diversity of freshwater molluscs is recorded in North America (USA 945 sp.; Bogan, 1998). There are currently 177 described molluscan species in Europe (excluding the Mediterranean basin and ex-USSR; Falkner et al., 2001). Data on freshwater mollusc richness are either lacking or scattered for South America and Asia.
Endemicity: Endemism of molluscan species occurs in river, spring and lake systems. Groombridge and Jenkins (1998) identified 27 areas of special importance for freshwater mollusc diversity world wide:

Table 1. Global diversity of freshwater molluscs

	North America[1]	South America	Europe[2]	Africa	Asia	Australia[4]	World
Bivalves	344		50[9]	172[5]		37	1000
Gastropods	601						4000[7]
Prosobranchs	439		133	250[3]		156	
Pulmonates	162		30	91[3]		c.67	
Total mollusks	945			513	285 SE Asia[8]	260	5000

[1] Bogan (1998).
[2] Excluding the mediterranean Basinand exUSSR (Falkner et al., 2001).
[3] Excluding Hydrobiidae from Northern Africa (Brown, 1994).
[4] Valid described taxa from Ponder (pers. com.) based on (Smith,1992) and published an unpublished data, there are at least 179 undescribed species and nine introduced species.
[5] Daget (1998).
[6] India, Sri lanka, Pakistan, Nepal, Bhutan, Bangladesh, Afghanistan and myanmar. *Source*: Naggs (in prep.).
[7] Bogan (1998) and Revenga & Kura (2002).
[8] Continental South East Asia: (Davis, 1982).
[9] Giller & Malmqvist (1998).

- habitats from ancient lakes: Lake Baikal, Lake Biwa and Lake Tanganyika respectively exhibit 70, 52 and 64% of endemic molluscan species (Table 2).
- habitats from lower river basins (Congo, Volta, Mekong, Mobile Bay, Uruguay River and Rio de la Plata): 93% of the total molluscan species found in the Mobile Bay basin (Tombigbee–Alabama rivers-USA) are endemic (Seddon, 2000). Another notable center of endemism is found in the Lower 500 km of the Mekong river basin with 92% of endemic molluscan species (Seddon, 2000).
- habitats from springs and underground aquifers (Australia, New Caledonia, the Balkans, western US, Florida and Cuatro Cienegas basin in Mexico).

The phylum Mollusca includes two classes: the class Bivalvia (mussels and clams), and the class Gastropoda (snail and slugs).

Class Bivalvia

Habitat: Bivalves are mainly marine organisms with less than 10% of the total number of species being found in freshwater (Bogan, 1998).

Diversity: Bivalves comprise approximately 15 000 species world wide, including around 1000, freshwater species which belong to 18 families, nine of which have radiated (Bogan, 1998). Most freshwater bivalves are included in two orders: Unionidae and Veneroidae.

Distribution: The most diverse freshwater bivalve fauna is found in North America and is dominated by Unionids with 300 described species (Bogan, 1998). Europe has about 48 bivalve species not including invasive species: 16 Unionoidae species and around 32 Veneroidae species (Nagel et al., 1998). The total number of freshwater bivalves for Asia is not known although it is expected that Asia, China and South Asia have a high diversity of Unionoidae species, second only to east and central North America (Banarescu, 1990). The Unionidae include 38 species in China (Liu Yueyin, 1979), 53 described species in India (Subba Rao, 1989), and 33 described species in Thailand (Brandt, 1974).

Ponder (1997) listed 37 described bivalve species in Australia: 18 Hyriidae, two Corbiculidae and 17 Sphaeriidae. Comprehensive data on freshwater bivalve diversity could not be reported for South America, where only local or national checklists exist and where data have not yet been compiled for the two main bivalve families: the Mycetopodidae and the Hyriidae. There are 172 bivalve species recorded in Africa, one family being restricted to the African continent: the Mutelidae (Daget, 1998).

Endemicity: Lakes Baikal, Victoria, Tanganyika and Biwa are amongst the main critical areas of

Table 2. Number of mollusc species in key areas of endemism: the major ancient lakes of the world and some major river basins (based on Gargomony et al[1]. Groombridge & Jenkins, 1998; Seddon, 2000)

Lakes	Gastropods (endemic)	Bivalves (endemic)	Total (endemic)
Baikal[2]	150(117)	31(16)	181(133)
Biwa	38(19)	16(9)	54(28)
Sulawesi[3]	ca. 50 (ca. 40)	4(l)	ca. 54 ca. 41)
Tanganyika	68(45)	15(8)	83(53)
Malawi	28(16)	9(1)	37(17)
Victoria	28(13)	18(9)	46(22)
Ohrid	72(55)		
Titicaca	24(15)		
River Basins			
Mobile Bay Basin	118(110)	74(40)	192(150)
Lower Uruguay River and Rio de la Plata	54(26)	39(8)	93(34)
Mekong River (lower 500 km)[4]	121(111)	39(5)	160(116)
Lower Congo Basin	96(24)		
Lower Zaire Basin	96(24)		

[1] Hotspots of freshwater molluscan diversity prepared by O. Gargominy, A Bogan, P. Bouchet, W. Ponder: draft discussion paper from the Mollusc Specialist Group IUCN-SSC.
[2] Timoshkin (1997).
[3] Lake Poso and the Malili lakes system.
[4] Davis (1982).

freshwater bivalve endemism with respectively 40, 50, 53 and 56% of endemic bivalve species (Table 2). Hot spots of endemism for freshwater bivalves are also identified in North America, i.e. Mobile Bay Basin which harbours 54% of endemic bivalves (Table 2).

Class Gastropods

Habitats: Gastropods are among the most ubiquitous organisms, they have widely colonized all marine, freshwater and terrestrial habitats. Freshwater gastropods are especially abundant in shallow littoral zones in lakes and streams.

Diversity and Endemicity: Gastropods include some 40 000 species of which about 4000 species inhabit freshwater habitats (Bogan, 1998). Gastropods are divided in two sub-classes: Prosobranchia and Pulmonata.

Most endemic species belong to the class Prosobranchia. The ancient lakes represent Key areas of gastropod endemicity, the rate of endemism ranging from 50% (Lake Biwa) to 80% (Lake Baikal; Table 2). Caves and springs also are largely colonised by endemic gastropods (mainly prosobranchs-hydrobiids). In addition, some major river basins are rich in endemic populations of gastropods: the lower Congo basin, the lower Mekong, the Mobile Bay basin or the Ohio–Tennessee rivers (Table 2). Isolated areas or islands are also susceptible to harbour endemic populations, i.e. Madagascar exhibits 12 endemic gastropod species of the 30 known species (Groombridge & Jenkins, 1998).[2]

Distribution: North America contains the highest diversity of freshwater gastropods with 601 species representing 14 families (Bogan, 1998). Freshwater gastropod diversity is also high in Africa with 341 known species (Brown, 1994), while in western Europe (excluding the Mediterranean basin and ex-USSR) there are only 166 known species (Falkner et al., 2001). We could not report comprehensive data on freshwater gastropod diversity for South America and Asia; however, local diversity estimates for major river basins (Uruguay River and Rio de La Plata, lower Mekong; Table 2) indicate that the diversity of freshwater gastropods on these continents might be very high.

Prosobranchia. Habitat: Most Prosobranchia species are marine, but there are also many freshwater taxa and a few terrestrial forms. The freshwater prosobranchs tend to occur in large and old lakes in tropical regions (Bogan, 1998).

Diversity: The two most diverse freshwater prosobranch families are Hydrobiidae with 228 known species world wide; and Pleuroceridae with 196 known species world wide (Revenga & Kura, 2002).

Distribution: The Afrotropical fauna differs from that of India and Southeast Asia in its richness of species in the families Ampullaridae, Bithyniidae, Thiaridae and Planorbidae. Pleurocidae are especially diverse in rivers of the Southeast United States (Revenga & Kura, 2002).

Endemicity: Prosobranchs generated numerous endemic populations. In particular, many Hyd-

[2]Contribution of IUCN Mollusk Specialist Group.

robiidae genera present in subterranean waters and springs radiated into endemic populations (Hersler et al., 1990; Kabat & Hershler, 1993). Among the hot spots of prosobranch (mainly hydrobiids) endemicity in springs and underground aquifers, we can cite: the Balkan region (180 endemic/190 known species), the arid and semi-arid western part of the USA (58/c. 100), Florida (43/84); the Great Artesian Basin in Australia, New Caledonia (65/81) and Western Tasmania (Groombridge-; Jenkins, 1998). Brown (1994) noted the lack of any hydrobioid radiation in the Afrotropical region, contrasting it with the radiations reported for the Mekong river (Davis, 1982).

Endemic prosobranch species are also found in large ancient lakes in tropical regions where they represent most of the endemic gastropod populations, i.e. the Thiaridae show large endemic radiation in Lake Tanganyika. (Banarescu, 1990; Michel, 1994; Mandaville, 2000).

Pulmonata. Habitat: The sub-class Pulmonata is mainly composed of terrestrial species (land snails). Freshwater pulmonate species are preferably found in muddy habitats of ponds and eutrophic lakes of small or moderate size (Brusca & Brusca, 1990). Many freshwater Pulmonata species are commonly found in ephemeral habitats (Brusca & Brusca, 1990).

Diversity and endemicity: The pulmonates did not generate as many endemic populations as the prosobranchs but the less numerous taxa are much more widely distributed throughout the world with a particularly high diversity at the northern latitudes. There are 28 000 species of pulmonates in terrestrial, marine and fresh waters (Smith, 2001). The freshwater representatives are all included in two orders: Archaepulmonata and Basommatophora (Smith, 2001). Pulmonates have spread throughout the world with a limited number of taxa resulting in few endemic species (Banarescu, 1990).

Distribution: Freshwater pulmonates reach their greatest size and diversity at northern latitudes (Smith, 2001). Data on freshwater pulmonate species richness are lacking for South America (56 species in Argentina; Ituarte, pers. com.) and Asia (285 species in South East Asia of which 200 species are from India; Naggs, in prep.).

PHYLUM ARTHROPODA

Class Crustacea

Most crustaceans are marine but some 10% of the extant species are found in freshwater (Smith, 2001). They spend all, or most of, their life in water, some species being diadromous.

Subclass Branchiopoda

Habitat: Most brachiopods live in fresh or brackish water and a few are found in marine habitats. They occur in stagnant temporary pools and permanent waters, from stagnant to weakly running (Maeda-Martinez et al., 1997).

Diversity: Branchiopods constitute a diverse class of Crustaceans including about 1000 species (Dumont & Negrea, 2002). Over 400 species belong to the 15 or more families of the 'Phyllopoda' (Anostraca, Notostraca, Laevicaudata, Spinicaudata and Cyclestherida), and more than 500 belong to the 15 families of 'Cladocera'.

Super-order Cladocera. Habitat: Cladocerans (or water fleas) inhabit almost any kind of freshwater habitats from large lakes to ponds. There are a few estuarine species but this group has not been successful in the oceans. The freshwater cladocerans are a widespread group, living on or near the bottom or on aquatic vegetation. Some species are planktonic in the open water of lakes.

Diversity and distribution: According to Dumont & Negrea (2002), the number of continental Cladocera species is slightly more than 500. The total number of cladoceran species includes 40 cosmopolitan species, 190 restricted to tropical and subtropical lowlands, 150 from cool-temperate and altitudinal zones, and 30 with an unclear status. It is now demonstrated that many species first considered to be cosmopolitan were, in fact, groups or complexes of morphologically similar species, each species having a more restricted distribution than the original group or complex.

Super-order Sarsostraca

Order Anostraca. Habitat: The fairy or brine shrimps are usually found in temporary ponds and pools; as well as in freshwater, alkaline or hypersaline lakes.

Diversity: There are 273 recorded species of Anostraca world wide (Belk & Brtek, 1995, 1997;

Dumont & Negrea, 2002). The most familiar members of the Anostraca are species of the genus *Artemia*.

Distribution and endemicity: In South Africa (Africa south of the Zambei and Kunene Rivers), 46 anostracan species are currently known, including 80% of endemic species (Hamer & Brendonck, 1997). Endemism of fairy shrimps is also abundant in South West Australia (12 endemic/19 total species), in Western USA (13 endemic/26 total species) and in the Southern part of South America (14 endemic/18 total species; Belk, pers. comm. in Groombridge & Jenkins, 1998). In Europe, Italy hosts seven endemic species on a total of 16 species (Belk, pers. comm. in Groombridge & Jenkins, 1998).

Super-order Conchostraca

Orders Laevicaudata, Spinicaudata and Cyclestherida. Habitat: Conchostracans are found exclusively in freshwater, usually on the bottom of temporary puddles or sometimes almost entirely dug into the mud. Conchostracans develop very quickly depending on temperature; they are thus well adapted to the extreme conditions of temporary water bodies.

Diversity: Conchostracans or 'clam shrimps' comprise about 130 species described world wide (Dumont & Negrea, 2002).

Super-order Notostraca and Haplopoda. Two other super orders of Branchiopoda: Calmanostraca (order Notostraca) and Leptodorida (Order Haplopoda) lack information about habitat, diversity and distribution. Notostraca (or 'tadpole shrimps') include the genera *Triops* with four species and the genera *Lepidurus* with five species (Dumont & Negrea, 2002). Haplopoda includes to date a single family with a single species, *Leptodora kindtii.*

Subclass Ostracoda

Habitat: The class Ostracoda (Seed shrimps or mussel shrimps) contains both marine and freshwater forms. Ostracods occur in practically every aquatic environment, from the abyssal depths to the shoreline. They inhabit estuaries, lagoons as well as freshwater lakes, ponds and streams, salt lakes, hot springs, and even damp vegetation. Ostracods may be free-swimming for all or part of their life-cycle, but, more commonly, they are benthic, living among aquatic plants or crawling on or through the sediment. A number of interstitial forms are known. Some are parasitic or commensal on other crustaceans, polychaete worms, echinoderms and even sharks.

Diversity: According to Kempf (1980, 1997), there are at present ca 30 000 ostracod species described, of which 8000 are freshwater species and 4500 ostracod genera, including 750 freshwater. However, these figures include nominal species (no synonymies are taken into account), while both living and fossil taxa are included here. From three checklists on recent non-marine ostracod faunas on Europe, Africa and South America (Limnofauna Europaea, Martens, 1984; Martens & Behen, 1994), it follows that presently some 500 described species occur in each of these continents and that they have few species in common. This would mean that there are ca. 3000 extant freshwater ostracod species, belonging to ca. 150 genera. In addition, about 5000 freshwater fossil species belonging to 600 genera have been described.

All freshwater Ostracods belong to the Podocopida, a group which is estimated to comprise around 5000 living species (Dole-Olivier et al., 2000). Most species are found in lakes, but recently it was shown that the diversity of subterranean (Danielopol & Rouch, 1991) and terrestrial ostracod faunas (Pinto et al., in press) is much higher than previously assumed.

Distribution and endemicity: There are few cosmopolitan freshwater ostracod species, most have more restricted distributions. All ancient lakes, for example, hold dozens of endemic species, for Lakes Baikal and Tanganyika these numbers could be up to 200 species for each lake (Martens, 1984). Surprisingly, the levels of endemicity reported for the African ancient lakes are only matched by the temporary pool faunas of South West Africa (Martens, 1998).

Subclass Copepoda

Habitat: Copepod habitats range from freshwater to hyaline conditions, from subterranean caves to streams, rivers, and lakes. Copepods may be free-living, symbiotic, or internal or external parasites on almost every phylum of animals in water (2000 described species are parasites of fish).

Table 3. Global diversity of freshwater crustaceans

Group	North America	Europe	Asia	Australia	South America	Africa	World
Branchiopoda							
Cladocera	140[1]						500[4]
Phyllopoda	67	72					420
Ostracoda	420[2]	400[6]			500[7]	500[8]	2000
Copepoda[5]	363	902	927	181	516	524	2085
Branchiura	23						
Malacostraca							4200

[1] Pennak (1989).
[2] Thorp & Covich (1991).
[3] Sternberg & Cumberlidge (2001).
[4] Dumont & Negrea (2002).
[5] D. Defaye (pers. com.).
[6] Giller & Malmqvist (1998).
[7] Martens (1984).
[8] Martens & Behen (1994).

Diversity: Among the ca. 13 000 species of copepods (free-living and parasites) more than 2900 species of free-living freshwater copepods are presently identified (Dussart & Defaye, 2002; D. Defaye, pers. com.) (Table 4).
Distribution: In freshwaters, three orders are dominant: Calanoida, Cyclopoida and Harpacticoida. Freshwater calanoid species have a relatively restricted distribution and the tropical and temperate species do not generally overlap (Dussart & Defaye, 2002). Two genera are dominant planktonic cyclopids in the tropics: *Mesocyclops* and *Thermocyclops* (Dussart & Defaye, 2002). Harpacticoids are the most numerous, but it is estimated that only 50% of the species have been described in interstitial waters and ground waters (Galassi, 2001).

Endemicity: some of the key hotspots of endemicity or copepods are: Lake Baikal (27 endemic/65 total described species; Timoshkin, 1997), and Lake Tanganyika (33/69; Coulter, 1991).

Subclass Malacostraca

The Malacostraca include about two thirds of all crustacean species, and contain all the larger forms such as shrimps, prawns, lobsters and crabs. There are about 4100 malacostracan species world wide (Table 5).

Order Mysidacea. Habitat: Mysids (opposum shrimps) inhabit coastal and open sea waters, as well as continental freshwaters, several taxa occur also in different groundwater habitats (Smith, 2001).

Table 4. Global Diversity of freshwater copepods (from Dussart & Defaye, 2002 updated by D. Defaye)

Order	North America	Europe	Asia	Austr. N. Zeal	South America	Mexico Central America	Africa	World
Calanoida	111	119	294	63	123	37	113	678
Cyclopoida	105	277	308	52	203	118	228	1045
Harpacticoida	147	504	325	66	190	61	183	1260
Gelyelloida		2						2
Total	363	902[1]	927[2]	181	516	216	524[3]	2080

[1] To the Urals, Caucasus not included.
[2] Plus Turkey, Philippines, Indonesia, Malaisia.
[3] Plus Madagascar.

Table 5. Global diversity of freshwater Malacostraca

Order	North America	Europe	Asia	Australia	South America	Africa	World
Mysidacea	3[1]	20[1]					43
Cumacea							20
Tanaidacea							2
Isopoda	130						> 660
Amphipoda	150[3]	350[3]					1700[2]
Decapoda							
Brachyura	1	3		30	234 + 90 Central Am	96	950[4]
Astacidae	342[5]						600
Caridea	15						221

[1] Pennak (1989).
[2] McAllister (1997).
[3] Giller & Malmqvist (1998).
[4] Sternberg & Cumberlidge (2001).
[5] Thorpe & Covich (2001).

Diversity and endemicity: Mysids include more than 1000 species (Abele, 1982). Only 25 mysid species occur in freshwaters and 18 additional species live in freshwater caves (Abele, 1982). The major family *Mysidae* shows the highest diversity in the basins of the Caspian, Azov and Black Seas, with 11 endemic genera.
Distribution: Mysids are widespread over all the continents. There are only three freshwater species in the US, and about 20 in Europe (Pennak, 1989).

Order Cumacea. Habitat: Cumaceans occur from tidal to abyssal depths in marine and brackish waters throughout the world (Pennak, 1989). Only one family has brackish and freshwater species that live in the Ponto-Caspian basin (Banarescu, 1990).
Diversity and endemicity: There are 20 species in the Caspian Sea (nine endemic species, seven endemic genera).

Order Tanaidacea. Habitat: While the vast majority of tanaidaceans are marine, a small number of species are found in brackish water.
Diversity and distribution: Two freshwater species are found in very distant locations. *Tanais stanfordi* on oceanic islands and *Nesotanais lacustris* on Rennell Island (Banarescu, 1990).

Order Isopoda. Habitat: Isopods or 'sow bugs' are common inhabitants of nearly all environments, aquatic and terrestrial.
Diversity: The isopods include approximately 10 000 described species, in 10 suborders. Freshwater species are mainly represented by Asellota with about 660 species world wide and 130 species in the US (Smith, 2001).
Diversity and endemicity: There are only a few isopod species in Lakes Tanganyika (three species, Coulter, 1991) and Baikal (five species, Hidding et al., 2003) but all of them are endemic.

Order Amphipoda. Habitat: A large number of amphipods is found in subterranean waters (Holsinger, 1993). These are stygobionts, and thus restricted to hypogean waters. However, the majority of the amphipods is epibiont.
Diversity and endemicity: According to McAllister et al. (1997) freshwater amphipods represent 24% of the 7000 known Amphipoda, which would amount to 1700 species world wide. Giller & Malmqvist (1998) provide lower estimates: 900 freshwater amphipod species are thought to be known world wide. The 740 species of subterranean Amphipoda. belong to the suborder Gammaridea (with the possible exception of the ingofiellids) an make up approximately 13% of the estimated 5700 gammaridean species (Holsinger, 1993). A large fraction (94%) of the stygobiont Amphipoda occur in only 12 families: the four most important ones are Niphargidae, Crangonyctidae, Hadziidae, and Bogidiellidae (Holsinger, 1993).

There are currently 345 amphipod species recorded from Lake Baikal, only one of these is not endemic to the lake (Timoshkin, 1997). The Baikalian amphipods are an extensive and at least partially adaptive radiation, with representatives having invaded almost every possible niche in the lake, even the pelagic. Kamaltynov (1999) has raised the level of endemicity to that of family by recognising four endemic families, while Väinölä & Kamaltynov (1999) recognised the existence of many cryptic (but reproductive isolated) species, so that the total specific amphipod diversity of Lake Baikal could exceed 1000 species.

Distribution: In Europe, the number of freshwater amphipods; is estimated to be around 350 species, while 150 described species are known from North America (Giller & Malmqvist, 1998). About 116 of the latter are groundwater amphipods (Thorp & Covich, 2001; Groundwater Amphipod Database). The current knowledge of the freshwater amphipod fauna in South America is neither up-to-date nor complete.

Order Decapoda

INFRA ORDER BRACHYURA (TRUE CRABS)

Habitat: Among the Brachyura, there are few freshwater crabs, all of these appear to be restricted to warmer waters.

Diversity and endemicity: There are about 950 freshwater crabs world wide (Sternberg & Cumberlidge, 2001). Three freshwater crab species can be found in Europe and 1 in North America (Giller & Malmqvist, 1998). In Africa, there are 96 species reported for the entire continent with 32 species in West Africa (Cumberlidge, 1999) and 12 species reported from southern Africa (Day et al., 2001). In Central America, about 90 species are recorded from Mexico to Panama including the Caribbean islands (Groombridge & Jenkins, 1998).[3] In South America, two families of freshwater crabs are present: the Trichodactylidae with 44 species and the Pseudothelphusidae with 190 species (Rodriguez, 1982, 1992). In Asia, the greatest diversity of freshwater crabs is found in Sumatra, Java, Borneo, Sulawesi and the Southern Philippines. Over a hundred species is found in Northeast India, Myanmar, Thailand, the Mekong basin in southern Indochina, to the Malaysian peninsula and Singapore (Groombridge & Jenkins, 1998). Thirty species are recorded from New Guinea and Australia (Groombridge & Jenkins, 1998).

Endemic populations of freshwater crabs are mostly found in Asia and Africa. In Asia, South China hosts 160 species, most of which are endemic; the west coast and south part of the Indian peninsula are known to harbour 20 endemic species, while there are 16 endemic species in Sri Lanka (Groombridge & Jenkins, 1998). In Africa, Lake Tanganyika hosts 8 endemic freshwater crab species (10 species in total, Coulter, 1991), a further 10 endemic species inhabit Madagascar, while also the Niger-Gabon area shows 10 endemic species (Cumberlidge, 1999).

Distribution: Freshwater crabs dominate tropical freshwaters and warm temperate zones of Central and South America, Southern Europe, Africa and Madagascar, South and Southeast Asia, China, Japan, The Philippines, New Guinea and Australia. They are absent from oceanic island in Atlantic and Pacific (Revenga & Kura, 2002).

INFRAORDER CARIDEA

Habitat: Most species of Caridea inhabit freshwater habitats or occasionally brackish waters, some species of the genus *Macrobrachium* being marine as juveniles (Jayachandran, 2001). A few troglobitic taxa are restricted to caves.

Diversity and endemicity: According to Jayachandran (2001), the freshwater representatives belong to three genera: *Palaemon* with one species, *Palaemonetes* with 20 species, and *Macrobrachium* with some 180 valid species and subspecies. There are also nine genera and some 20 troglobitic species. Lake Tanganyika harbors 15 species of which 14 are endemic (Coulter, 1991).

Distribution: Ten *Palaemonetes* and five *Macrobrachium* species are reported in North America (Thorp & Covich, 1991). Most species of *Macrobrachium* are pantropical and subtropical. Other freshwater representatives of Caridea belong to the family *Athyidae* represented by three endangered species in North America and by four species in Brazil (Jayachandran, 2001).

[3]Based on the contribution of N. Cumberlidge and R. von Sternberg.

INFRA ORDER ASTACIDEA

Habitat: Crayfish and true lobsters are represented in freshwater and marine environments. Three families of Astacidea are restricted to freshwater habitats: Astacidae, Cambaridae and Parastacidae.

Diversity and Endemicity: There are 590 freshwater species recognised world wide of which 342 native species of crayfishes have been identified from USA and Canada, (Thorp & Covich, 2001; Hobbs, 1989). More than 65 taxa are known from a single locality or a single river or drainage (Taylor et al., 1996).

There are two well identified centres of species endemism for freshwater crayfishes: the first is located in the southeastern United States (including the eastern and southern Mississippi drainage) were some 80% of the cambarid species can be found (Horwitz, 1990); the second is centred on Victoria (Australia) and Tasmania and houses a large proportion of the parastacid species (K Crandall pers. com. in Groombridge & Jenkins, 1998).

Distribution: In the Northern Hemisphere, the astacids are widespread in Europe and the Western United States while the cambarids occur in Eastern North America and parts of Asia, the latter representing 70% (404 sp.) of all known freshwater crayfish. In the Southern Hemisphere, only the parastacids (156 sp.) are found (Crandall, 2003).

Class Arachnida

Habitat: Arachnids (spiders, mites and ticks, scorpions) are mainly terrestrial invertebrates. There are a few freshwater spiders and a large number of water mites in the sub-class Acari. Water mites can be parasitic, for example on insects, as larvae and are predatory as deutonymphs and adult. A few taxa are parasitic on bivalves or crayfish in their post-larval stages (Walter & Proctor, 1999).

Diversity and endemicity: The class Arachnida is represented in freshwater by few genera of semi-aquatic true spiders (*Dolomedes, Argyroneta*) and a large number of water mites. The freshwater mites belong to five unrelated groups that have independently invaded freshwaters. The Hydracarina or Hydrachnidia is the most diverse group with more than 5000 named species world wide in 300 genera (Smith & Cook, 1991; Giller & Malmqvist, 1998).

Distribution: Over 1500 species are currently estimated to occur in North America, but only half of these have been formally described. In Australia, there are 413 described species of Hydracarina in 89 genera, representing 22 families (Harvey, 1998).

Class Entognatha (Order Collembola)

Habitat: Collembols or springtails are primarily terrestrial, soil and litter dwelling, preferring however wet or damp surroundings (Hopkin, 1997). The majority of species associated with aquatic habitats are accidentals, often as temporary inhabitants of the water surface. However, virtually all collembolans have an affinity for areas marginal to aquatic habitats because of their requirement of high humidity. A few species are restricted to aquatic habitats and exhibit a high degree of specialisation for an aquatic existence (Heckman, 2001).

Diversity: There are ca 9000 described species world wide (Resh & Carde, 2003). Thirty species (4 families) are associated with European waters (Gisin, 1978 in Ward, 1992), and 50 aquatic species are found in North America (Waltz & McCafferty, 1979 in Ward, 1992).

Class Insecta

Habitat: Insects are mainly terrestrial but some groups are denominated as 'aquatic insects' because they spend at least part of their lives in aquatic environments. These 'aquatic insects' most likely evolved as terrestrial organisms and later on adapted to freshwater; they are thus secondary aquatic organisms. They occur in all freshwater environments, sometimes in brackish waters, but they almost never colonised marine systems.

Diversity: More than one million insect species have been described, that is over 50% of all known organisms. Only 2% of all insect species exhibit aquatic stages representing a total of about 50 000 species (Table 6).

Order Ephemeroptera. Habitat: Mayflies are hemimetabolous insects; the nymphs are aquatic and the adults are terrestrial. Nymphs are found in virtually all types of freshwaters throughout the world. (Williams & Feltmate, 1992). The most diverse fauna occurs in warm lotic habitats.

Table 6. Estimates of the number of aquatic insects in the world and for continents or large biogeographic areas. (adapted and completed from Hutchinson 1993)

	Afrotropic	Neartic	Paleartic oriental	Europe	Neotropic	Oriental	Australian	World
Ephemeroptera	295[1]	670[6]		350[6]			84[4]	>3000
Odonata	699[1]	> 650[5]		150[7]			302[4]	5500
Plecoptera	49[1]	578[4]		423[7]			196[4]	2000
Megaloptera	8[1]	43[4]		6[4]	63[3]		26[4]	300
Trichoptera	>1000[1]	1524[1]	1228[1]	1724[1]	2196[2]	3522[1]	1116[1]	>10000
Hemiptera		404[4]		129[4]			236[4]	3300
Coleoptera		1655[4]		1077[4]			730[4]	>6000
Diptera		5547[4]		4050[4]			1300[4]	>20000
Orthoptera		ca 20		0				ca 20
Neuroptera		6[4]		9[4]			58[4]	ca 100
Lepidoptera		782[8]		5[4]				ca 1000
Hymenoptera		55[4]		74[4]				>129

[1] Elouard & Gibon (2001).
[2] Flint et al. (1999).
[3] Contreras-Ramos (1999).
[4] Hutchinson (1993).
[5] Ward (1992).
[6] Resh (2003).
[7] Limonofauna Europaea (2003).
[8] Lange (1996).

Diversity: The Ephemeroptera either comprise a little over 2100 described species (Williams & Feltmate, 1992) or more likely 3000 species in 375 genera (Resh & Carde, 2003).
Distribution and endemicity: A high diversity of Ephemeroptera is found at northern latitudes and especially in North America (McCafferty, 1996). The rate of endemism is high (99%) for the 111–172 known ephemeropteran species from Madagascar, because ephemeropterans are poor flyers and do not disperse easily (Elouard & Gibon, 2001). The most diverse ephemeropteran fauna occurs in warm lotic habitats (Wiggins & Mackay,1978; Ward & Berner, 1980). The mayfly faunas of Australia, New Zealand and temperate South America are similar, suggesting a Gondwanean origin for most Australian families: Leptophlebiidae, Siphlonuridae, Oniscigastridae, Ameletopsidae, and Coloburiscidae (Peters & Campbell, 1991).

Order Odonata. Habitat: Also Odonata are hemimetabolous insects with aquatic nymphs and terrestrial adults. There are a few truly marine species, several that live in brackish water, and many that survive in arid regions where the larvae can develop quickly in the warm waters of temporary ponds before they dry up.
Diversity: Extant Odonata species belong to three suborders: Anisozygoptera containing only two species, Zygoptera or damselflies, and Anisoptera or true dragonflies. A total of around 5500 species has been described in the world (Ward, 1992; Williams & Feltmate, 1992). According to Corbet (1999), there are about 2500 species of Zygoptera and considerably more than 2500 described species of Anisoptera.
Distribution and endemicity: Odonates are distributed from the tropics to the tree line in polar regions. In Australia, there are 314 known species of dragonflies, half of these are endemic and appear to be Gondwanean relics (Watson et al., 1991). The two species of Anisozygoptera are restricted to Japan and the Himalayas (Tsuda, 2000). Endemism at species level is high in Madagascar for Zygoptera (94.4%) but lower for Anisoptera, which are the better fliers (52%; Elouard & Gibon, 2001). The level of endemism is also high for Australian odonates in cold water genera present

in Tasmania and southwest Australia (Peters & Campbell, 1991).

Order Plecoptera. Habitat: Stoneflies are hemimetabolous insects. The nymphs are aquatic and usually occur in cool or cold running waters typically below 25 °C and with a high oxygen content.
Diversity and Endemicity: Some 2000 species have been described world wide (Williams & Feltmate, 1992; Ward, 1992).
Distribution: Stoneflies are adapted to low temperatures so they are mostly restricted to higher latitudes and altitudes. Most extant families of plecopterans are confined to the temperate zones of either the Southern or the Northern Hemispheres (Illies, 1965; Zwick, 1980). Consequently the highest number of plecopteran species are encountered in North America (478 sp.) and in Europe (423 sp.).

Order Hemiptera – Heteroptera. Habitat: Heteroptera are hemimetabolous insects. Most species of Hemiptera or true bugs are terrestrial. All species of the sub-order Homoptera are terrestrial with a few semi-aquatic species associated with vegetation on the intertidal zone or freshwater margins (Foster & Treherne, 1976; Polhemus, 1984). Within the suborder Heteroptera, which is primarily terrestrial, there are several truly aquatic or semi-aquatic families. According to Hutchinson (1993) two separate groups of Heteroptera are associated closely with water in two different ways: one group consists of the Gerromorpha, which live primarily on the water surface and floating vegetation though, at least, a few can dive; the other group consists of the Nepomorpha that spend most of their life under water (water bugs).
Diversity: The world fauna of aquatic and semi-aquatic heteropterans comprises over 3300 species in 16 families (Ward, 1992). Williams & Feltmate (1992) mentioned 3200 hydrophilic species of Heteroptera.

Order Orthoptera. Habitat: The grasshoppers, locusts and crickets constitute an almost exclusively terrestrial order. Orthopterans are hemimetabolous insects. They are not usually thought of as being aquatic, nor even semi-aquatic, but several species live in association with water (hydrophilous species). Semi-aquatic species frequent wet margins of freshwater bodies or are found on emergent littoral vegetation (Ward, 1992).
Diversity: There are 20 described semi-aquatic species of Orthoptera world wide (Hutchinson, 1993).

Order Trichoptera. Habitat: The caddisflies are holometabolous insects: the larvae and pupae live in most types of water bodies including both cold and warm springs, temporary waters, etc. (Williams & Feltmate, 1992).
Diversity: About 10 000 described species of caddisflies are known world wide (Williams & Feltmate, 1992). Flint et al. (1999) suggested that the real number of trichopteran species is around 50 000.
Distribution and endemicity: Trichopteran species are distributed over most of the globe, except Antarctica. However, while some families occur world wide, others are restricted to either the northern or southern hemisphere (Ward, 1992; Williams & Feltmate, 1992). Africa has the lowest diversity in Trichoptera (Elouard & Gibon, 2001). The trichopteran diversity is higher in the tropical regions, but many trichoptera species can also be found in high elevation areas of temperate regions (Elouard & Gibon, 2001). In Australia, most trichopteran species are endemic with hotspots located in Tasmania and in the Southwest of Australia (Peters & Campbell, 1991).

Order Megaloptera. Habitat: The Megaloptera is a small order of holometabolous insects: all species are aquatic as larvae and terrestrial as eggs, pupae and adults.
Diversity and Endemicity: About 250–300 species of Megalopterans are described world wide (Ward, 1992; Williams & Feltmate, 1992).
Distribution: All living species belong to one of the two super-families. the Corydalidae are widely distributed throughout temperate regions but they are absent from Europe with a few species in the tropics; the Sialidae are confined to temperate latitudes (Ward, 1992; Williams & Feltmate, 1992).

Order Coleoptera. Habitat: Coleoptera is a holometabolous order with primarily terrestrial species, but about 10% of the coleopteran families have

aquatic or semi-aquatic species. Aquatic beetles inhabit freshwater, brackish-water and some marine environments. The most diverse and abundant fauna occurs in well-vegetated freshwater habitats (Williams & Feltmate, 1992). Unlike many of the preceding groups, in aquatic beetles both larvae and adults can be aquatic.
Diversity: The Coleoptera are one of the most diverse groups of living organisms with over 500 000 estimated species. The world fauna contains about 6000 aquatic and semi-aquatic beetles (Hutchinson, 1993; Williams & Feltmate, 1992). The most important families of aquatic beetles are the Dysticidae (4000 species), the Gyrinidae (700 species) and the Noteridae (150 species; Hutchinson, 1993). Revenga & Kura (2002) suggested much higher estimates for aquatic beetles, at approximately 10% of all Coleoptera, which would account for 35 000 aquatic species of Coleoptera.
Distribution: Most aquatic beetles are thought to be cosmopolitan (Dysticidae & Gyrinidae), or widespread (Noteridae; Hutchinson, 1993).

Order Diptera. Habitat: Although most species are terrestrial, those with aquatic larvae may be predominant insects in many freshwater habitats. Dipterans are holometabolous and are found in every conceivable aquatic environment and are often the only insects in freshwater habitats with extreme environmental conditions. At least 30 families have aquatic or semi-aquatic representatives (Williams & Feltmate, 1992; Hutchinson, 1993).
Diversity: Over half of all known aquatic insects are dipterans. The order contains an estimated 200 000 species world wide, although only just over half have been described (Hutchinson, 1993). The number of dipteran species breeding in water is estimated to be >20 000 (Resh, pers. com.).

Lane & Crosskey (1993) estimated the number of species in groups of medical importance: Tipulidae, 14 000 sp.; Culicidae, 3450 sp.; Anophelidae, 420 sp.; Simulidae, 1570 sp.; Ceratopogonidae, 5000 sp.; Tabanidae, 4000 sp.
Distribution: The Tipulidae are distributed world wide, although their greatest diversity is in the humid tropics. Stone et al. (1965) recorded 1458 species in America north of Mexico. The Simulidae or black flies occur on all the major landmasses, apart from Antarctica. The Tabanidae or Horse flies are distributed throughout the world. The Chironomidae are non-biting midges whose distribution extends to both the northern and southern limits of land, and they are the dominant group in the Arctic.

Order Lepidoptera. Habitat: The butterflies and moths are traditionally considered a terrestrial group, but there are nevertheless some aquatic or semi-aquatic species. Lepidopters are holometabolous insects. With few exceptions, aquatic and semi-aquatic Lepidoptera are intimately associated with aquatic vascular plants, and thus most species occur in ponds and the littoral of lakes. Most of the truly aquatic species occur in the family Pyralidae. All pre-imaginal stages of aquatic pyralids occur in the water while adults are terrestrial.
Diversity and distribution: 'The total number of aquatic and semi-aquatic described species of Lepidoptera has not been compiled. However, over 782 species of Lepidoptera with aquatic and semi-aquatic stages are found in North America (Lange, 1996). There are 148 pyralids in North America and 1670 pyralids in Australia (Lange, 1996; Nielsen, 1999).

Order Neuroptera. Habitat: Neuropterans (also called planiplenns) are holometabolous insects with mostly terrestrial species. Three families contain truly aquatic members: Sisyridae or spongillaflies, Osmylidae, and Neurorthidae, whose larvae are semi-aquatic and live in wet margins along freshwater bodies (Williams & Feltmate, 1992). The species of Sisyridae are associated to freshwater sponges. This constitutes a rare example of a single family of an otherwise terrestrial insect group depending for its existence on a small freshwater family (Spongillidae) of an otherwise exclusively marine phylum (Ward, 1992).
Diversity: There are some 4300 species of Neuroptera world wide, of which ca. 100 species are considered as aquatic or semi-aquatic (Hutchinson, 1993). There are 45 species of spongillaflies distributed throughout the world. (Ward, 1992).
Distribution and endemicity: Most of the known aquatic species of Neuroptera are found in Australia (58 sp.) while the diversity is much lower in Europe (9 sp.) or North America (6 sp). We cannot report comprehensive data on neuropteran endemicity.

Order Hymenoptera. Habitat: Hymenoptera is a holometabolous order with a few aquatic or semi-aquatic species. Some species are only marginally aquatic while others might spend most of their life stages under water (Ward, 1992). Several families contain species that are, in some way, associated with water such as the parasitic wasps depending on aquatic hosts. The adult enters the water to attack the hosts, usually the aquatic stage of other insects, except Collembola, Ephemeroptera and Plecoptera. The specific habitat of aquatic Hymenoptera is dictated by the habitat of the hosts that can occur in both running and standing freshwaters (Williams & Feltmate, 1992; Ward, 1992).
Diversity and distribution: There is currently a total of ca. 100 parasitic wasps with aquatic hosts world wide. There are 10 North American wasp families and 55 species of parasitic wasps are considered aquatic (Hagen, 1996 in Merritt & Cummins, 1996).

PHYLUM CHORDATA (SUBPHYLUM VERTEBRATE)

Class Teleostomi – Subclass Actinopterygii (Fish)

Habitat: Fish are well represented in both marine and freshwater habitats. Some species also migrate between salt and freshwater (diadromous species)
Diversity: An estimated 24 600 valid fish species have been described world wide. Nelson's (1994) estimate of a total of 28 500 fish (marine and freshwater) is probably reasonable. Although fish are among the best-known freshwater groups, the rate of discovery of new species was on an average of 309 species a year for the period 1976–1994, indicating that many species are still to be discovered. Presently, ca. 10 000 species of fish live principally in freshwaters, and an additional 500 are diadromous.
Endemicity: Hot-spots of endemicity of freshwater fishes occur in tropical freshwaters, which are currently the least investigated areas but also the most threatened biota. According to our present knowledge, there are 'flocks' of several hundreds of cichlid species in the largest East Africal Lakes (Victoria, Malawi, Tanganyika) (Lévêque, 1997; Lévêque & Paugy, 1999). In Lake Tanganyika evolution has led to the occurrence of species flocks within a few families : seven Mastacembelid species, six species of the Bagrid *Chrysichthys*, seven species *Synodontis* and four species of the Centropomi. The remarkable diversity of the large barbs (genus *Barbus*) in Lake Tana (Ethiopia) constitutes a potential species flock (Nagelkerke et al., 1994). A totally endemic fish flock, comprising 29 species (11 genera) of sculpins (Cottotoidei) occurs in Lake Baikal (Asia) (Timoshkin, 1997). In lake Titicaca (South America) 24 endemic *Orestias* species (Cyprinodontidae) are presently recognised (Lauzanne, 1982). These species-flocks are sometimes considered to be a world heritage which is endangered and has to be preserved from destruction by human activities such as over fishing or introductions of aquatic species.

While the high fish species endemicity in lakes has promoted a considerable interest among biologists, the endemicity in river systems is less known.
Distribution: The ichthyofaunas of Europe west of the Ural mountains (ca. 360 sp.), of North America (ca. 1050 sp.) and Australia-New Guinea (ca.500 sp.) are the most thoroughly documented to date (Lundberg et al., 2000). The current estimate for Africa is ca. 3000 species (Lévêque, 1997) but the number of some families such are Cichlidae is likely to be greatly underestimated. The neotropical ichthyofauna (Central and South America) is estimated to include from 3500 species (Stiassny, 1999) to more than 5000 species (Lundberg et al., 2000). For tropical Asia,

Table 7. Distribution of fish species diversity at continental scale

Zones	Species number	FAO areas fishbase
Europe + (USSR)	360[1]	393 + 448
Africa	3000[2]	3042
North America	1050[1]	1542
South America	5000+[1]	3731
Asia	3500+[3]	3443
Australia–New Guinea	500[1]	
Australasia		616
Total	13 400	13 215

[1] Lundberg et al. (2000).
[2] Lévêque (1997).
[3] Kottelat & Whitten (1996).

extending from the Indus basin eastward to South China, an estimate of 3000 species is proposed by Lundberg et al. (2000). Kottelat & Witten (1996) gave a figure of more than 3500 species for Asia. In China, about 900 freshwater fishes are recorded (He & Chen, 1996) (Table 7).

Class Amphibia

Habitat: Amphibians are strictly freshwater animals and do not tolerate salt water. The great majority of amphibian species have aquatic larval stages and therefore depend on inland waters for implementing their life cycle and for continued survival of populations. However, relatively few species are fully aquatic.

Diversity: A total of 5504 amphibian species have currently been recorded world wide (World Database 'Amphibian species of the world'), including 4837 Anura (frogs and toads), 502 Caudata (newts and salamanders) and 165 Gymnophiona (caecilians).

Distribution and endemicity: Corbett (1999) reported a total of 74 amphibians known in Europe, but Gasc et al. (1997) listed only 62 species of European amphibians. Approximately 230 species of amphibians occur in the continental United States. In Japan a total of 56 amphibians have been described to date, many of which are salamanders. Most endemic species in the western US are widely dispersed while endemics in the eastern and southeastern US tend to be clustered in centres of endemism.

Class Reptilia

The class Reptilia includes turtles, crocodiles, lizards, and snakes. All crocodilians and many turtles inhabit freshwaters but nest on land. Many lizards and snakes occur along water margins; a few snakes are entirely aquatic.

Subclass Parareptila (Turtles). Habitat: There are two suborders of living turtles: the Pleurodira or side-necked turtles which are found in the south-hemisphere and are semi-aquatic turtles; the Cryptodira, or hidden-necked turtles including all turtles of the world, marine, freshwater and terrestrial.

Diversity: There are around 200 species of freshwater turtles throughout the warm temperate and tropical regions of the world (IUCN/ SSC Tortoise and Freshwater Turtle Specialist group 1991).

Distribution: The side-necked turtles include two families, the Chelidae with 36 species distributed in Australia, New Guinea, and South America and the Pelomedusidae with 23 species, found in Africa and South America (Ernst & Barbour, 1989).

The freshwater and semi-aquatic hidden-necked turtles include three species of Chelydridae (snapping turbes) found in North America and China; 22 species of Trionychidae (soft-shelled turtles) distributed in Africa, Asia, Indonesia, Australia and North America; 22 species of Kinosternidae (mud and musk turtles) exclusively on the American continent; 91 species of Emydidae (pond, box or water turtles) represented in all parts of the world except for Australia and Antarctica.

Subclass Diapsida

Lepidosauromorph

Crocodilians

Diversity and Distribution: There are 23 living crocodilian species widespread throughout tropical and subtropical countries. Alligators and caimans (family Alligatoridae) are found almost exclusively in North, Central and South America. The sole exception is the Chinese alligator, which occurs in Eastern China. A few species of the family Crocodylidae (true crocodiles) is present on the American continent, but most Crocodylidae are found throughout Africa, India and Asia. The single member of the family Gavialidae (Indian gavial) is found in India and adjacent countries.

Fresh water snakes

Habitat: Two species belonging to the family Acrochordidae are strictly adapted to freshwater habitats: the file snake (*Acrochordus arafurae*) and the Javan wart snake (*A. javanicus*). In addition, there are a few species of semi-aquatic snakes, such as the green anaconda (*Eunectes murinus*), one of the largest snakes in the world.

Diversity and distribution: The file snake and the Javan wart snake are both found in the Indo-Pacific region. Revenga & Kura (2002) reported an estimate of seven species of semi-aquatic snakes. Kottelat & Whitten (1996) mentioned 24 species of freshwater semi-aquatic snakes in Asia.

Saurishia – Class Aves (Birds)

Habitat: No birds are entirely aquatic, because avian eggs cannot survive a prolonged immersion in water. It is impossible to clearly separate inland water species from primarily terrestrial ones. Aquatic birds may be defined as wading, swimming and diving birds of either fresh or salt water. However, many birds are closely associated with wetlands and water margins; relatively few including divers, grebes and ducks are restricted to river and lake systems. There are also a number of non-wading birds that feed largely on fish and other aquatic animals and are adapted to diving and surface-snatching. Among them are kingfishers families, the fish owls, fish eagles and a few other raptors.

Diversity and distribution: Among the 420 North American breeding birds, Sauer et al. (1997) identified a wetland-open water group including 86 species, that is about 1/5 of the breeding species. There are 253 European bird species affiliated to freshwaters including all the charadriiform wading birds (Roselaar, pers. com.). So for Europe, about 27% out of a total of 700 European bird species depend on inland waters. Using a conservative value of 20% for the 9000 species of birds currently known in the world, there might be about 1800 bird species world wide that are more or less dependant on freshwaters.

Class Therapsida – Mammalia

Many mammals live in close proximity to freshwater ecosystems on which they depend for their survival. However, only a small number of representatives of several mammal orders are considered to be aquatic or semi-aquatic spending most of their life in slow-flowing rivers (Revenga & Kura, 2002). The best known representatives are the beavers, otters, hippopotamus, and river dolphins.

The beaver is found in northern boreal regions with two species, one in Europe, and one in North America. There are also rice rats, swamp rats, muskrats and water voles in North and South America. However the best known aquatic rodents in south America are the capybara. (*Hydrochaeris hydrochaeris*) and the myocastor (*Myocastor coypus*). In Africa there are several species of water rats and creek rats (Kingdon, 1997). Four species of beaver rats occur in Oceania.

There are four species of otters in Africa (Nel & Somers, 2002), one in North America, one in Europe, three in South and Central America, and four in Asia. There are four water dependent water shrews (genus *Sorex*) in North America (Whitaker, 1991) and ten species in Europe and Asia (Stone, 1995). In Africa, three species of otter shrews (Potamogale) occur in the Congo basin, and a few species of Tenrecs in Madagascar. (Kingdon, 1997).

The hippopotamus (*Hippopotamus amphibious*) only occurs in Africa as well as the pigmy hippopotamus (*Hexaprotodon liberiensis*). In addition, several semi-aquatic ungulates are known from Africa (the sitatunga, *Tragelaphus spekei*), South America (the marsh deer, *Blastocerus dichotomus*), Asia (water buffalo, *Bubalus bubalis*, and two deer species).

Five species of river dolphins and one of freshwater purpoises live in large rivers in South America and South Asia. In Asia there are also five species of cetaceans, among which the Yangtse River dolphin or baiji (*Lipotes vexillifer*) and the Gange River dolphin (*Orcaella brevirostris*) (Reeves et al., 2003).

Other representatives are:

- the duck-billed platypus (*Ornithorhynchus anati nus*) from Australia (Nowak, 1991),
- the water opossum (*Chironectes minimus*) from South America,
- the mink (*Mustela vison*) in North America,
- two species of seal in Europe and Siberia (Reijnders et al., 1993),
- three species of manatees (*Trichechus*),
- three species of desmans present in Europe (Nowak, 1991),
- the otter civet (*Cynogale bennettii*) in Asia,
- the aquatic genet (*Osbornictis piscivora*) in Africa.

Discussion and conclusion

Importance of freshwater habitats for animal groups

It is commonly recognised that the first life forms originated in the sea and have later colonised freshwater and terrestrial habitats (Barnes & Mann, 1991). The adaptation to terrestrial life occurred independently in various lineages of aquatic animals, and some terrestrial lineages re-adapted secondarily to freshwaters (i.e. insects). Hence, two main groups of freshwater animals can be identified (Banarescu, 1990):

A group of marine origin (primary aquatic animals with no terrestrial ancestors) that invaded freshwaters directly from the sea: lower metazoans, branchiate molluscs (mussels and prosobranchiates), crustaceans, lampreys and fishes. Among these groups, some species live indifferently in saline, brackish and freshwater, others live almost exclusively in brackish waters. Numerous fish and prawn species are diadromous with a life cycle including both marine and freshwater stages. Marine biodiversity has remained higher at higher taxonomic levels than freshwater biodiversity because many important marine invertebrate groups such as Echinodermata, Ctenophora, or Chaetognatha, have failed to colonise freshwater habitats. The osmotic challenge of life in freshwaters probably prevented colonisation by several marine invertebrate groups.

A group of terrestrial origin (secondary aquatic taxa) that probably underwent major co-evolution on land and subsequently invaded freshwaters. For example, many pulmonate snails with a primitive kind of lung are thought to have first colonised the land, and secondarily returned to freshwater where they can still breathe air (Barnes & Mann, 1991). The origin of terrestrial insects is speculative, but most likely they had their main adaptive radiation on open land at the same time as flowering plants.

However, a significant proportion of present-day insects spend part of their life history in freshwater. No equivalent re-invasion of marine habitats could be observed, maybe because insects failed to compete with the abundant crustaceans found in the fringing marine habitats and occupying similar niches to those occupied by insects in freshwaters.

The different origins of freshwater species may explain why the group of 'freshwater dependant species' occupies a wide range of habitats (Table 8): some species may be in the on-going process of adapting to freshwater, while others are only colonising the ecotones, while they are physiologically or biologically unable to live in freshwater.

Global assessment of freshwater animal diversity

This paper is an attempt to provide a quantitative evaluation of the global animal freshwater species diversity with regard to data availability (Table 9). The overall order of magnitude of presently described freshwater species is 100 000, half of which are represented by the very speciose class of Insecta. However, the true number is definitely much higher than this.

Even doubling the present number of described species, the order of magnitude remains low compared to the species diversity of coral reefs or tropical rainforests. It is assumed that tropical forests contain more than half of the, world's species estimated around 10 million (1.4 million named species; Revenga & Kura, 2002). For instance, it is expected that as many as 30 million arthropod species may exist in tropical.forests (Revenga & Kura, 2002). However, it should be stressed that some 20 000 species of vertebrates (35–40% of the known vertebrates) are true or water dependant freshwater species while inland water habitats occupy around 1% of the planet's surface not submerged in seas and oceans.

The present preliminary study highlighted gaps in the basic knowledge of freshwater species richness:

- as expected, most information is available from North America and Europe. The tropical areas are still poorly investigated while their species richness is widely recognised to be higher than in temperate zones. South America and Asia are especially lacking global comprehensive estimates of species richness for many groups, even the usually well-known ones such as molluscs or insects.
- Some animal groups have been less studied than others, and data on their diversity and distribution is scarce. The vertebrates are the best known groups, whereas in many cases knowledge on invertebrates that are not useful or posing threats to humans (e.g. disease vectors) is still very poor.

Table 8. Relationships of the 'true freshwater' and the 'freshwater dependent' taxa (phyla and classes) to marine, freshwater and terrestrial habitats (adapted from Smith 2001)

Phyla	Classes	M/FW$^-$	M/FW$^+$	M = FW	FW/M$^-$	P	FW/LC	T/FW
Porifera		X						
Cnidaria								
	Hydrozoa	X						
Nemertea		X						
Platyhelminthes								
	Turbellaria			X				
	Cestoda					X		
	Trematoda					X		
Gastrotricha				X				
Rotifera					X			
Nematoda				X				
Annelida								
	Polychaeta	X						
	Oligochaeta				X			
	Hirudinae				X			
Bryozoa		X						
Tardigrada				X			X	
Mollusca								
	Bivalvia		X					
	Gastropoda		X					
Amphipoda								
Branchiopoda	Cladocera				X			
	Anostraca				X			
	Notostraca				X			
	Conchostraca				X			
	Haplopoda				X			
Ostracoda				X				
Copepoda				X				
Malacostraca								
	Mysidacea	X						
	Cumacea							
	Tanaidacea	X						
	Isopoda							
	Amphipoda		X					
	Decapoda		X					
Arachnida								X
Entognatha	Collembola							X
Insecta							X	
Vertebrata								
	Teleostimi				X			
	Amphibia						X	
	Reptilia							X
	Aves							X
	Mammalia	X						X

M/FW$^-$ = mostly marine with few freshwater species; M/FW$^+$ = Mostly marine with many freshwater species; FW/M$^-$ = Mostly freshwater with few marine species; P = parasites; FW/LC = complete part of their life cycle in freshwater; T/FW = Terrestrial but need fresh water for food & habitat.

Table 9. Global species richness of freshwater animal groups: Current state of knowledge

Phyla	Classes/orders	Species numbers
Porifera		197
Cnidaria		30
	Hydrozoa	ca 20
Nemertea		12
Plathelminthes		ca. 500
Gastrotricha		ca. 250
Rotifera		1817
Nematoda		3000
Annelida		
	Polychaeta	?
	Oligochaeta	700
	Hirudinae	ca. 300
Bryozoa		70–75
Tardigrada		
Mollusca	Bivalvia	ca. 1000
	Gastropoda	ca. 4000
Arthropoda		
Crustacea		
Branchiopoda		
	Cladocera	>400
	Anostraca	273
	Notostraca	9
	Conchostraca	130
	Haplopoda	1
Amphipoda		
Ostracoda		3000
Copepoda		2085
Malacostraca		
	Mysidacea	43
	Cumacea	20
	Tanaidacea	2
	Isopoda	ca. 700
	Amphipoda	1700
	Decapoda	1700
Arachnida		5000
Entognatha	Collembola	
Insecta		
	Ephemeroptera	>3000
	Odonata	5500
	Plecoptera	2000
	Megaloptera	300
	Trichoptera	>10 000
	Hemiptera	3300
	Coleoptera	>6000
	Diptera	>20 000
	Orthoptera	ca. 20
	Neuroptera	ca. 100
	Lepidoptera	ca. 100
	Hymenoptera	ca. 100
Vertebrata		
	Teleostomi	13 400
	Amphibia	5504
	Reptilia	ca.250
	Aves	ca. 1800
	Mammalia	ca. 100

Table 9. (Continued)

Assessment of freshwater biodiversity and the Global Freshwater System

The attempt to address freshwater biodiversity at a global scale is related to the development of a new global approach integrating data and knowledge from local over regional to global scales. This global perspective is motivated by the global scale reached by freshwater biodiversity changes and drivers of change.

Freshwater biodiversity changes occur at global scale

Due to a general lack of data, it is difficult to assess the status of the inland water biodiversity. However, evidence of biological impoverishment is pervasive in aquatic systems and when available the data are very disturbing. For example in USA, where the status of aquatic biodiversity is relatively well-documented compared to other areas, 34% of fish, 65% of crayfishes and 75% of unionid mussels are classed rare to extinct (Master, 1990). Of 214 stocks of Pacific salmon, 74% have a high or moderate risk of extinction (Nehlsen et al., 1991). As a whole, the native fishes of North America are in serious decline and currently 364 taxa are listed by the American Fisheries Society as endangered, threatened or of special concern (Williams et al., 1991). In California, Móyle et al. (1995) reported a rising trend of extinction and endangerment in the region's native fishes from 24% in 1988 to 43% in 1992. Nearly half of Mexico is arid or semiarid with scarce waters. At least 92 springs and 2500 km of river have dried in this area, This has affected nearly 200 species of freshwater fishes, 120 under some threat 15 extinct through human impact. As of 1985, an average of 68% of species was eradicated in local fish faunas (Contreras & Lozano, 1994).

Freshwater biodiversity and global drivers of change
The magnitude of the response of freshwater systems to global change will depend on a complex array of physical and biological factors. These changes are likely to directly affect the survival, reproduction and growth of organisms, as well as the distribution, persistence and diversity of species. Global change may affect principally the climate through temperature and precipitation regime (amount, annual distribution, etc.). Bot may have consequences on the overall distribution of freshwater biota (Tonn, 1990), as well as on the physiology and biology of individual species (Regier et al., 1990).

Several anthropogenic factors are also responsible for the erosion of freshwater biodiversity. Until recently, these factors had impacts at local or regional scales. However, some of them are spreading throughout the world and can now be considered as global trends:

- Habitat alteration is a major cause of aquatic biodiversity loss, and degradation or destruction of habitats is particularly threatening in rivers. Flow regulation occurs on almost all large river and alters the natural flood regime in fresh waters. Man-made lakes associated with hydroelectric dams or built for irrigation purposes prevent fish migrations, and alter the flow pattern downstream. Catchments are affected by large-scale land-use practices often associated with deforestation. These practices generate rapid headwater erosion and increase of sediment load in river waters, as do mining-related activities.
- Human activities generate excessive inputs of nutrients (nitrogen and/or phosphor-us) from point or non-point sources resulting in river and lake eutrophication. The eutrophication process leads to changes in pelagic communities and benthic invertebrate biodiversity, mostly through depletion of available oxygen, and abundance can be modified as well.
- Introduction of alien species into aquatic systems has been frequent. Welcomme (1988) listed 237 inland animal species that have been introduced into 140 countries world wide, but the total number is now much higher. One of the major problems of freshwater species introductions is its irreversibility, at least at the human lifetime scale. Once introduced and established, it is impossible, given current technology, to eradicate a fish, mollusc or plant species from a large natural water body. As a consequence, we are likely to see a continued impoverishment of native aquatic biodiversity and an increased homogenisation of the world's freshwater biota.

We need to develop tools and methodology to address the problem of scale. In climate and hydrologic studies, data and processes are commonly integrated from local to global scale. In ecology, studies still address biodiversity changes usually at a local scale. The challenge is also to provide a global assessment of freshwater biodiversity that could be related to global data on climate and water cycle. Such a global assessment is also critical for decision makers to address issues related to the global water system including the biological component. Meantime, the efforts to provide a global assessment would also open the path for a macro-ecological approach of freshwater biodiversity.

Acknowledgements

We would like to thank for their input, expertise and reviews: Christian Amblard, Maria Balsamo, Penny Berents, Tom Bongers, Georgina Bond-Buckup, Danielle Defaye, Yusbelly Diaz, Marie-José Dole-Olivier, Anno. Faubel, Olivier Gargominy, Raymond Gibson, Cristian Ituarte, Renata Manconi, Alberto Martin, Jos. A. Massard, Alessandro Minelli, Thierry Oberdorf, Didier Paugy, Winston Ponder, Vince Resh, Kees Roselaar, Hendrik Segers, Cristiana Serejo, Bernard Statmer, Walter Traunspurger, Ronald Vonk, Georges Wilson.

References

Abele, L. G. (sci. ed), 1982. Biology of the Crustacea, Vol. 1 and 2. Academic Press, New York.

Andrássy, I., 1978. Nematoda. A checklist of the animals inhabiting European inland waters with accounts of their distribution and ecology. In Illies (ed.), Limnofauna Europacea. 2nd edn. Gustav Fisher Verlag, Stuttgart, German: 98–117.

Balmford, A., A. Bruner, P. Cooper, R. Costanza, S. Farber, R. E. Green, M. Jenkins, P. Jefferiss, V. Jessamy, J. Madden, K. Munro, N. Myers, S. Naeem, J. Paavola, M. Rayment, S. Rosendo, J. Roughgarden, K. Trumper & R. K. Turner, 2002. Why conserving wild nature makes economic sense. Science 297: 950–953.

Banarescu, P., 1990. Zoogeography of Fresh Waters, Vol. 1, General distribution and dispersal of freshwater animals. AULA – Verlag, Weisbaden; 511 pp.

Banarescu, P., 1995. Zoogeography of Fresh Waters. Vol. 3, Distribution and dispersal of freshwater animals in Pacific areas and South America, Verlag, Weisbaden: 524 pp.

Barnes, R. S. K. & K. H. Mann, 1991. Fundamentals of Aquatic Systems. Blackwell Scientific Publications.

Belk, D. & Brtek, J., 1995. Checklist of the Anostraca. Hydrobiologia 298: 315–353.

Belk, D. & J. Brtek, 1997. Supplement to 'Checklist of the Anostraca'. Hydrobiologia 359: 243–245.

Boaden, P. J. S., 1985. Why is a gastrotrich? In Conway Morris S., J. D. George, R. Gibson, & H. M. Platt (eds), The Origins and Relationships of Lower Invertebrates. The Systematics Association and Clarendon Press, Oxford, England: 248–260.

Bogan, A. E., 1998. Freshwater Molluscan conservation in North America: problems and practices. In Killeen, I. J., M. B. Seddon & A. Holmes (eds), Molluscan Conservation: a Strategy for the 21th century. Journal of Conchology Special Publication No. 2: 223–230.

Bouchet, P. (ed.), 1995. Résultats des Campagnes MUSORSTOM. Vol. 14. MMNHN Zoologie Tome 167: 654 pp.

Brandt, R. A. M., 1974. The non-marine aquatic Mollusca of Thailand. Archiv für Molluskenkunde 105(1–4): 423 pp.

Brinkhurst, R. O. & M. J. Wetzel. 1984. Aquatic oligochaeta of the world: Supplement. A catalogue of new freshwater species, descriptions and revisions. Canadian technical report of Hydrography and Ocean Science 44: 101 pp.

Brinkhurst, R. O. & S. R. Gelder, 1991. Chap 12 Annelida: Oligochaeta and Branchiobdellida. In Thorp and Covich (eds), Ecology and classification of North American freshwater invertebrates. Academic Press, San Diego: 428–435.

Brown, D. S., 1994. Freshwater snails of Africa and their medical importance. Revised 2nd edn. Taylor & Francis Ltd. London: 489 pp.

Brusca, R. C. & G. J. Brusca. 1990. Invertebrates. Sinauer Associates, Sunderland, Massachusetts: 922 pp.

Contreras, B. S. & V. M. L. Lozano, 1994. Water, endangered fishes, and development perspectives in arid lands of Mexico. Conservation Biology 8: 379–387.

Contreras-Ramos, A. 1999. List of species of Neotropical Megaloptera (Neuropterida). Proceedings of the Entomological Society of Washington. 101: 274–284.

Coomans, A. 2000. Nematode systematics: past, present and future. Nematology 2: 3–7.

Corbet, P. S., 1999. Dragonflies. Behavior and Ecology of Odonata. Cornell University Press.

Coulter, G. W.(ed), 1991. Lake Tanganyika and its life. Oxford University Press, London. 83–89.

Crandall, K., 2003. Crayfish World Checklist. http://bioag.byu.edu/mlbean/crayfish/crayhome.htm.

Cumberlidge, N., 1999. The freshwater crabs of West Africa Family Potamonautidae. Faune et Flore tropicale, 36, 384 p. IRD, Paris.

Daget, J., 1998. Catalogue raisonné' des mollusques bivalves d'eau douce africains. Backhuis Publishers, Orstom, Leiden, Paris: 329 pp.

Daly, H. V., J. T. Doyen & A. H. Purcell III. 1998. Introduction to Insect Biology and Diversity, 2nd edn. Oxford University Press, Oxford.

Danielopol, D. & R. Rouch, 1991. L'adaptation des organismes au milieu aquatique souterrain Réfflections sur l'apport de recherches écologiques récentes. Stygologia 6(3): 129–142.

Davies, D. A. L. & P. Tobin, 1985. The dragonflies of the world: a systematic list of the extant species of Odonata. Vol. 2. Anisoptera. Societas Internationalis Odonatologica rapid Comm. 5 (Suppl.) 1–151.

Davis, G. M., 1982. Historical and ecological factors in the evolution, adaptive radiation, and biogeography of freshwater mollusks. American Zoologist 22: 375–395.

Day, J. A., B. A. Stewart, I. J. de Moor & A. E. Louw, 2001. Guides to the Freshwater Invertebrates of Southern Africa – Vol. 4: Crustacea III Bathynellacea, Amphipoda, Isopoda, Spelaeogriphacea, Tanaidacea and Decapoda. Water Research Commission Report TT141/01, Pretoria.

Day, J. A. & I. J. de Moor, 2002. Guides to the Freshwater Inverebrates of Southern Africa Vol. 5. Non Arthropods: The Protozoans, Porifera, Cnidaria, Platyhelminthes, Nemertea, Rotifera, Nematoda, Nematomorpha, Gastrotricha, Bryozoa, Tardigrada, Polychaeta, Oligochaeta, and Hirudinea. Water Research Commission Report TT167/02, Pretoria.

Dole-Olivier, M.-J., D. M. P. Galassi, P. Marmonier & M. Creuzé des Châtelliers, 2000. The biology and ecology of lotic microcrustaceans. Freshwater Biology 44: 63–91.

Dudgeon, D., 1999. Tropical Asian Streams Zoobenthos, Ecology and Conservation. Hong Kong University Press, Hong Kong, 830 pp.

Duker, L. & L. Borre, 2001. Biodiversity conservation of the World's Lakes: a preliminary framework for identifying priorities. Lake Net Reports Series 2. Monitor International, Annapolis, Maryland, USA.

Dumont, H. J. & S. V. Negrea, 2002. Introduction to the class Branchiopoda. In Dumont, H. J. F. (ed), Guides to the Identification of the Microinvertebrates of the Continental Waters of the World. Vol. 19: Backhuys Publishers, Leiden, The Netherlands, 398 pp.

Dumont, H. J., 1994a, On the diversity of Cladocera in the tropics. Hydrobiologia 272: 27–38.

Dumont, H. J., 1994b. The distribution and ecology of the fresh- and brackish water medusae of the world. Hydrobiologia 272: 1–12.

Dumont, H. J. F. (ed.), 1993–2003. Guides to the identification of the Microinvertebrates of the Continental Waters of the

World.Vol. 1–20, SPB Academic Publishing, The Hague, The Netherlands.

Dussart, B. & D. Defaye, 2002. World directory of the Crustacea Copepoda of inland waters. I. Calaniformes. SPB Academic Publishing, The Netherlands, 285 pp.

Elouard, J. M. & F. M. Gibon, 2001. Biodiversité et Biotypologie des eaux continentales de Madagascar. IRD, Montpellier, France, 446 pp.

Ernst, C. H. & R. W. Barbour, 1989. Turtles of the World. Smithsonian Institution Press, Washington, D.C: 313 pp.

Falkner, G., R. A. Bank & T. Von Proschwitz, 2001. CLECOM-PROJECT check-list of the non-marine Molluscan Species-group taxa of the States of Northern, Atlantic and Central Europe (CLECOM I) Heldia, 4(1/2): 1–76.

Fauna Europaea, 2002. Eurpean database on freshwater and terrestrial organisms. http://www.faunaeur.org/PUBLIC/EMAIL/EMAIL1.html. Zoological Museum Amsterdam, The Netherlands.

Flint, O. S., R. W. Holzenthal & S. C. Harris, 1999. Catalog of the Neotropical caddisflies (Insecta, Trichoptera). Ohio Biological Survey, Columbus, Ohio, USA.

Gasc, J. P., A. Cabela, J. Crnobrnja-Isailovic, D. Dolmen, K. Grossenbacher, P. Haffner, J. Lescure, H. Martens, J. P. Martínez Rica, H. Maurin, M. E. Oliveria, T. S. Sofianidou, M. Veith & Zuiderwijk. A. (eds), 1997. Atlas of amphibians and reptiles in Europe. Collection Patrimoines Naturels, 29, Paris, SPN / IEGB / MNHN, 496 pp.

Gibson, R., 1995. Nemertean genera and species of the world. An annotated checklist of original names and description citations, synonyms, current taxonomic status, habitats and recorded zoogeographic distribution. Journal of Natural History 29: 271–562.

Giller, P. S. & B. Malmqvist, 1998. The biology of streams and rivers. Oxford University Press , New York, 272 pp.

Gisin, H., 1978. Collembola. In Illies, J. (ed), Limnofauna Europaea. Fisher Stuttgart. 254–255.

Groombridge, B. & M. Jenkins, 1998. Freshwater biodiversity: a preliminary global Assessment. Biodiversity Series No. 8. World Conservation Monitoring Centre. (ISBN 1899628002), 140 pp.

Hagen, K. S., 1996. Chapter 21: Aquatic Hymenoptera. In Merritt, R. W., K. W. Cummins (ed.), An Introduction to the Aquatic Insects of North America. 3rd edn. Kendall/Hunt Publishing Co. Dubuque, Iowa.

Hamer, M. L. & L. Brendonck., 1997. Distribution, diversity and conservation of Anostraca (Crustacea: Branchiopoda) in southern Africa. Hydrobiologia 359: 1–12.

Harvey, M. S., 1998. The Australian water mites: a guide to families and genera. CSIRO Publishing Collingwood, Victoria.

He, S. & Y. Chen, 1996. The status of endangered freshwater fishes in China, and the analysis of the causes of their endangerment? In Conserving China's Biodiversity, China's Environmental Science Press, Beijing: 169–174.

Heckman, C. W., 2001. Encyclopedia of Southern American Aquatic Insects: Collembola. Kluwer Academic Publishers, Boston, 418 pp.

Hershler, R., J. M. Pierson & R. S. Krotzer, 1990. Rediscovery of Tulotoma magnifica (Conrad) (Gastropoda: Viviparidae). Proceedings of the Biological Society of Washington 103(4): 815–824.

Hidding, B., E. Michel, A. V. Natyaganova & D. Y. Sherbakov, 2003. Molecular evidenc reveals a polyphyletic origin and chromosomal speciation of Lake Baikal's endemic asellid isopods. Molecular Ecology 12(6): 1509–1514.

Hobbs, H. H., Jr. 1989. An illustrated checklist of the American Crayfishes (Decapoda: Astacidae, Cambaridae, and Parastacidae). Smithsonian Contributions to Zoology 480: 1–236.

Holsinger, J. R., 1993. Biodiversity of subterranean amphipod crustaceans: global patterns and zoogeographic implications. Journal of Natural History 27: 821–835.

Hopkin, S. P., 1997. Biology of the Springtails (Insecta: Collembola). Oxford University Press, Oxford, England: 330 pp.

Horwitz, P. 1990. The conservation status of Australian freshwater Crustacea with a provisional list of threatened species, habitats and potentially threatening processes. Report Series No. 14. Australian National Parks and Wildlife Service.

Hutchinson, G. E., 1993. A Treatise on Limnology. Vol. IV, The Zoobenthos. Y.H. Edmondson (ed), John Wiley & Sons, Inc., 944 pp.

Illies, J., 1965. Phylogeny and zoogeography of Plecoptera. Annals et Revue d' Entomologie 10: 117–140.

Jacobs, L. J., 1984. The free-living inland aquatic nematodes of Africa: a review. Hydrobiologia 113: 259–291.

Jayachandran, K. V., 2001. Palaemonid Prawns: Biodiversity, Taxonomy, Biology and management. Sciences Publishers, Inc., 624 pp.

Kabat, A. & R. Hershler, 1993. The prosobranch snail family Hydrorbiidae (Gastropoda: Rissoidea): Review of classification and supraspecific taxa. Smithsonian Contributions to Zoology 547: 1–94.

Kamaltynov, R. M., 1999. On the higher classification of Lake Baikal amphipods. Crustaceana 72: 933–944.

Kempf, E. K., 1980. Index and bibliography of non-marine Ostracoda. 1, Index A. Geologisches Institut der Universitaet zu Koeln, Sonderveroeffentlichungen, 35, 188 pp.

Kempf, E. K., 1997. Index and bibliography of non-marine Ostracoda. 6, Index A, supplement 1. Gdologisches Institut der Universitaet zu Koeln, Sonderveroeffentlichungen. 109: 142 pp.

Khlaphake, A., W. Scheumann & R. Scliep, 2001. Biodiversity and international water policy. International agreements and experiences related to the protection of freshwater ecosystems. Institute of Management in Environmental Planning, Technical University of Berlin.

Kingdon, J., 1997. The Kingdon field guide to African Mammals. Academic Press, London and New York: Natural World.

Kottelat, M. & T. Whitten. 1996. Asia-wide assessment of freshwater biodiversity. World Bank Technical Paper Number 28 1. The World Bank, Washington, DC, USA.

Lane, R. & R. W. Crosskey, 1993. Medical Insects and Arachnids. Chapman & Hall, London, 723 pp.

Lange, W. H., 1996. Chapter 19: Aquatic and semi-aquatic Lepidoptera. In Merritt R. W. & K. W. Cummins. An Introduction to the Aquatic Insects of North America. 3rd edn. Kendall/Hunt Publishing Co., Dubuque, Iowa: 387–398.

Lauzanne, L., 1982. Les Orestias (Pisces, Cypronodontidae) du Petit Lac Titicaca. Revue d' Hydrobiologie Tropicale 15(1): 39–70.

Lévêque, C. & D. Paugy, 1999. Les poissons des eaux continentales afficaines Diversité, écologie et utilisation par l'homme. Editions IRD, Paris.

Lévêque, C. & Mounolou J. C., 2003. Biodiversity. Wiley & Sons Ltd: 284 p.

Lévêque, C. 2002. Biodiversity in Freshwaters. pp. 146–152, In Encyclopedia of Global Environmental Change, Vol. 2. The Earth system: Biological and Ecological Dimensions of Global Environmental Change. In Harold, A. Mooney & Josep, G. Canadell (eds), John Willey & Sons, Ltd, Chichester:

Lévêque, C., 1997. Biodiversity Dynamics and Conservation: The Freshwater Fish of Tropical Africa. Cambridge University Press, Cambridge, USA, 486 p.

Liu Yueyin (ed), 1979. Economic Fauna of China: Freshwater Mollusks. China Scientific Book Service, 134 pp.

Lundberg, J. G., M. Kottelat, G. R. Smith, M. I. Stiassny & A.C. Gill, 2000. So many fishes, so little time: an overview of recent ichthyological discovery in continental waters. Annals of the Missouri Botanical Garden 87: 26–62.

Maeda-Martinez, A. M., D. Belk, H. Obregón-Barboza, & H. J. Dumont, 1997. Large branchiopod assembalges common to Mexico and the United States. Hydrobiologia 359: 45–62.

Malakhov, V. V., 1994. Nematodes. Structure, Development, Classification and Phylogeny. Smithsonian Institution Press, Washington, D.C.

Manconi, R. & R. Pronzato, 2002 Spongillina n. subord. Lubomirskiidae, Malawispongiidae, Metaniidae, Metschnikowiidae, Palaeospongillidae, Potamolepiidae, Spongillidae. In Hooper, J. N. A. & R. W. M. Van Soest (eds), Systema Porifera: a guide to the classification of sponges. Kluwer Academic/Plenum Publishers, New York, Vol. 1: 921–1019.

Mandaville, S. M., 2000. Limnology-Eutrophication and Chemistry, Carrying Capacities, Loadings, Benthic Ecology, and Comparative Data. Project F-1, Soil & Water Conservation Society of Metro Halifax. xviii, Synopses 1, 2, 3, 13, and 14. 210 p.

Martens, K., 1984. Annotated checklist of non-marine ostracods (Crustacea, Ostracoda) from African inland waters. Koninklijk Museum voor Midden-Afrika, Zoologische Dokumentatie (Tervuren) 20: 1–51.

Martens, K., 1998. Diversity and endemicity of Recent non-marine ostracods (Crustacea, Ostracoda) from Africa and South America: a faunal comparison. Verhandlungen der Internationalen Vereinigung für Limnologie 26(4): 2093–2097.

Martens, K. & F. Behen, 1994. A checklist of the non-marine ostracods (Crustacea, Ostracoda) from South-American inland waters and adjacent islands. Travaux Scientifiques du Musée d'Histoire Naturelle de Luxembourg 22, 81 pp.

Master, L., 1990. The imperiled status of North American aquatic animals. Biodiversity Network News 3: 1–2, 7–8.

May, R. M., 1994. Biological diversity: differences between land and sea. Philosophical Transactions of the Royal Society of London, Series B 343: 105–111.

McAllister, D. E., A. L. Hamilton & B. Harvey, 1997. Global Freshwater Biodiversity: Striving for the Integrity of Freshwater Ecosystems. Special issue of Sea Wind, Bulletin of Ocean Voice International, 11(3).

McCafferty, W. P., 1996. The Ephemeroptera species of North America and index to their complete nomenclature. Transactions of the American Entomological Society 122 (1): 1–54.

Merritt, R. W. & K. W. Cummins (eds), 1996. An Introduction to the Aquatic Insects of North America 3rd edn. Kendall/Hunt Publishing Co. Dubuque, Iowa, 862 pp.

Michel, E., 1994 Why snails radiate: a review of gastropod evolution in long-lived lakes, both recent and fossil. Advances in Limnology 44: 285–317.

Moore, J. & R. Gibson, 1985. The evolution and comparative physiology of terrestrial and freshwater nemerteans. Biological Reviews. Cambridge Philosophical Society 60: 257–312.

Moyle, P. B., R. M. Yoshiyama, J. E. Williams & E. D. Wikramanayake, 1995. Fish Species of Special Concern in California. Second Edition. Department of Fish and Game. Final Report for Contract 2181F.

Nagel, K. O., G. Badino & G. Celebrano, 1998. Systematics of European naiades (Bivalvia: Margaritiferidae and Unionidae): a review and some, new aspects. Malacological Review Supplement 7 (Bivalvia I, Ann Arbor, Mich): 83–104.

Nagelkerke, L. A. J., F. A. Sibbing, J. G. M. van den Boogaart & E. H. R. R. Lammens, 1994. The barbs (Barbus spp.) of Lake Tana: a forgotten species flock? Environmental Biology of fishes 39: 1–22.

Naggs, F. (in prep.). South Asia. In Gargominy, O. & Bouchet, P. (eds), Action Plan for the Conservation of Land and Freshwater Molluscs. IUCN, Gland, Switzerland.

Nehlsen, W., J. E. Williams & J. A. Lichatowich, 1991. Pacific salmon at the crossroads: stocks at risk from California, Oregon, Idaho, Washington. Fisheries 16(2): 4–21.

Nel, J. A. J. & M. J. Somers, 2002. The status of otters in Africa: an assessment. In Dulfer, R, J. Conroy, J. Nel & A. Gutieb, Otter Conservation – An example for a sustainable use of wetlands. Proceedings of the VIIth International Otter Symposium. IUCN Otter Specialist Group Bulletin, 19A (Special Issue). 258–266.

Nelson, D. R & S. J. McInnes, 2002. Tardigrada. In Rundle, S. D., A. L. Robertson & J. M. Schmid-Araya (eds). Freshwater Meiofauna: Biology and Ecology. Backhuys Publishers, Leiden, The Netherlands: 177–215.

Nelson, J. S., 1994. Fishes of the World. 3rd edn. John Wiley & Sons, New York, NY.

Nielsen, E. S., 1999. Systematics and Conservation. In Ponder, W. & D. Lunney. The Other 99%: The Conservation and Biodiversity of Invertebrates. Transactions of the Royal Zoological Society of theNew South Wales 1999: 166–173.

Nowak, R. M. 1991. Walker's Mammals of the World. The John Hopkins University Press, Baltimore, 1629 pp.

Pennak, R.W., 1989. Freshwater invertebrates of the United States. 3rd edn. Wiley and Sons, New york, 628 pp.

Peters, W. L. & I. C. Campbell, 1991. Ephemeroptera. In Naumann, I. D., P. B. Carne, J. F. Lawrence, E. S. Nielsen, J. P. Spradberry, R. W. Taylor, M. J. Whitten & M. J. Littlejohn (eds), Insects of Australia: A Textbook for Students and Research Workers. Vol. I. 2nd edn. Melbourne University Press, Carlton, Australia: 279–293.

Pinto, R. L., C. E. F. Roacha & K. Martens (in press). On new terrestrial ostracods (Crustacea, Ostracoda) from Brazil, primarily from Sao Paulo State. Zoological journal of the Linnean Society.

Polhemus J. T., 1984. Aquatic and Semiaquatic Hemiptera. In Merrit and Cummins (eds), An Introduction to the Aquatic Insects: 231–260.

Ponder, W. F., 1997. Conservation status, Threats and habitat requirements of. Australian terrestrial and freshwater molluscs. Memoirs of the Museum of Victoria 56(2): 421–430.

Postel, S., 2001 Troubled waters In Wilson, E. O. & B. Bilger (eds), Best American Science and Nature Writing 2001 Houghton Mifflin.

Preston, H. B., 1915. The Fauna of British India, including Ceylon and Burma: Mollusca (Freshwater Gastropoda and Pelecypoda). Today and Tomorrow's Printers and Publishers, New Delhi, 244 pp. (reprint)

Pronzato, R. & R. Manconi, 2002. Atlas of European freshwater sponges. Annali Museo Civico Storia Naturale Ferrara: 4: 3–64.

Reeves, R. R., B. D. Smith, E. A. Crespo & G. Notarbartolo di Sciara (compilers). 2003. Doliphins, Whales and Porpoises: 2002–2010 Conservation Action Plan for the World's Cetaceans. IUCN/SSC Cetacean Specialist Group, IUCN, Gland, Switzerland and Cambridge, UK.

Regier H. A., J. A. Holmes & D. Payly, 1990. Influence of temperature changes on aquatic ecosystems: an Interpretation of empirical data. Transactions of the American Fisheries Society 119: 374–389.

Reijnders, P., S. Brasseur, J. van der Toorn, P. van der Wolf, J. Harwood, D. Lavingne, & L. Lowry, (eds), 1993. Seals, Fur Seals, Sea Lions and Walrus. Status Survey and Conservation Action Plan. IUCN, Gland, 88 pp.

Resh, V. & R. T. Carde (eds), 2003. Encyclopedia of insects. Academic Press, 900p.

Revenga, C. & Y. Kura, 2002. Status and trends of inland water biodiversity. WRI publications, Washington, 90 pp.

Ricci, C. & M. Balsamo, 2000. The biology and ecology of lotic rotifers and gastrotrichs. Freshwater Biology 44: 15–28.

Ricciardi, A. & J. B. Rasmussen, 1999. Extinction rates of North American freshwater fauna. Conservation Biology 13(5): 1220–1222.

Rodriguez, G., 1982. Les crabes d'eau douce d'Amerique. Famille des Pseudothelphusidae. Faune tropicale XXII, ORSTOM, Paris, 224 pp.

Rodriguez, G., 1992. The freshwater crabs of America. Family Trichodactylidae and supplement to the family Pseudothelphusidae. Faune tropicale XXXI, ORSTOM, Paris: 189 pp.

Rouse, G. W. & F. Pleijel, 2001. Polychaetes. Oxford University Press, London.

Rundle, S. D., A. L. Robertson & Schmid-Araya (eds), 2002. Freshwater Meiofauna: Biology and Ecology. Backhuys Publishers, Leiden, The Netherlands, 369 pp.

Ruppert, E. E. & R. D. Barnes, 1995. Sponges and Placozoans. In Invertebrate Zoology: 6th edn. Saunders College Publishing, New York: 73–94.

Sauer, J. R., J. E. Hines, G. Gough, I. Thomas & B. G. Peterjohn, 1997. The North American Breeding Bird Survey. Results and Analysis. Version 96.3. Laurel, Maryland, USA: Patuxent Environmental Science Center.

Sawyer, R. T., 1986. Leech biology and behaviour. Vol. 2. Feeding Biology, Ecology and Systematics. Clarendon Press, Oxford, England, 793 pp.

Seddon, M. B., 2000. Molluscan biodiversity and the impact of large dams. Contributing Paper for Thematic Review II.1: Dams, ecosystem functions and environmental restoration for the World Commision on Dam's final report: "Dams and Development: A New Framework for Decision-Making." http://www.dams.org/.

Segers, H., 1996. The biogeography of littoral Lecane Rotifera. Hydrobiologia 323: 169–197.

Segers, H., 2001. Zoogeography of the southeast Asian Rotifera. Hydrobiologia 446/447: 237–246.

Segers, H., 2002. The nomenclature of the Rotifera: annoted checklist of valid family – and genus-group names. Journal of Natural History 36: 631–640.

Segers, H., 2003. A biogeographical analysis of rotifers of the genus Trichocerca lamarck, 1801 (Trichocercidae, Monogononta, Rotifera), with notes on taxonomy. Hydrobiologia 500: 103–114.

Shiel, R. J. & W. Koste, 1986. Australian Rotifera: Ecology and biogeography. In De Deckker, P. & W. D. Williams (eds), Limnology in Australia. CSIRO, Melbourne, Dr W. Junk BV, Dordrecht: 142–150.

Shoshin, 1999. The Baikal Lake nematodes fauna: the present and prospects. General problems of Nematology. http://www.zin.ru/conferences/ns/abstracts.htm#Intr_lec.

Shumway, C., 1999. Forgotten waters: freshwater and marine ecosystems in Africa. Biodiversity support program and New England Aquarium. http://www.neaq.org.

Simovich, M. A., C. Sassaman & D. Belk (eds), 1997. Studies on Large Branchiopod Biology and Conservation. Kluwer Academic Publishers, Belgium.

Smith, B. J., 1992. Zoological Catalogue of Australia Vol. 8, Non-marine Mollusca-Zoological Catalogue of Australia Series. CSIRO PUBLISHING/Aust. Biological Resources Study, 412 pp.

Smith, D. G., 2001. Pennak's freshwater invertebrates of the United States: 4th edn. Porifera to Crustacea. John Wiley and Sons Inc. New York. 648 pp.

Smith, I. M. & D. R. Cook, 1991. Water mites. In Thorp, J. H. & A. P. Covich (eds), Ecology and Classifiation of North American Freshwater Invertebrates, Academic Press, San Diego: 523–592.

Sternberg, R. V. & N. Cumberlidge, 2001. Notes on the position or the true freshwater crabs within the brachyrhychan

Eubrachyura (Crustacea: Decapoda: Brachyura). Hydrobiologia 449: 21–39.

Stiassny, M., 1999. The medium is the message: Freshwater biodiversity in peril. In Cracraft, J. & F. Griffo (eds), The Living Planete in Crisis: Biodiversity Science and Policy. Columbia University Press, New York, 53–71.

Stone, A., C. W. Sabrosky, W. W. Wirth, R. H. Foote & J. R. Coulson (eds), 1965. A catalog of the Diptera of America north of Mexico. Agriculture Handbook 276: 1–1696.

Stone, D. (compiler), 1995. Eurasian Insectivores and Tree Shrews: Status, Survey and Conservation Action Plan. IUCN/SSC Insectivore, Tree Shrew and Elephant Shrew Specialist Group. IUCN, Gland, Switzerland and Cambridge UK.

Subba Rao, N. V., 1989. Handbook freshwater molluscs of India. Zoological Survey of India, Calcutta, 289 pp.

Taylor, C. A., M. L. Warren, Jr., J. F. Fitzpatrick, Jr., H. H. Hobbs III, R. F. Jezerinac, W. L. Pflieger & H. W. Robison, (1996). Conservation status of crayfishes of the United States and Canada. Fisheries 21(4): 25–38.

Taylor, D. W., 1988. Aspects of freshwater molluscs ecological biogeography. Palaeogeography, palaeoclimatology, palaeoecology 62: 511–576.

Thorp, J. H. & A. P. Covich (eds), 1991. Ecology and classification of North American Freshwater invertebrates. Academic Press.

Tim, T., 1985. Distribution of aquatic oligochaetes. In Brinkhurst, R. O. & D. G. Cook (eds), Aquatic Oligochaetes Biology. Plenum Press, New York and London: 55–77.

Timoshkin, O. A., 1997. Biodiversity of Baikal fauna: state-of-the-art (Preliminary analysis). Wada, Timoshkin, Fujita & Tanida (eds), New Scope on the Boreal Ecosystems in East Siberia/Proceedings of International Symposium, Kyoto, Nov. 23–25 1994, (DIWPA Ser. No.2). Novosibirsk: SB RAS, 1997a. – P. 35–76.

Tonn, W. M., 1990. Climate change and fish communities: a conceptual framework. Transactions of the American Fisheries Society 119: 337–352.

Tsuda, S., 2000. A distributional list of world Odonata 2000. Privately published. Osaka.

Turgeon, D. D., A. E. Bogan, E. V. Coan, W. K. Emerson, W. G. Lyons, W. L. Prat, C. F. E. Roper, A. Sheltema, F. G. Thompson & J. D. Williams, 1988. Common and scientific names of aquatic invertebrates from the United States and Canada: Mollusks. American Fisheries Society Special Publication 16, 277 pp.

Väinölä & R. M. Kamaltynov, 1999. Species diversity and speciation in the endemic amphipods of lake Baikal: molecular evidence. Crustaceana 72: 945–956.

Walter, D. E. & H. C. Proctor, 1999. Mites: Ecology, Evolution and Behaviour. University of New South Wales Press, Sydney, New South Wales.

Waltz, R. D. & W. P. Mc Cafferty, 1979. Freshwater Springtails (Hexopoda: Collembola) of North America. Purdue University Aricultural Experimental Station Research Bulletin: 960.

Ward, J. V., 1992. Aquatic insect ecology. 1 – Biology and habitat. John Wiley and Sons.

Ward, J. V. & L. Berner, 1980. Abundance and altitudinal distribution of Ephemeroptera in a rocky mountain stream. In Flannagen, I. F. & K. E. Minshall (eds), Advances in Ephemeroptera Biology. Plenum Press, New York and London: 169–177.

Watson, J. A. L., G. Theischinger & H. M. Abbey, 1991. The Australian Dragonflies. A guide to the identification, distributions, and habitats of Australian Odonata. CSIRO: Canderra and Melbourne.

Welcomme, R. L., 1988. International introductions of inland aquatic species. United Nations Food and Agricultural Organization, Rome, 318 pp.

Whitaker, J. O. Jr., 1991. Audubon Society Field Guide to North American Mammals, Seventh printing. Alfred A. Knopf, Inc. New York, 745 pp.

Wiggins G. B. & R. J. Mackay, 1978. Some relationships between systematic and trophic ecology in Nearctic aquatic insects, with special reference to Trichoptera. Ecology 59: 211–220.

Williams, D. & B. Feltmate, 1992. Aquatic Insects. CAB International. Wallingford, UK, 358 p.

Williams, R. B., P. F. S. Cornelius, R. G. Hughes & E. A. Robson (eds), 1991. Coelenterate biology: Recent research on Cnidaria and Ctenophora. (Proceedings of the Fifth International Conference on Coelenterate Biology, 1989). Kluwer Academic Publishers, Dordrecht and other cities, 742 pp.

Zhadin, V. I., 1952. Mollusks of fresh and brackish waters of the U.S.S.R. Academy of Sciences of the Union of Soviet Socialist Republics. [Translation, 1965, by the Israel Program for Scientific Translations].

Zhuge, Y., X. Huang & W. Koste, 1998. Rotifera recorded from China 1893–1997. International Review of Hydrobiology 183(3): 217–232.

Zwick, P., 1980. Plecoptera. Handbuch der Zoologie. 4(2) 2/7: 1–115.

Hydrobiologia (2005) 542:69–75
H. Segers & K. Martens (eds), Aquatic Biodiversity II
DOI 10.1007/s10750-004-5736-8

Does inland aquatic biodiversity have a future in Asian developing countries?

Brij Gopal
School of Environmental Sciences, Jawaharlal Nehru University, New Delhi 110067, India
(Tel.: +91-11-26704324, Fax: +91-11-26172438, E-mail: brij@nieindia.org; birj0400@mail.jnu.ac.in)

Key words: water policy, socio-cultural factors, globalisation, water resource management, aquaculture

Abstract

Inland aquatic ecosystems and their biodiversity in Asia represent a wide spectrum along a complex continuum of interacting ecological, economic, socio-cultural and political gradients all of which determine their present and future. Whereas the diversity of biophysical environments ensures a rich inland aquatic biodiversity, their present status has been greatly influenced by human societies that have depended on them for millennia. Besides high population densities and developmental pressures, socio-cultural factors, economic considerations and various policies concerning land and water resources are major factors responsible for the degradation of habitats and loss of biodiversity. The looming global climate change may only worsen the situation unless remedial measures are taken on a large scale and urgently. The future of aquatic biodiversity in Asian countries will depend upon a radical change in national policies on water, and upon research that can support the development of appropriate policies.

Introduction

Less than 0.01% of the Earth's total water resources occur in liquid form covering approximately 3% of the land surface. Extreme variability in the spatial and temporal distribution of this small amount of water, together with the large geomorphological variability, has resulted in a variety of water bodies – from ephemeral pools to large lakes and mighty rivers. A significantly large proportion of the Earth's biodiversity representing nearly all kinds of organisms – from microorganisms to mammals – inhabits these inland waters (Revenga & Kura, 2003). During the past decade or so, there has been great concern over the rapid loss of biodiversity in general, and the causes and consequences of the growing loss have been discussed in numerous publications. The biodiversity of inland aquatic ecosystems has also received some attention in recent years and several factors responsible for its decline are well recognized (Groombridge & Jenkins, 1998, Revenga et al., 2000). The present state and future of all kinds of aquatic ecosystems – inland, coastal and marine – has been examined in some detail only recently (see a series of reviews in *Environmental Conservation* vols. 29 and 30). This Symposium in honour of Prof. Henri Dumont, and coincidentally in the International Year of Freshwater (2003), provides another opportunity to discuss the past, present and future of inland aquatic biodiversity. Here I focus attention on those factors which threaten today the aquatic biodiversity the most and are critical to its 'survival' in the future. Further, I confine my discussion to the situation in Asian countries which is in my view quite different from that in other continents. Some of these issues have also been discussed elsewhere in the context of tropical wetlands (Gopal et al., 2004).

Aquatic ecosystems and their biodiversity in Asia

Asia is not only the most populous region of the world (more than 60% of the world's population living over about 30% land area), but has had a long human history. Geologically, the Asian landmass is highly varied as it includes areas representing Pre-Cambrian to the most recent periods. Geomorphologically, it includes the world's highest mountain ranges, some of the large rivers and their vast floodplains, and numerous lakes. Climatically, Asia includes the most dry, most rainy, coldest and hottest places. Both hot and cold deserts as well as tropical rainforests occur in Asia. Therefore, practically all kinds of aquatic ecosystems – from temporary seasonal ponds to large lakes, ephemeral streams to mighty rivers, cold and hot water springs, as well as most types of wetlands (from bogs to mangroves) occur in different parts of Asia. Tibet and other cold desert areas at high altitude have hundreds of saline lakes and associated marshes.

It is therefore only natural that Asia's inland aquatic ecosystems are very rich in their biological diversity. The available information shows that a significant proportion of total biodiversity of a country or region resides in a disproportionately small area of water bodies. On a national level, 15–20% of plant and animal diversity in India occurs within the less than 6% area occupied by inland water bodies (Anonymous, 1991; Gopal, 1997). Similarly, more than 25% of Malaysia's plant biodiversity occurs in peat swamps. The riverine systems are highly rich in various taxa (Dudgeon, 2000a, b); there are more than 930 fish species in the Indochinese Peninsula (Kottelat, 1989), over 500 species of fish (or probably more) in the Mekong Basin (Zakaria-Ismail, 1994; Rainboth, 1996), 262 species in the Zhujiang (Pearl) River, 150 species in the Salween River, 290 species in the Kapuas River, 147 species in the Mahakam River, and 115 species in the Baram River (see Dudgeon, 2000a, b). Tropical Asia is richer in freshwater fish families (>105) than Africa (74 families) and South America (only 60). Yet, these figures do not reflect the true state of biodiversity because numerous habitats such as the saline lakes of Tibet have not yet been explored and many groups of organisms, particularly the benthic invertebrates, are very poorly known and only rarely identified to the level of species.

Factors affecting biodiversity

The paleoclimatic and evolutionary history play an important role in determining the present biodiversity of a region. Biodiversity is then regulated by a large number of biophysical factors that operate over a range of temporal and spatial scales (see Barbault and Sastrapradja, 1995). One must add to these factors the socio-cultural, economic and political influences (Fig. 1). The biodiversity of aquatic ecosystems depends upon and is determined primarily by their hydrological characteristics (Gopal, 1990, in press; Mitsch & Gosselink, 2000), and to a great extent by the nutrient status (Gopal et al., 1990; Harding et al., 1998; Keddy, 2000). Spatial and temporal variations in the

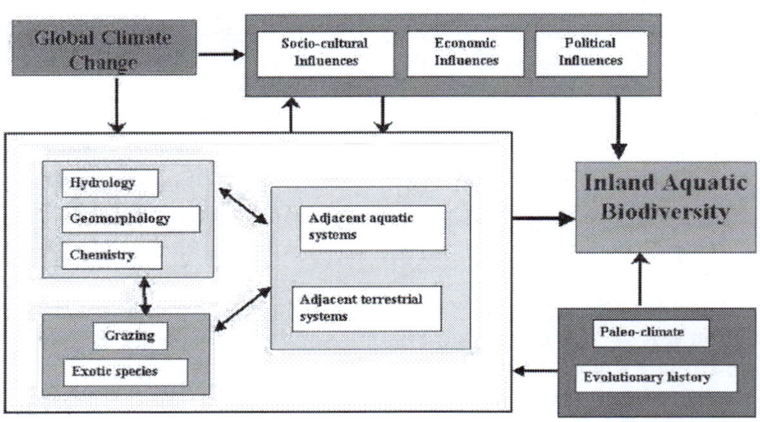

Figure 1. Major factors that influence biodiversity of inland aquatic ecosystems.

amount of water and its source (precipitation, runoff or marine) generate the habitat heterogeneity and thereby affect the biodiversity. The hydrological processes are not only influenced by climate and geomorphology but, in turn, also regulate them to a great extent (Petts and Amoros, 1996; Gopal, in press).

The biodiversity of inland aquatic systems is further influenced significantly by their terrestrial watersheds, and the interactions between adjacent aquatic ecosystems also play an important role in each other's biodiversity.

The current state and the future of aquatic biodiversity in Asian region may be examined in this background, Most of the Asian countries are influenced by the monsoon and experience seasonal rains followed by prolonged dry periods (Robinson, 1976). The situation is most dramatic in south Asia (the Indian subcontinent) where monsoon behaviour is nearly unpredictable (Gopal and Chauhan, 2001). Therefore, since historical times, rivers have been regulated by diverting flows for irrigation and domestic use. In recent years large dams have been constructed creating multipurpose reservoirs. Surface runoff is impounded to create thousands of reservoirs and tanks, particularly in the arid and semi-arid regions (see Gopal & Krishnamurthy, 1992; Gopal, 2000). Natural wetlands such as floodplains, marshes and swamps (including mangroves) have been eliminated or converted to agriculture and aquaculture while innumerable small and large water bodies were created. Today, a large proportion of aquatic habitats is man-made (e.g., extensive network of canals, wastewater drains, reservoirs, fishponds and paddy fields) and intensively managed.

Second, humans have maintained a very close relationship with rivers, lakes and wetlands. They have depended heavily, since historical times, upon the aquatic systems particularly wetlands for food, fiber, fuel, fodder, timber, and numerous other natural resources. Agriculture in wetlands is the most common practice ranging from cultivating cereals, fodder and vegetables on river beds (or islands) exposed during the dry period to intensive paddy cultivation by clearing the swamp vegetation. There are also 'wetlander' communities, such as the Marsh Arabs, that have lived on the lakes and wetlands. Interesting examples come from India, Myanmar, China and southeast Asia where floating islands are constructed for cultivation, fishing and shelter (van Duzer, 2004).

Threats to biodiversity

The biodiversity of inland aquatic ecosystems, and in some cases the very existence of the ecosystems, is increasingly threatened by a variety of factors all of which are related to humans. Ever increasing human population places a direct pressure on the limited resources through increasing demand for meeting the basic needs (e.g., for drinking and agriculture) as well as the cultural needs (e.g., waste disposal). However, real threats come from the anthropogenic activities that cause biophysical changes in and around aquatic ecosystems, and the policies that govern these activities aimed at economic development.

Biophysical changes

The loss and degradation of the habitat are the greatest threats to the biodiversity of a region. Alterations in the hydrological regime constitute the most important cause of loss or degradation of habitats in the aquatic ecosystems. The regulation and diversion of river flows by constructing dams and weirs (barrages) directly affects both upstream and downstream habitats (Poff et al., 1997). The construction of embankments to reduce flood hazards directly eliminates floodplain habitats. Surface runoffs are altered in many ways by human activities in the watershed, and are accompanied by changes in the amount of sediments and nutrients transported to the aquatic systems. Extensive deforestation, agriculture and urbanization of the watersheds right up to the headwaters of most rivers in Asia contribute further to the already high sediment load of the rivers (Milliman, 1997; Milliman and Meade, 1983). These changes lead to the modification of the habitats through siltation, nutrient enrichment, and consequent biological changes. The impacts of flow regulation on riverine biodiversity in Asia have been discussed in several recent papers (Dudgeon, 1999; 2000c, 2002).

Extensive degradation of habitats has occurred throughout Asia due to the conversion of a variety of inland aquatic systems to paddy fields and aquaculture farms (Boyd & Clay, 1998; Ahmed, 2001; Thornton et al., 2003). These converted systems retain some of their wetland characteristics but only at the cost of greatly reduced biodiversity (see Gopal, 2002). Aquatic habitats are lost also directly by drainage and landfill to reclaim land for other land uses, particularly agriculture and extension of urban settlements. Eutrophication and pollution from toxic substances are the next most important threats to aquatic biodiversity. In most of the Asian countries, domestic sewage and industrial effluents are discharged into rivers, lakes and wetlands with only partial or no treatment. The intensive agriculture dependent upon heavy use of agrochemicals (fertilizers and pesticides) is a major source of eutrophication and pollution as the agrochemicals find their way into surface waters with the runoff, subsurface flows and periodic floods.

Exotic invasive species, such as the widely distributed water hyacinth, are another major threat to the aquatic biodiversity in Asia (Gopal, 1987). During less than a century of its invasion and spread in Asia, water hyacinth has become an agent of habitat destruction as well, and has greatly altered the natural biodiversity. Several other weeds (*Salvinia molesta*, *Ipomoea fistulosa*, etc.) and introduced fishes have become nuisance in numerous waterbodies throughout Asia, affecting native biodiversity (Fernando and Holcik, 1991; Mitchell and Gopal, 1991).

High sediment loads in rivers and siltation of water bodies directly affect the biodiversity in various ways such as smothering of the organisms, reduction in primary production, burial of seeds and propagules (Keddy, 2000; Gopal and Chauhan, 2001).

Overexploitation of biological resources (fishing, plant harvest, grazing) is another common and widespread factor impinging on biodiversity of different aquatic ecosystems. Rapid growth in human population exerts increasingly greater pressure on the resources of both land and water, and the aquatic ecosystems are invariably affected directly or indirectly.

While the threats from biophysical changes are well known and most discussed, more serious threats come from the socio-cultural changes, inappropriate economic development and policies, and the growing pressures of globalization.

Socio-cultural changes

Traditional human societies throughout Asia had developed a bond of relationship with their environment, particularly the aquatic environment, based on the highly variable and periodic availability of resources. Traditional societies treated water bodies (as also many land based resources such as forests and grazing lands) as 'common property resources' that were managed with a sense of responsibility. It promoted sustainable use and led to the development of a conservation ethic based around sustainability and survival. However, with the economic development, the availability of alternate resources and access to them from other areas, often distant, disrupted the social structures, cultural practices and the human–environment relationships resulting in the neglect and abuse of water bodies which became 'open access' resources. Changes in the land use practices in the watershed also had their impact on the aquatic ecosystems and their biodiversity.

Inappropriate policies

National policies that are oriented towards rapid economic development, often driven by the urge to earn foreign exchange (hard currencies) through increased exports in a competitive market, and guided by the forces of globalisation, are most responsible for the environmental degradation and loss of biodiversity. Recently, Armitage (2002) discussed the institutional and policy issues related to mangroves in Indonesia and demonstrated the need for 'strategies that contest existing policy narratives and challenge entrenched economic interests and power relationships'. The large-scale water resource development projects aimed at irrigation for increased agricultural production and/or hydropower generation for industrial development, directly affect the riverine ecosystems. Pollution from agricultural runoff, and discharge of domestic and industrial wastewaters is a natural consequence. However, it is the policy of intensification of agriculture in water-stressed regions and promotion of cash crops dependent upon excessive use of water that has resulted in turning the rivers dry and depleted of their biodiversity. For example, rice and sugarcane are now grown in the arid and semi-arid regions of India. The consumption of water in agriculture is further promoted by policies supporting large subsidies on agrochemicals and energy that do not necessarily increase production.

The declining fish catches from dry, polluted rivers are sought to be compensated by intensive aquaculture which also degrades water quality and reduces the biodiversity. The aquaculture in most Asian countries does not help the local communities, either nutritionally or economically, because it is controlled by commercial interests and most production is exported for consumption by the affluent sections of the society in other parts of the country or in the developed countries. Aquaculture is being promoted not merely to meet the local needs but the demand of international markets. Shrimp industry in southeast Asia depends upon the European and American markets at the cost of degradation of aquatic environments in the region (Primavera, 1994). More than a decade ago, I had raised the issue of mismanagement of lake Kolleru on the eastern coast of India (Gopal, 1991). The Government-promoted aquaculture by cooperative societies has converted a shallow lake with high biodiversity into a series of large intensive aquaculture farms. These farms are now unable to support even the three major Indian carps due to high organic matter loading, blooms of blue green algae, and frequent hypoxic to anoxic conditions (personal observations in January, 2003).

However, the greatest threat to aquatic biodiversity comes from the policies related to the management of a country's water resources. All national water policies accord highest priority to, apart from drinking water supplies, irrigational use of water for increasing agricultural production. Other policies targeted at agriculture sector only complicate the problem by further increasing the demand and inefficient use of water. After rivers have been drained completely, they are turned into sewers by discharging all kinds of wastewaters with little or no treatment. Despite the existence of various laws concerning control of water pollution, and considerable efforts to install wastewater treatment facilities, the pollution from both point and non-point sources continues unabated. The policy- and decision makers fail to appreciate the long-term consequences of their short-sighted policies as the technocrats continue to suggest technological solutions to all kinds of problems. The requirements of water for the maintenance of goods and services provided by the aquatic ecosystems are currently totally ignored. Large scale projects such as the Three Gorges dam in China and the Narmada Sarovar in India, and the ambitious plans for interbasin transfers (linking rivers of different basins) cannot be expected to leave adequate flows or mimic the flow regimes of the recent past.

Finally, the future of aquatic biodiversity is also linked to the global climate change that is likely to occur over the next few decades. There is a general consensus among all researchers that the climate change will result in greater spatial and temporal variability (IPCC, 2001). Given the present policies, the response of the policy makers to this climate change can only be expected to aggravate the situation with respect to the biodiversity of aquatic ecosystems because there will be greater demand for water in certain areas and there will be efforts to transfer the available water to water-starved areas.

Under the circumstances, the future of aquatic biodiversity remains at least uncertain, if not entirely bleak. In conclusion, I may only emphasise that there is an urgent need for the adoption of national policies for the conservation of aquatic ecosystems in a watershed perspective and for providing environmental flows in all rivers through efficient utilization of the available water resources. Unfortunately, the interplay of poor facilities, lack of expertise and the failure to properly identify and prioritise research issues has resulted in a very weak scientific base upon which the policy makers and resource managers can depend. In most countries of Asia, the current understanding of the inland aquatic ecosystems is extremely poor and the ongoing research on biodiversity is grossly inadequate as well as inappropriate for the policy requirements. The future of biodiversity, therefore, hinges upon the research that can provide answers to policy relevant questions, and at the same time, research and training need to be strengthened to meet the challenge.

Acknowledgement

I gratefully acknowledge the financial support provided by Kluwer Academic Publishers towards travel for my participation in the workshop.

References

Ahmed, N., 2001. Socio-economic aspects of freshwater prawn culture development in Bangladesh. Ph.D. thesis. University of Stirling, UK.

Anonymous, 1991. Animal Resources of India: Protozoa to Mammalia. State of the Art. Zoological Survey of India, Calcutta.

Armitage, D., 2002. Socio-institutional dynamics and the political ecology of mangrove forest conservation in Central Sulawesi, Indonesia. Global Environmental Change 12: 203–217.

Barbault, R. & S. D. Sastrapradja, (Coordinators) 1995. Generation, maintenance and loss of biodiversity. In Heywood, V. H. (ed), Global Biodiversity Assessment. Cambridge University Press, Cambridge, UK, 193–274

Boyd, E. & J. W. Clay, 1998. Shrimp aquaculture and the environment. Scientific American 278(6): 58–65.

Dudgeon, D., 1999. The future now: prospects for the conservation of riverine biodiversity in Asia. Aquatic Conservation – Marine and Freshwater Ecosystems 9(6): 497–501.

Dudgeon, D., 2000a. Riverine biodiversity in Asia: a challenge for conservation biology. Hydrobiologia 418: 1–13.

Dudgeon, D., 2000b. Riverine wetlands and biodiversity conservation in tropical Asia. In Gopal, B., W. J. Junk and J. A. Davis (eds), Biodiversity in Wetlands: Assessment, Functions and Conservation, Vol. 1. Backhuys Publishers, Leiden: 35–60.

Dudgeon, D., 2000c. Large-scale hydrological changes in tropical Asia: prospects for riverine biodiversity. BioScience 50: 793–806.

Dudgeon, D., 2002. The most endangered ecosystems in the world? Conservation of riverine biodiversity in Asia. Verhandlungen der internationale Vereinigung der Limnologie 28: 59–68.

Fernando, C. H. & J. Holcik, 1991. Some aspects of fish introductions into tropical freshwaters. In Ramakrishnan, P. S. (ed.), Ecology of Biological Invasion in the Tropics. International Scientific Publications, New Delhi.

Gopal, B., 1987. Water Hyacinth. Elsevier Science, Amsterdam, 471 pp.

Gopal, B. (with a contribution by Masing, V.) 1990. Biology and ecology. In Patten, B. C., S. E. Jorgensen, H. J. Dumont, B., Gopal, P. Koryavov, J. Kvet, H. Löffler, Y. Sverizhev & J. G. Tundisi (eds), Wetlands and Shallow Continental Water Bodies. Vol. 1. Natural and Human Relationships. SPB Academic Publishing, The Hague: 91—239.

Gopal, B., 1991. Wetland (mis)management by keeping people out: two examples from India. Landscape and Urban Planning 20: 53–59.

Gopal, B., 1997. Biodiversity in inland aquatic ecosystems in India: an overview. International Journal of Ecology and Environmental Sciences 23: 305–313.

Gopal, B., (with contributions from Bose, B. & Goswami, A.B.) 2000. River conservation in the Indian subcontinent. In Boon, P.J., B.R. Davies & G.E. Petts (eds), Global Perspectives on River Conservation: Science, Policy and Practice. John Wiley, London: 233–261.

Gopal, B., 2002. Conserving biodiversity in Asian wetlands: issues and approaches. In: Ali, Ahyaudin, Che Salmah Md Rawi, Mashhor Mansor, Reiko Nakamura, Sundari Ramakrishna, & Taej Mundkur (eds), The Asian Wetlands: Bringing Partnerships into Good Wetland Practices. Penerbit Universiti Sains Malaysia, Pulau Pinang, Malaysia: ID15–ID22.

Gopal, B., in press. Wetlands and biodiversity. In E. Maltby, (ed), Wetlands Handbook. Blackwell, Oxford.

Gopal, B. & M. Chauhan, 2001. South Asian wetlands and their biodiversity: the role of monsoons. In Gopal, B., W. J. Junk & J.A. Davis (eds), Biodiversity in Wetlands: Assessment, Functions and Conservation, Vol. 2. Backhuys Publishers, Leiden: 257–275.

Gopal, B. & W. J. Junk, 2001. Assessment, determinants, function and conservation of biodiversity in wetlands: present status and future needs. In Gopal, B., W. J. Junk & J. A. Davis (eds), Biodiversity in Wetlands: Assessment, Functions and Conservation, Vol. 2. Backhuys Publishers, Leiden: 277–302.

Gopal, B., W. J. Junk, C.M. Finlayson & C. M. Breen, 2004. Present state and the future of tropical wetlands. In Polunin, N. (ed), Aquatic Eco. systems: Trends and Global Prospects. Cambridge University Press, Cambridge.

Gopal, B. & K. Krishnamurthy, 1993. Wetlands of south Asia. In Whigham, D. F., D. Dykyjova, & S. Hejny (eds), Wetlands of the World. I. Inventory, Ecology and Management. Handbook of Vegetation Science 15/2). Kluwer Academic Publisher, Dordrecht: 345–414.

Gopal, B., J. Kvet, H. Löffler, V. Masing & B. C. Patten, 1990. Definition and classification. In Patten, B. C., S. E. Jorgensen, H. J. Dumont, B. Gopal, P. Koryavov, J. Kvet, H. Löffler, Y. Sverizhev & J. G. Tundisi (eds), Wetlands and Continental Shallow Waterbodies. Vol. 1. Natural and Human Relationships. SPB Academic Publishing, Amsterdam: 9–15.

Groombridge, B. & M. Jenkins. 1998. Freshwater Biodiversity: a Preliminary Global Assessment. WCMC-World Conservation Press. Cambridge, UK.

Harding, J. S., E. F. Benfield, P. V. Bolstad, G. S. Helfman, & E. B. D. Jones, 1998. Stream biodiversity: the ghost of land use past. Proceedings of the National Academy of Sciences of the United States of America 95(25): 14843–14847.

IPCC, 2001. Technical summary, and summary for policymakers. Third Assessment Report of Working Group I of the Intergovernmental Panel on Climatic Change [www document]. URL http://www.ipcc.ch.

Keddy, P. A., 2000. Wetland Ecology: Principles and Conservation. Cambridge University Press, Cambridge, UK 614 pp.

Kottelat, M., 1989. Zoogeography of the fishes from Indochinese inland waters with an annotated checklist. Bulletin of the Zoological Museum, University of Amsterdam 12: 1–56.

Milliman, J. D., 1999. Flux and fare of fluvial sediment discharge to the ocean with particular emphasis on southern Asia and Oceania. Mitteilungen aus dem Geologisch-Paläontologischen Institut der Universität Hamburg, SCOPE Sonderband, 82: 33–34.

Milliman, J. D. & R. H. Meade, 1983. World-wide delivery of river sediment to the oceans. Journal of Geology 91: 1–21.

Mitchell, D. S. & B. Gopal, 1991. Invasion of tropical freshwaters by alien aquatic plants. In Ramakrishnan, P. S. (ed.), Ecology of Biological Invasion in the Tropics. International Scientific Publications, New Delhi: 139–155.

Mitsch, W. J. & J. G. Gosselink, 2000. Wetlands, Third Edition. John Wiley, New York; 920 pp.

Petts, G. E. & C. Amoros, 1996. Fluvial Hydrosystems. Chapman & Hall, London. 322 pp.

Poff, N. LeRoy, J. D. Allan, M. B. Bain, J. R. Karr, K. L. Prestegaard, B. D. Richter, R. E. Sparks & J. C. Stromberg, 1997. The natural flow regime: a paradigm for river conservation and restoration. Bioscience 47: 769–784.

Primavera, J. H., 1994. Shrimp farming in the Asia-Pacific: environmental and trade issues and regional Cooperation. Paper presented at the Nautilus Institute Workshop on Trade and Environment in Asia-Pacific: Prospects for Regional Cooperation, 23–25 September 1994, East–West Center, Honolulu, USA. Available at http://www.nautilus.org/papers/enviro/trade/shrimp.html.

Rainboth, W. J., 1996. Fishes of the Cambodian Mekong. Food and Agriculture Organization of the United Nations (FAO), Rome, 265 pp.

Revenga, C., J. Brunner, N. Henninger, K. Kassem & R. Payne, 2000. Pilot Analysis of Global Ecosystems: Freshwater Systems. Washington, DC: World Resources Institute. Online at: http://www.wri.org/wr2000/freshwater_page.html.

Revenga, C. & Y. Kura, 2003. Status and trends of biodiversity of inland water ecosystems. CBD Technical Series no. 11. Secretariat of the Convention on Biological Diversity, Montreal, 120 pp.

Robinson, H, 1976. Monsoon Asia. Third Edition. McDonald & Evans, Estover, UK, 510 pp.

Thornton, C., M. Shanahan & J. Williams, 2003. From wetlands to wastelands: Impacts of shrimp farming. SWS Bulletin 2003; 48–53.

Van Duzer, C., 2004. Floating islands: a global bibliography, with an edition and translation of G. C. Munz' Exercitatio academico de insulis natantibus (1711) manuscript, 234 pp.

Zakaria-Ismail, M., 1994. Zoogeography and biodiversity of the freshwater fishes of Southeast Asia. Hydrobiologia 285: 41–48.

Hydrobiologia (2005) 542:77–93
H. Segers & K. Martens (eds), Aquatic Biodiversity II
DOI 10.1007/s10750-004-3951-y

Recovery in diversity of fish and invertebrate communities following remediation of a polluted stream: investigating causal relationships

S. M. Adams*, M. G. Ryon & J. G. Smith
Environmental Sciences Division, Oak Ridge National Laboratory, Oak Ridge, TN, USA
*(*Author for correspondence: Tel.: +1-865-574-7316, Fax: +1-865-576-8543, E-mail: Adamssm@ornl.gov)*

Key words: recovery, restoration, fish and macroinvertebrate communities, causal relationships

Abstract

Spatial and temporal responses of biota to anthropogenic disturbance were measured over a 15 year period in a contaminated stream undergoing remediation and recovery. Along the spatial gradient of the stream, levels of contaminants decreased downstream along with improved responses of instream biota at several levels of biological organization. Recovery of the biota in this stream over the 15 year study period is demonstrated by the temporal relationships between levels of decreasing contaminants and the concomitant responses of the periphyton, macroinvertebrate, and fish communities and changes in the various bioindicators of individual fish health. Decreases in contaminants over a temporal scale were followed closely by an improvement in physiological and organismal-level indicators, increases in the diversity of macroinvertebrate and fish communities, and rapid increases in the chlorophyll *a* biomass and photosynthesis rate of the periphyton community. These results emphasize that field studies designed to assess and evaluate the effectiveness of restoration activities on stream recovery should incorporate a variety of response endpoints ranging from sensitive and short-term responses to long-term but ecological relevant indicators of change. The close spatial and temporal relationships observed between changes in physicochemical factors and positive responses in various components of the stream biota over the 15-year study period suggest a strong cause and effect relationship between remediation activities and stream recovery. Understanding causal relationships and the mechanistic processes between environmental stressors, stress responses of biota, and the recovery process is important in the effective management and restoration of aquatic ecosystems.

Introduction

Many aquatic ecosystems are subjected to some level of environmental stress either from natural causes, anthropogenic sources, or both. The aim of restoration ecology and associated recovery studies is to remediate impacted ecosystems to some level of integrity or health that is acceptable to society and to environmental managers. There has been much debate, however, relative to the criteria that defines acceptable states of recovery or restoration in aquatic ecosystems. Recovery has generally been considered as a return to a pre-disturbance steady-state, termed the nominal state, as a result of the operation of homeostatic control mechanisms operating in the ecosystem. Recovery can also be defined as the rate and manner in which an ecosystem subsequently returns to its unstressed condition or follows a chronological sequence of change that coincides with an unstressed reference condition (Kelly & Harwell, 1990). Studies of recovery in aquatic systems are important for establishing exceedance criteria for water quality standards (Plafkin, 1988), for testing current ecology theory (Resh et al., 1988, Yount & Niemi, 1990), and for evaluating the effectiveness of restoration activities and remedial actions (Detenbeck et al., 1992; Adams et al., 2002).

There are currently no widely accepted approaches or criteria that are used to evaluate when a disturbed aquatic system has 'recovered' to an acceptable state. Conventional approaches for monitoring the status and recovery of aquatic systems include measures of biomass, abundance, species diversity and richness, or biotic indices based on combinations of these parameters (Ford, 1989). The more conventional viewpoint is that structural attributes of the community should be the goals of restoration success while recently a vocal minority has advocated functional endpoints such as ecosystem processes which could serve as indicators of restored ecosystems (Muotka & Laasonen, 2002; Nienhuis et al., 2002). In spite of a well developed theory for recovery and an abundance of studies on the effects and subsequent recovery of aquatic systems, important issues remain to be addressed relative to the dynamics of the recovery process within the framework of restoration ecology (Power, 1999). For example, what are the most important ecosystem parameters for determining ecosystem recovery from stress? Does information on recovery at one level of biological organization reflect or provide reliable predictions of recovery at other levels of organization?

The effects and consequences of stress on aquatic ecosystems may be manifested at many levels of biological organization. Also, the rate and extent of recovery may differ at different levels of biological organization. Subtle changes in community composition, the genotypic make-up of populations, and the extent to which physiological, cellular, and biochemical systems are impaired in organisms following stress may have a profound effect on how a recovered ecosystem responds to subsequent environmental stressors. Ecosystems that are thought to have recovered may, in fact, be more vulnerable than they were originally due to losses in both intraspecific and interspecific biodiversity, such that the ability to adapt to new environments is compromised (Depledge, 1999).

According to Maltby (1999) there is not a 'right' level to study stress and recovery, but different levels of biological organization provide different types of information that, in combination, provide insight into the effects of stress, their mechanistic bases, and their ecological significance. Population and community studies can provide a description of the effects of stress but do not, in themselves, provide information on how effects may be caused (Maltby, 1999). For example, it is relatively straightforward to document that some change has occurred at the population or community level, but identifying the mechanisms or causes of such changes are difficult. Community-level measures, such as biotic and diversity indices, provide useful descriptions of community structure but they are generally insensitive to sublethal levels of stress (Gray et al., 1990; Dawson-Shepherd et al., 1992). In contrast, individual and sub-organismal measures of stress can be used to detect sublethal effects because of their short response times, sensitivity and specificity to stressors, and their potential to provide information on causal agents (Depledge, 1999; Adams et al., 2002). Knowledge of organism-level responses to stress and recovery is essential for understanding how stressors cause adverse effects at higher levels of biological organization, and it follows, therefore, that the effects of stressors on communities could be predicted from knowledge of the effects of stressors at the individual and population level (Maltby, 1999). In this regard, when assessing the effects of stressors on aquatic systems, the effects on the individual should also be understood in order to help explain the ecological relevance of the results (Heinonen et al., 1999).

There are, however, advantages and limitations of using either individual organism metrics or community-level metrics alone for assessing the effects of stressors and the subsequent recovery process of aquatic systems. For example, Detenbeck et al. (1992) reported that autecological factors alone were inadequate for explaining recovery and that recovery of individual species densities was much slower than recovery based on community parameters such as species richness or total density. Community-based metrics are useful for assessing the degree of recovery but these measures provide little insight into the causal mechanisms or underlying processes responsible for recovery (Adams et al., 2002). One of the main concerns in using population and community level responses alone for assessing stress and recovery in aquatic systems is that, by the time an effect is observed using these endpoints, it may be too late to initiate effective environmental management or mitigation. Thus, in addition to these ecologically relevant endpoints, responses at lower levels of biological organization should also be measured to serve as sensitive and early-warning indicators of impending changes in ecosystem health (Adams et al., 2002).

When assessing the effects of environmental stressors on aquatic systems and the recovery of these systems once restoration activities have been initiated, studies should include a variety of response endpoints representing several levels of biological organization. Different endpoints typically represent a variety of specificities, sensitivities, variability, and ecological relevance to environmental stressors (Adams et al., 2002). One of the main reasons that environmental assessment and recovery studies should involve the use of a variety of responses at different levels of biological organization is that disturbance can be unequivocally identified and associated with various specific ecological levels of organization. Disturbance may affect each major level of organization, from the individual to ecosystem and landscape, and the consequences and mechanisms of disturbance are different at each hierarchal level (Rykiel, 1985). Analyses of disturbance at each level of organization are vital to understand the importance of disturbance and the dynamics of the recovery process (Pickett et al., 1989). Given this background, therefore, the principle objectives of this study are (1) assess changes in community diversity following remediation of a contaminated stream, (2) investigate potential causal relationships between changes in community diversity and other abiotic and biotic components of a stream ecosystem, and (3) provide some basic guidance for the design of field studies related to restoration and recovery of stream ecosystems.

Methods and approach

Study sites

Both spatial and temporal responses of a stream ecosystem to environmental contamination were examined in a disturbed stream undergoing remediation and restoration over a 15 year period. Levels of contaminants in water and biota along with various measures of periphyton, benthic macroinvertebrate, individual fish, and fish community health were measured in East Fork Poplar Creek (EFPC), a third order stream in East Tennessee whose headwaters historically have received point-source discharges of various contaminants from a nuclear weapons production facility. This system is typical of most ridge and valley streams in East Tennessee which are characterized by alternating pool and riffle areas, riparian vegetation along both banks, silt and sand substrates, and annual mean flows generally ranging between 5 and 15 cfs. The headwaters of EFPC originate a few hundred meters upstream of this facility and then flows through the city of Oak Ridge and the Oak Ridge Department of Energy reservation before confluence with Poplar Creek 24 km downstream (Fig. 1). The reference stream, Brushy Fork Creek, which is located in a nearby watershed (Fig. 1), was selected for its hydrological and physicochemical similarity to EFPC except this stream has no measurable levels of contaminants or other anthropogenic disturbances. From the mid-1940s until the late 1980s, EFPC received discharges of relatively high levels of various contaminants including heavy metals, chlorinated organics, PAHs, and residual chlorine. During this 50 year period the stream was characterized by a distinct longitudinal gradient in contaminant loading as evidenced by decreasing downstream concentrations of Polychlorineted biphenyl (PCBs) and mercury in fish, sediment, and water (Southworth, 1990; Peterson et al., 1994; LMES, 1997). In addition to point-source contaminant discharges, this stream has also been subjected to multiple disturbances such as sediment inputs, altered hydrodynamic flows, thermal loading, and habitat modification. For the past 15 years, however, EFPC has experienced extensive remediation primarily through water treatment processes and abatement of contaminant sources at the site facility along with the physical removal of contaminated sediment from various areas within the floodplain.

To monitor and assess spatial patterns and temporal changes in stream recovery during and following remediation activity, sampling for physicochemical parameters and instream biota including periphyton, benthos, and fish was

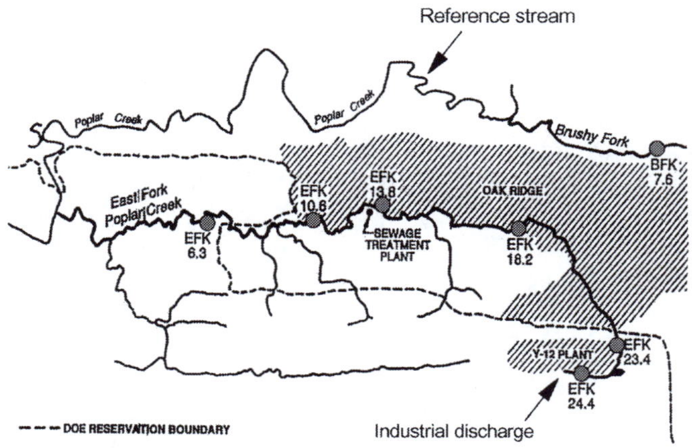

Figure 1. Location of sample sites at the contaminated stream, EFPC, and at the reference stream, Brushy Fork. Sample site numbers represent the distance in kilometers from the stream mouth with EFK 24 located closest to the industrial discharge.

conducted annually from 1988 through 2002 at three sites in EFPC and at the reference stream. Assessment of periphyton, benthic macroinvertebrate communities, and fish communities along with individual bioindicators of fish health were conducted in the spring from April to June of each year. Sample sites in the contaminated stream were located at East Fork Kilometer (EFK) 24 and 23 in the upper reaches of the stream near the sources of industrial discharge and at EFK 14, located 9 km below the industrial discharge (Fig. 1). The reference site, Brushy Fork Creek (BFC), is located in an adjacent watershed and it was sampled each year for the same chemical and biological parameters as were measured for the contaminated stream.

Contaminant analysis

Contaminants were measured in EFPC and in the reference stream for mercury in water and PCBs and mercury in fish tissue. In addition, heavy metals were analyzed annually in periphyton samples collected from the two upper EFPC sites and the reference site. For metals in periphyton, two to three composite samples were collected from each site with each sample consisting of periphyton harvested from 5 to 10 large cobbles. Samples were dried to constant mass at 60 °C before being extracted with hydrochloric and nitric acid and analyzed by ICP or ICP/MS.

To evaluate spatial and temporal trends in mercury and PCB contamination in fish, redbreast sunfish, *Lepomis auritus*, were collected twice annually from three sites in EFPC and the reference area from 1988 to 2002. Samples were collected and processed according to project-specific standardized technical procedures developed by the Biological Monitoring and Abatement Program (BMAP) at this laboratory to ensure sample quality and integrity (Peterson & Phipps, 1995). In general, 6–8 adult (50–150 g) sunfish were collected by electrofishing from each site at each sample period and placed on ice for transport to the processing laboratory. Total mercury analysis on filets was conducted by cold vapor atomic adsorption spectrometry (AA). PCBs were quantified against standard commercial mixtures (Aroclor 1254 and 1260) by capillary column gas chromatography using electron capture detection (EPA, 1980; EPA, 1984).

Periphyton analysis

Biomass of periphyton (chlorophyll *a*) and photosynthesis were measured seasonally (four times a year) at three sites in EFPC and at one reference site in Brushy Fork Creek. Sampling occurred at EFK 24 and 23 and the reference site from 1988 to 2002 and at EFK 14 during 1988–1995. Four rocks per site were collected at each sampling time and taken to the laboratory where photosynthesis was measured by ^{14}C uptake in glass chambers with recirculating flow (Hill & Boston, 1991). Uptake occurred at photosaturating irradiance (300–400 μmol quanta m^{-2} s^{-1}) and at ambient stream temperature. Rocks were removed from the chambers, rinsed in stream water, and placed overnight in individual jars containing DMSO which extracted both chlorophyll *a* and ^{14}C-labeled photosynthate (Palumbo et al., 1987).

Benthic macroinvertebrates

Five randomly selected benthic samples were collected during October each year from designated riffles at each of the three sites in EFPC and from the reference stream with a Hess stream bottom sampler (0.1 m^2) fitted with a 363-μm-mesh collection net. In the field, samples were placed in polyurethane coated glass jars, preserved with 95% ethanol, and returned to the laboratory for processing. In the laboratory, samples were placed in a US Standard No. 60-mesh (250-μm-mesh) sieve and washed with tap water. Organisms were sorted from the sample debris in a white tray at a magnification of 2×, identified to the lowest practical taxon (usually genus), and enumerated. Annelids were identified to class and chironomids were identified to subfamily or tribe. Additional details of the field and laboratory procedures are provided in Smith & Smith (1995).

Fish communities

For assessment of community-related metrics, quantitative sampling of fish was conducted by electrofishing in April through May each year at each of the EFPC sites and at the reference site. Information obtained from this quantitative sampling was used to estimated population size, calculate fish biomass per unit area, determine species richness, and evaluate community composition characteristics, such as the abundance of sensitive species (Karr et al., 1986), All sampling reaches were approximately 100 m in length and were isolated at the top and bottom of each reach by block-nets to prevent fish movement in or out of the area during sampling. A sampling event consisted of 5–8 investigators using two to three backpack electrofishers making three upstream passes, and collecting all stunned fish. Data obtained from this progressive removal method was use to calculate fish community structure at each stream site (Zippin, 1956). Following collection, fish were anesthetized, identified, measured (total length), and weighed on site. After processing, fish were allowed to fully recover and returned to the stream. Species population estimates were calculated using the method of Carle & Strub (1978). Biomass was estimated by multiplying the population estimate by the mean weight per individual per size class. To calculate biomass per unit area, total numbers and biomass were divided by the surface area (m^2) of each study reach. For each sampling date, surface area was estimated by multiplying the length of the reach by the mean width based on measurements taken at 5-m intervals.

Individual fish health

A suite of bioindicators reflecting individual fish health were measured for redbreast sunfish (*Lepomis auritus*), the dominant higher trophic level fish species collected at the EFPC sites and the reference stream during the spring of each year. At each site, 14–17 adult sunfish of each sex were collected by electrofishing techniques for assessing biochemical, physiological, and overall condition. Immediately after collection, a blood sample was taken from each fish by puncture of the caudal vessels, centrifuged immediately to obtain serum, and frozen in liquid nitrogen for subsequent analysis of biochemical and physiological parameters. Total length (mm) and mass (g) were recorded for each fish, and the liver and spleen removed for additional analysis. The liver (LSI) and spleno-somatic (SS1) indices were calculated as the mass of these respective organs divided by total body mass. Condition factor was calculated as $K = 100(W/L^3)$, where W = body mass (g) and L = total length (cm). Blood samples from each fish were analyzed for indicators of organ dysfunction and nutritional status. Creatinine (Rock et al., 1986) was used as an indicator of kidney damage and serum triglycerides (Bucolo & David 1973) as a short-term indicator of feeding or nutritional status. Growth and age structure of the redbreast sunfish population at each site were determined by analyzing scales taken from fish of all sizes. Scales were measured and annulus marks identified following the procedures of Jearld

(1983). Data from the scale analysis were used to assign fish to age groups. Growth for each age-class was estimated by back-calculation of size at age (Bagenal & Tesch, 1978).

Statistical procedures

Temporal patterns in chemical and biological parameters were tested with a least squares regression analysis (SAS, 1996) to determine significance of long-term changes in the data. Our assumption was that long-term trends in the various biological metrics at the reference site would show neither a persistent increase or decrease while at the impacted sites there would be a significant increase or decrease in metric values through time as demonstrated by significant ($p \leq 0.05$) deviations of the regression slope from zero. Data were first checked for heteroscadasticity using Levene's test (Johnson & Wichern, 1992). In some cases data were log transformed to meet the heterosadasticity assumption. Each regression analysis utilized all the individual observation points within a temporal data set which represented a sample size range of from 20 to 25 in the case of some chemical measurements up to 250–300 observations in a data set for some of the biological parameters. Fish community data were compiled and analyzed by a comprehensive Fortran 77 program developed by Railsback et al. (1989). The benthic macroinvertebrate data were transformed using a square root transformation (Elliott, 1977), and temporal changes in various aspects of the benthic invertebrate data such as EPT richness were analyzed with regression analysis.

Results

Spatial patterns

Distinct spatial gradients in contaminants and various biological components were observed in EFPC. Concentrations of PCBs in sediment and fish, levels of mercury in water, and heavy metal loading in periphyton displayed a distinct downstream gradient from the upper to lower reaches of EFPC (Fig. 2). The downstream pattern in various components of the stream biota also tracked the spatial gradient in contaminants within the stream (Fig. 3). Biomass of periphyton was highest upstream and species richness of the benthic macroinvertebrate and fish communities where lowest upstream but increased downstream (Fig. 3).

Temporal patterns

In addition to distinct spatial patterns of contaminants and biota in EFPC, there were also clear temporal patterns in contaminant levels and in the temporal response of the periphyton, macroinvertebrate, and fish communities over the 15 year study period.

Contaminants

Mercury concentrations in sunfish gradually declined from an annual mean peak of about 1.2 mg/l in 1993–1994 at the upper sections of EFPC to annual mean values approaching 0.5 mg/l in the late 1990s (Fig. 4a). Mercury levels were not only higher for fish at the upper sites of EFPC compared to downstream, they also demonstrated a more dramatic decline in the upper sites over the study period. Among all three EFPC sites, there were no evident temporal patterns in PCB levels in fish even though there were occasional high spikes in body burden levels particularly in 1994 and 1996 at the upper sites. There was also a tendency for PCB levels in fish to be lower in the downstream reaches compared to upstream areas which were closer to the contaminant source (Peterson et al., 1994). Levels of PCBs in reference fish were typically below 0.1 mg/kg and exhibited no temporal variation (Peterson et al., 1994). In the upper reaches of EFPC, total mercury concentrations in water steadily declined from high values of about 1.7 μg/l in 1989–1990 to current levels of about 0.5 μg/l (Fig. 4a). This significant temporal decrease in water concentrations of mercury (significance of regression line slope, $p < 0.05$) generally followed that of reductions in levels of mercury in fish tissue in upper EFPC after 1994.

Periphyton are known to efficiently sequester metals from the water, thus analysis of metals in periphyton provides an informative method to evaluate the dynamics of metal loading in aquatic systems. Zinc, cadmium, and chromium concen-

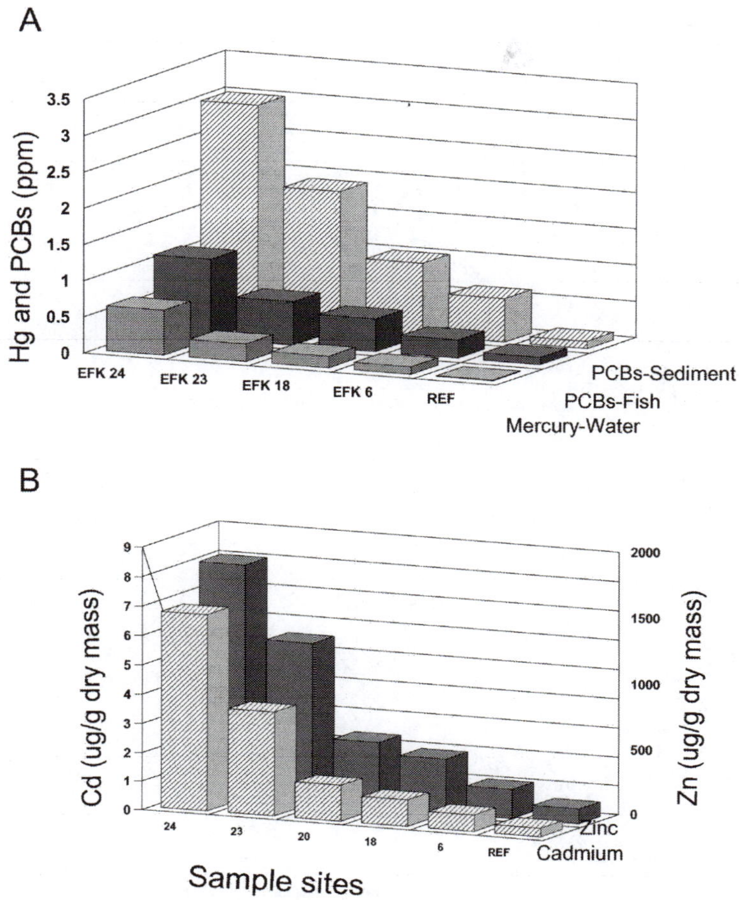

Figure 2. Spatial gradient in the concentration of PCBs in sediment and fish tissue and levels of mercury in water (A) and heavy metals in periphyton (B) at four sites in EFPC and in the reference stream.

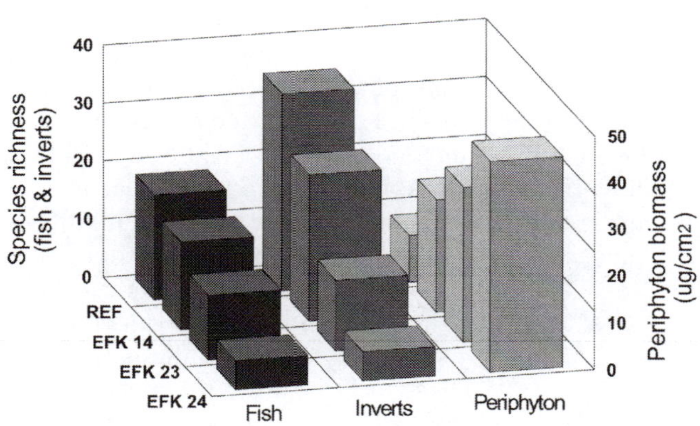

Figure 3. Spatial gradient in species richness of the fish and invertebrate communities and biomass (chl *a*) of the periphyton community at three sites in EFPC and at the reference site.

trations in periphyton generally decreased over time in upper EFPC (Fig. 4b). From 1996 to 2002, chromium levels in the upper reach of the stream declined approximately 50% , cadmium about 80%, and zinc 90% over this time period. Zinc and cadmium declined rapidly from 1996 to 1998–1999

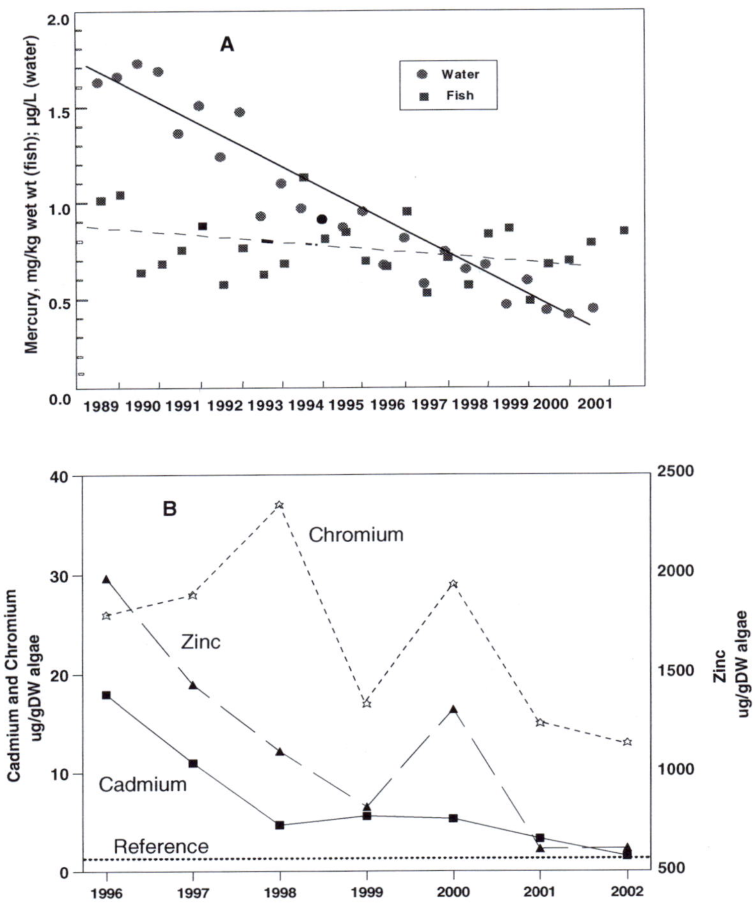

Figure 4. Temporal changes of contaminants in the upper sections of EFPC for mercury in water and in fish tissue (A) and three heavy metals in periphyton (B) over the 15 year study period.

and then exhibited slower decreases over time. Over the period of the study, concentrations of metals in reference site periphyton ranged from 2 to 10 times lower for cadmium and chromium and 2–3 times lower for zinc than levels in periphyton from EFPC sites, In addition, no temporal trends were evident for metals in periphyton from the reference site.

Periphyton

Biomass of periphyton as reflected by chlorophyll (chl) *a* increased over time in upper EFPC (Fig. 5) (regression; $p < 0.05$). Chlorophyll *a* levels increased from lows around 15 $\mu g/cm^2$ of rock surface in the late 1980s to peaks of 45–50 $\mu g/cm^2$ in the late 1990s, a threefold increase (Fig. 5). No significant temporal trends in chl *a* levels were apparent at the lower EFPC site (EFK 14) and at the reference stream. In the upper reaches of EFPC, photosynthesis generally tracked the temporal trend in algal biomass with the lowest levels occurring at initiation of the remediation period (1988–1990), after which photosynthesis increased rapidly during the early and mid-1990s, reaching maximum values by 1997 (regression: $p < 0.05$). Photosynthesis was similar to reference values during the first 3 years of the study but increased 3–4 times over the reference by the late 1990s.

Macroinvertebrate community diversity

Both total taxonomic richness and EPT (Ephemeroptera, Plecoptera, Trichoptera) richness

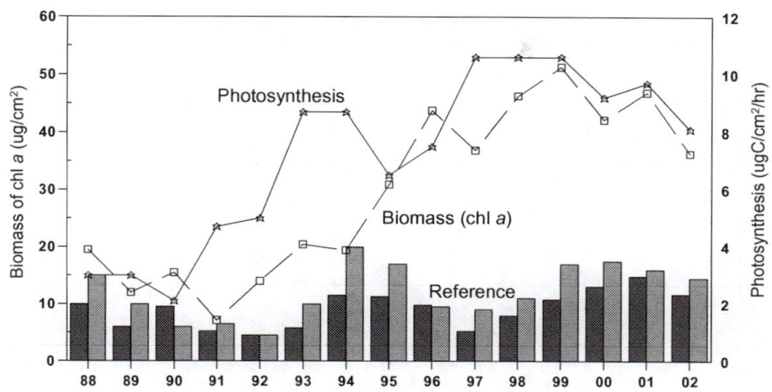

Figure 5. Temporal changes in biomass (chlorophyll *a*) and photosynthesis of the periphyton community in the upper sections of EFPC over the 15 year study period. Darker histograms are reference values for chlorophyll *a* and lighter histograms are reference levels for photosynthesis.

were depressed at the upper two sites in EFPC (EFK 24 and 23) even though a significant positive trend (regression: $p < 0.05$) in temporal recovery was indicated over the 15 year study period (Fig. 6). Total number of taxa per sample increased from a low of 5 and 8 species in 1988 to 21 and 21 species in 2002 at EFK 24 and 23, respectively. In general, EFK 14 and the reference site had similar values for total richness over the study period and both of these sites had 2–3 times the total number of taxa than the upper two EFPC sites. Even though there was no significant temporal increase in total richness at EFK 14 (regression: $p > 0.05$) from 1988 to 2002, there was a higher inter-annual variability at this site than was evident for total taxa at the upper two EFPC sites. Total EPT richness increased from less than 1 taxa at EFK 24 and 23 in 1988 to 5 taxa at both sites in 2002 and did not demonstrate the significant temporal increase in taxa as did total richness (Fig. 6).

A comparison of temporal changes in the major taxonomic groups of benthic macroinvertebrates at the upper two EFPC sites indicates a substantial increase in EPT taxa after the early 1990s, particularly at EFK 23, and a gradual reduction of the pollution-tolerant Chironomidae over the same time period (Fig. 7). The Oligochaeta and other taxa, however, do not appear to change in a consistent pattern over this period at any of the sites. There were no apparent temporal trends in any of the major taxonomic groups at EFK 14 or the reference site. Both of these sites are characterized by a relatively large portion (10–60% at EFK 14 and 10–25% at the reference) of EPT taxa.

Fish community diversity

Compared to the benthic macrobenthos, the total number of fish species increased very little in the upper two EFPC sites over the 15 year study period (Fig. 8a). Before 1988, no fish were present at EFK 24 and the total number of species at this site increased to only 3–4 by 2002. There was, however, a significant temporal trend in recovery at EFK 23 with the number of species increasing significantly (regression: $p < 0.05$) from a low of 7 during 1986–1989 to 14–15 in the early 2000s (Fig. 8a). Temporal species trends at EFK 14, however, exhibited more inter-annual variability, peaking at highs of 23 species in 1995 and 20 in 1997. The species most responsible for recolonization at the EFPC sites were northern hogsucker (*Hypentelium nigricans*), rock bass (*Ambloplites rupestris*), and snubnose darter (*Etheostoma simoterum*).

In contrast to total taxonomic richness, the biomass of sensitive fish species in upper EFPC experienced a rather rapid explosion (Fig. 8b). Biomass of sensitive species (i.e., northern hogsucker, snubnose darter, rock bass) was generally higher at the reference sites than in EFPC (Fig. 8b). In fact 50–60% of the total biomass at the reference site consists of sensitive species whereas in EFPC only 5–10% and 10–20% of the biomass is composed of sensitive species at the upper two sites and at EFK 14, respectively.

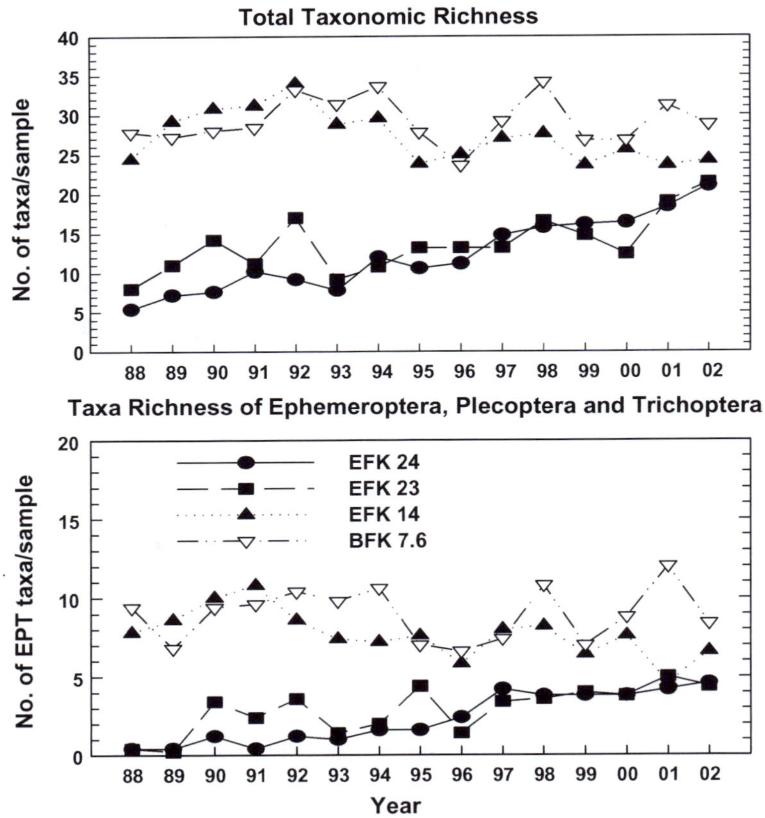

Figure 6. Total species richness (A) and total EPT (Ephemeroptera, Plecoptera, Trichoptera) richness (B) of the benthic macroinvertebrate community at three sites in EFPC and at the reference site.

Individual fish health

Indicators of physiological and bioenergetic status were also measured to determine the health and condition of individual fish in EFPC. Physiological indicators consisted of creatinine (an indicator of kidney dysfunction) and serum triglycerides, an indicator of short-term nutritional and bioenergetic status. Individual health condition indices were represented by the SSI, (general indicator of immune system status) (Goede & Barton, 1990), the LSI (energy storage and metabolism), and individual fish growth (tissue elaboration).

Creatinine, and indicator of kidney dysfunction, was elevated above reference values in upper EFPC from 1988 to about 1991 then declined significantly ($p < 0.05$) to within range of reference values before spiking in 1996 and 1997 (due to an episodic input of mercury into upper EFPC), after which levels again declined to reference values (Fig. 9a). In upper EFPC, serum triglycerides increased significantly ($p < 0.05$) over the study period reaching levels 2–3 times the values observed in 1988 (Fig. 9b). Reference fish generally had higher levels of serum triglycerides that did those from EFPC but showed no clear temporal pattern (Fig. 9b). The SSI, a general indicator of disease or immune system competence, declined significantly ($p < 0.05$) over the study period in upper EFPC (Fig. 9c), approximating reference levels by the mid 1990s. Fish in the upper reaches of EFPC displayed a significant increase ($p < 0.05$) in the LSI, or energy storage capacity, over the study period (Fig. 9d). Reference values for the LSI were relatively low until 1993 after which they increased and approximated those from the upper reaches of EFPC. Reduced reference values for the

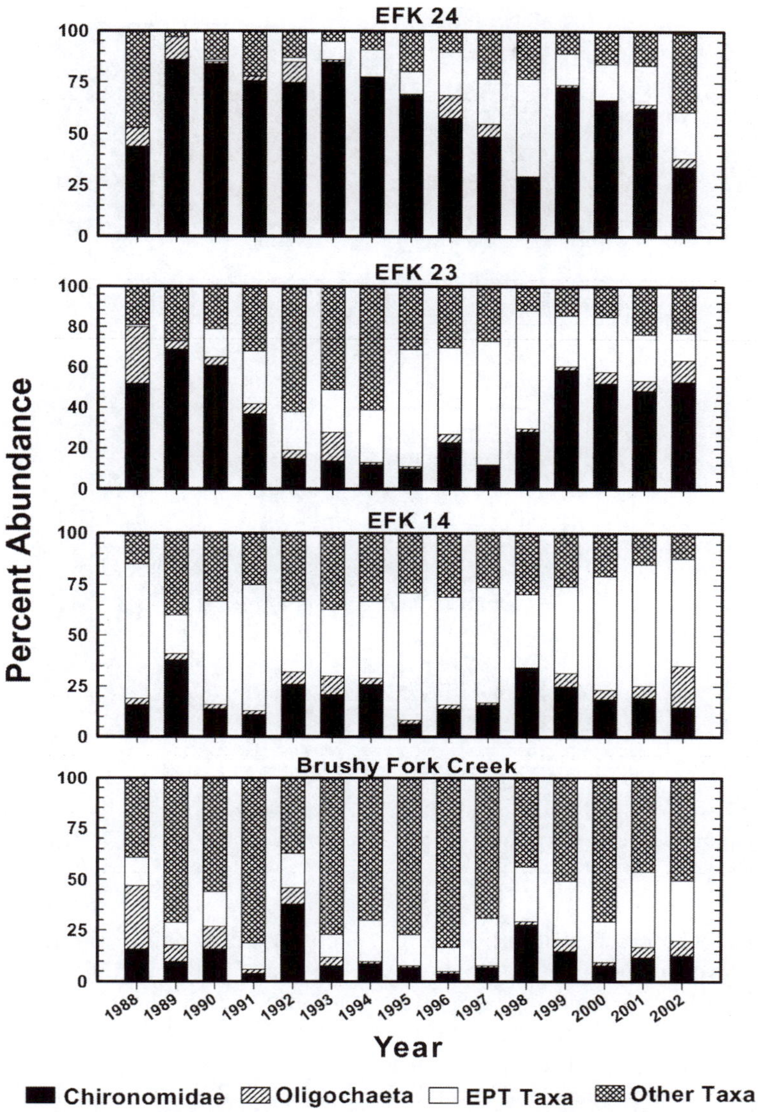

Figure 7. Temporal patterns of change in the percentage of abundance in major benthic macroinvertebrate groups in EFPC and the reference site. EPT = Ephemeroptera, Plecoptera, Trichoptera.

LSI in the first 5–6 years of the study were primarily due to a widespread drought in the Tennessee Valley region which reduced stream flows and consequently available food for growth. Growth rate of the redbreast sunfish population displayed a gradual but significant ($p < 0.05$) temporal reduction over the 15 year study period approaching reference values near the end of the study (Fig. 9e).

Discussion

Spatial responses

Contaminant loading into EFPC had a pronounced effect on the response of instream biota as demonstrated by the spatial patterns in biological components which tracked the downstream gradient in contaminants. The elevated levels of

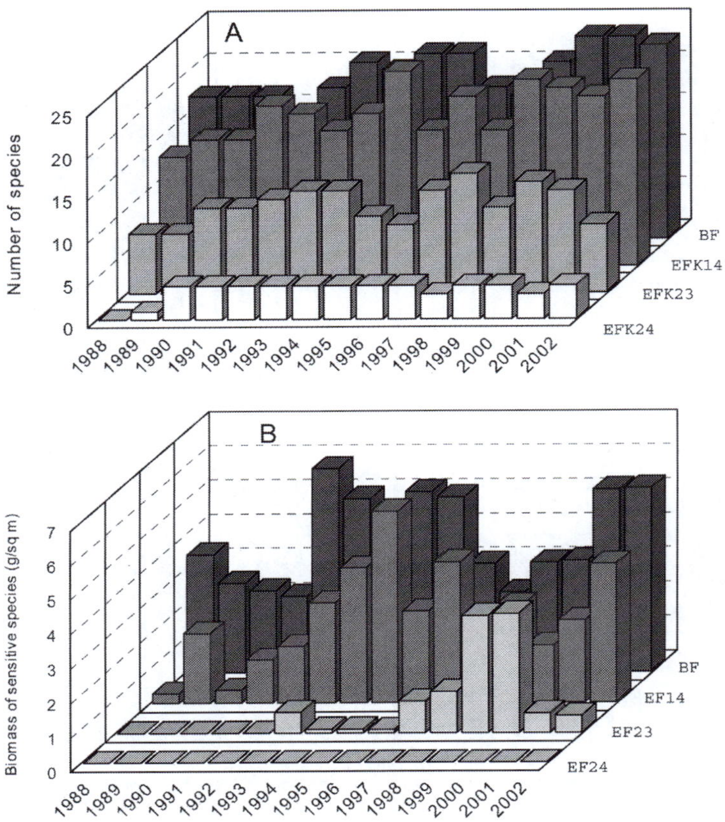

Figure 8. Annual variability in the total species richness (A) and biomass of sensitive species (B) for the fish community at three sites in EFPC and at the reference site.

contaminants in the upper reaches of the stream reflected the highest levels of periphyton biomass, the lowest levels of fish and macroinvertebrate community diversity (species richness), and the poorest condition of individual fish as demonstrated by the various physiological and bioindicators of fish health. The higher levels of periphyton biomass upstream simply reflected the higher nutrient loading upstream which decreased along the downstream gradient of the stream. As the levels of contaminants in the water, sediment, and biota decreased downstream, the health of the individual fish improved in conjunction with diversity (species richness) of the benthic and fish communities.

Other environmental factors, in addition to the decreasing contaminants along the gradient of the stream, could also have contributed to the improved condition of the fish and macroinvertebrate communities downstream. The role of confounding and modifying ecological factors such as food availability, competition, and habitat quality on the health and recovery of biological systems is well documented (Yount and Niemi 1990; Dickson et al., 1992; Power, 1999). Along the spatial gradient of EFPC, the physical habitat changes from relatively low quality in the upstream reaches to higher quality downstream. Upstream sections are primarily channelized with portions of the streambank being relatively steep and stabilized by sections of rip-rap. Habitat diversity in upper EFPC is relatively low due to scarcity of submerged physical structures in pool and riffle areas, there is a higher sinuosity in flow, and higher and more variable temperatures regimes generally occur. In contrast, the lower sections are characterized by a higher diversity and quality of habitats and more moderate and less variable temperature and flow regimes. In addition, food resources for secondary consumers

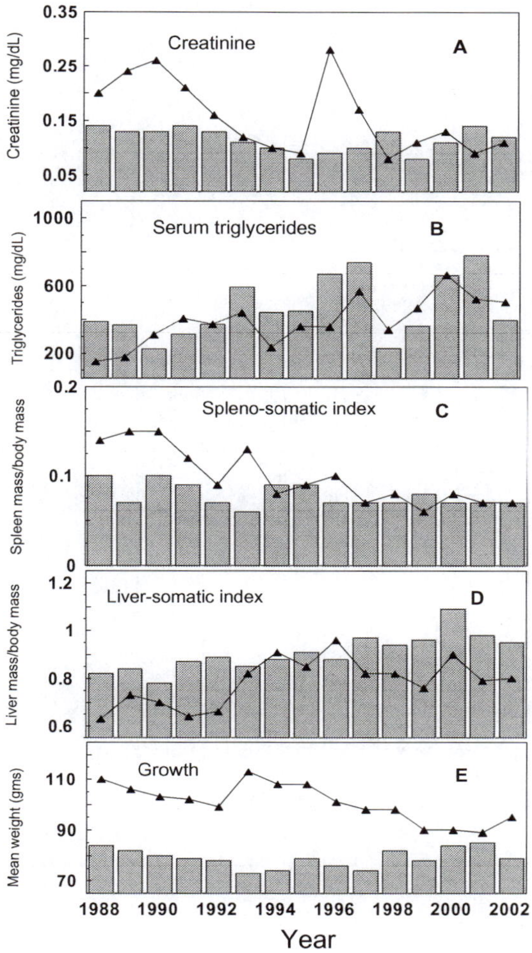

Figure 9. Temporal responses in five bioindicators of individual fish health including creatinine an indicator of kidney dysfunction (A), triglycerides an indicator of bioenergetic status (B), the SSI, a gross indicator of immune system competence (C), the LSI an indicator of metabolic status (D), and growth (mean weight at age 3) which is an overall reflection of individual fish health (E) for redbreast sunfish collected from the upper reaches of East Fork Poplar Creek. Histograms represent temporal responses of sunfish sampled from the reference site.

appears to be more plentiful downstream which would also favor and sustain the higher trophic groups. Episodic variations in physicochemical factors are also more dramatic in the upper sections of the stream which may contribute to the biological responses observed along the spatial gradient of the stream. High variabilities in temperature regimes typically occur during the summer months in upper EFPC. Although these temporal variations in temperature are not at levels which normally cause mortality in fish or other stream biota, nevertheless, these wide variations can cause physiological stress in organisms ultimately resulting in poor condition and fitness of stream biota. In addition, those episodic events which influence the upstream sections of the stream more than at downstream sites can destabilize communities, hinder the ability of populations and communities to realize their maximum recovery potential, and extend recovery time of instream biota (Resh et al., 1988; Death & Winter-bourn, 1995; Brant et al., 1999).

Temporal responses

In general, temporal response patterns in biota tracked the decreases in stream contaminants over time. Contaminants including heavy metals, mercury, and chlorine declined over the 15 year study period particularly in the upper reaches of the stream. Decreases in mercury levels in fish and heavy metals in periphyton followed similar temporal patterns as did the reductions of mercury in stream water. In upper EFPC, increases in chl *a* of periphyton tracked the general temporal decrease of residual chlorine and mercury in water over the study period. Likewise, key metrics of the benthic macroinvertebrate community improved at the upper two EFPC sites which may have been in response to the contaminant decreases in the stream and the increasing trend in chl *a* levels in periphyton. Similar temporal patterns in recovery were found by Watanabe et al. (2000) for a macrobenthic community previously experiencing stress from acid mine drainage. In the upper reaches of EFPC, several bioindicators of individual fish health also demonstrated the same general temporal patterns as did other biological components such as the periphyton, macroinvertebrate, and fish communities.

Recovery of the biota in EFPC over the 15 year study period is demonstrated by the temporal relationships between levels of decreasing contaminants in the stream and the concomitant responses of the various biological components including periphyton, macroinvertebrates, and fish. A summary of these temporal relationships is shown in Fig. 10 for the upper reach of EFPC where recovery was the most dramatic. Decreases

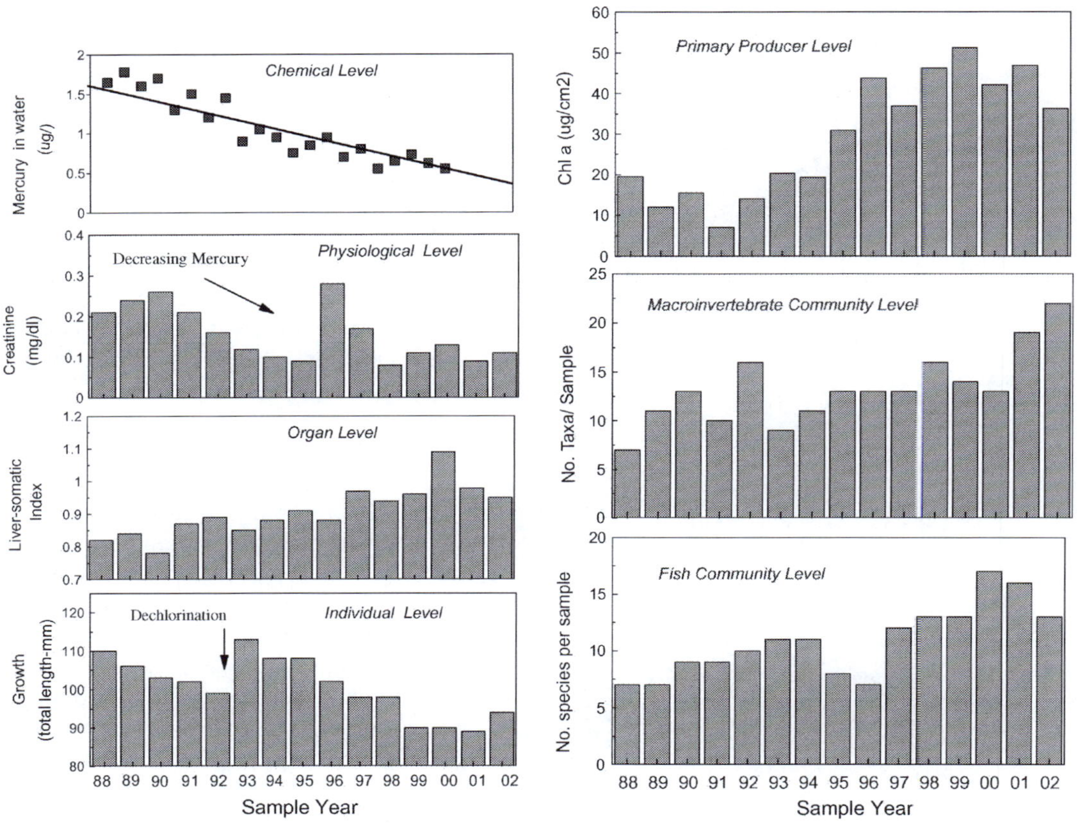

Figure 10. A summary of temporal relationships between remediation of contaminants in EFPC and biological responses at increasing levels of biological organization. The temporal patterns at the higher levels of organization track those temporal changes observed at the lower levels of organization.

in contaminants over a temporal scale was followed closely by an improvement in physiological and organismal-level indicators of individual fish, increases in the diversity of macrobenthic and fish communities, and the rapid response of structural and function properties of the periphyton community. Such responses in the recovery process of stream biota can be due to both direct and indirect factors. In the case of EFPC, direct effects of a stressor such as contaminants can impact organisms by impairing metabolic pathways. Stressors acting through indirect pathways, however, can ultimately affect the health and condition of organisms through bioenergetic pathways via the food chain. For example in EFPC, decreases of chlorine in the late 1980s and early 1990s in EFPC could have resulted in a reduction of toxic effects on the periphyton community, ultimately resulting in increased availability of plant biomass for primary consumers such as benthic macroinvertebrates. Consequently, the fish community would benefit from an increase in the quality (e.g., EPT taxa) and quantity (i.e., biomass) of the macrobenthos (i.e., see Fig. 7b).

Causal relationships

The close spatial and temporal relationships observed between changes in physicochemical factors and positive responses in various components of the stream biota in this study suggest a cause and effect relationship between remediation activities and recovery of this stream ecosystem. When such a relationship appears to exist in an aquatic system, then a credible case for causality between these components can be established (Fox, 1991; Adams, 2003). Results of this study have demonstrated that, on both a spatial and temporal scale, there was a distinct gradient of contaminant levels in EFPC and subsequent responses of biota. Based, therefore, on the causal criteria given by Hill (1965) and Fox (1991), there appears to be a strong relationship between reduction of contaminants and positive responses in the biota of EFPC.

Causality between environmental stressors and biological effects cannot reliably be established, however, based on one or just a few measured response endpoints. Measurement of community-level responses alone such as species richness or biomass would indicate that a change has occurred in an ecologically relevant parameter but the cause(s) of such a change would not be immediately evident. Conversely, if only suborganism or organism-level changes were measured there would be a high fidelity between the cause (stressor) and lower-level effect (i.e., a biomarker response) but the ecological significance of such an observation could not easily be established. In this study, however, we observed concurrent and concomitant changes among the contaminant stressors and several trophic groups (Fig. 10) and also among different levels of biological organization, with changes at the lower levels of organization (physiological, bioenergetic status of individuals) preceding observations at the higher levels (i.e., community diversity) of biological organization. Such relationships provide a 'weight of evidence' approach in establishment of causality in field studies. Establishment of causal relationships are particularly challenging in field studies because of the complex nature of ecological systems, the many biotic and abiotic factors which can influence or modify responses of biological systems to stressors (McCarry & Munkittrick, 1996; Wolfe, 1996), the orders of magnitude involved in extrapolation of effects over different spatial and temporal scales (Holdway, 1996), and compensatory mechanisms operating in populations (Power, 1997).

Understanding causal relationships between environmental stressors and effects on biota is important in the management and restoration of aquatic ecosystems. One of the primary goals of restoration ecology is to help identify the underlying mechanisms or processes responsible for observed changes in biota, and particular those mechanisms that may ultimately influence the status of ecologically-relevant endpoints such as population and community-level parameters. By understanding the factors or mechanisms responsible for responses in biological parameters and the subsequent recovery process of aquatic systems, environmental managers and regulators have a more informed basis for making decisions relative to environmental remediation and restoration activities. Recognizing, however, that efforts to establish relationships between causal factors (i.e., mechanisms) and various levels of biological response may challenge available biomonitoring and assessment resources, the experimental design of field studies should incorporate a selected set of biological responses that represent a wide range of sensitivities, specificities, and response time scales to stressors. Such a design would necessarily include a suite of biological responses represented by different levels of biological organization (Adams et al., 2000). Multiple endpoints are needed in such studies because no single response variable is sufficient to assess the extent of damage to an ecological system or to help in the identification of the factors responsible for damage (Karr, 1993). Therefore, within the framework of this experimental design, the focus of field studies should not only be on the individual organism level but also include (1) a few selected measures at the lower levels of biological organization (represented by high response sensitivity and specificity), and (2) some responses at the higher levels of biological organization (represented by high ecological relevance). The later category would include community-level parameters such as species diversity, richness, biomass of sensitivity species, and integrative community-level indices (e.g., the index of biotic integrity). Inclusion of a variety of biotic response endpoints in environmental effects and restoration studies would lend higher credibility, for example, to the ecological risk assessment process that incorporates community-level attributes as the basis for important decisions relevant to ecological and societal issues.

Acknowledgements

This research was conducted in support of the Y-12 Plant Biological Monitoring and Abatement Program, sponsored by the Environmental Safety and Health Organization, Environmental Compliance

Organization. The Oak Ridge Y-12 Plant is managed by BWXT-Y12 for the U.S. Department of Energy under contract DE-AC05-00-OR22800. The Oak Ridge National Laboratory is managed by UT-Battelle, LLC, for the U.S. Department of Energy under contract DE-AC05-00-OR22725.

References

Adams, S. M., 2003. Establishing causality between environmental stressors and effects on aquatic ecosystems. Human and Ecological Risk Assessment 9: 17–35.

Adams, S. M., M. S. Greeley & M. G. Ryon, 2000. Evaluating effects of contaminants on fish health at multiple levels of biological organization: extrapolating from lower to higher levels. Human and Ecological Risk Assessment 6: 15–27.

Adams, S. M., W. R. Hill, M. J. Peterson, M. G. Ryon, J. G. Smith & A. J. Stewart, 2002. Assessing recovery in a stream ecosystem: applying multiple chemical and biological endpoints. Ecological Applications 12: 1510–1527.

Bagenal, T. B. & F. W. Tesch, 1978. Age and growth. In Bagenal, T. B. (ed), Methods for Assessment of Fish Production in Fresh Waters, 3rd ed. Blackwell Sci. Pubs., Oxford, England: 101– 136.

Brant, P., M. Urban, N. Goodman, S. Bissell & I. Spiegal, 1999. Stability and resilience in benthic macroinvertebrate assemblages. Hydrobiologia 403: 123–133.

Bucolo, G. & H. David, 1973. Quantitative determination of serum triglycerides by the use of enzymes. Clinical Chemistry 19: 476–482.

Carle, F. L. & M. R. Strub, 1978. A new method for estimating population size from removal data. Biometrics 34: 621–630.

Dawson-Shepherd, A., R. M. Warwick, K. R. Clarke & B. E. Brown, 1992. An analysis of fish community responses to coral mining in the Maldives. Environmental Biology of Fishes 33: 367–380.

Death, R. G. & M. J. Winterbourn, 1995. Diversity patterns in stream benthic invertebrate communities:the influence of habitat stability. Ecology 76: 1446–1460.

Depledge, M. H., 1999. Recovery of ecosystems and their components following exposure to pollution. Journal of Aquatic Ecosystem Stress and Recovery 6: 199–206.

Detenbeck, P. W., DeVore, G. J. Niemi & A. Lima, 1992. Recovery of temperate-stream fish communities from disturbances: a review of case studies and synthesis of theory. Environmental Management 16: 33–53.

Dickson, K. L., W. T. Waller, J. H. Kennedy & L. P. Ammann, 1992. Assessing the relationship between ambient toxicity and instream biological response. Environmental Toxicology and Chemistry 11: 1307–1322.

Elliott, J. M., 1977. Some methods for the statistical analysis of samples of benthic invertebrates. Scientific Publication No. 25, Freshwater Biological Association of the United Kingdom.

Environmental Protection Agency (EPA), 1980. Interim methods for the sampling and analysis of priority pollutants in sediments and fish tissues. EPA 600/4-81-055. Environmental Monitoring and Support Laboratory, US Environmental Protection Agency, Cincinnati, Ohio.

Environmental Protection Agency (EPA), 1984. Extraction and analysis of priority pollutants in biological tissue, Method PPB 12/83. Environmental Services Division, Region IV, Analytical Support Branch, US Environmental Protection Agency, Athens, Georgia.

Ford, J., 1989. The effects of chemical stress on aquatic species composition and community structure. In Levine, S.A., M. A. Harwell, J. R. Kelley & K. D. Kimball (eds), Ecotoxicology: Problems and Approaches. Springer-Verlag, New York, NY: 99–179.

Fox, C., 1991. Practical causal inference for ecoepidemiologists. Journal of Toxicology and Environmental Health 33: 359–373.

Goede, R. W. and B. A. Barton, 1990. Organismic indices and an autopsy-based assessment as indicators of health and condition of fish. American Fisheries Society Symposium 8: 93–108.

Gray, J. S., K. R. Clarke, R. M. Warwick & G. Hobbs, 1990. Detection of initial effects of pollution on marine benthos: an example from the Ekofish and Eldfish oilfields, North Sea. Marine Ecology Progress Series 66: 285–299.

Heinonen, J. J., K. Kurronen & I. J. Holopainen, 1999. The effects of parasites and temperature on the accumulation of xenobiotics in a freshwater clam. Ecological Applications 9: 475–481.

Hill, A. B., 1965. The environment and disease: association or causation? Proceedings of the Royal Society of Medicine 58: 295–300.

Hill, W. R. & H. L. Boston, 1991. Community development alters photosynthesis-irradiance relations in stream periphyton. Limnology and Oceanography 36: 1375–1389.

Holdway D. A., 1996, The role of biomarkers in risk assessment. Human and Ecological Risk Assessment 2: 263–267.

Jearld, A. 1983. Age determination. In L. A. Nielsen & D. L. Johnson (eds), Fisheries Techniques. Southern Printing Co., Blacksburg, Virginia: 301–324.

Johnson, R. A. & D. W. Wichern, 1992. Applied multivariate analysis for biologists. John Wiley & Sons, New York, NY.

Karr, J. R., 1993. Defining and assessing ecological integrity: beyond water quality. Environmental Toxicology and Chemistry 12: 1521–1531.

Karr, J. R., K. D. Fausch, P. L. Angermeier, P. R. Yant & I. J. Schlosser, 1986. Assessing biological integrity in running waters: a method and its rationale. Illinois Natural History Survey Special Publication 5.

Kelley, J. R. & M. A. Harwell, 1990. Indicators of ecosystem recovery. Environmental Management 14: 527–545.

LMES (Lockheed Martin Energy Systems), 1997. Mercury abatement report for the US Department of Energy Oak Ridge Y-12 Plant for fiscal year 1997, Oak Ridge, Tennessee. Report No. Y/ER-297. Lockheed Martin Energy Systems, Inc., Oak Ridge, Tennessee, USA.

Maltby, L., 1999. Studying stress: the importance of organism-level responses. Ecological Applications 9: 431–440.

McCarty, L. S. & K. R. Munkittrick, 1996. Environmental biomarkers in aquatic toxicology: fiction, fantasy, or functional? Human and Ecological Risk Assessment 2: 268–274.

Muotka, T. & P. Laasonen, 2002. Ecosystem recovery in restored headwater streams: the role of enhanced leaf retention. Journal of Applied Ecology 39: 145–156.

Nienhuis, P. H., A. D. Buijse, R. S. E. W. Leuven, A. J. M. Smits, R. J. W. de Nooij & E. M. Samborska, 2002. Ecological rehabilitation of the lowland basin of the river Rhine (NW Europe). Hydrobiologia 478: 53–72.

Palumbo, A. V., P. J. Mulholland & J. W. Elwood, 1987. Extraction with DMSO to simultaneously measure periphyton photosynthesis, chlorophyll, and ATP. Limnology and Oceanography 32: 464–471.

Peterson, M. J. & T. L. Phipps, 1995. Bioaccumulation monitoring-aquatic. In Rev. 1: Biological Monitoring and Abatement Program Quality Assurance Plan, QAP-X-90-ES-063. Oak Ridge National Laboratory, Oak Ridge, Tennessee.

Peterson, M. J, G. R. Southworth & K. D. Ham, 1994. Effects of sublethal chlorinated discharges on PCB accumulation in transplanted asiatic clams (*Corbicula fluminea*). Water, Air, and Soil Pollution 73: 169–178.

Pickett, S. T. A., J. Kolasa, J. J. Armesto & S. L. Collins, 1989. The ecological concept of disturbance and its expression at various hierarchial levels. Oikos 54: 129–136.

Plafkin, J. L., 1988. Water quality-based controls and ecosystems recovery. In Cairns, J. Jr. (ed), Rehabilitating damaged ecosystems, Vol. II. CRC Press, Boca Raton, Florida: 87–96.

Power, M., 1997. Assessing the effects of environmental stressors on fish populations. Aquatic Toxicology 39: 151–169.

Power, M., 1999. Recovery in aquatic ecosystems: an overview of knowledge and needs. Journal of Aquatic Ecosystem Stress and Recovery 6: 253–257.

Railsback, S. F., B. D. Holcomb & M. G. Ryon, 1989. A Computer Program for Estimating Fish Population Sizes and Annual Productions Rates. ORNL/TM-11061. Oak Ridge National Laboratory, Oak Ridge, Tennessee, USA.

Resh, V. H., A. V. Brown, A. P. Covich, M. E. Gurtz, H. W. Li, G. W. Minshall, S. R. Reice, A. L. Shelton, J. B. Wallace & R.C. Wissmar, 1988. The role of disturbance in stream ecology. Journal of the North American Benthological Society 7: 433–455.

Rock, R. C., M. G. Walker & C. D. Jennings, 1986. Nitrogen metabolites and renal function. In Norbert, W. T. (ed), Textbook of Clinical Chemistry. W. B. Sanders Publishers, Philadelphia, Pennsylvania: 1271–1273.

Rykiel, E. J., 1985. Towards a definition of ecological disturbance. Australian Journal of Ecology 10: 361–364.

SAS Institute, 1996. SAS/STAT User's Guide, version 6.1. SAS Institute, Gary, North Carolina.

Smith, M. R. & J. G. Smith, 1995. Biological Monitoring and Abatement Program (BMAP) Benthic Macroinvertebrate Community Studies Quality Assurance Plan. QAP-90-ES-068. Oak Ridge National Laboratory, Oak Ridge, Tennessee, USA.

Southworth, G. R., 1990. PCB concentrations in stream sunfish (*Lepomis auritus* and *L. macrochirus*) in relation to proximity to chronic point sources. Water, Air, and Soil Pollution 51: 287–296.

Watanabe, N. C., S. Harada & Y. Komai. 2000. Long-term recovery from mine drainage distrubance of a macroinvertebrate community in the Ichi-Kawa River, Japan. Hydrobiologia 429: 171–180.

Wolfe, D. A., 1996. Insights on the utility of biomarkers or environmental impact assessment and monitoring. Human and Ecological Risk Assessment 2: 245–250.

Yount, J. D. & G. J. Niemi (eds), 1990. Recovery of lotic communities and ecosystems following disturbance: theory and application. Environmental Management 14: 515–762.

Zippin, C., 1956. The removal method of population estimation. Journal of Wildlife Management 22: 82–90.

Hydrobiologia (2005) 542:95–102
H. Segers & K. Martens (eds), Aquatic Biodiversity II
DOI 10.1007/s10750-004-2488-4

Ecological remarks on Mastigodiaptomus nesus Bowman, 1986 (Copepoda: Calanoida) in a Mexican karstic sinkhole

Adrian Cervantez-Martínez[1], Manuel Elías-Gutiérrez[1,*], Martha A. Gutiérrez-Aguirre[2] & Alexey A. Kotov[3]
[1]*Ecologia y Sistematica Acuatica, El Colegio de al Frontera Sur (ECOSUR), Km 2 Carretera Chetumal-Bacalar, Av. Centenario Km 5.5, Chetumal, Quintana Roo 77900, Mexico*
[2]*Lab. de Bioquímica. Instituto Technológico de Chetumal (ITCH) A.P. 267. Chetumal, Quintana Roo 77013, México*
[3]*A.N. Severstov Institute of Ecology and Evolution, Leninsky Prospect 33, Moscow 119071, Russia*
*(*Author for correspondence: Tel./Fax: +52-983-8350440 × 274, E-mail: eliasgm@ecosur-qroo.mx)*

Key words: body size, distribution, diaptomidae, Yucatan Peninsula, vertical migration, Chlorophyll-α

Abstract

Mastigodiaptomus nesus (Bowman, 1986) was previously described from the West Indies, and its distribution was considered restricted to these islands. Recently it has been detected in several systems from the continental Yucatan Peninsula in Mexico. In this study, we contribute to the body size variability, abundance, distribution and vertical migration of *Mastigodiaptomus nesus* dwelling in a small, deep (264 m^2, 47 m deep) karstic beaker-shaped sinkhole, located in this Peninsula. Here, this species is an important component of the zooplankton community during all the year. We analyzed the relationship between its abundance and concentration of nutrients (NO^{-3}), and Chlorophyll α. Differences in body size by sex ($F = 67.56$, $p < 0.0001$; $n = 180$), and sampled season ($F = 6.59$, $p < 0.01$; $n = 180$) were found. When animals from this sinkhole were compared with other localities, size difference was detected. Bigger animals were found in the West Indies, and smaller ones in continental localities. The nauplii were the most abundant development instars, followed by copepodids, and adults. In general, all developmental instars had a normal migration, with higher abundance at night time in upper layers (0.5 and 5 m), and diminishing towards daylight time. Direction of migrations seems to be towards the walls of the sinkhole (horizontal migration) instead of the deeper water layers, suggesting that these copepods are hidden in the irregular calcareous walls of the sinkhole. Abundance of total copepodid developmental stages were correlated with the Chl-α concentration. In addition, we found a highly significant difference between Chl-α concentration ($F = 24.172$, $p < 0.001$, $n = 72$), and NO^{-3} concentration at all strata among seasons ($F = 25.46$, $p < 0.001$, $n = 60$). Our results suggest that the body size of *M. nesus* is influenced by the nutrient concentration, and its vertical abundance and distribution may be explained by food availability.

Introduction

Mastigodiaptomus nesus (Bowman, 1986) is a species described from San Salvador Island. It is distributed in several Caribbean islands as Cayman, Bahamas, Cuba, and continentally in several localities from the north to the center of Yucatan Peninsula (Bowman, 1986; Suárez-Morales et al., 1996; Suárez-Morales, 2003). This species is the most important zooplankter in a small sinkhole named Minicenote (located in the center of the Yucatan Peninsula, near the village of Carrillo Puerto) (see Cervantes-Martinez et al., 2002) coexisting with several rotifers as *Keratella lenzi* (Hauer, 1953), *K. americana* Carlin, 1943, *Brachionus havanaensis* Rousselet, 1911, *and B. falcatus*

Zacharias, 1898; cyclopoids as *Thermocyclops inversus* Kiefer, 1936, and *Tropocyclops prasinus* (Fischer, 1860); and cladocerans as *Bosmina hagmanni* Stingelin, 1904 (pers. obs.).

As it is well known, studies on tropical zooplankton are still meager. Although some contributions to the taxonomy of zooplankters (Suárez-Morales et al., 1996; Elías-Gutiérrez & Suárez-Morales, 1999) and general limnology (Cervantes-Martínez et al., 2002; Schmitter-Soto et al., 2002) have been added, the basic structure of zooplankton populations from these freshwater bodies is still unknown (Torres-Orozco & Estrada-Hernandez, 1997).

The aim of this work is to contribute (with some ecological remarks) to the knowledge on *M. nesus*, as its vertical distribution in a 24 h cycle, abundance, body size variability, and relationship with some environmental factors.

Methods

The studied sinkhole (Minicenote) is located at 19° 36′ 23″ N, 87° 59′ 18″ W. It was described in detail by Cervantes-Martínez et al. (2002). This water body is characterized by small size (water surface about 264 m^2), deep (47 m), and oligotrophic water. In this system the diaptomid *Mastigodiaptomus nesus* is the most abundant zooplankton species along the year, and it is the largest limnetic crustacean. All our samples were collected with a plankton net of 50 μm mesh, by filtering a known volume, and fixed with buffered 4% formaldehyde.

Body size: 179 individuals (females and males) of *M. nesus* were measured in three different seasons, according to Schmitter-Soto et al. (2002); dry = March–May, 2001; rainy = June–October, 2003; and winter storm = November–February, 2003. These measurements were compared with data of Bowman (1986), as well as with our data obtained during the study of material on this species from the Grand Cayman collected in 1997, and Chichancanab lake (Yucatan, 19° 57′ N, 88° 15′ W), collected in the same year. Differences in the body size among seasons in the same sinkhole were tested by one-way ANOVA. Tukey's test was performed in order to determine in which season this difference in length was present.

Vertical migration: in order to determinate the diel vertical distribution of *M. nesus*, we collected samples by duplicate with a transparent 2.5l Van Dorn bottle in October 24–25th, 2003 from different vertical strata (0.5, 5, 10, 15, 20, and 30 m), in the limnetic zone of the system. The water samples were collected at 3 h interval, and filtered in a 50 μm mesh. Different development instars (nauplii, copepodids and adults) were counted in the whole samples to determine the abundance of organisms per one liter. Transparency with a Secchi disk, temperature, and dissolved oxygen with a Horiba U-10 water quality checker, were determined from the environment.

The food availability was estimated through the concentration of Chlorophyl α (Chl-α) (Williamson et al., 1996; Conde-Porcuna et al., 2002). The concentrations of Chl-α in mg m^{-3}, nitrates (NO_3) and ortophosphates (PO_4^{3-}) were measured each month by duplicate (at surface level in the three seasons) by spectrophotometric methods (A.P.H.A. 1989). Differences between Chl-α concentration among seasons were tested by one-way ANOVA.

Also, we measured the Chl-α concentrations in the same strata as copepods, only in rainy season. A Pearson's product-moment correlation was utilized to evaluate if the account of Chl-α is related with distribution and abundance of each development instar of *M. nesus*. All statistical tests were performed in the software Statgraphics, DOS version 7.0.

Results

All variables measured in the sinkhole were in the same range as previously detected by Cervantes-Martínez et al. (2002) for this system. Secchi disk transparency was 4.3 ± 0.86, 8.0 ± 4.95 and 6.6 ± 1.4 m for rainy, winter storm and dry seasons respectively. A vertical profile of dissolved oxygen and temperature by seasons is given in Figure 1. Oxygen was present in the whole column of water; in average, the lower values were found in winter storm season (2.6 ± 0.4 mg l^{-1}), whereas the higher values were observed during dry season (9.0 ± 0.4). The maximun temperature was found in rainy (30.1 °C) and minimum in winter storm season (24.6 °C). In general, all seasons had similar values of temperature (Figure 1).

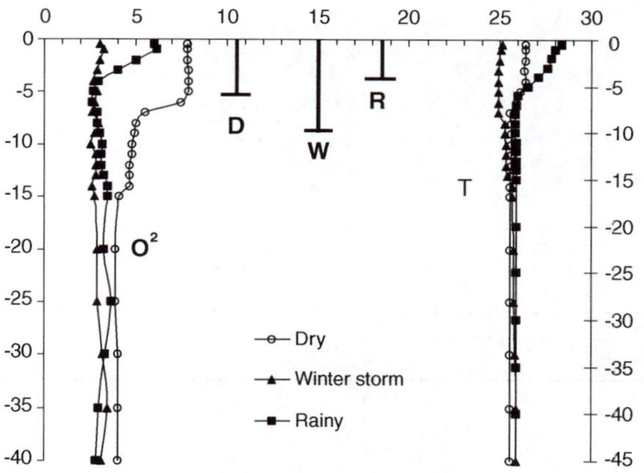

Figure 1. Vertical profile of dissolved oxygen (O_2 in mg l^{-1}), temperature (*T* in °C) and Secchi disk (m). Each line represents an average of each season. Secchi: D = dry, W = winter storm, R = rainy season.

Mastigodiaptomus nesus was a dominant component of the zooplankton community along all the year in this system. According to Bowman (1986), it is a close relative of *M. albuquerquensis*, but easily recognizable because the former lacks of a butterfly-like sclerotization on the basipodite 2 of the right male leg 5. In the Figure 2 SEM micrographs of the male and female from Minicenote are represented to confirm our determination of this species.

In addition to the typical differences in length of the adult males and females ($F = 67.56$, $p < 0.0001$; $n = 180$), we found a significant difference in body size by season ($F = 6.59, p < 0.01$; $n = 180$) (Figure 3). A Tukey test demonstrated that the body size of *M. nesus* in dry season was higher, with an average of 1071 ± 60 μm for males and 1111 ± 53 μm for females. In the rainy season it was 1020 ± 38 and 1102 ± 46 μm, and in the winter storm season it was 1043 ± 22 and 1091 ± 46 μm, respectively. Size of the animals from the Minicenote population was intermediate, when it was compared with populations from Chichancanab and Grand Cayman. Figures 3 and 4 allows to compare the size differences from different localities where this species was studied.

The study of vertical migration of the adults, copepodids and nauplii revealed that (Figure 5):

(1) the nauplii were the most abundant, followed by copedids and adults;
(2) in general, all development instars had major abundances in the night time, between 18:00–6:00 h and diminished towards 9:00–12:00 h (daylight hours);
(3) all development instars of *M. nesus* were always present in the layer from the surface to 10 m deep, but a clear differential distribution of these instars was observed. Nauplii were about 150 times more abundant in the surface layer (0.5 m), whereas copepodids and adults were located predominantly at depth of 5 m. Maximum activity in the 0–5 m was detected at 21:00 h for the copopodids and adults, and at 18:00 for nauplii. The deepest zone of the Minicenote seems to be a vacant environment for these calanoids (abundance in 15, 20, and 30 m strata, was minor to 7 individuals/1, even in the night time). They did not appear in the bottom layers during the day or immediately near the bottom.

Vertical distribution of Chl-α (Figure 6) was similar to the profile of copepod distribution and abundance: major values were found in the upper layers (0.5–5 m), and diminished towards deeper strata. In fact, the total abundance of copepodids was significantly correlated with the Chl-α concentration (Table 1).

In addition, we found a highly significant difference between Chl-α concentration among layers ($F = 24.172$, $p < 0.001$, $n = 72$), *sensu* Tukey test in the 0.5 and 5 m strata. The average of 0.25–0.33 is higher than the 0.03–0.12 values recorded in deeper layers, but no differences were found between seasons ($F = 0.304$, $p = 0.74$, $n = 30$).

We found significant differences between seasons ($F = 25.46$, $p < 0.001$, $n = 60$) in the nitrate concentration; *the posteriori* test, demonstrated that the higher values took place during the rainy season (52.58 ± 26.9 μM l^{-1}) in comparison to dry, and winter storm seasons (26.9 ± 12.9 and 38.3 ± 27.2 μM l^{-1}, respectively). Concentration of orthophosphates was higher in winter storm season and lower in dry season (2.45 ± 2.2, 0.005 ± 0.002 μM l^{-1}, respectively).

Discussion

From the environmental point of view, this water body has a slight thermic gradient and oxycline.

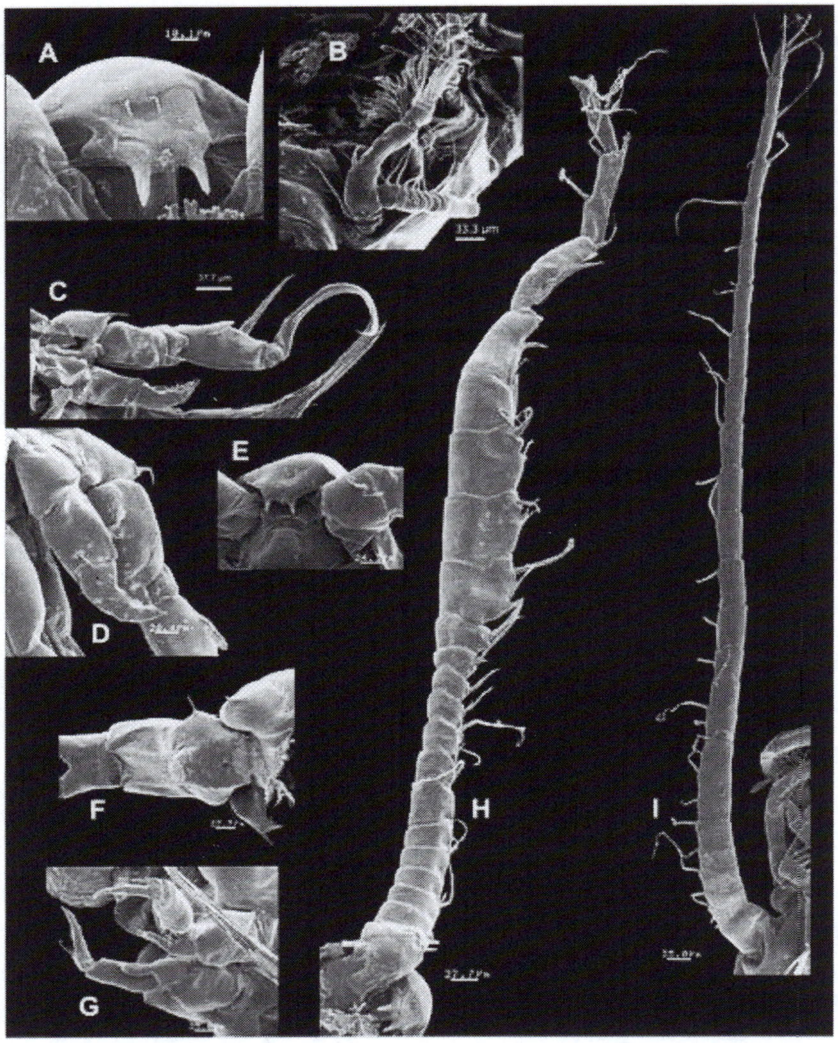

Figure 2. Scanning electron microscope photographs of *M. nesus* from Minicenote. (A) male rostrum, (B) first antenna, (C) male fifth leg, (D) male fifth leg, left side, (E) female rostrum, (F) genital segment, (G) female fifth leg, (H) male antennule, and (I) female antennule.

The temperature above 25 °C during the whole year in all the water column indicates a typical tropical system (Umaña & Collado, 1990; Lewis, 1996). The situation is the same in other karstic systems of Yucatan (Herrera-Silveira et al., 1998). As a result of the thermic stability in the environment, even during winter storm season, *M. nesus* was present during all seasons. Its abundance in the strata, expressed as org l^{-1} was similar with other diaptomids in Mexican lakes (see Lugo et al., 1999).

A clear differential distribution of development instars of *M. nesus* was observed: earlier developmental stages were much more abundant in the most superficial layers (0.5 m), whereas later developmental stages, remained in a deeper layer (5 m). This phenomena could be due to avoid intraspecific competition or to avoid predation. Laboratory experiments are necessary to investigate the food preferences of each developmental stage, as well as to analyze the gut content of all predators, to understand the shackle of the food web in these kind of systems.

Differences in body size of *M. nesus* between the water bodies is difficult to explain at this stage

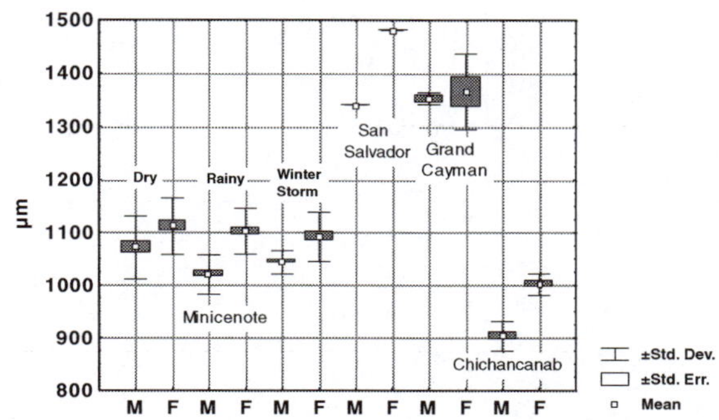

Figure 3. Body size of adult males and females of *Mastigodiaptomus nesus* in different seasons and localities. M = Males F = Females.

of knowledge, due to absence of environmental information from other localities where the species was studied. Among possible explanations, predation is one of the most important points to determine in future works. Chichancanab lake is the locality where the smallest animals appeared. Here, special conditions of isolation allowed the speciation of five Cyprinodontidae fishes (Schmitter-Soto, 1998). In particular *Cyprinodon simus* Humpries & Miller, 1981 is an active zooplanktivore. In the case of Minicenote, the main planktivore species of fishes are '*Cichlasoma*' *salvini* (Günther, 1862), *Thorichthys meeki* Brind, 1918, *Poecilia orri* Fowler, 1943, *P. mexicana* Steindachner, 1863, and *Gambusia yucatana* Regan, 1914 (pers. obs.). Size of *M. nesus* here was intermediate as compared with the other systems. It seems that this copepod is smaller in both continental systems because predator pressure, mainly the fishes, but this hypothesis should be tested.

Nevertheless, the variation in size between seasons in Minicenote could be related with the nutrient concentrations in the water, as it has been recorded in temperate zooplankton, where nutrient limitation can reduce the body size (Plath & Boersma, 2001). The larger body size of *M. nesus* during dry season in this study, was correlated with the slight increasing of Chl-α concentration, and a decreasing of Nitrate concentration, but it should be taken in account that we did not find any significant difference in Chl-α concentration among seasons in the surface of the water body. These data are similar to other observations where significant high values of Chl-α have been found during the dry season in other karstic systems from the region (Flores-Nava, 1994, Díaz-Arce, et al., 2000). Orthophosphates were scarce in Minicenote, as it has been detected previously, because they react with Ca and Mg and precipitate to the sediments as Ca_2PO_4 (Flores-Nava, 1994; Cervantes Martínez, et al., 2002).

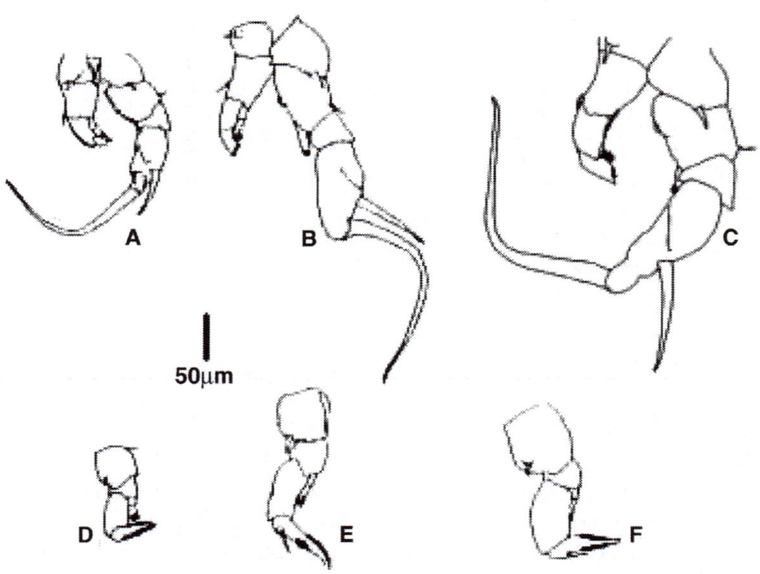

Figure 4. Fifth leg size variability in three localities where *M. nesus* dwells A–C Males, D–E Females, (A, D) Chichancanab lake, (B, E) Minicenote, (C, F) Grand Cayman, Pease Bay Pond. Scale bar is the same for all figures.

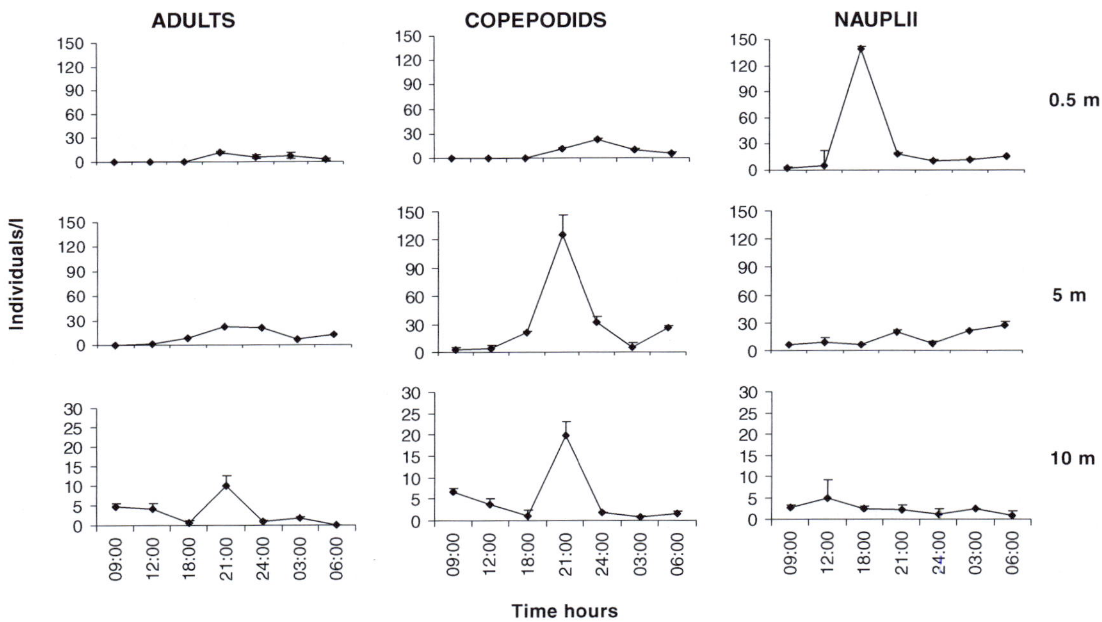

Figure 5. Diel vertical migration patterns for development instars *of Mastigodiaptomus nesus,* on October 24 and 25, 2003. Notice the modified abscissa scale for 10 m depth.

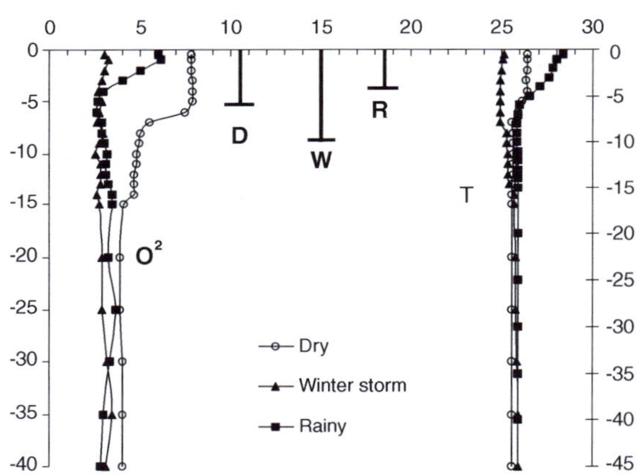

Figure 6. Vertical profile of Chl-α (mg m^{-3}) in rainy season.

It is accepted that physical and chemical features, that eventually affect the zooplankton community, are quite similar in all tropical systems (Lewis, 1996). However, as in temperate lakes, we found significant differences in the vertical profile, with major copepod abundance and Chl-α concentration in the superficial layers. Therefore, a good amount in the variability of the abundance of developmental stages of *M. nesus* is explained by food availability in this study, as well as has been recorded for diaptomids dwelling template and tropical lakes (Zaret, 1976; Williamson et al., 1996; Caramujo & Boavida, 2000).

Mastigodiaptomus nesus is absent in the water column during the day. It did not appear even in vertical tows from the bottom to the surface. It was not associated with the bottom also, we think, that they moved towards the walls of the sinkhole, suggesting that these copepods are hidden in the irregular calcareous walls on it. This kind of

Table 1. Correlation between Chl-α concentration and abundance in each strata, for the three developmental instars of *M. nesus* in Minicenote, in rainy season ($n = 60$ for each instar)

Instar	Pearson correlation and significance level	Equation
Adults	0.73***	$y = 0.09 + (0.012)(x)$
Copepodids	0.69***	$y = 0.11 + (0.0004)(x)$
Nauplii	0.60***	$y = -0.33 + (0.32)(x)$

behavior, known as horizontal migration has been observed in arctic and temperate lakes (De Stasio, 1993; Laurensen et al., 2001). However, a further study is needed to determine if this kind of movement occur in the surveyed system. Probably this behavior is to avoid predators, as fishes. A preliminary study of the gut content from *G. yucatana* indicated a preference for adult calanoids and copepodids (42.8%), followed by rotifers (37.08%), and cladocerans (20.12%). As it was noticed by Mischke & Wise (2003), the nauplii are not consumed, even by the small fry. Because of this reason they appear in the surface layer during the dusk.

In summary, after this work some interesting questions remain to be answered. Among them, we consider to study why nauplii are not consumed as other copepod instars, and to test if really there is a movement of adults and juveniles towards the walls of this system. Also remains to be studied the circulation of water on it and the effect on its thermal structure, just to mention some queries arising from these unique karstic tropical water bodies.

Acknowledgements

We thank E. Tovar-Juarez, S. Morales-Gutiérrez, I. Sánchez Álvarez, M.P. Chi López, G. Rodríguez Martínez, Johannes Hechenbichler and O. Sánchez-Aké from ECOSUR-Chetumal and Instituto Tecnológico de Chetumal (ITCh), for assisting in the collection of samples. Water samples were analyzed by A. Zavala for nutrients. ECOSUR Chetumal provided financial support for the sampling. This work is part of the graduate research study, sponsored by CONACYT to A. Cervantes-Martínez. AAK thanks the CONACYT for supporting his stay in Mexico, through the Catedras Patrimoniales program. Guadalupe Nieto kindly assisted us with SEM session.

References

A.P.H.A. American Public Health Association, 1989. Standard methods for the examination of water and wastewater. APHA, 17th edn. Washington, DC:1527 pp.

Bowman, T. E., 1986. Freshwater calanoid copepods of the West Indies. Syllogeus 58: 237–246.

Caramujo, M. J. & M. J. Boavida, 2000. The seasonal dynamics of *Copidodiaptomus numidicus* (Guraey, 1909) and *Thermocyclops dybowskii* (Lande, 1890) in Coastelo-do-Bode Reservoir. Aquatic Ecology 34: 143–153.

Cervantes-Martinez, A., M. Elías-Gutiérrez & E. Suárez-Morales, 2002. Limnological and morphometrical data of eight karstic systems "cenotes" of the Yucatan Peninsula, Mexico, during the dry season (February–May, 2001). Hydrobiologia 482: 167–177.

Conde-Porcuna, J. M., E. Ramos-Rodríguez & C. Pérez-Martínez, 2002. Correlations between nutrient concentrations and zooplankton populations in a mesotrophic reservoir. Freshwater Biology 47: 1463–1473.

De Stasio, B. Jr., 1993. Diel vertical and horizontal migration by zooplankton: population budget and the diurnal deficit. Bulletin of Marine Sciences 53: 44–64.

Díaz-Arce, V., J. Herrera-Silveira & F. A. Comín, 2000. Limnological characteristics of two types of cenotes of Yucatan. Verhandlungen der Internationalen Vereinigung für Theoretische und Angewandte Limnologie 27: 1–4.

Elias-Gutiérrez, M. & E. Suárez-Morales, 1999. (Crustacea; Anomopoda; Chydoridae) a new cladoceran from Southeast *Alona pectinata* Mexico. Revista de Biología Tropical 47: 105–112.

Flores-Nava, A., 1994. Some limnological data from five water bodies of Yucatan as a basis for aquaqulture development. Anales del Instituto de Ciencias del Mar y Limnología, Universidad Nacional Autonoma de México 21: 87–98.

Herrera-Silveira, J. A., F. A. Comín, S. López & I. Sánchez, 1998. Limnological characterization of aquatic ecosystems in Yucatan Peninsula (SE México). Verhandlungen der Internationalen Vereinigung für Theoretische und Angewandte Limnologie 26: 1348–1351.

Laurensen, T., E. Jeppesen, F. Landkildehus & M. Søndergaard, 2001. Horizontal distribution of cladocerans in arctic

Greenland lakes-impact of macrophytes and fish. Hydrobiologia 442: 107–116.
Lewis, M. W., 1996. Tropical lakes: how latitude makes a difference. In Schiemer, F. & K. T. Boland (eds), Perspectives in tropical limnology. Academic Publishing, The Netherlands: 27–42.
Lugo, A., M. E. González, M. del R. Sanchez & J. Alcocer, 1999. Distribution of *Leptodiaptomus novamexicanus* (Copepoda: Calanoida) in a Mexican hyposaline lake. Revista de Biología Tropical 47: 141–148.
Mischke, C. C. & D. J. Wise, 2003. Zooplankton size and taxonomic selectivity of channel catfish fry. North American Journal of Aquaculture 65: 141–146.
Plath, K. & M. Boersma, 2001. Mineral limitation of zooplankton: stoichiometric contraints and optimal foraging. Ecology 82: 1260–1269.
Schmitter-Soto, J. J. 1998. Catálogo de los peces continentales de Quintana Roo. Guías Cientificas ECOSUR. El Colegio de la Frontera Sur. San Cristóbal de las Casas, Chiapas: 239 pp.
Schmitter-Soto, J. J., F. A. Comín, E. Escobar-Briones, J. Herrera-Silveira, J. Alcocer, E. Suárez-Morales, M. Elías-Gutierrez, V. Díaz-Arce, L. E. Marin & B. Steinich, 2002. Hydrogeochemical and Biological characteristics of cenotes in the Yucatan Peninsula (SE Mexico). Hydrobiologia 467: 215–228.
Suárez-Morales, E., 2003. Historical biogeography and distribution of the freshwater calanoid copepods (Crustacea: Copepoda) of the Yucatán Peninsula, México. Journal of Biogeography 30: 1851–1859.
Suárez-Morales, E., J. W. Reid, T. M. Iliffe & F. Fiers, 1996. Catálogo de los copépodos (Crustacea) continentales de la Península de Yucatán, México. Comision Nacional para el Conocimiento y Uso de la Biodiversidad (CONABIO) & El Colegio de la Frontera Sur (ECOSUR), Chetumal, México: 296 pp.
Torres-Orozco, B. R. E. & M. Estrada-Hernández, 1997. Patrones de migración vertical en el plancton de un lago tropical. Hidrobiológica 7: 33–40.
Umaña, G. & C. Collado, 1990. Asociación planctónica en el Embalse Arenal, Costa Rica. Revista de Biología Tropical 38: 311–321.
Williamson, C. E., R. W. Sanders, R. E. Moeller & P. L. Stutzman, 1996. Utilization of subsurface food resources for zooplankton reproduction: implications for diel vertical migration theory. Limnology and Oceanography 41: 224–233.
Zaret, M. T., 1976. Vertical migration in zooplankton as a predator avoidance mechanism. Limnology and Oceanography 21: 804–813.

Hydrobiologia (2005) 542:103–111
H. Segers & K. Martens (eds), Aquatic Biodiversity II
DOI 10.1007/s10750-004-2411-z

Primary Research Paper

Consumptive and non-consumptive effects of turbellarian (*Mesostoma* sp.) predation on anostracans

Els R. M. De Roeck[1,*], Tom Artois[2] & Luc Brendonck[1]
[1]*Department of Biology, KULeuven, Charles de Bériotstraat 32, Leuren 3000, Belgium*
[2]*Limburgs Universitait Centrum, Centre for Environmental Sciences, Research group Biodiversity Phylogeny and Population Studies, Dept. SB6, Universitaire Campus Gebouw D, B-3590 Diepenbeek, Belgium*
(**Author for correspondence: Tel.: +32-16-32-4574, Fax: +32-16-32-4575, E-mail: els.deroeck@bio.kuleuven.ac.be*)

Key words: large branchiopods, turbellarians, predation, temporary pools, resting eggs, hatching characteristics

Abstract

We assessed the predatory impact of *Mesostoma* sp. (Rhabditophora) on hatching and survival of *Branchipodopsis wolfi* (Anostraca) under laboratory conditions. The hatching percentages of resting eggs of *B. wolfi* were negatively influenced to a significant degree by the presence of *Mesostoma* sp. or the medium in which the worms were kept before, but only when prey and predator originated from the same pool. This relationship was also not observed when eggs of the American anostracan *Thamnocephalus platyurus* were exposed to the Turbellaria from Botswana. There was, however, no indication of predation of *B. wolfi* eggs by *Mesostoma* sp., either by eating them totally or by sucking them out. When the same number of anostracans of different ages was offered simultaneously to *Mesostoma* sp., the 1- and 3-day-old specimens were eaten significantly faster than the 5-day-old juveniles and the adult *B. wolfi*. We conclude that *Mesostoma* sp. has a potentially strong influence on the population dynamics and life history characteristics of *B. wolfi* in temporary aquatic systems. Additionally, the feeding behaviour of *Mesostoma* sp. was described and the ultra structure of its rostral area was investigated.

Introduction

Predation is an important factor structuring zooplankton communities in both permanent and ephemeral water bodies (Zaret, 1980; Sih et al., 1985; Fryer, 1986; Kerfoot & Lynch, 1987; Williams, 1987; Blaustein, 1990; Blaustein & Dumont, 1990; Brendonck et al., 2002). Periodic droughts ensure that the temporary systems usually lack fish, but other organisms like invertebrates and amphibians still remain as predators. Predators in general can have a major impact on morphology and life history traits of their prey. Prey organisms can often 'sense' the predators via predator-derived kairomones. The reaction of the prey is assumed to reduce predation risk and is therefore considered adaptive (Lass & Spaak, 2003). Weber (2003) suggested differences in fish kairomones not related to concentration effects alone, since she detected differences in the effect of the medium of two fish species on life history traits of *Daphnia galeata*. Different kairomones could therefore add a possible factor of complexity to zooplankton ecology (Weber, 2003).

Dodson (1988) suggested that an animal only responds to those predators that have been a source of mortality in nature, not just to predators in general, and that a correspondence between response and coexisting predator suggests an evolutionary history. On the other hand, Binckley & Resetarits (2003) found that in contrast with most fish predators of *Hyla chrysoscelis*, *Aphred-*

oderus sayanus (Aphredoderidae) did not induce a behavioural response in this tree frog. Possibly the frog could never escape this fish predator because of its large gape. If predation is important in an island-like habitat, it is expected that prey populations are adapted to the predation pressure in their local habitat (Boersma et al., 1999). Such local adaptation could occur if selection for heritable traits differs from environment to environment and if gene flow between environments is restricted (Knight & Miller, 2004). Predation can as such cause rapid evolutionary changes in prey communities (Lass & Spaak, 2003).

Some of the most important predators of temporary aquatic systems generally are turbellarians, notonectids, diving beetles and dragonfly larvae (Williams, 1987; Blaustein, 1990; Blaustein & Dumont, 1990; Blaustein et al., 1995). This is also the case in very ephemeral rock pool systems in south-eastern Botswana (Brendonck et al., 2002). *Branchipodopsis wolfi* Daday is the only anostracan inhabiting these pools. Population densities of these fairy shrimps were frequently observed to decrease drastically from one day to another, probably predation from dragonfly larvae and Turbellaria (Brendonck et al., 2002).

At least five species of yet undescribed Turbellaria, including a *Mesostoma* species, occurs in these rockpools. A paper with the descriptions of these species is submitted for publication. In the course of this paper we will refer to this species as *Mesostoma* sp. As they also produce drought-resistant resting stages, they are probably the only predators of *B. wolfi* living permanently in these pools. *Mesostoma* may have a strong impact on pool-communities due to its high population growth rate (Blaustein & Dumont, 1990) and a relatively high predation rate (Schwartz & Hebert, 1982; Blaustein & Dumont, 1990; Dumont & Schorreels, 1990; Brendonck et al., 2002).

Several strategies of turbellarian predation have been described in literature, like producing traps, sit-and-wait predation, release of toxins and active prey searching (Schwartz & Hebert, 1982; MacIsaac & Hutchinson, 1984; Dumont & Carels, 1987; Blaustein & Dumont, 1990; Cash et al., 1995). Kolasa (2001) reported a learning ability for tasks that might be advantageous under natural conditions, such as recognition of prey. So far it is not clear how *Mesostoma* sp. preys on *B. wolfi* and whether there is a selection for different life stages.

Besides the active anostracan population, their egg bank may also suffer from turbellarian predation. As such, predation may have an impact on the chances for persistence of the prey populations relying on such dormant stages. However, in general, little is known of the impact of predation on egg banks. Cáceres & Hairston (1998) described predation of amphipods on the resting eggs of the water flea *Daphnia*, but this was not reported for turbellarians. Dumont et al. (2002) suggested that the large variation in surface structure of the resting stages of the anostracan *Chirocephalus* is an adaptive, maybe inducible response to predation by *Mesostoma* sp.

The aim of this study was to evaluate the impact of *Mesostoma* sp. on *B. wolfi*, both of which coexist in ephemeral rock pools in Botswana. Firstly, the impact on hatching fractions of dormant propagules of *B. wolfi* has been studied in the presence of turbellarian predators. By simultaneously exposing an anostracan species from America, possible prey predator co-evolution and local adaptation has been tested. Additionally, turbellarian-feeding preference has been assessed under laboratory conditions. Finally, predation on resting stages has been studied. We have also observed the predatory behaviour of the Turbellaria and searched for possible morphological adaptations used for predation.

Materials and methods

Anostraca and Turbellaria were collected in small rock pools in the semi-arid environment of Botswana. The sampled pools were situated on two isolated granite hill sites in the south-eastern Hardveld region (Kgale Hill and Thamaga, respectively). These sites are about 50 km apart.

Branchipodopsis wolfi is the only anostracan species inhabiting ephemeral rock pools in south-eastern Botswana. This species is adapted to the harsh conditions of periodically filling and quick, often unpredictable drying of the pools. They cope with this by growing and reproducing very rapidly and like all large freshwater branchiopods by the production of resting stages, which can adapt their hatching percentages to environmental cues

(Brendonck & Riddoch, 2000, 2001). *Mesostoma* sp. lives permanently in the Botswana rock pools, by also making resting stages that can survive long periods of drought. Like *B. wolfi*, they hatch within the first days after filling.

The resting egg bank of both study organisms was sampled by removing some surface sediment from the rock pools. Part of the sampled sediment was inundated with distilled water in 2-l aquaria under standardised laboratory conditions (at a temperature of 20 °C and at 24 h of light) after which *Mesostoma* sp. and *B. wolfi* emerged. Resting stages of both species were also isolated from the second part of the sediment by means of a sugar flotation method (Onbé, 1978; Marcus, 1990).

Hatching experiments

Influence of adult Mesostoma *sp. on hatching fractions of* B. wolfi

The hatching response of *B. wolfi* and *Thamnocephalus platyurus* Packard eggs to the presence of caged Turbellaria and Turbellaria-conditioned medium was investigated. *Mesostoma* sp. was held in a plastic tube with fine gauze of 120 μm at the end to assess for non-consumptive effects of their presence, while predation on freshly hatched anostracan larvae was avoided. The Turbellaria-conditioned medium consisted of diluted (20 μS/cm) EPA-medium (0.096 g/l $NaHCO_3$, 0.06 g/l $CaSO_4 \cdot 2H_2O$, 0.06 g/l $MgSO_4$, 0.004 g/l KCl, distilled water) (cf. US-EPA/600/4-85/013, 1985) in which turbellarians had been present for one day at a density of two adults per 10 ml. The medium used for control is 20-μS/cm EPA. In the respective control treatments for both species, hatching fractions were studied in pure medium without Turbellaria. Each treatment was replicated four times. Eggs were incubated at 20 °C and 24 h of light. For all replicates, 10 ml medium containing 20 *B. wolfi* eggs was used. The medium was refreshed daily in all treatments. In the treatment with Turbellaria-conditioned medium this should compensate for potential de-activation of info-chemicals. Hatchlings were counted daily for a period of 3 days and a hatching fraction was calculated. To test for the generality of any response to predator kairomones, the hatching response was also studied by using turbellarian populations from another rock pool site (Thamaga), hence not coexisting with the *B. wolfi* population from Kgale Hill and by studying resting eggs of *T. platyurus*, an anostracan originating from the USA and Mexico (Maeda-Martínez et al., 1997).

Predation on B. wolfi *eggs*

Twenty *B. wolfi* eggs were offered to five starved (3 days without food) and non-starved specimens of *Mesostoma* sp. in 10-ml petri dishes. The medium used was diluted (20 μS/cm) EPA. The study was conducted for 7 h only, in order to exclude predation on any larvae that might start hatching. The remaining eggs were tested for their viability by checking for the presence of a healthy embryo by pinching them with a pair of tweezers (method adopted after Brendonck et al., 1998; Brendonck & Riddoch, 2000). Six replicates were used.

Predation rate on different stages of B. wolfi

The feeding experiments of *Mesostoma* sp. for various life cycle stages of *B. wolfi* were studied in 200-ml plastic vials under a photoperiod of 12L/12D and at 25 °C. Three 1-day, 3-day, 5-day and adult Anostraca were added simultaneously to each of seven vials filled with 50 ml diluted (20 μS/cm) EPA medium and holding three starved Turbellaria. The length of the 1-day-old larvae was on average 0.73 mm, of the 3-day-old juveniles 1.15 mm, the 5-day-old juveniles measured 2 mm and the adult Anostraca 10.5 mm. The Turbellaria had an average length of 3 mm. The number of each anostracan life cycle stage eaten was counted during about 21 h. The electivity index (Chesson, 1983) was calculated to detect whether the predator prefers one or several prey age-class(es). An electivity index of 0 indicates that the predator has no preference for a specific prey type; a positive index indicates a prey preference for that specific age class.

Behaviour and morphology

The predation behaviour (encounter, attack, capture and ingestion of prey) by *Mesostoma* sp. was observed using a dissecting microscope. As prey organisms we used both *B. wolfi* and the cladoceran *Daphnia magna* Straus. Eggs and adult specimens of *Mesostoma* sp. were furthermore studied under a Scanning Electron Microscope (type XL

30 ESEM FEG and type XL 30 ESEM TMP) to detect any devices that could play a role in predation. The eggs were dried at 30 °C in an incubator, placed on a stub, critical point dried and covered with gold. The adults were fixed in warm (50 °C) formaldehyde (4%), dehydrated in an alcohol-series (50%, 75%, 90% en 100% alcohol), critical point dried, placed on a stub, coated with gold, placed in the SEM and photographed.

Statistical analysis

ANOVA tests were applied to compare the effects of the different treatments in each experiment, after meeting the requirement for homoscedasticity as assessed with Cochran's test. The statistical analysis was performed with Statistica 6.0.

Results

Hatching experiments

Influence of adult Mesostoma *sp. on hatching fractions of* B. wolfi

The mean hatching percentages of *Branchipodopsis wolfi* (originating from Kgale Hill) in absence or presence of *Mesostoma* sp. (originating from Kgale Hill and Thamaga, respectively) and the hatching fractions of similarly exposed *Thamnocephalus platyurus* are presented in Figure 1.

Overall, the different treatments had a significant influence on the hatching percentages of *B. wolfi* originating from the same pools as *Mesostoma* sp. (Kgale Hill) (ANOVA; $F = 14.8237$, 2df, 9df error, $p = 0.001$), but not on those of *T. platyurus* (ANOVA; $F = 0.8937$, 2df, 9df error, $p = 0.443$). The mean hatching percentages of *B. wolfi* eggs incubated in pure EPA (24.1 ± 7.5%, $n = 4$) was significantly higher (Tukey post-hoc, $p < 0.05$) than those incubated in Turbellaria-medium (4.1 ± 5.6%, $n = 4$) or in the presence of caged Turbellaria (6.7 ± 3.0%, $n = 4$). The hatching percentages of those last two treatments did not differ significantly from each other (Tukey post-hoc, $p > 0.05$).

For *B. wolfi* originating from another pool (Kgale Hill) as *Mesostoma* sp. (Thamaga), there was no overall significant effect on its hatching fractions (ANOVA; $F = 1.1673$, 2df, 9df error, $p = 0.354$) and neither on these of *T. platyurus* (ANOVA; $F = 0.8937$, 2df, 9df error, $p = 0.443$). The mean hatching percentages of *B. wolfi* eggs incubated in diluted EPA (33.8 ± 16.5%, $n = 4$) was not significantly different (Tukey post-hoc, $p > 0.05$) from those incubated in Turbellaria-medium (21.2 ± 4.8%, $n = 4$) or from those incubated in the presence of caged Turbellaria from another pool (33.8 ± 15.5%, $n = 4$). The hatching percentages of those last two treatments also did not differ significantly from each other (Tukey post-hoc, $p > 0.05$).

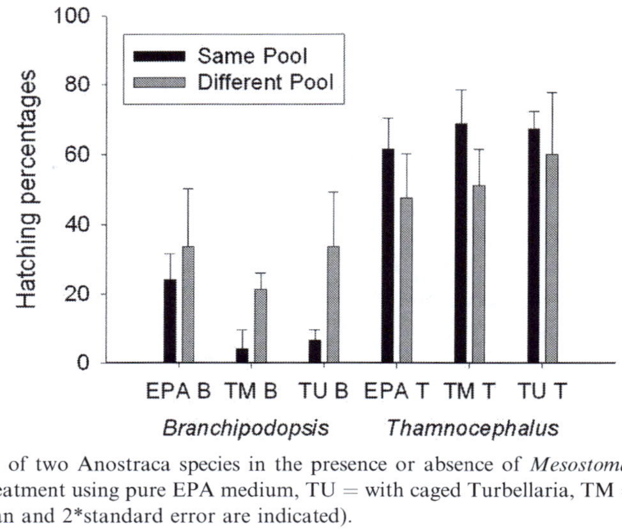

Figure 1. Hatching percentages of two Anostraca species in the presence or absence of *Mesostoma* sp. from the same and from a different site (EPA = control treatment using pure EPA medium, TU = with caged Turbellaria, TM = in Turbellaria medium, B = *B. wolfi* and T = *T. platyurus*; mean and 2*standard error are indicated).

Predation on B. wolfi *eggs*

None of the *B. wolfi* eggs were eaten within 7 h, neither by the saturated nor by the unsaturated *Mesostoma* specimens. We did, however, observe that the turbellarians ate freshly hatched larvae.

Predation rate and preference

The electivity index for the different age classes of *B. wolfi* after 150 min of exposure to the predators are shown in Figure 2. An overall significant difference in the electivity index was observed between the different age categories (ANOVA; $F = 26.1879$, 3df, 24df error, $p = 0.000$).

There was no difference between the electivity index of 1-day-old larvae (0.43 ± 0.21, $n = 7$) and of 3-day old juveniles (0.31 ± 0.16, $n = 7$) (Tukey post-hoc, $p > 0.05$), and between the electivity index of 5-day old juveniles (-0.80 ± 0.35, $n = 7$) and that of adults (-0.63 ± 0.49, $n = 7$) (Tukey post-hoc, $p > 0.05$). For all other pairwise comparisons between the studied stages there were significant differences in electivity index. The number of *B. wolfi* eaten by *Mesostoma* sp. over time is presented in Figure 3.

Feeding and swimming behaviour

In a series of phases to catch *B. wolfi*, the first step is the encounter. *Mesostoma* sp. first gave a strike with the frontal region to the anostracans swimming by. When this action was successful, the prey was apparently paralysed and showed uncontrolled movements. In this state, the anostracans were less successful in avoiding the predator. This paralysis was usually of short duration (less than a minute). In the mean time, the predator tried to attach itself to its prey. Hereafter, *Mesostoma* sp. extended its pharynx and sucked out the anostracans contents. Meanwhile, other Turbellaria also attempted to eat the prey specimen. It took only a few minutes to suck out the entire prey.

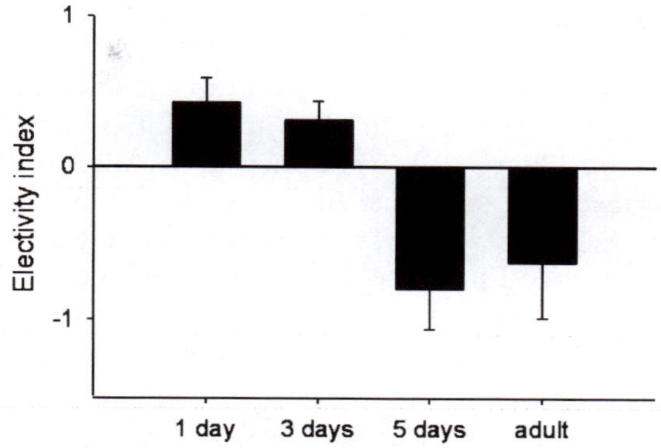

Figure 2. The electivity indices of different anostracan life cycle stages eaten after 150 min by *Mesostoma* sp. (1 day = 1-day-old larvae; 3 days = 3-day-old juveniles; 5 days = 5-day-old juveniles, adult = adult Anostraca; mean and 2*standard error are indicated).

Mesostoma sp. is also very successful in catching and eating *Daphnia magna*. This prey gets stuck in the secreted mucus and thus forms an easy immobilised prey. In contrast to the situation of catching Anostraca, this mucus was clearly visible on the caught *Daphnia*-bodies. *Mesostoma* sp. subsequently crawls partly in the carapace of

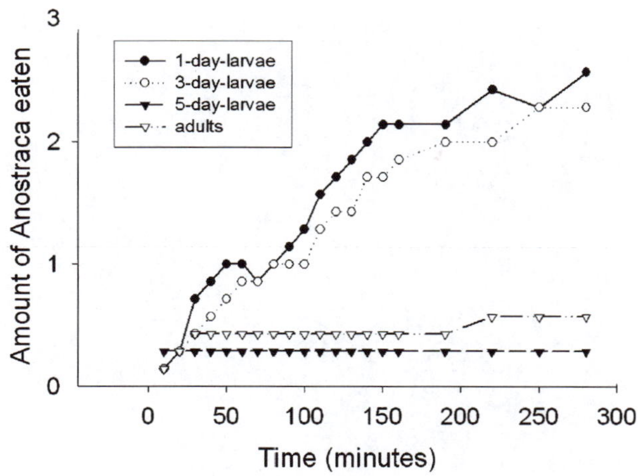

Figure 3. Number of Anostraca life cycle stages eaten by *Mesostoma* sp. in function of time (in minutes).

Daphnia and sucks its contents out with the emerged pharynx.

In general, an adult *Mesostoma* sp. mainly glides on the sides of the aquarium or on the bottom. Once in a while they glide towards the surface and stay there. Juveniles are mainly found in the open water. In measuring cylinders they often glide up to the surface via the side of the cylinder, where after they lose grip and sink slowly to the bottom whilst twisting around.

Detailed morphology of predatory structures in adults and of the egg surface

The rostrum of *Mesostoma* sp. consists of four parts, which can be folded towards each other. In this rostral invagination very long thick cilia are present (Fig. 4), arranged in groups that have a fixed composition. In the middle there is a thick hair (A in Fig. 4), flanked by two thinner cilia (B in Fig. 4). Behind these three cilia there is a group of three or four curved hairs (C in Fig. 4). Apart from these groups, mushroom-shaped structures that are probably sensory organs can also be observed (D in Fig. 4).

Discussion

There was a significant difference in hatching fractions of dormant eggs of the anostracan *Branchipodopsis wolfi* when incubation took place either in the presence of caged *Mesostoma* sp. or in a medium probably loaded with its kairomones. The hatching percentages were about 20% lower in these treatments than in the control. This is an indication that developing eggs of *B. wolfi* can detect kairomones of *Mesostoma* sp. and adapt hatching fractions accordingly. We therefore expect that in the presence of active *Mesostoma* sp., fewer anostracans will hatch in the rock pools to avoid predation. Only *B. wolfi* eggs from a population that originally thrived in the same pool as the predator showed this reaction. When eggs of *B. wolfi* from a far distant pool or from an American species (*Thamnocephalus platyurus*) were exposed to *Mesostoma* sp., no impact on hatching fractions were found. This might be a result of co-evolution between *B. wolfi* and *Mesostoma* sp. in relatively isolated pools (reviewed in Abrams, 2000). The fact that prey species may detect kairomones of different predators is confirmed by Weber (2003). In this study, two fish species had a different effect on life history traits of *Daphnia galeata*. To our knowledge, however, it is the first time that an indication is found of intra-specific differences in the action of predator kairomones. Larger experiments are needed to investigate whether local adaptation has indeed occurred in our case.

It is not known what type of chemical the Turbellaria-kairomone could be. Dumont & Carels (1987) have demonstrated that *Mesostoma* cf. *lingua* secretes toxins with which they can paralyse or kill their prey. Possibly anostracan resting eggs can react to such a substance. It is also not known how the kairomones reach the

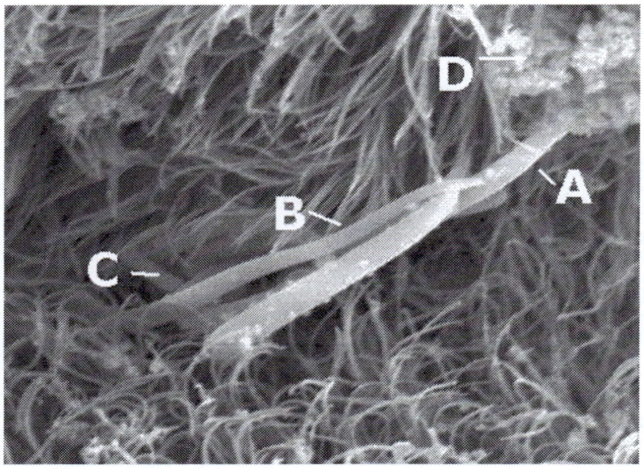

Figure 4. Long hair like structures on the rostrum of an adult *Mesostoma* sp.

developing anostracan embryo. Possibly, the infochemicals enter the eggs through pores on the surface of anostracan eggs as described by Brendonck & Coomans (1994). There, they may bind with or penetrate the larval membranes.

The sensitivity of zooplankton resting eggs towards environmental information is also confirmed by Blaustein (1997) who has suggested a reduction in the hatching fraction in several crustacean species (*Arctodiaptomus similis*, *Ceriodaphnia quadrangula* and *Cyzicus* sp.) when salamander larvae were present. Later, Spencer & Blaustein (2001) have demonstrated this reduction in one species to be due to hatching inhibition. Changes in egg hatching in response to the presence of turbellarians are also known from other organisms. Sih & Moore (1993), for example, have documented a shift in the timing of hatching in salamander eggs when the size selective predatory flatworm *Phagocotus gracilis* or its chemicals were present. This delayed hatching was considered adaptive because a larger and stronger developmental stage hatched from the eggs when the flatworm was present.

Mesostoma sp. apparently preferred different development stages of *B. wolfi* when offered simultaneously. Younger (1–3 days old) stages were removed significantly faster than older individuals. This seeming preference for larvae and young juveniles may be explained by their swimming behaviour and by the fact that it takes longer before satiation of the predator occurs. Five-day-old larvae are already much more developed and swim faster than 3-day-old juveniles. Hence our results largely conform to the general statement of Brooks & Dodson (1965) that an animal chooses its prey on the basis of its size, abundance, edibility and ease with which it can be caught. In our experiment, abundance of the prey stages was comparable so that selection was probably largely based on size and ease of capture.

Brendonck et al. (2002) state that *Mesostoma* sp. can probably perform fast (within 24 h) and massive predation on *B. wolfi* in the Botswana rock pools. By means of enclosure experiments they demonstrated that one *Mesostoma* sp. could eat about 1/4 to 1/3 adult *B. wolfi* per worm per day. Our experiments have shown a predation rate of about 1/5 *B. wolfi* per worm per day. Of *Mesostoma lingua* it is known that it obtains maximal growth when they eat 4 *Daphnia* per day (Schwartz & Hebert, 1986) and that it shows an increased predation rate when plants, used as ambush, were available (MacIsaac & Hutchinson, 1984). Although the observed predation rate of *Mesostoma*. sp. is lower than that of other predators in the Botswana rock pools (dragon fly larvae: 2–3 *B. wolfi* in less than 8 h; notonectids: 4 anostracans in less than 3 h) (Brendonck et al., 2002), it nevertheless could have a major influence on the *B. wolfi* population. Flatworms are indeed the only known predators permanently present in these temporary aquatic systems, they easily feed on larvae and juveniles, have a high population growth rate and are effective predators (Schwartz & Hebert, 1982; Blaustein & Dumont, 1990; Dumont & Schorreels, 1990).

When an anostracan is approaching, *Mesostoma* sp. quickly bulges out its rostrum in the direction of the passing prey. A comparable behaviour was previously described by Blaustein & Dumont (1990) and Wrona & Koopowitz (1998) for *Mesostoma*. Possibly *Mesostoma* detects its prey with the mushroom-like structures on the rostrum (Fig. 4), which are probably sensory organs. After this action, the anostracan is apparently paralysed, often hanging head-down in the water. In this condition it does not avoid the predator as much as before. This phenomenon of loss of orientation has also been described for *Daphnia* in the presence of *Mesostoma* cf. *lingua* (Dumont & Carels, 1987) and other species of the genus by several other authors (Adams, 1980; Blaustein & Dumont, 1990; Wrona & Koopowitz, 1998). Dumont & Carels (1987) assume that both *Mesostoma* cf. *lingua* and its mucus webs are toxic and can immobilize the prey. After attaching itself to *B. wolfi*, the flatworm empties the prey by the sucking action of the pharynx. Meanwhile, other Turbellaria are attracted to the caught and hurt prey, a behaviour that was previously observed by Brendonck et al. (2002) in the field. Also for other Turbellaria, communal feeding occurs (Young, 1977; Kolasa, 1984; Ogren, 1995). Maybe some substance derived from the hurt prey attracts turbellarians. The studied *Mesostoma* sp. also easily captured and ate *Daphnia* in the laboratory. Water fleas are a slower and weaker prey than the relatively fast and big Anostraca.

No indication was found for predation of *Mesostoma* sp. on intact *B. wolfi* eggs. Probably the eggshell is too thick to penetrate and maybe the golf ball-like external sculpturing of the eggs offers additional protection. Dumont et al. (2002) also found that other *Mesostoma* sp. hardly predate on intact cysts of *Chirocephalus diaphanus* (Anostraca), in contrast with decapsulated cysts, which were easily predated. Protection of the cysts shell against predation is a necessary survival strategy as the persistence of the population depends largely on the size and hatching characteristics of the resting egg bank. As the time between egg laying and hatching is often quite long (months or even years), the predation risk is indeed potentially high and predation defence mechanisms are needed (Dumont et al., 2002). Cáceres & Hairston found that turbellarians were not capable of ingesting ephippia of *Daphnia* in the laboratory, neither were gastropods, chironomids or zebra mussels. Amphipods on the other hand, did consume ephippia.

Adult *Mesostoma* sp. were mainly crawling or gliding on the sides or bottom of the aquarium, a behaviour also noted by Dumont & Carels (1987). Sometimes they swam towards the surface where they stayed for a while. Juveniles mainly occurred in the open water, making them more visible than adults, a behaviour also reported by Wrona & Koopowitz (1998). In measuring cylinders they often glided up to the surface, after which they sunk again whilst twisting around. This behaviour, which probably raises the chance of encountering swimming prey, corresponds with observations by others on several species of *Mesostoma* (Dumont & Carels, 1987; Blaustein & Dumont, 1990).

Due to their permanent presence in temporary pools, their direct consumptive predatory impact (especially on early stages), and their non-consumptive effect on the hatching behaviour of the egg bank, we can conclude that predatory turbellarians potentially have a significant impact on the size and dynamics of anostracan populations and on their chances for persistence in shallow ephemeral rock pools.

Acknowledgments

This work was supported by FWO grant G.011803 'Ecological biogeography of rockpools' and by OT/00/14 'Evolutionary ecology of resting egg banks'. E. De Roeck is a fellow of the Institute for the Promotion of Innovation through Science and Technology in Flanders. Special thanks go to Julien Cilis (KBIN) and the FEI-Company for being able to use the SEM under excellent guidance. We also wanted to thank Merlijn Jocqué and Kelle Moreau for helping with preparation of the experiments.

References

Abrams, P. A., 2000. The evolution of predator-prey interactions: theory and evidence. Annual Review of Ecology and Systematics 31: 79–105.

Adams, J., 1980. The feeding behaviour of *Bdellocephala punctata*, a freshwater triclad. Oikos 35: 2–7.

Binckley, C. A. & W. J. Resetarits, 2003. Functional equivalence of non-lethal effects: generalized fish avoidance determines distribution of gray treefrog, *Hyla chrysoscelis*, larvae. Oikos 102: 623–629.

Blaustein, L., 1990. Evidence for predatory flatworms as organisers of zooplankton and mosquito community structure in rice fields. Hydrobiologia 199: 179–191.

Blaustein, L., 1997. Non consumptive effects of larval *Salamandra* on crustacean prey: can eggs detect predators? Oecologia 110: 212–217.

Blaustein, L. & H. J. Dumont, 1990. Typhloplanid flatworms (*Mesostoma* and related genera): Mechanisms of predation and evidence that they structure aquatic invertebrate communities. Hydrobiologia 198: 61–77.

Blaustein, L., B. P. Kotler & D. Ward, 1995. Direct and indirect effects of a predatory backswimmer (*Notonecta maculata*) on community structure of desert temporary pools. Ecological Entomology 20: 311–318.

Boersma, M., L. De Meester & P. Spaak, 1999. Environmental stress and local adaptation in *Daphnia Magna*. Limnology and Oceanography 44: 393–402.

Brendonck, L. & A. Coomans, 1994. Egg morphology in African Streptocephalidae (Crustacea: Branchiopoda: Anostraca) Part 1: South of Zambezi and Kunene rivers. Archives für Hydrobiology 3: 313–334.

Brendonck, L. & B. J. Riddoch, 2000. Egg bank dynamics in anostracan desert rock pool populations (Crustacea: Branchiopoda). Archives für Hydrobiology 148: 71–84.

Brendonck, L. & B. J. Riddoch, 2001. Hatching characteristics of the fairy shrimp *Branchipodopsis wolfi* in relation to the stochastic nature of its habitat, desert rock pools. Verhandlungen der Internationale Vereinigung für Theoretische und Angewandte Limnologie 27: 3931–3935.

Brendonck, L., B. J. Riddoch, V. Van de Weghe & T. Van Doornen, 1998. The maintenance of egg banks in very short-lived pools – a case study with anostracans (Branchiopoda). Archives für Hydrobiology 52: 141–161.

Brendonck, L., E. Michels, L. De Meester & B. J. Riddoch, 2002. Temporary pools are not enemy free. Hydrobiologia 486: 147–159.

Brooks, J. L. & S. I. Dodson, 1965. Predation, body size and composition of plankton. Science 150: 28–35.

Cáceres, C. E. & N. G. Hairston, 1998. Benthic pelagic coupling in planktonic crustaceans: the role of the benthos. Archives für Hydrobiology 52: 163–174.

Cash, K. J., F. J. Wrona & G. J. Scrimgeour, 1995. The effect of group size on per capita ingestion in flatworms. Freshwater Biology 34: 477–483.

Chesson, J., 1983. The estimation and analysis of preference and its relationship to foraging models. Ecology 64: 1297–1304.

Dodson, S. I., 1988. The ecological role of chemical stimuli for the zooplankton: predator avoidance behaviour in *Daphnia*. Limnology and Oceanography 33: 1431–1439.

Dumont, H. J. & I. Carels, 1987. Flatworm predator (*Mesostoma* cf. *lingua*) releases a toxin to catch planktonic prey (*Daphnia magna*). Limnology and Oceanography 32: 699–702.

Dumont, H. J. S. Nandini & S. S. S. Sarma, 2002. Cyst ornamentation in aquatic invertebrates: a defence against egg-predation. Hydrobiologia 486: 161–167.

Dumont, H. J. & S. Schorreels, 1990. A laboratory study of the feeding of *Mesostoma lingua* (Schmidt) (Turbellaria: Neorhabdocoela) on *Daphnia magna* Straus at four different temperatures. Hydrobiologia 198: 79–89.

Fryer, G., 1986. Enemy free space: a new name for an ancient ecological concept. Biological Journal of the Linnean Society 27: 287–292.

Kerfoot, W. C. & M. Lynch, 1987. Branchiopod communities: associations with planktivorous fish in space and time. In W. C. Kerfoot & A. Sih (eds), Predation: Direct and Indirect Impacts on Aquatic Communities. University Press, New England.

Kolasa, J., 1984. Predation on mosquitoes by juveniles of *Mesostoma* spp. (Turbellaria). Freshwater Invertebrate Biology 3: 42–46.

Kolasa, J., 2001. Flatworms: Turbellaria and Nemertea. In Throp H. J. & A. P. Covich (eds), Ecology and classification of North American Freshwater Invertebrates. Academic Press, 155–180.

Knight, T. M. & T. E. Miller, 2004. Local adaptation within a population of *Hydrocotyle Bonariensis*. Evolutionary Ecology Research 6: 103–114.

Lass, S. & P. Spaak, 2003. Chemically induced anti-predator defences in plankton: a review. Hydrobiologia 491: 221–239.

MacIsaac, H. & T. Hutchinson, 1984. The influence of benthic tundra pond vegetation and prey behavior on zooplankton predation by the flatworm *Mesostoma lingua*. Canadian Journal of Zoology 63: 1617–1621.

Maeda-Martínez, A. M., D. Belk, H. Obregón-Barboza & H. J. Dumont, 1997. Large branchiopod assemblages common to Mexico and the United States. Hydrobiologia 359: 45–62.

Marcus, N. H., 1990. Calanoid copepod, cladoceran, and rotifer eggs in sea-bottom sediments of Northern Californian coastal waters – identification, occurrence and hatching. Marine Biology 105: 413–418.

Ogren, R. E., 1995. Predation behavior of land planarians. Hydrobiologia 305: 105–111.

Onbé, T., 1978. Sugar flotation method for sorting the resting eggs of marine cladocerans and copepods from sea-bottom sediment. Bulletin of the Japanese Society of Scientific Fisheries 44: 1411.

Schwartz, S. & P. Hebert, 1982. A laboratory study of the feeding behavior of the rhabdocoel *Mesostoma ehrenbergii* on pond Cladocera. Canadian Journal of Zoology 60: 1305–1307.

Schwartz, S. & P. Hebert, 1986. Prey preference and utilization by *Mesostoma lingua* (Turbellaria, Rhabdocoela) at a low arctic site. Hydrobiologia 135: 251–257.

Sih, A. & R. D. Moore, 1993. Delayed hatching of salamander eggs in response to enhanced larval predation risk. American Naturalist 142: 947–960.

Sih, A., P. Crowley, M. Mcpeek, J. Petranka & K. Strohmeier, 1985. Predation, competition, and prey communities – a review of field experiments. Annual Review of Ecology and Systematics 16: 269–311.

Spencer, M. & L. Blaustein, 2001. Hatching responses of temporary pool invertebrates to signals of environmental quality. Israel Journal of Zoology 47: 397–417.

Weber, A., 2003. More than one 'fish kairomone'? Perch and stickleback kairomones affect *Daphnia* life history traits differently. Hydrobiologia 498: 143–150.

Williams, D. D., 1987. The Ecology of Temporary Waters. Croom Helm, London.

Wrona, F. J. & H. Koopowitz, 1998. Behavior of the Rhabdocoel flatworm *Mesostoma ehrenbergii* in prey capture and feeding. Hydrobiologia 383: 35–40.

Young, J. O., 1977. An ecological study of *Phaenocora unipunctata* (Örsted) (Turbellaria; Rhabdocoela): predator–prey interrelationships. Hydrobiologia 60, 2: 145–149.

Zaret, T. M., 1980. The effect of prey motion on planktivore choice. In Kerfoot, W. C. (ed.), Evolution and Ecology of Zooplankton Communities. University Press New England: 594–603.

Hydrobiologia (2005) 542:113–120
H. Segers & K. Martens (eds), Aquatic Biodiversity II
DOI 10.1007/s10750-004-8771-6

Changes in functional biodiversity in an invaded freshwater ecosystem: the Moselle River

Simon Devin*, Jean-Nicolas Beisel, Philippe Usseglio-Polatera & Jean-Claude Moreteau
Universite dé Metz, UFR SciFA, Laboratoire Biodiversité et Fonctionnement des Ecosystémes, Campus Bridoux, Av. du Géneral Delestraint, 57070 Metz, France
*(*Author for correspondence: Tel.: +33-3-8737-8431, Fax: +33-3-8737-8423, E-mail: devin@sciences.univ-metz.fr)*

Key words: invasive species, freshwater macroinvertebrates, bio/ecological traits, functional diversity, biomonotony concept

Abstract

The spread of non-indigenous species and the decline of autochthonous ones are leading to a homogenization of freshwater fauna in terms of systematic units, but the functional consequences are poorly documented. We studied the peculiar case of the lower, French section of the Moselle River where 20 invertebrate species have been introduced since 1854, with a rate increasing exponentially with time. Dredge sampling performed in 1994, 1996, 2000 and 2001 at four sampling stations allowed for an evaluation of faunal changes in terms of composition, structure and function. During this period, no structural changes were recorded in spite of multiple, new and successful introductions. The evaluation of functional modifications was based on a typology of taxa exhibiting homogeneous biological/ecological traits. Functional diversity, measured as the diversity of taxa distribution among functional groupings, revealed a significant increase between 1994/1996 and 2000/2001 because those species that were over-represented during the former period reached more equilibrated densities during the latter. The major, indirect implications of these functional changes are discussed.

Introduction

The fauna in freshwater ecosystems is being modified at unprecedented rates by human activities, resulting in a decline of local species and a spread of exotic or invasive ones. Navigable sectors of large rivers are the most affected or the most threatened aquatic ecosystems due to their multiple uses and the resulting stresses. While many studies have focused on the characteristics of successful invaders (Williamson, 1996; Ricciardi & Rasmussen, 1998; Kolar & Lodge, 2001), the impact of multiple introductions on recipient ecosystems is still poorly documented (Van der Velde et al., 2000).

In this context, the main, inter-related aspects of faunal changes in a community – composition, structure and functional diversity – have to be evaluated. Composition (i.e. a list of species) and taxonomic structure (usually described with biotic indices or distribution models) are often studied, but the functional diversity of a given community, much more difficult to evaluate, is scarcely approached. However, an evaluation of functional modifications can be based on a recent, innovative functional classification of benthic macroinvertebrates assembling taxa with similar suites of biological and ecological traits in increasingly larger faunal groups (Usseglio-Polatera et al., 2001). Species belonging to a given group may be considered as 'functionally redundant' for the major processes occurring in the ecosystem, considering that the functional effects of a taxon are output variables or consequences of its combination of traits and use of resources (Rosenfeld, 2002). From this viewpoint, three different factors were taken into account in the analysis of 22 biological and ecological traits performed by Usseglio-Polatera et al. (2001): (i) morphological and physiological features of taxa related to ecological processes (described with traits such as 'feeding habits', 'respiration mode' or 'maximal size'), (ii) demographic characteristics of populations (traits such as 'life cycle duration' or 'potential number of reproduction cycles per year'), and (iii) ecological preferences influencing the functionally relevant distribution of a taxon (traits such as 'longitudinal distribution' or 'salinity preferendum'). If each taxon in a given community is replaced by its functional group, the assessment of 'group diversity' has the potential to provide a useful, simple measure of that community's functional diversity. In this context, we aimed at analysing recent faunal changes in the Moselle River, in terms of composition, structure and functioning, following a succession of uncontrolled introductions of macroinvertebrates. After a short description of the invasion pattern we focused on the consequences of repeated introductions on functional diversity.

Materials and methods

Study sites

The lower part of the Moselle River is canalized to allow for a high shipping traffic between the Rhine, the Meuse and the Saône catchments. The large number of weirs induces the presence of several diverted stretches flowing parallel to the canalized one. The present study was carried out at four sites (Fig. 1): Dieulouard, Metz, Cattenom and Sierck-les-Bains, covering approximately 100 km of the lower Moselle. The sites were chosen because they represent the main environmental conditions encountered along the river, and are typical of the canalized sections (except for the Metz sampling site located 400 m upstream from a navigable section). All four sampling sites were characterised by a mineral substratum ranging from coarse sand to pebbles.

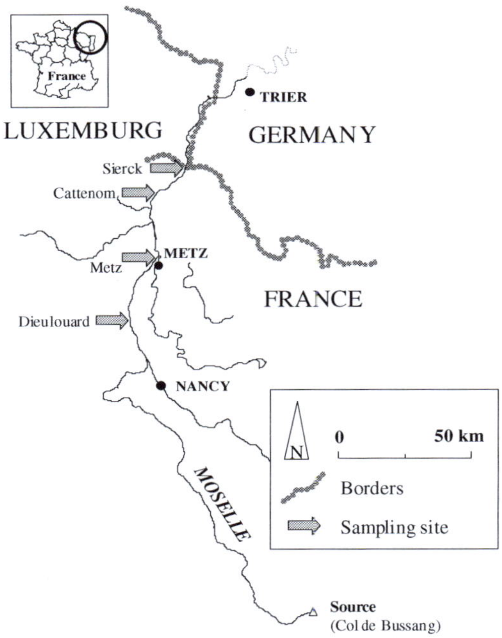

Figure 1. Location of the four sampling sites along the Moselle River.

Sampling protocol

Benthic samples were collected in 1994, 1996, 2000 and 2001, at a depth of about 4 m, by dredging the bottom in a direction perpendicular to one transect. On each transect, the left side, the right side (each one at 4 m from the bank) and the middle of the river were sampled. This method is designed to collect the main populations living in the channel and thus representing the major part of the macrobenthic biomass. On each sampling occasion, the surface of the sampled area was evaluated according to (1) the length of the dredge draught estimated from the volume of dredged material, (2) the dredge section and (3) the sediment depth (Berly, 1989). Macroinvertebrate densities were then expressed per square metre (Table 1 for detailed information on the sampling protocol).

The samples collected in 1994, 1996 and 2000 were preserved in 4% formaldehyde, but those from 2001 were frozen (−20 °C) shortly (<4 h) after collection. Macroinvertebrates were sorted, identified and counted under a stereomicroscope. Most taxa were identified to the genus level, except for some exotic species identified to species level,

Table 1. Basic data for the four sites at each sampling date

Sampling date	Station	Number of samples	Surface (m^2)	Mean total density ($ind\ m^{-2}$)	Individual number ($ind\ m^{-2}$)		Taxa number	
					Exotic	Autochthonous	Exotic	Autochtonous
	Sierck	3	0.232	26,791	9711	17,081	4	15
1994	Cattenom	3	0.046	272,456	135,503	136,952	3	19
	Metz	3	0.293	209,292	190,685	18,607	4	22
	Dieulouard	3	0.084	169,077	110,021	59,056	3	20
	Cattenom	1	0.144	17,325	270	17,055	5	14
1996	Metz	3	0.526	5292	4204	1088	5	20
	Dieulouard	3	0.938	12,280	9444	2836	6	27
	Sierck	3	0.238	7306	3601	3705	6	23
2000	Cattenom	3	0.155	12,959	3213	9745	7	18
	Metz	3	0.114	19,679	7390	12,290	8	22
	Dieulouard	3	0.352	5158	642	4516	6	16
	Sierck	3	0.310	26,103	25,865	238	8	7
2001	Cattenom	3	0.669	11,547	11,138	409	8	14
	Metz	3	0.747	4300	3465	834	8	14
	Dieulouard	3	0.498	32,122	24,633	7489	8	22

and Diptera identified only at family level. Oligochaeta and Nematoda were recorded as such.

Statistical analyses

The functional diversity of each community was calculated using the Shannon–Weaver index (Shannon & Weaver, 1963) according to the distribution of individuals (and taxa) among functional groups (Usseglio-Polatera et al., 2001). The significance of temporal variations in diversity indices was tested using the non-parametric Friedman ANOVA ($\alpha = 0.05$) test. The assessment of changes in functional diversity was performed using the Mann–Withney test ($\alpha = 0.05$).

Results

Biological invasion is a common phenomenon in the Mosellè River, with 20 species of macroinvertebrates introduced since 1854, most of them belonging to Crustacea and Mollusca (Table 2). Seven exotic species were recorded before the first half of the 20th century and thirteen thereafter. The trend in cumulative numbers of successful introductions is best described by an exponential function (Fig. 2), which explains 7% more variance than a linear one.

The taxonomic composition of the fauna reveals an increase in the relative abundances of autochthonous individuals between 1994 (34.2%) and 2000 (67.1%), followed by a severe decrease in 2001 (12.1%), due mainly to a decrease in relative abundances of 'insects' and 'other' autochthonous taxa (Cnidaria, Achaeta, Nematoda and Oligochaeta). Conversely, an increase in 'other' exotic species can be observed, linked to the introduction of the freshwater Polychaeta *Hypania invalida*. A decline in exotic molluscs occurred in relation with a massive decrease in *Dreissena polymorpha* between 1994 and 1996. Finally, only one autochthonous crustacean *Asellus aquaticus*, present at Dieulouard, still remained in the Moselle River in 2001, while the proportion of exotic crustaceans increased with the establishment of two new species, *Dikerogammarus villosus* and *Jaera istri* (Fig. 3).

During the period from 1994 to 2001, the major, structural characteristics of macrobenthic communities did not change in terms of taxonomic

Table 2. Introduced species in the French part of the Moselle River

Species	Systematic unit	Introduction date	Origin	Functional group	Reference
Dreissena polymorpha	Mollusca: Bivalvia	1854	Ponto-Caspian	ζ	Dhur & Massard (1995)
Orchestia cavimana	Crustacea: Amphipoda	First half of the 20th century	Ponto-Caspian and Mediterranean		Dhur & Massard (1995)
Echinogammarus berilloni	Crustacea: Amphipoda	First half of the 20th century	South-Western Europe	ζ	Dhur & Massard (1995)
Atyaephyra desmarestii	Crustacea: Decapoda	1920s	Mediterranean	$\gamma 2$	Dhur & Massard (1995)
Physa acuta	Mollusca: Gasteropoda	1920s	Mediterranean	$\gamma 1$	Dhur & Massard (1995)
Branchiura sowerbyi	Oligochaeta	1930s	South-Eastern Asia	ζ	Dhur & Massard (1995)
Lithoglyphus naticoides	Mollusca: Gasteropoda	1931	Ponto-Caspian	$\gamma 1$	Dhur & Massard (1995)
Dugesia tigrina	Turbellaria: Tricladida	1950s	North America	$\gamma 2$	Dhur & Massard (1995)
Potamopyrgus jenkinsi	Mollusca: Gasteropoda	1976	New-Zealand	ζ	Dhur & Massard (1995)
Craspedacusta sowerbyi	Cnidaria: Hydrozoa	About 1978	Asia	$\gamma 2$	Dhur & Massard (1995)
Orconectes limosus	Crustacea: Decapoda	1978	North America	$\gamma 2$	Dhur & Massard (1995)
Cordylophora caspia	Cnidaria: Hydrozoa	1987	Ponto-Caspian	ζ	Dhur & Massard (1995)
Gammarus tigrinus	Crustacea: Amphipoda	1991–1992	North America	β	Dhur & Massard (1995)
Chelicorophium curvispinum	Crustacea: Amphipoda	1994	Ponto-Caspian	ζ	Bachmann (2000)
Corbicula fluminea	Mollusca: Bivalvia	1994	Asia	ζ	Bachmann (2000)
Corbicula fluminalis	Mollusca: Bivalvia	1994	Asia	ζ	Bachmann (2000)
Dikerogammarus villosus	Crustacea: Amphipoda	1998	Ponto-Caspian	ζ	Devin et al. (2001)
Limnomysis benedeni	Crustacea: Mysidacea	2001	Ponto-Caspian	ζ	Personal observation
Jaera istri	Crustacea: Isopoda	2001	Ponto-Caspian	ζ	Personal observation
Hypania invalida	Polychaeta: Amphareti-dae	2001	Ponto-Caspian	ζ	Personal observation

Functional groups were reported from Usseglio-Polatera et al. (2001) and from unpublished results.

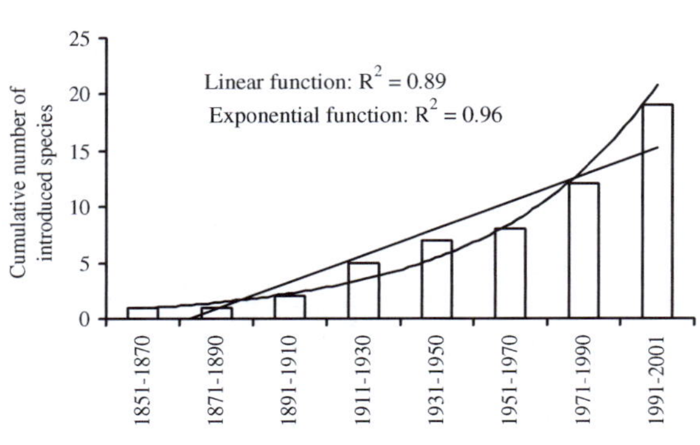

Figure 2. Cumulative number of introduced species in the Moselle River since the mid-19th century. The final number of exotic species (19) is lower than that presented in the Table 2 (20) because an introduced gammarid (*E. berilloni*) disappeared from the river before 2001. Linear and exponential adjustments were applied to describe the trend.

richness or taxonomic diversity (non-parametric Friedman ANOVA, $p = 0.34$ for taxonomic richness and $p = 0.054$ for taxonomic diversity; Fig. 4). Conversely, in terms of functional characteristics, the functional diversity increased significantly between 1994–1996 and 2000–2001 ($p = 0.005$, Fig. 5). During the period from 1994 to 1996, functional diversity was very low due to the high relative abundance of taxa (such as *D. polymorpha* and *C. curvispinum*) belonging to the same bio/ecological group (ζ). In 2000–2001, a decrease in the dominance of this group was observed (Simpson dominance, $\lambda = 0.82$ in 1994, $\lambda = 0.66$ in 2001; Mann–Withney test, $p = 0.04$), that was not counterbalanced by the establishment of several new exotic species also belonging to the

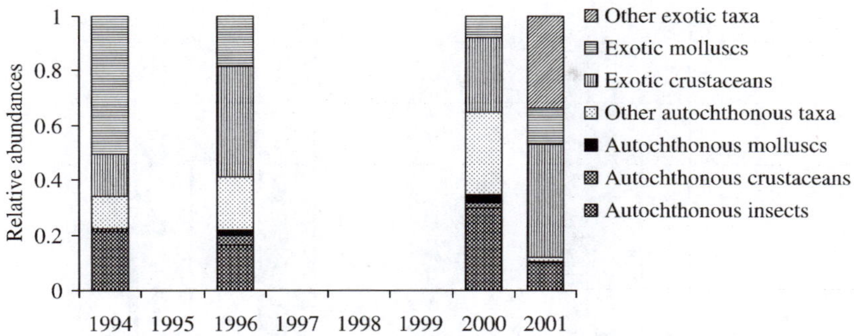

Figure 3. Relative abundances of the main taxonomic groups in the Moselle River sections studied between 1994 and 2001. See Table 1 for the abundances.

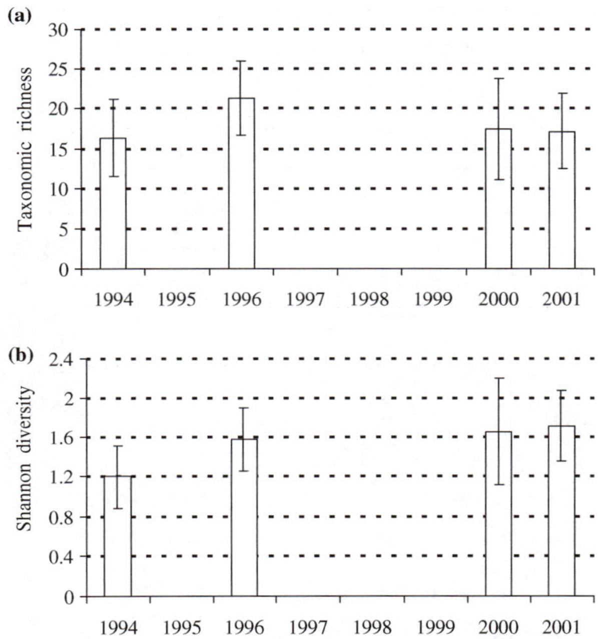

Figure 4. Variation of the mean (±standard deviation) taxonomic richness per sample (a) and diversity (b) in the Moselle River section studied between 1994 and 2001. See Table 1 for the number of samples per year.

group ζ (*D. villosus*, *J. istri*, *H. invalida*). The major decrease in *D. polymorpha* density in the Moselle River (Bachmann et al., 2001) seemed to play a key role in this pattern. We can also note that the increase in functional diversity was far more significant upstream than downstream, even with a decrease in the section furthest downstream at Sierck. Conversely, the other functional groups became increasingly represented along the temporal gradient (Fig. 6), involving the increase in functional diversity.

Discussion

The exponential trend in the cumulative, successfully introduced species over time in the Moselle River highlights an accelerating accumulation of exotic species during the last decade. This increase may also reflect a growing interest in biological invasions and increased knowledge of freshwater fauna. Otherwise, several hypotheses can be proposed potentially to explain this increasing rate over time: (1) an increase of propagule pressure

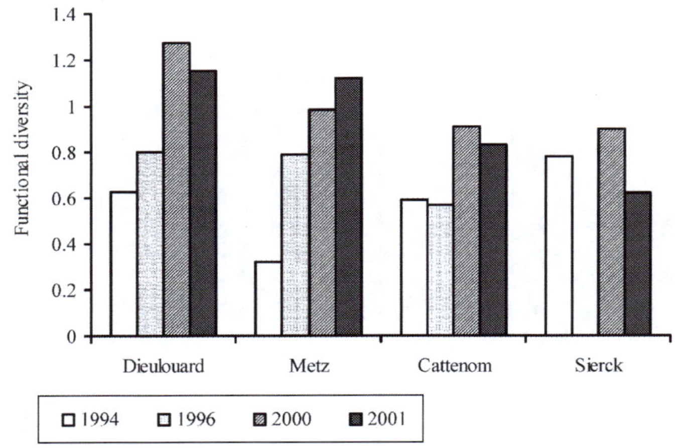

Figure 5. Functional diversity of four Moselle River sections between 1994 and 2001. (Shannon–Weaver index was calculated on distribution of individuals among functional groups defined by Usseglio-Polatera et al., 2001). Note: The Sierck station was not sampled in 1996.

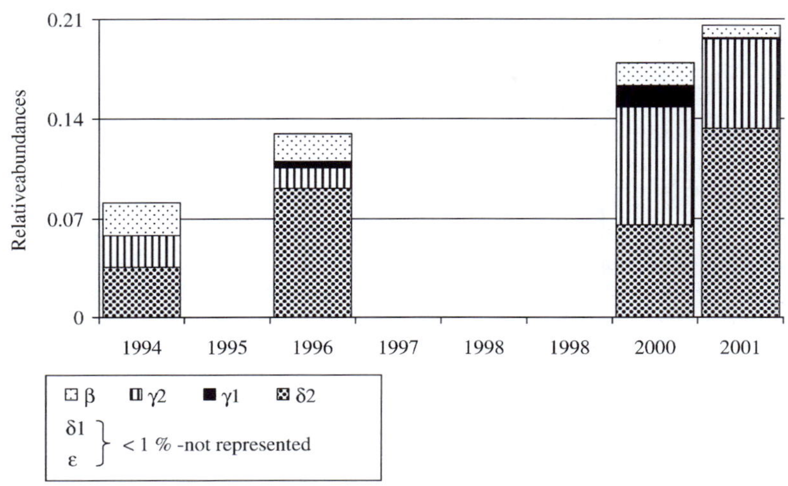

Figure 6. Relative abundances of functional groups (excluding the ζ group) in the benthic community of four Moselle River sections. Two groups were not shown on this graph because of their low relative abundance (see Usseglio-Polatera et al., 2001 for further details on mean bio-ecological profiles of groups).

due to the increasing development of shipping traffic, (2) a process of facilitative interactions which favors new exotic species from the same biogeographic area than previously successful introduced species (Simberloff & Von Holle, 1999) and (3) a decline in biotic resistance following the alteration of the environment and the decline of autochthonous species.

Most introduced species in the Moselle River belong to Crustacea and Mollusca, a fact already observed in other freshwater ecosystems (Morton, 1997; Van der Velde et al., 2000; Grigorovich et al., 2002). Such taxonomic clumping of invaders can result in an over-representation of specific traits which enhance transport, dispersal, establishment and/or massive development in a recipient ecosystem.

Most of the major, recent invasive species in lowland sections of large, western-European rivers belong to the ζ group, including organisms with a unique combination of features (large-size, asexual reproduction, thermophilous, polysaprobic, affinity for silt and sand, mainly filter or deposit feeders, burrowers or permanently attached), which distinguish them clearly from taxa of other functional groups (Usseglio-Polatera et al., 2001, unpublished results).

Assessing the functional consequences of a given introduction is a stimulating, challenging part of biological invasion research. Functional diversity translates the bio/ecological response of organisms to multiple environmental constraints. Its interest lies in the condensed expression of major ecological processes such as nutrient cycles or energy flows throughout different ecosystem compartments. This method, as applied to the Moselle River, is based on the study of groups of taxa with similar suites of bio/ecological traits. This approach highlights functional changes over a very short period of time following a succession of successful introductions. In 1994–1996, functional diversity was very low, illustrating the strong degradation of the river community after the demographic explosion of several new exotic species introduced in the beginning of the 1990s (Bachmann et al., 1995). Recent temporal variations in the density of exotic species probably led to the more equilibrated faunal structure observed in 2000–2001, due in particular to the decline of *Dreissena polymorpha* (Bachmann et al., 2001), which exhibited a very high density in the Moselle during the 1994–1996 period. As a counterpoint to the decrease in relative abundances of group ζ, an increase in relative abundance of groups β and $\gamma 2$ was observed. This phenomenon allowed a better representation of various ecological functions in the ecosystem, and contributed to the increase in the community's functional diversity between 1994–1996 and 2000–2001. The migratory corridor of invasive species, following the downstream–upstream gradient, could explain the pattern of changes in functional diversity within the four sections. For example, *J. istri*, one of the most recent invaders, was only present in Sierck in 2001, thus increasing the relative abundance of the ζ group and decreasing the functional diversity on this section.

Functional diversity takes into account the relative abundances of taxa and gives the best condensed representation of the diversity of bio/ecological trait combinations in an ecosystem. The use of evenness to track faunal changes also allows to highlight functional modifications in benthic communities before the total disappearance of a set of species with the same bio/ecological profile is observed (Chapin et al., 2000).

An increase in functional diversity following several, successful invasions is not so surprising. Indeed, species introductions can increase community diversity locally, whereas they are always involved in a decrease in diversity at a global scale (Sax & Gaines, 2003). However, even at a local scale, this increase is often counterbalanced by the unpredictable impact of non-indigenous species on the ecosystem, such as non-endemic pathogens. For example, the presence of *Orconectes limosus*, the American crayfish introduced in 1978 in the Moselle River, spread the crayfish plague (*Aphanomyces astaci*) which infested the very susceptible indigenous crayfish *Astacus astacus* (Van der Velde et al., 2000). Moreover, most introduced species are distributed worldwide and have achieved the status of cosmopolitan species.

Navigated sectors of large rivers in Europe are more and more homogeneous both in terms of physical attributes and in terms of macrobenthic fauna. This homogenization of freshwater fauna, already reported for fishes (Rahel, 2002) and birds (McKinney & Lockwood, 1999), could lead to a decrease in the resilience of ecosystems exposed to disturbance. The spread of cosmopolitan species involves a decrease in the diversity of bio/ecological traits combinations in communities. As a result, the range of potential responses of a given community to disturbance becomes narrower. The community is destabilized and weakened, and the ecological impacts it suffers intensify (Olden et al., 2004).

Acknowledgements

The authors wish to thank V. Bachmann for contribution to data collection from 1994 to 1996, P. Wagner for efficient assistance during field work and Anna Matheny-Cartier for revision of the manuscript. This study was supported by the French Ministry of Ecology and Sustainable Development, as part of the 2003–2005 INVABIO – Biological Invasions Program.

References

Bachmann, V., 2000. Dynamique spatio-temporelle d'espéces invasives, particulièrement *Corbicula* spp. et *Dreissena polymorpha* (Mollusca: Bivalvia) en hydrosystème fluvial: évolution des populations et effets de l'artificialisation sur les peuplements macrobenthiques. Ph.D. Thesis, University of Metz, 229 pp.

Bachmann, V., E. Cegielka, P. Wagner, P. Usseglio-Polatera & J. C. Moreteau, 1995. Establishment of the amphipod *Corophium curvispinum* and the asiatic clam *Corbicula* sp. in the French part of the River Moselle. Hydroécologie Appliquée 7: 185–190.

Bachmann, V., J. N. Beisel, P. Usseglio-Polatera & J. C. Moreteau, 2001. Decline of *Dreissena polymorpha* in the River Moselle: biotic and abiotic key factors involved in dynamics of invasive species. Archiv für Hydrobiologie 151: 263–281.

Berly, A., 1989. Distribution spatio-temporelle des peuplements macrobenthiques prélevés par dragage dans une station du Haut-Rhône. Ph.D. Thesis, University of Lyon I, 309 pp.

Chapin, F. S. III, E. S. Zavaleta, V. T. Eviners, R. L. Naylor, P. M. Vitousek, H. L. Reynolds, D. U. Hooper, S. Lavorel, O.E. Sala, S. E. Hobbie, M. C. Mack & S. Diaz, 2000. Consequences of changing biodiversity. Nature 405: 234–242.

Devin, S., J. N. Beisel, V. Bachmann and J. C. Moreteau 2001. *Dikerogammarus villosus* (Amphipoda: Gammaridae): another invasive species newly established in the Moselle River and French hydrosystems. Annales de Limnologie–International Journal of Limnology 37: 21–27.

Dhur, G. & J. A. Massard 1995. Etude historique et faunistique des Invertébrés immigrés ou introduits dans la Moselle luxembourgoise et ses affluents. Bulletin de la Société Naturelle luxembourgeoise 127–156.

Grigorovich, I. A., H. J. MacIsaac, N. V. Shadrin & E. L. Mills 2002. Patterns and mechanisms of aquatic invertebrate introductions in the Ponto-Caspian region. Canadian Journal of Fisheries and Aquatic Sciences 59: 1189–1208.

Kolar, C. S. & D. M. Lodge, 2001. Progress in invasion biology: predicting invaders. Trends in Ecology and Evolution 16: 199–204.

McKinney, M. L. & J. L. Lockwood, 1999. Biotic homogenization: a few winners replacing many losers in the next mass extinction. Trends in Ecology and Evolution 14: 450–453.

Morton, B. 1997. The aquatic nuisance species problem: a global perspective and review. In: D'Itri, F. (ed.), Zebra Mussels and Aquatic Nuisance Species Ann Arbor Press, Chelsea, Michigan: 1–53.

Olden, J. D., N. LeRoy Poff, M. R. Douglas, M. E. Douglas & K. D. Fausch, 2004. Ecological and evolutionary consequences of biotic homogenization. Trends in Ecology and Evolution 19: 18–24.

Rahel, F. J., 2002. Homogenization of freshwater fauna. Annual Review of Ecology and Systematics 33: 291–315.

Ricciardi, A. & J. B. Rasmussen, 1998. Predicting the identity and impact of future biological invaders: a priority for aquatic resource management. Canadian Journal of Fisheries and Aquatic Sciences 55: 1759–1765.

Rosenfeld, J. S. 2002. Functional redundancy in ecology and conservation. Oikos 98: 156–162.

Sax, D. F. & S. D. Gaines, 2003. Species diversity: from global decreases to local increases. Trends in Ecology and Evolution 18: 561–566.

Shannon, C. E. & W. Weaver, 1963. The Mathematical Theory of Communication. University of Illinois Press, Urbana, 117 pp.

Simberloff, D. & B. Von Holle, 1999. Positive interactions of non-indigenous species: invasional meltdown? Biological Invasions 1: 21–32.

Usseglio-Polatera, P., P. Richoux, M. Bournaud & H. Tachet, 2001. A functional classification of benthic macroinvertebrates based on biological and ecological traits: application to river condition assessment and stream management. Archiv für Hydrobiologie 139 (Suppl.): 53–83.

Van der Velde, G., S. Rajagopal, B. Kelleher, I.B. Musko and A. bij de Vaate, 2000. Ecological impacts of crustacean invaders: general considerations and examples from the river Rhine. Crustacean Issues 12: 3–33.

Williamson, M., 1996. Biological Invasions, Chapman & Hall, London.

Hydrobiologia (2005) 542:121–129
H. Segers & K. Martens (eds), Aquatic Biodiversity II
DOI 10.1007/s10750-004-4951-7

The impact of stocking on the genetic structure of European grayling (*Thymallus thymallus*, Salmonidae) in two alpine rivers

Nina Duftner[1], Stephan Koblmüller[1], Steven Weiss[1], Nikolaus Medgyesy[2] & Christian Sturmbauer[1,*]
[1]*Department of Zoology, University of Graz, Universitätsplatz 2, 8010 Graz, Austria*
[2]*Department of Zoology and Limnology, Technikerstrosse 25, University of Innsbruck, Austria*
*(*Author for correspondence: Tel: +43-316-380-5595, Fax: +43-316-380-9875,*
E-mail: christian.sturmbauer@uni-graz.at)

Key words: mtDNA sequences, control region, molecular phylogeny, conservation biology

Abstract

The European grayling (*Thymallus thymallus*) is considered to be threatened in several European regions. In recent decades fishery managers have increasingly turned to stocking programs as one way to combat the negative effect that human influence has had on population densities. The present study surveyed the genetic structure of two Danube drainage populations at the Inn and Drau rivers, in Austria, on the basis of mtDNA sequences. Data were placed in the context of the phylogeographic structure of European *Thymallus* and thus could reveal unexpected geographical mixing due to stocking with allochthonous individuals. Our analyses revealed that regular stocking of fish not originating from their natal Rivers has left genetic traces in both systems surveyed. These traces may be classified as marginal for the Inn river and its tributaries in which 97% of the graylings investigated carried haplotypes belonging to the northern alpine lineage, corresponding to the region through which the Inn flows. In contrast, the genetic composition of the Drau population, situated in the southern Alps, has been seriously altered through the stocking of fish belonging to the northern alpine mtDNA lineage as only 62% of the fishes sampled carried haplotypes representing the native southern alpine lineage.

Introduction

A series of studies on the ecology, biology, and molecular genetics of European grayling (*Thymallus thymallus*) populations point to the acute endangerment of this species in several water bodies (Uiblein et al., 2000, 2001; Koskinen et al., 2001, 2002a, b). The causes are pollution, the construction of hydroelectric facilities that both inhibit migration and dispersal as well as alter flow regimes, river bed channelization, overfishing, and in some specific localities bird predation or competition from introduced rainbow trout, *Oncorhynchus mykiss* (Weiss et al., 2002). Human influence has so greatly affected grayling populations that fishery managers have recently turned to intensive stocking efforts in an attempt to restore and maintain stocks. Numerous molecular based studies have demonstrated a clear phylogeographic structure for European salmonids, primarily reflecting current drainage systems but also the past effects of Pleistocene glacial cycles (Bernatchez et al., 1992; Bernatchez 2001; Koskinen et al., 2002a; Duftner et al., 2003). For *Thymallus*, however, this structure is exceedingly complex, particularly in the upper Danube where multiple divergent lineages are found, most likely reflecting isolated glacial refugia (Weiss et al., 2002). The occurrence of such a pattern is assumed to be a consequence of grayling's limited dispersal ability and relatively narrow ecological niche. This is further supported by the fine-scaled population substructuring of grayling reported in Scandinavian populations (Koskinen et al., 2001, 2002b). Thus, stocking with individuals from non-autochthonous populations might have severe effects on the genetic structure of local populations in *Thymallus* and seriously threaten their maintenance through the deleterious effects of inbreeding and outbreeding resulting in reduced fitness to local conditions (Avise & Hamrick, 1996).

The phylogeographic structure of European grayling was recently elucidated in a study of Weiss et al. (2002) including 44 populations throughout the western-European range of the species. The authors demonstrated the presence of multiple divergent mtDNA clades in European rivers, most of which were assigned to particular sections of drainage systems. This finding supports the notion that European grayling have had a long history in Western Europe, pre-dating Pleistocene glacial cycles. Their intraspecific phylogenetic tree revealed two well-supported clades within the Danube basin, one highly divergent clade in the Adriatic basin, and one large, diverse group representing most other central- and western-European populations. A deeply divergent haplotype fixed in the Loire basin in Central France, more groups of haplotypes from distinct Danubian tributaries, and a relatively ancestral haplotype fixed in former tributaries of the Elbe in Denmark all suggest a complex pattern of inter- and post-glacial expansions originating from disjunct refugia throughout Central Europe. Despite some evidence for human-mediated stock transfers, the authors suggest by means of nested clade analysis (Templeton, 1998) that corridors across drainage systems such as the lower Rhine (Moselle) and Elbe served as sources for expansion into the Baltic to the north, as well as the upper Rhine and Danube to the south. Moreover, specific Rhine populations (Doller, Orbe, and Reuss) served as sources of colonization of the Rhone. Concerning the distribution of mtDNA lineages in the Danube system, Weiss et al. (2002) found that the Danube and its tributaries north of the Central Alps comprise two mtDNA lineages, and that the Danube and its tributary south of the Central Alps is inhabited by a third distinct mtDNA lineage, as well as the Danube in Slovenia, which again comprises another mtDNA lineage. In consequence, it was argued that the current trends to trans-locate brood stocks for rearing and release in response to the decline of natural populations must be reconsidered in order to protect the natural diversity of this species.

Here we report the results of a mtDNA based study on the haplotype distribution of European grayling in two Austrian river systems, the Inn river, belonging to the northern alpine Danube drainage, and the Drau River, belonging to the southern alpine Danube drainage. Both rivers are known to receive regular inputs of hatchery grayling, the origin of which is traceable for the majority of stocks, at least for the most recent past (5–10 years). The records for the Inn River covering our sample sites document the release of ca. 100 000 yearling grayling per year within the last 6 years, all originating from a single hatchery that uses only brood stock captured from the Inn itself. In the Drau River system several thousand grayling of Inn origin were stocked 20 years ago, and during the last 5 years the water body has been managed through the release of 25 000 yearlings per year raised from brood stock of Drau origin. However, we note that both river systems are quite large and contain many independently managed reaches in which fish of unknown origin might have been released.

We aimed to evaluate grayling from the Inn and Drau river systems in the context of the mtDNA phylogeographic structure of European *Thymallus* and thereby assess the extent of influence that stocking has had on the occurrence of expected mtDNA lineages.

Materials and methods

In total, 94 individuals were collected from nine localities in two Austrian Danube river tributaries – Inn and Drau – north and south of the Central Alps. Detailed sample information is given in Table 1 (for a map of sampling localities see Fig. 1). For all 94 specimens we sequenced the second section of the mitochondrial control region (457 bp, base positions 631–1087 of the complete control region), and for 51 individuals we se-

Table 1. Characterization of samples including river, sampled locality, major river drainage, haplotype, number of individuals, and GenBank accession number

River	Locality	System[a]	Haplotype[b]	R[d]	N[e]	GenBank Accession No.
Isel	Ainet	Drau	Da1	NA	1	AF522395[f]
Isel	Ainet	Drau	Da4	NA	1	AF522398[f]
Große Drau	Lienz	Drau	Da1	NA	2	AF522395[f]
Große Drau	Lienz	Drau	Da5	SA	5	AF522399[f]
Große Drau	Lienz	Drau	Da7	SA	2	AF522401[f]
Große Drau	Lienz	Drau	Da11	NA	1	AF522405[f]
Große Drau	Lienz	Drau	Da15	SA	1	AF522409[f]
Große Drau	Lienz	Drau	-[c]	SA	1	AY594183
Große Drau	Nikolsdorf	Drau	Da1	NA	2	AF522395[f]
Große Drau	Nikolsdorf	Drau	Da4	NA	1	AF522398[f]
Große Drau	Nikolsdorf	Drau	Da5	SA	6	AF522399[f]
Große Drau	Nikolsdorf	Drau	Da7	SA	4	AF522401[f]
Große Drau	Nikolsdorf	Drau	Da11	NA	2	AF522405[f]
Kohlenbach	Kössen	Inn	Da1	NA	6	AF522395[f]
Kohlenbach	Kössen	Inn	Da11	NA	6	AF522405[f]
Kohlenbach	Kössen	Inn	new	M	1	AY594181
Inn	Pfunds	Inn	Da1	NA	4	AF522395[f]
Inn	Pfunds	Inn	Da11	NA	1	AF522405[f]
Inn	Pfunds	Inn	-[c]	NA	5	AY594184
Inn	Haiming	Inn	-[c]	NA	8	AY594185
Inn	Zirl	Inn	Da1	NA	2	AF522395[f]
Inn	Hall	Inn	Da7	SA	2	AF522401[f]
Inn	Hall	Inn	new	NA	1	AY594182
Inn	Hall	Inn	-[c]	NA	16	AY594186
Inn	Hall	Inn	-[c]	M	1	AY594187
Inn	Kirchbichl	Inn	-[c]	NA	12	AY594188

[a] River system to which river sampled belongs to. According to Weiss et al. (2002) the Inn-system is assigned to the Danube drainage (Northern Alps) and the Drau-system is assigned to the Danube drainage (Southern Alps).
[b] Haplotypes of complete mitochondrial control region sequences as defined by Weiss et al. (2002).
[c] Only the second part of the mitochondrial control region was available.
[d] mtDNA lineage to which the samples were assigned to based on the complete or the second part of the mitochondrial control region: NA, Danube drainage (Northern Alps); SA, Danube drainage (Southern Alps); M, Mixed Rhine/Danube drainage.
[e] Number of individuals per haplotype and sampling locality.
[f] GenBank accession number according to Weiss et al. (2002).

quenced the entire mitochondrial control region (1087 bp).

DNA-extraction, polymerase chain reaction (PCR), and automatic sequencing followed standard protocols (Bruford et al., 1998; Weiss et al., 2001). The primers for PCR-amplification and chain termination sequencing for the first section of the mitochondrial control region were LRBT-25, 5′-AGA GCG CCG GTC TTG TAA TC-3′ (Uiblein et al., 2001) and the newly designed primer INT-R, 5′-CGA CCT TAT TAG TTC TTC TTA G-3′. For the second section of the mitochondrial control region we used the primers INT-F, 5′-AGC CGG GCG TTC TCT CCT ATA T-3′, and LRBT-1195, 5′-GCT AGC GGG ACT TTC TAG GGT C-3′ (Uiblein et al., 2001). DNA fragments were visualized on an ABI 3100 automatic sequencing system. DNA sequence alignment was performed using the program CLUSTAL W (Thompson et al., 1994). The 51 DNA sequences of the whole mitochondrial control region were added to the data set of Weiss et al. (2002) to assign them to

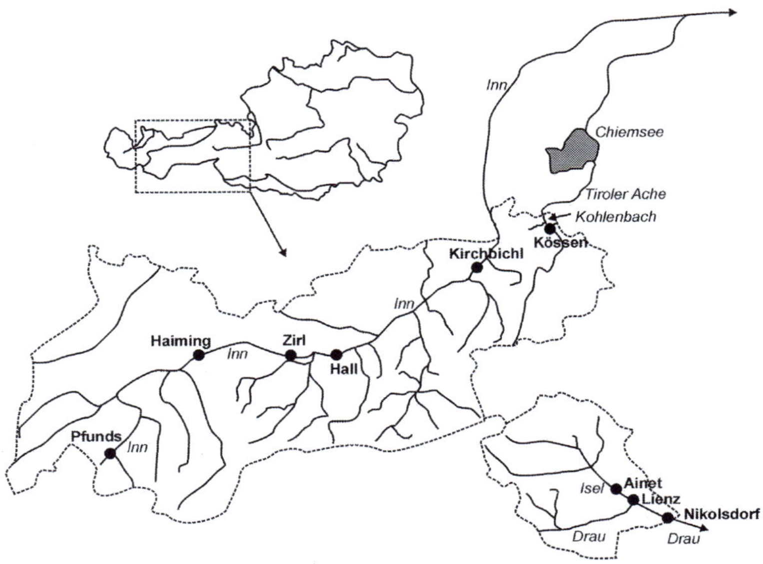

Figure 1. Map of the Inn- and Drau-river system in the Tyrol, Austria, showing the sampling localities.

previously identified European grayling lineages. Sequences of all haplotypes, described herein, are available from GenBank (see Table 1).

First, a phylogenetic tree was constructed based upon the entire control region by applying the neighbor joining algorithm (NJ) implemented in PAUP*4.0b3 (Swofford, 2000) (Fig. 2). For NJ, the substitution model HKY + I + Γ (Hasegawa et al., 1985; base frequencies A, 0.3203; C, 0.2127; G, 0.1475; T, 0.3195; gamma shape parameter α, 0.9043; proportion of invariable sites I, 0.7761; transition/transversion ratio, 2.6238) was used, since it was chosen as the most appropriate model of sequence evolution by Modeltest (Hierarchical Likelihood Ratio Tests; Posada & Crandall, 1998). *Thymallus arcticus*, *T. brevirostris*, and *T. grubii* were declared as outgroup taxa. The base positions 603–639 were excluded from the analysis due to ambiguities in the alignment. As a measure of confidence we performed bootstrapping (Felsenstein, 1985) with 1000 pseudo-replicates. In a second step of analysis a NJ analysis based upon the second section of the control region (substitution model HKY; base frequencies: A, 0.3351, C, 0.1944, G, 0.1527, T, 0.3178; transition/transversion ratio, 1.4625) was conducted for 94 specimens, again declaring *Thymallus arcticus*, *T. brevirostris*, and *T. grubii* as outgroup taxa.

For this data set, a minimum spanning network was constructed in order to visualize the number of base substitutions separating the mtDNA haplotypes and lineages (Fig. 3). The network is based on one most parsimonious tree [un-weighted maximum parsimony analysis; tree length, 66 base substitutions; consistency index excluding uninformative characters, 0.73; retention index, 0.83; rescaled consistency index, 0.69] that was most similar to the NJ tree.

Results

The NJ tree, based on the data set of Weiss et al. (2002) plus 51 new sequences of the complete mitochondrial control region from the Drau- and Inn-system, shows that all individuals sequenced can be assigned to the three previously identified mtDNA haplotype groups present in Austrian rivers: the northern alpine Danube drainage clade, the southern alpine Danube drainage clade, and the mixed clade from the Rhine and Danube (Fig. 2). In total, nine haplotypes were found in the newly sequenced samples from the Inn- and the Drau-system. Most individuals could be assigned to previously identified haplotypes (Da1, Da4, Da5, Da7, Da11, Da15; Weiss et al., 2002), but two individuals, one from the Kohlenbach River at

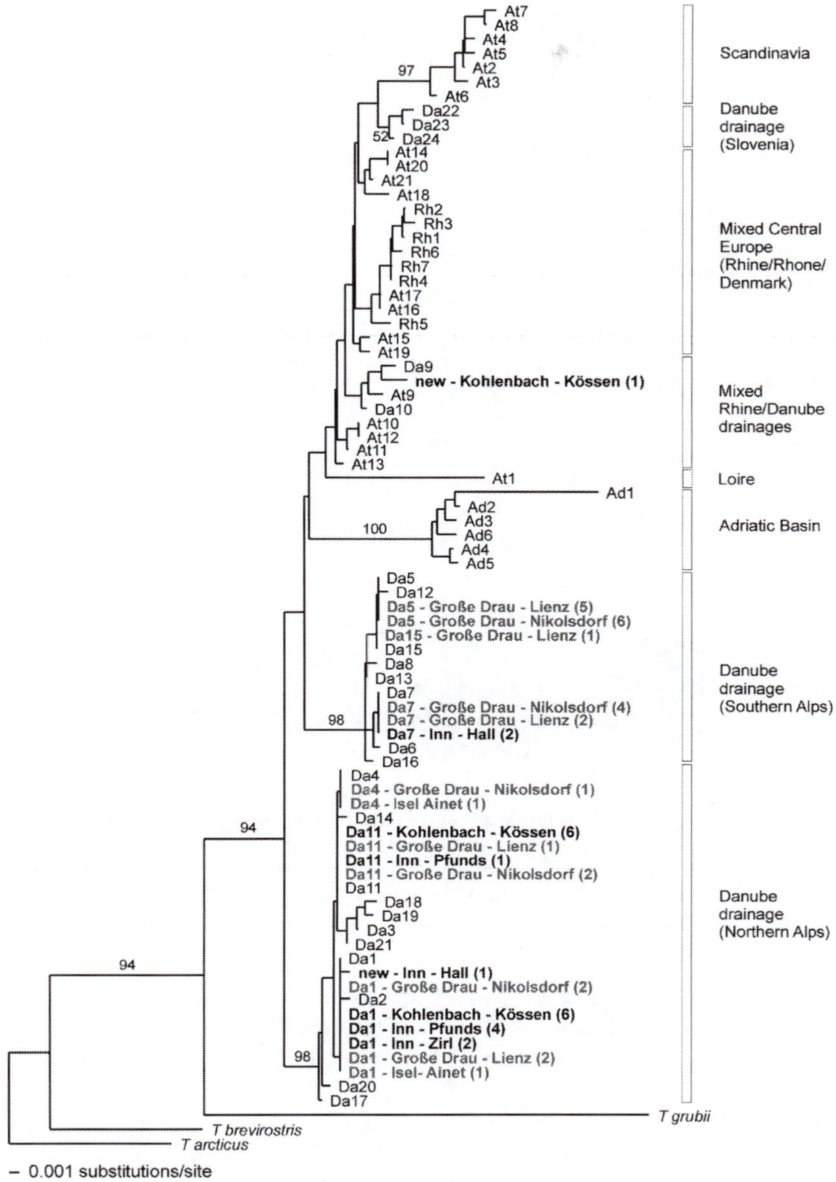

Figure 2. NJ phylogram based on the data set of Weiss et al. (2002) including 58 haplotypes plus 52 new sequences of the complete mitochondrial control region from the Drau- and Inn-system using the HKY + I + Γ model of molecular evolution, to assign the haplotypes found in the Inn-and Drau River to major clades defined by Weiss et al. (2002). Two individuals, one from the Kohlenbach at Kössen and one from the Inn River at Hall, represented new haplotypes, denoted by 'new'. Node support for the major clades defined by Weiss et al. (2002) is shown in the form of bootstrap values (1000 replicates) for NJ at the branches. Bold taxa represent newly sampled haplotypes (grey, Drau-system; black, Inn-system).

Kössen and one from the Inn River at Hall, represented new haplotypes. While the new haplotype from Inn clustered within the clade comprising haplotypes from the northern alpine Danube drainage system, the new haplotype from Kohlenbach was assigned to the mixed group containing haplotypes from the Danube- and Rhine drainage system.

Haplotypes could be unambiguously assigned to the three distinct clades regardless of whether

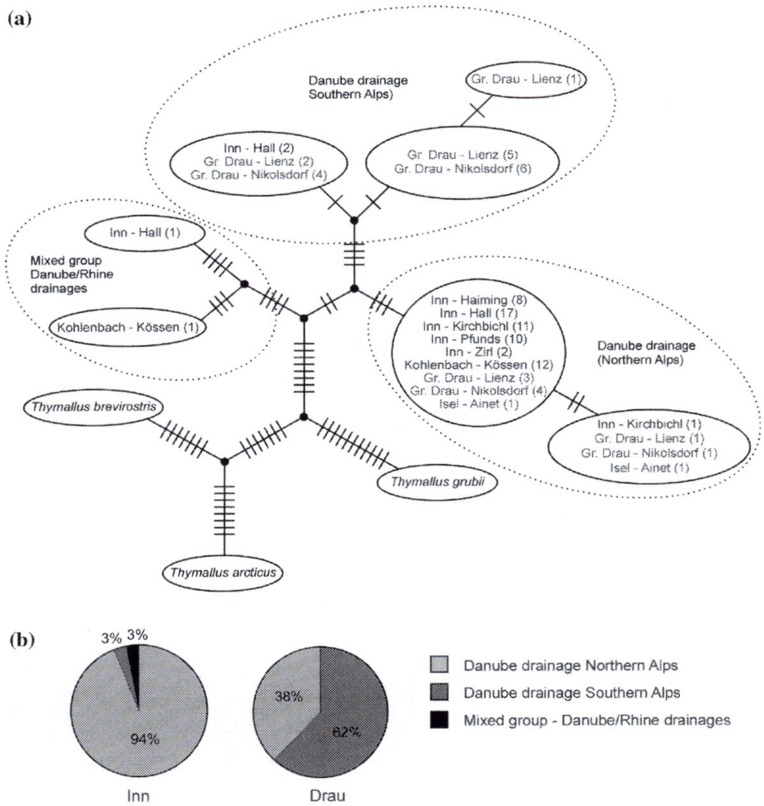

Figure 3. (A) Minimum spanning tree based on 94 DNA sequences (plus three outgroup taxa: *Thymallus arcticus*, *T. brevirostris*, and *T. grubii*) of the second section of the mitochondrial control region (457 bp, base positions 631–1087 of the complete control region). Bars indicate the number of base substitutions among haplotypes assigned to three mtDNA lineages – Danube lineage north of the central Alps, Danube lineage south of the central Alps, and Mixed lineage Danube and Rhine (grey, Drau-system; black, Inn-system). Origin and frequency (in parentheses) of haplotypes are given. (B) Pie diagrams depicting frequencies of mtDNA lineages for the Inn- and Drau river.

the complete or partial control region sequence was used. Thus, for summary purposes we consider the partial sequence in order to encompass the total sample (94 individuals). From the 65 individuals sampled in the Inn drainage, 61 (94%) were assigned to the northern alpine clade, 2 (3%; from Kohlenbach at Kössen and Inn at Hall) were assigned to the mixed clade of the Rhine- and Danube system, and two individuals (3%; Inn at Hall) were assigned to the southern alpine clade. Out of the 29 individuals collected from the Drau system, 18 specimens (62%) were assigned to the southern alpine clade, whereas the remaining 11 individuals (38%) were attributed to the northern alpine haplotype group.

The minimum spanning tree (Fig. 3), based on 94 sequences of the partial control region shows the number of substitutions between haplotypes and highlights the high genetic divergence among the three mtDNA lineages: A minimum of 12 base substitutions separate haplotypes of the northern alpine Danube drainage clade from the mixed clade of the Rhine-and Danube drainage, a minimum of 14 substitutions were counted between the southern alpine Danube drainage haplotype group and the mixed clade of the Rhine- and Danube drainage, and at least eight substitutions between the two Danube drainage lineages north and south of the Alps.

Discussion

The population decline of European grayling throughout central Europe is considered to be in a crisis phase, which has led to an increase in management activities involving the rearing and stocking of fish, with little input from biologists in terms of genetic or ecological considerations. A conservation-oriented management framework must consider the genetic architecture of the natural populations, the source of the stocking material, and a proper breeding design that maintains genetic variation. To achieve this framework, basic genetic data are needed on the current status of both natural and hatchery stocks. This study provides an overview of the existing lineages of grayling in two major Austrian river systems subject to intense management activities. The introduction of allochthonous material in the Drau system has clearly left a genetic imprint in that 38% of the individuals screened carried a mtDNA control region haplotype belonging to the northern alpine clade. The mean sequence divergence between the northern and southern alpine clades has been shown to be 1.4% (Weiss et al., 2002). Based on a substitution rate calibration specific for *Thymallus* (Koskinen et al., 2002), this divergence corresponds to at least one million years. This is a considerable time period for both local adaptations and (historical) chance events to occur and thus we must assume that these two lineages are evolutionarily distinct and should be managed as separate units (Moritz, 1999). Thus, we conclude that the upper Drau system has been seriously affected by the stocking of a highly divergent albeit Danubian lineage, in agreement with the stocking record documenting a massive introduction event about 20 years ago, albeit local mother fish are used since 5 years. The long-term presence of mtDNA haplotypes from an allochthonous lineage in such percentages strongly implies that introgression has taken place and the original genetic architecture of the native Drau river stocks has been compromised.

Even if a similar level of stocking has been carried out in the Inn drainage, it is not reflected in the frequency distribution of mtDNA haplotypes. At least 94% of the individuals screened can be assigned to the northern alpine mtDNA clade, whereby two additional individuals (then 97%) belong to another clade that may in fact represent an autochthonous lineage in the region. This so-called 'mixed Rhine-Danube' clade was first defined based on a small sample of individuals found in the Saalach and Vöckla rivers in Austria, various samples from the Rhine catchment in Switzerland, as well as former tributaries of the Elbe system in Denmark (Weiss et al., 2002). Following the authors' biogeographic interpretation, this clade represents a distinct central European refuge from which fish have expanded post-glacially into distinct regions of the upper Danube system. The presence of two individuals (3%) from the southern alpine clade in the Inn drainage reflects a minimal influence of stocked allochthonous lineages. While the complete stocking history in large Austrian systems is nearly impossible to track, it appears from our results that the Inn system has received stocked fish primarily from broodstocks stemming from the northern alpine mtDNA lineages, in agreement with the stocking record. However, it is also possible that fish stocked into the Inn have simply not survived or interbred with native stocks and thus introgression has not occurred (hybrids among lineages cannot be detected with mtDNA due to its maternal pathway of inheritance). Moreover, it should be mentioned that mtDNA data reflect a regional phylogeographic perspective, but do not allow to address more fine-scale genetic relationships within and among local populations.

It is noteworthy to mention that despite the historical records of extensive stocking activities, no haplotypes of geographically distant origins have been detected. This is in contrast to the documentation of distant lineages appearing in rivers of France, for example (Weiss et al., 2002), and our knowledge of the use of imported material in some Austrian hatcheries. Thus, it appears that Danube drainage lineages have been almost exclusively used in Austrian management activities.

As sustainable conservation-oriented strategy for Austrian grayling populations will require both additional population-level genetic data as well as assessment of the ecological conditions and degree of natural reproduction. Aquatic habitats must be adequate to support the full life-history cycle of the organism for any management strategy beyond a 'put-and-take' fishery to make sense. From the genetic perspective, it is not only necessary to use stocking material of local origin, but also the phylogeographic integrity and the level of genetic variation in brood stocks must be monitored, or more ideally, the use of 'broodstocks' completely avoided. As an alternative, gametes can be collected in the wild and yearling fish raised and released into the habitats from which the gametes came.

While the screening of mtDNA lineages provides a valuable overview of the mixture of geographically distinct lineages, an expanded role for genetics in the future management of Austrian salmonids must be based on multi-locus data (e.g. microsatellites; Weiss et al., 2002). Since cross-breeding (or natural reproduction) cannot be directly assessed on the basis of mtDNA, only nuclear DNA markers allow to directly estimate the level of introgression among distinct lineages. Such data are also needed to assess the fine-scale level of genetic variation, the degree of genetic substructure within large systems such as the Drau and Inn, or genetic loss within hatchery stocks. These data can also be used for choosing brood fish. This approach is presently being pursued on the Soca River, Slovenia, where a similar problem has been created in that Danube drainage (Sava) strains have been stocked into the Soca (Adriatic drainage) where a highly distinct lineage of grayling exists (Sušnik et al., 2001). Here, the future management strategy is to selectively rear fish exhibiting the least degree of introgression from allochthonous lineages.

While genetic considerations are playing an increasingly important role in conservation-oriented management schemes it can not be understated that a broad range of primarily ecological factors are the primary cause of the decline of salmonid populations in Austria (Uiblein et al., 2000; Weiss et al., 2001) and until these causes are addressed, the sustainability of any management plan must be held in question.

Acknowledgements

We thank the Tiroler Fischereiverband – Tyrolean Fisheries Association – for financial support of this study and for providing information about the stocking record. We are further grateful to G. Steiner, Alpenzoo Innsbruck, for some of the samples. N.D., S.K., and C.S. were further supported by the Austrian Science Foundation (Grant P15239).

References

Avise, J. C. & J. L. Hamrick, 1996. Conservation genetics: case histories from nature. Chapman & Hall, New York.

Bernatchez, L., R. Guyomard & F. Bonhomme, 1992. DNA sequence variation of the mitochondrial control region among geographically and morphologically remote European brown trout, *Salmo trutta* populations. Molecular Ecology 1: 161–173.

Bernatchez, L., 2001. The evolutionary history of brown trout (*Salmo trutta* L.) inferred from phylogeographic, nested clade, and mismatch analysis of mitochondrial DNA variation. Evolution 33: 351–379.

Bruford, M. W., O. Hanotte, J. F. Y. Brookfield & T. Burke, 1998, Multilocus and single-locus DNA fingerprinting. In Hoelzel, A. R. (ed.), Molecular genetic analysis of populations: a practical approach. Oxford University Press, Oxford, New York, Tokyo: 287–336.

Duftner, N., S. Weiss, N. Medgyesy & C. Sturmbauer, 2003. Enhanced phylogeographic information about Austrian brown trout populations derived from complete mitochondrial control region sequences. Journal of Fish Biology 62: 427–435.

Felsenstein, J., 1985. Confidence limits on phylogenies: an approach using the bootstrap. Evolution 39: 783–791.

Hasegawa, M., H. Kishino & T. Jano, 1985. Dating of the human-ape splitting by a molecular clock of mitochondrial DNA. Journal of Molecular Evolution 22: 160–174.

Koskinen, M. T., J. Piironen & C. R. Primmer, 2001. Interpopulation genetic divergence in European grayling (*Thymallus thymallus*, Salmonidae) at a microgeographic scale: implications for conservation. Conservation Genetics 2: 133–143.

Koskinen, M. T., J. Nilson, A. J. Veselov, A. G. Potutkin, E. Ranta & C. R. Primmer, 2002a. Microsatellite data resolve phylogeographic patterns in European grayling, *Thymallus thymallus*, Salmonidae. Heredity 88: 391–401.

Koskinen, M. T., P. Sundell, J. Piironen & C. R. Primmer, 2002b. Genetic assessment of spatiotemporal evolutionary relationships and stocking effects in European grayling, (*Thymallus thymallus*, Salmonidae). Ecology Letters 5: 193–205.

Moritz, C., 1999. Conservation units and translocations: strategies for conserving evolutionary processes. Hereditas 130: 217–228.

Posada, D. & K. A. Crandall, 1998. Modeltest: testing the model of DNA substitution. Bioinformatics 14: 817–818.

Schneider, S., D. Roessli & L. Excoffier, 2000. Arlequin ver. 2000: a software for population genetics analysis. Genetics and Biometry Laboratory, University of Geneva, Switzerland.

Sušnik, S. A., Snoj & P. Dovc, 2001. Evolutionary distinctiveness of grayling (*Thymallus thymallus*) inhabiting the Adriatic river system, as based on mtDNA variation. Biological Journal of Linnean Society 74: 375–385.

Swofford, K., 2000. PAUP*: phylogenetic analysis using parsimony (and other methods) version 4.0. Sinauer Associates, Sunderland, MA.

Thompson, J. D., D. G. Higgins & T. J. Gibson, 1994. CLUSTAL W: improving the sensitivity of progressive multiple sequence alignment through sequence weighting, position-specific gap penalties, and weight matrix choice. Nucleic Acids Research 22: 4673–4680.

Templeton, A. R., 1998. Nested clade analysis of phylogeographic data: testing hypotheses about gene flow and population history. Molecular Ecology 7: 381–397.

Uiblein, F., A. Jagsch, W. Honsig-Erlenburg & S. Weiss, 2000. Status, habitat use, and vulnerability of the European grayling (*Thymallus thymallus*) in Austrian waters. Journal of Fish Biology 59(A): 223–247.

Uiblein F., A. Jagsch, G. Kössner, S. Weiss, P. Gollman & E. Kaniz, 2000. Untersuchungen zu lokaler Anpassung, Gefährdung und Schutz der Äsche (*Thymallus thymallus*) in drei Gewäsern in Oberösterreich. Österreichs Fischerei 53(4): 87–162.

Weiss, S., C. Schlötterer, H. Waidbacer & M. Jungwirth, 2001a. Haplotype (mtDNA) diversity of brown trout *Salmo trutta* L. in tributaries of the Austrian Danube: massive introgression of Atlantic basin fish – by man or nature? Molecular Ecology 10: 1241–1246.

Weiss, S., F. Uiblein & A. Jagsch, 2001b. Conservation of salmonid fishes in Austria: current status and future strategies. In Collares-Pereira, M. J. M. M Coelho & I. G Cowx (eds). Freshwater Fish Conservation: options for the future. Fishing News Books, Blackwell Science, Oxford.

Weiss, S., H. Persat, R. Eppe, C. Schlöterer & F. Uiblein, 2002. Complex patterns of colonization and refugia revealed for European grayling *Thymallus thymallus*, based on complete sequencing of mitochondrial DNA control region. Molecular Ecology 11: 1393–1407.

Hydrobiologia (2005) 542:131–136
H. Segers & K. Martens (eds), Aquatic Biodiversity II
DOI 10.1007/s10750-004-5495-6

Connectivity and nestedness of the meta-community structure of moss dwelling bdelloid rotifers along a stream

Diego Fontaneto*, Giulio Melone & Claudia Ricci
Dipartimento di Biologia, Università degli Studi di Milano, via Celoria 26, Milano 20133, Italy
*(*Author for correspondence: Tel.: +39-02-50314810, E-mail: diego.fontaneto@unimi.it)*

Key words: Rotifera Bdelloidea, Italy, meiofauna, lotic environment, benthic freshwater

Abstract

We analyzed meta-community structure of bdelloid rotifers colonizing mosses along an 80 meter section of Rio Valnava in NW Italy. Bdelloid rotifers are small animals living associated with a substratum; colonization in bdelloids can be produced by active animals moving along the riverbed, or by passive dormant propagules, moved by wind. To detect which kind of colonization might be stronger at different spatial scales, we designed a spatially nested sampling experiment at three hierarchical levels: (1) single sample, (2) 10 communities inside each pool, (3) complete section of 10 pools. Assessing species richness and species similarity of communities, and coherence and nestedness of bdelloid meta-communities, we found that different forces may drive species composition at different spatial scales: at the largest scale, colonization of propagules may over-ride direct dispersal between pools, while at the scale of the single pool, differential movements of species give a nested structure to the meta-communities. The number of species increased as the level of analysis increased, even though this study was carried out along only a small stream section.

Introduction

Classical analysis of freshwater community structure usually has attempted to recognize environmental features shaping species composition as if such communities were spatially and temporally isolated entities. The connectivity of communities has been considered as a relevant factor influencing species structure only recently (Forbes & Chase, 2002; Cottenie & De Meester, 2003; Cottenie et al., 2003; Kneitel & Miller, 2003). Naturally, with the addition of this new perspective, the scale over which communities are potentially connected – local vs. regional, for example – has also become important (Amarasekare, 2000; Taneyhill, 2000; Mouquet & Loreau, 2002). One of the major innovations arising from these recent studies is the concept of the meta-community, that is, a set of ecological communities at different sites that are potentially connected by dispersal (Leibold and Mikkelson, 2002) and for which species composition is at least partially structured by the interplay of colonization and selective local extinction (Lomolino, 1996; Worthen et al., 1998; Wright et al., 1998; Davidar et al., 2002; Mac Nally et al., 2002, Chase, 2003).

To date, no evaluation of meta-community interaction has been based on freshwater benthic meiofaunal communities. Such communities are very widespread in freshwater habitats of many kinds, including those physically connected by water courses that could serve as effective conduits for species migrations. Hence, meiofaunal communities are excellent systems to evaluate the applicability of meta-community theory. The current work assesses the significance of community connectivity for a series of meiofaunal communities numerically dominated by bdelloid rotifers associated with submerged mosses of a freshwater stream (Linhart et al., 2002a, b).

Bdelloids are small (<1000 μm) animals that live in close association with the substrate and which can move actively by swimming or 'creeping', or be passively transported downstream by drift (Schmid-Araya, 1998a, b; Bilton et al., 2001; Elliott, 2003). Bdelloids can withstand prolonged drought by entering dormancy and can be dispersed in the dormant state to colonize distant areas not directly connected by water (Cáceres & Soluk, 2002; cf. Jenkins & Underwood, 1998 and Bohonak & Jenkins, 2003). Moreover, their small size and low vagility allow study of different communities in a relatively small area. Restricting study to a local area in a single environment can minimize differences in species composition caused by extensive differences in environmental features, disturbance processes or geographic history. Because of these unique characteristics, bdelloid communities are a particularly useful field model for investigation of different meta-community structures.

Here we analyze a series of bdelloid rotifer communities from submerged mosses in pools in a short section of a mid-elevation stream. We propose the following alternative hypotheses: (1) if individual pool community structure is driven primarily by aerial colonization by dormant propagules from distant source communities (which should be rare and random events), inter-community comparisons should show stochastic meta-community structure; (2) if direct (i.e. aquatic) connectivity dominates over only occasional aerial colonization, nearby communities are expected to be more similar in species composition than more distant ones.

Material and methods

Study area

Moss samples were collected in June, July, and August 2003 along a 80 m long section of Rio Valnava (approx. coordinates: 4° 11′ E, 45° 39′ N). The sampling site was near Bercovei cave at an elevation of about 400 m, in the municipality of Sostegno, Biella province, Piedmont Region, NW Italy. The transect was characterized by a sequence of 10 pools, with low flow rate, linearly connected by segments of riffles, each with a higher flow rate. Each pool measured from 1 to 4 m^2 (Table 1) with shoreline edges almost fully covered with submerged mosses; mosses were absent in the riffle segments. Distance between any two adjacent pools varied from 3 to 20 m. Except for this section of 10 pools, mosses were not observed along the stream.

Temperature was about 16 °C, pH around 7 and the rocky substratum at all sampled sites was made of limestone. Moss species was the same too (*Brachythecium* sp.). Macrobenthos or other biotic components at the different sites were not investigated, and we assumed that their effect on bdelloids could have been equal throughout the different sampling sites, due to the high vagility of macrobenthos at the scale of the length of the analyzed transect (Bilton et al., 2001).

We considered 10 continuous pools (called, from upstream to downstream, A–J), and we designed a cumulative spatial analysis at three different nested levels. The first level is for the individual moss sample (local level; single community), the second level is for each pool (intermediate level; 10 communities, one from each moss sample), and the third level is at the stream level (largest level; 10 communities, one from each pool). This spatially nested design should have allowed us to detect hierarchical differences in the structure of the meta-communities at different spatial scales.

Sampling

Each sample of moss had a surface area of 5 cm^2. Samples were kept in small plastic bottles and were directly taken to the laboratory, where they were maintained at 6 °C to keep the animals alive. As soon as possible, but in no case longer than 3 days, living bdelloids were isolated and identified from each sample.

Statistical analyses

The meta-community analysis performed on the data followed Leibold & Mikkelson (2002). This analysis deals with presence of species and disregards the frequency of individuals. The data sets for each of the three scales analysis were arranged

Table 1. Number of sites, number of bdelloid species, and area of each analyzed matrix, with *p*-values from 200 simulated matrices, counting number of embedded absences for coherence and number of species replacements for nestedness, as described in Leibold & Mikkelson (2002)

Pool	Number of sites	Number of species	Area (m^2)	Coherence *p* values	Nestedness *p* values
A	10	9	1.5	*0.000*	0.054
B	10	9	1.0	*0.000*	*0.012*
C	10	7	2.0	*0.002*	*0.010*
D	10	7	1.8	*0.009*	*0.013*
E	10	7	3.0	*0.038*	0.052
F	10	10	2.2	*0.000*	*0.021*
G	10	9	4.0	*0.000*	*0.038*
H	10	8	3.0	*0.002*	0.050
I	10	7	1.2	*0.005*	0.082
J	10	9	3.8	*0.000*	*0.044*
System	10	15		*0.000*	0.696

Cells with '0.000' mean a lower value than '0.001'. Significant *p* values are in italics.

in ordered presence/absence matrices listing species in columns, and communities in rows. The ranking of rows and columns was obtained by reciprocal averaging (also known as correspondence analysis), so that communities (rows) with the closest lists of species were close each other, and species (columns) with the closest distributions were close each other.

We tested each matrix for coherence and for nestedness. A matrix of species distribution is considered to be coherent when the presences are arranged along the diagonal line of the matrix; coherence is the pre-requisite to detect nestedness (Leibold & Mikkelson, 2002). A pattern of species distribution is considered to be nested when the communities poorer in species are subsamples of the richest ones (Patterson & Atmar, 1986; Fisher & Lindenmayer, 2002).

Coherence was detected through the number of embedded absences in the matrix, while nestedness was detected through the number of absence/presence changes; that is, how many species are replaced in the sequence of communities. Each data set was tested by comparing the absences or replacements with those present in 200 matrices randomly generated by Monte Carlo simulations. Each generated matrix had 2 constraints: (1) the number of presences equal to that of the observed data matrix, and (2) no row or column empty.

First, we tested the nestedness of the whole meta-community of the 10 pools at the system level, and then the nestedness of each pool, at the intermediate level.

Results

At the stream level (largest level), a total of 15 species were identified. At this largest spatial level (10 pools), the meta-community described by the analysis had a coherent pattern of species distribution, but revealed a random spatial turnover of species (Table 1). No nested structure could be recognized (Fig. 1). The rank order of pools in the ordered matrix (Fig. 1) was not related to the downstream position of pools in the field (Spearman $r = 0.563$, n.s.), indicating that close pools were not more similar than distant pools. Mean number of species per pool was 8.2 ± 1.13 (SD), with no significant difference between communities in the downstream position in the field (Spearman $r = 0.061$, n.s.). The number of species present in each pool was independent from the area of the pool (Spearman $r = 0.23$, n.s.).

The analysis at the intermediate level revealed that the species in the 10 meta-communities from each of the 10 pools were distributed non-randomly. Significant values of nestedness were evident. Communities of A, E and H pools were slightly nested (*p*-value ~ 0.05), and *p* value for I pool resulted a bit higher than this level. (Table 1),

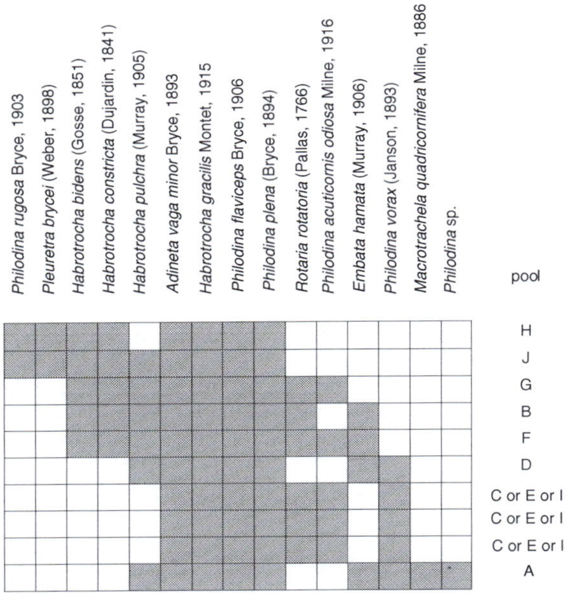

Figure 1. Bdelloid species presence in 10 analyzed pools in the ordered matrix. Communities are ordered in rows and species in columns.

but these values could be due to the low number of cells of our matrices. Nevertheless, all matrices resulting from ordination were typically nested (Fig. 2), except matrix from C pool.

At the local level, the average number of species per sample was 5.7 ± 1.56, without significant differences between pools (ANOVA test: $F_{9,90} = 1.602$, $p = 0.127$).

Discussion

The 15 species found along the stream were not equally present in the 10 pools, and so obviously not all present in the 10 moss samples for each pool. On average 8 species were present in each pool, and adjacent pools did not show same species assemblages. In three cases the pools had the same species, but these pools were not directly connected. Bdelloid rotifers can actively move on the river bed and be spread downstream by drift. Thus the similarity between the pools could be due either to chance or to sampling bias. Thus, and although bdelloid species were apparently randomly distributed, a species structure of the communities along the stream was revealed by application of meta-community analysis.

The analysis run at different scale levels showed that the structure of the bdelloid meta-communities changes according to the scale of investigation, with stochastic distribution at the largest level and nestedness at the intermediate level. At the largest level, each pool could be recognized as a separate system, sharing only few species with other pools. Passive dispersal of bdelloids by wind or rain is possible, and this process should be independent of the connectivity between pools. If this were the case, all pools should present the same composition in species. In contrast, we found a random spatial substitution of species. This could be explained hypothesizing that new invaders could be precluded from successful colonization of a given area by the already established community

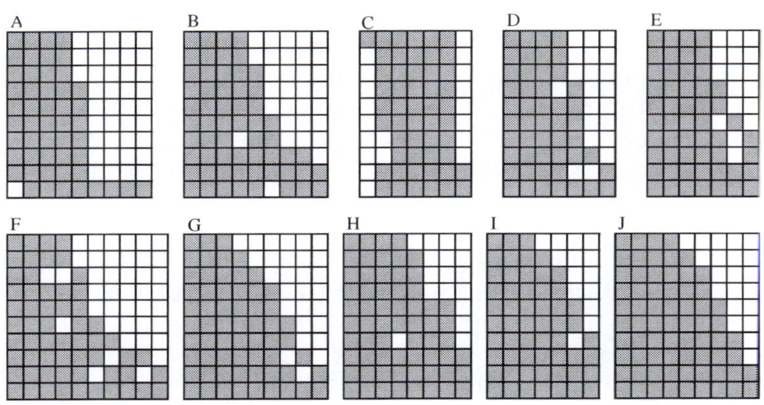

Figure 2. Reciprocal averaging ordered matrices with data from the 10 communities in each pool. Communities are ordered in rows and species in columns.

(Shurin, 2000; Rundle et al., 2002). And a strong priority effect of early founders may drive the species assemblages of the pools to differ. As founder effect may be rare and random, the meta-community structure was random, as well (De Meester et al., 2002).

In contrast, at intermediate level, all pools revealed nested structures. Nestedness could be produced by (1) differential colonization (Loo et al., 2002), or (2) sampling artifact (Cutler, 1994; Fischer & Lindenmayer, 2002). In both cases, some species should be present in every sample from a same pool, while others should be present only in the richest communities. In our system this pattern could be related to different motility. Locally rare species in each pool system were *Embata hamata* and *Rotaria rotatoria*, that are those species that can move more easily by swimming. More common within each pool were the species that move mainly by inching or creeping on the bottom, such as *Adineta* and *Habrotrocha*. In the attempt of relating different locomotory behavior to colonization capability, we could have expected that the swimming species should have been the most widespread inside each pool, while the opposite was found. If swimming of bdelloids is not related to their colonization capability, as seems to be the case from our results, active colonization should be expected through water beds, only.

As noted earlier, we considered two alternative hypotheses, one predicting dominance of colonization by passive dispersal, and the other predicting the structuring influence of active animal movements. In the former case the meta-community would show a random character, and in the latter a statistically significant structure. The hypotheses seem to fit the observed meta-communities if the analysis is run at the largest level or at the intermediate level, respectively.

The meta-community approach for the study of benthic meiofauna evidences patterns and provides interpretation for species distribution that could not be approached through other ecological analysis. Although this study was carried out along a small stream section, 80 m long, the number of species for the three different levels of analysis increased (single moss sample = 5.7, pools = 8.2, stream = 15), analogously to the general feature of scale-dependent diversity found in community ecology at larger geographical scales (Shorrocks & Sevenster, 1995; Cornell, 1999; Lawton, 1999; Gaston, 2000; Arita & Rodriguez, 2002). The characteristics of poor vagility, colonization capability, widespread distribution and minute size make bdelloid rotifer a well suited organism model for these analyses.

Acknowledgements

We wish to thank Mathew A. Leibold for providing us the meta-community analysis tool, Graziella Rodondi for moss identification, and Peter Starkweather for improving the English text.

References

Amarasekare, P., 2000. The geometry of coexistence. Biological Journal of the Linnean Society 71: 1–31.

Arita, H. T. & P. Rodríguez, 2002. Geographic range, turnover rate and the scaling of species diversity. Ecography 25: 541–550.

Bilton, D. T., J. R. Freeland & B. Okamura, 2001. Dispersal in freshwater invertebrates. Annual Review of Ecology and Systematics 32: 159–181.

Bohonak, A. J. & D. G. Jenkins, 2003. Ecological and evolutionary significance of dispersal by freshwater invertebrates. Ecology Letters 6: 783–796.

Cáceres, C. E. & D. A. Soluk, 2002. Blowing in the wind: a field test of overland dispersal and colonization by aquatic invertebrates. Oecologia 131: 402–408.

Chase, J. M., 2003. Community assembly: when should history matter? Oecologia 136: 489–498.

Cornell, H. V., 1999. Unsaturation and regional influences on species richness in ecological communities: a review of the evidence. Ecoscience 6: 303–315.

Cottenie, K. & L. De Meester, 2003. Connectivity and cladoceran species richness in a metacommunity of shallow lakes. Freshwater Biology 48: 823–832.

Cottenie, K., E. Michels, N. Nuytten & L. De Meester, 2003. Zooplankton metacommunity structure: regional vs. local process in highly interconnected ponds. Ecology 84: 991–1000.

Cutler, A. H., 1994. Nested biotas and biological conservation-metrics, mechanisms and meaning of nestedness. Landscape and Urban Planning 28: 73–82.

Davidar, P., K. Yogananad, T. Ganesh, & S. Devy, 2002. Distribution of forest birds and butterflies in Andaman islands, Bay of Bengal: nested patterns and process. Ecography 25: 5–16.

De Meester, L., A. Gómez, B. Okamura & K. Schwenk, 2002. The monopolization hypothesis and the dispersal-gene flow paradox in aquatic organisms. Acta Oecologica 23: 121–135.

Elliott, J. M., 2003. A comparative study of the dispersal of 10 species of stream invertebrates. Freshwater Biology 48: 1652–1668.

Fischer, J. & D. B. Lindenmayer, 2002. Treating the nestedness temperature calculator as a 'black box' can lead to false conclusions. Oikos 99: 193–199.

Forbes, A. E. & J.N. Chase, 2002. The role of habitat connectivity and landscape geometry in experimental zooplankton metacommunities. Oikos 96: 433–440.

Gaston, K. J., 2000. Global patterns in biodiversity. Nature 405: 220–227.

Jenkins, D. G. & M.O. Underwood, 1998. Zooplankton may not disperse readily in wind, rain, or waterfowl. Hydrobiologia 387/388: 15–21.

Kneitel, J. M. & T. E. Miller, 2003. Dispersal rates affect species composition in metacommunities of *Sarracenia purpurea* inquilines. American Naturalist 162: 165–171.

Lawton, J. H., 1999. Are there general rules in ecology? Oikos 84: 177–192.

Leibold, M. A. & G. M. Mikkelson, 2002. Coherence, species turnover, and boundary clumping: elements of meta-community structure. Oikos 97: 237–250.

Linhart, J., M. Fiuraskova & V. Uvira, 2002a. Moss- and mineral substrata-dwelling meiobenthos in two different low-order streams. Archiv für Hydrobiologie 154: 543–560.

Linhart, J., S. Vlckova & V. Uvira, 2002b. Bryophytes as a special mesohabitat for meiofauna in a rip-rapped channel. River Research and Applications 18: 321–330.

Lomolino, M. V., 1996. Investigating causality of nestedness of insular communities: Selective immigrations or extinctions? Journal of Biogeography 23: 699–703.

Loo, S. E., R. Mac Nally & G. P. Quinn, 2002. An experimental examination of colonization as a generator of biotic nestedness. Oecologia 132: 118–124.

Mac Nally, R., G. Horrocks, & A. F. Bennett, 2002. Nestedness in fragmented landscapes: birds of the box-ironbark forests of south-eastern Australia. Ecography 25: 651–660.

Mouquet, N. & M. Loreau, 2002. Coexistence in a competitive metacommunity: the regional similarity hypothesis. American Naturalist 159: 420–426.

Patterson, B. D. & W. Atmar, 1986. Nested subsets and the structure of insular mammalian faunas and archipelagos. Biological Journal of the Linnean Society 28: 65–82.

Rundle, S. D., A. Foggo, V. Choiseul & D. T. Bilton, 2002. Are distribution patterns linked to dispersal mechanism? An investigation using pond invertebrate assemblages. Freshwater Biology 47: 1571–1581.

Schmid-Araya, J. M., 1998a. Small-sized invertebrates in a gravel stream: community structure and variability of benthic rotifers. Freshwater Biology 39: 25–39.

Schmid-Araya, J. M., 1998b. Rotifers in interstitial sediments. Hydrobiologia 387/388: 231–240.

Shorrocks, B. & J. G. Sevenster, 1995. Explaining local species diversity. Proceedings of the Royal Society of London Series B – Biological Sciences 260: 305–309.

Shurin, J. B., 2000. Dispersal limitation, invasion resistance and the structure of pond zooplankton communities. Ecology 81: 3074–3086.

Taneyhill, D. E., 2000. Metapopulation dynamics of multiple species: the geometry of competition in a fragmented habitat. Ecological Monographs 70: 495–516.

Worthen, W. B., M. T. Jones & R. M. Jetton, 1998. Community structure and environmental stress: desiccation promotes nestedness in mycophagous fly communities. Oikos 81: 45–54.

Wright, D. H., B. D. Patterson, G. M. Mikkelson, A. Cutler & W. Atmar, 1998. A comparative analysis of nested subset patterns of species composition. Oecologia 113: 1–20.

Hydrobiologia (2005) 542:137–150
H. Segers & K. Martens (eds), Aquatic Biodiversity II
DOI 10.1007/s10750-004-1452-7

Implications of taxonomic modifications and alien species on biological water quality assessment as exemplified by the Belgian Biotic Index method

W. Gabriels*, P. L. M. Goethals & N. De Pauw
Laboratory of Environmental Toxicology and Aquatic Ecology, Ghent University, J. Plateaustraat 22, 9000 Ghent, Belgium
*(*Author for correspondence: E-mail: w.gabriels@vmm.be)*

Key words: biological assessment methods, macroinvertebrates, surface water, exotic species

Abstract

In this paper, some important problems related to taxonomic resolution in water quality assessment by means of macroinvertebrates are discussed. Most quality indices based on macroinvertebrates only require identification up to genus or family level. Although this can be seen as a practical trade-off between taxonomic precision and time constraints and financial resources, it can result in biased assessment scores for certain stream types. An additional difficulty of identification levels other than species is caused by possible changes in taxonomy over time. A given genus may indeed have been split up into two or more genera or a species could be assigned to a different genus. These changes may alter biotic index values calculated over time, due to a change in number of taxa or replacement of one taxon by another one having a different tolerance class. An additional problem is caused by the invasion of exotic species. The genus *Corbicula* for instance is currently invading Belgian watercourses in increasing numbers. Since no Belgian Biotic Index (BBI) tolerance class is defined for *Corbicula*, this may cause inconsistencies in index calculations as well. In order to eliminate these, a semi-fixed taxa list, including a tolerance class for each taxon, for BBI calculation is proposed.

Introduction

In biomonitoring, two approaches can be distinguished: the bioconservation approach, where biodiversity and species conservation are the key incentives for sampling an aquatic habitat, and the bioassessment approach, where the focus is on water quality assessment and hence, insight in the biological community is a means rather than a goal. The former approach implies a species-level identification of the sampled organisms, while the latter usually involves a trade-off between a higher level of identification with lower costs but a less precise outcome, and a species-level identification with higher costs but a higher precision. The objectives of a sampling campaign should therefore be decisive for the choice of identification level. In this paper, some problems related to taxonomic resolution in water quality assessment by means of macroinvertebrates are discussed.

Various authors recommend identification to species level to ascertain a detailed insight in the community composition, avoiding information loss due to lumping of taxa, and showing a strong assemblage–environment relationship (e.g. Resh & McElravy, 1993; Stubauer & Moog, 2000; Verdonschot, 2000; Lenat & Resh, 2001; King & Richardson, 2002; Adriaenssens et al., 2004). On the other hand, species identification is time-consuming and expensive. On top of that, information loss when identifying to genus or even family level is often small, and according to several authors it is therefore not necessary to descend to the species level (e.g. Warwick, 1988; Bowman & Bailey, 1997; Ghetti, 1997; Olsgard et al., 1998; Dolédec et al., 2000; Gayraud et al., 2003). Another problem associated with species level identifications is the increasing uncertainty that arises with an increasing level of detail. Ellis (1985) acknowledged this when defining taxonomic sufficiency as the level to which the organisms should be identified in order to balance the need to indicate the biological community versus accuracy of the identifications. When deciding upon the taxonomic level, all aspects mentioned above should be taken into consideration. According to Guérold (2000) and Roach et al. (2001) family level is sufficient for detecting perturbations on the macroinvertebrate community, but a more detailed level of identification is necessary for ecological interpretation. Williams & Gaston (1994) proposed the use of higher-taxon categories as surrogates for species in rapid biodiversity surveys. Karr & Chu (1999) consider genus level to be sufficient for developing a multimetric index and also family level to be acceptable in case of limited time and/or financial resources.

Whatever taxonomic level is used for a biotic water quality index, the level should be fixed with the method description because (1) many methods can only be calculated when using the predefined level, e.g. when taxon-specific tolerance values are defined and (2) taxonomic level can affect index calculation (e.g. Guérold, 2000; Schmidt-Kloiber & Nijboer, 2004).

An additional difficulty with identification levels other than species, however, is caused by possible changes in taxonomy over time, giving rise to inconsistencies in index calculation. A given genus may be split up into two or more genera or a species can be assigned to a different genus. These changes may alter the value of the biotic indices calculated based on the given taxa, respectively because the number of taxa (of a level higher than species) has changed or a taxon is replaced by another one (having a different tolerance class). This is demonstrated by a simple example of Belgian Biotic Index (BBI) calculation of a virtual sample.

Similar problems are due to the invasion of exotic species. Newly occurring taxa raise discussions whether or not to include them in the existing index, which may imply defining a tolerance class for the new taxon, as used in most biotic index methods. This problem has risen for at least one exotic genus in Belgium, as will be discussed later.

Calculation of the BBI

The BBI method is a standardised method to assess biological quality of watercourses based on the macroinvertebrate community. The method was proposed by De Pauw & Vanhooren (1983) and has been adopted as a standard method by the Belgian Institute of Normalisation (IBN, 1984). Since its first publication, the method has been extensively used to assess water quality in Belgium but also abroad (De Pauw & Hawkes, 1993). Since 1989, the Flemish Environment Agency (VMM) assesses around thousand sites throughout Flanders (Belgium) each year by means of the BBI.

The BBI combines characteristics of the indices proposed by Woodiwiss in the UK (1964, Trent Biotic Index) and Tuffery and Verneaux in France (1968, Indice Biotique). The method is based on aquatic macroinvertebrates sampled with a standard handnet, as in the method of Woodiwiss (1964) and the calculation of the biotic index using the table as proposed by Tuffery & Verneaux (1968). Some adaptations were made concerning the sampling method and the taxonomic level of identification. Table 1 summarises the taxonomic levels of identification for the BBI as proposed by De Pauw & Vanhooren (1983). Only taxa of which at least two individuals are found in the sample, are taken into account. The calculation of the BBI is based on a combination of the highest tolerance class encountered, the class frequency within the highest tolerance class and the total number of taxa (Table 2). For instance, a sample containing 9 taxa, 2 of which having a tolerance class of 3 (being the lowest tolerance class encountered in the sample) would be assigned a BBI of 5. The column with indicator groups in Table 2 contains some modifications, which will be discussed further in this paper. BBI values correspond to water quality classes with their associated formal valuation, which are summarised in Table 3 (De Pauw & Vanhooren, 1983).

Bervoets et al. (1989) proposed, along with some modifications in sample processing, to

Table 1. Identification levels of macroinvertebrate taxa for calculating the BBI (De Pauw & Vanhooren, 1983)

Taxonomic group	Determination level of systematic units
Plathelminthes	Genus
Oligochaeta	Family
Hirudinea	Genus
Mollusca	Genus
Crustacea	Family
Plecoptera	Genus
Ephemeroptera	Genus
Trichoptera	Family
Odonata	Genus
Megaloptera	Genus
Hemiptera	Genus
Coleoptera	Family
Diptera	Family, excl. Chironomidae (Chironomidae *thummi-plumosus*, Chironomidae *non-thummi-plumosus*)
Hydracarina	Presence

include taxa represented by only one individual in BBI calculation, but this modification was never incorporated into routinely monitoring schemes of the VMM.

Table 3. Water quality classes corresponding to the BBI values (De Pauw & Vanhooren, 1983)

Quality class	BBI	Colour code	Valuation
I	9–10	Blue	Lightly polluted or unpolluted
II	7–8	Green	Slightly polluted
III	5–6	Yellow	Moderately polluted
IV	3–4	Orange	Heavily polluted
V	0–2	Red	Very heavily polluted

Inconsistencies due to taxonomic modifications

De Pauw & Vannevel (1991) published keys in Dutch for identification of aquatic macroinvertebrates, for each group up to the appropriate BBI level. Since the publication of these identification keys, taxonomy of some groups of macroinvertebrates was changed, resulting in genera splitting up into more than one genus. Examples are the gastropod genera *Lymnaea*, *Stagnicola*, *Radix* and *Galba*, formerly all considered as *Lymnaea* species; the gastropod *Physella*, formerly belonging to the genus *Physa*; and *Aquarius najas* (De Geer, 1773), formerly belonging to the genus *Gerris*. As a result,

Table 2. Calculation of the BBI, based on the highest tolerance class encountered, the class frequency within the highest tolerance class and the total number of taxa (De Pauw & Vanhooren, 1983)

Tolerance class Indicator groups	Class Frequency	Number of taxa				
		0–1	2–5	6–10	11–15	≥16
1. Plecoptera; Heptageniidae	≥2	–	7	8	9	10
	1	5	6	7	8	9
2. Cased Trichoptera	≥2	–	6	7	8	9
	1	5	5	6	7	8
3. Ancylidae; *Acroloxus*; Ephemeroptera (excl. Heptageniidae)	> 2	–	5	6	7	8
	1–2	3	4	5	6	7
4. *Aphelocheirus*; Odonata; Gammaridae; Mollusca (excl. **Ancylidae**, *Acroloxus*, Sphaeriidae & *Corbicula*)	≥1	3	4	5	6	7
5. Asellidae; Hirudinea; Sphaeriidae; Hemiptera (excl. *Aphelocheirus*)	≥1	2	3	4	5	–
6 .Tubificidae Chironomidae *thummi-plumosus*	≥1	1	2	3	–	–
7. Syrphidae-Eristalinae	≥1	0	1	1	–	–

Proposed modifications of indicator groups after Gabriels et al., this paper (in bold).

two samples containing the same species and the same number of individuals for each species could result in a different index depending on whether the current state-of-the-art in taxonomy is followed for identifying the organisms or the taxonomic levels *sensu* De Pauw & Vannevel (1991) are used.

This is demonstrated with a simple example of a BBI calculation for two virtual samples (Table 4). The two approaches produce different BBI values in both examples. Table 4 (panels A and B) gives a list of species with their respective abundances and tolerance classes. Subsequently, the BBI is calculated following both approaches. In the first example (Table 4, panel A), identification of the sample following the keys of De Pauw & Vannevel (1991) will result in a decrease of taxa richness with two units, and a decrease of the BBI with one unit, because the genera *Aquarius* and *Radix* are assigned to other genera (*Gerris* and *Lymnaea*, respectively). In the second example (Table 4, panel B) the actual taxa richness decreases with one unit, but for BBI calculation it increases with a unit because two individuals are only counted when representing the same taxon since two is the minimal abundance for inclusion in BBI calculation. As a result, the BBI increases with two units in this case.

Both approaches can be justified since the original publication of the BBI (De Pauw & Vanhooren, 1983) only indicates the levels of identification (Table 1). Application of the BBI *sensu stricto* today would therefore imply using the current levels of identification, although only using the same taxonomic identification keys at all time would lead to stable results, i.e. a time-independent calculation of BBI values.

An estimation of the percentage of actual samples for which both approaches provide different results was not possible since the identifications of the VMM are only recorded at the lumped levels (e.g. *Lymnaea* including *Stagnicola*, *Radix* and *Galba*). In order to obtain a rough indication, both approaches were compared for *Anisus*, a genus that was split before the publication of the identification keys of De Pauw & Vannevel (1991) and hence all actual taxa are recorded in the VMM data set. The recorded taxa are *Anisus*, *Armiger*, *Bathyomphalus*, *Gyraulus*, *Hippeutis*, *Planorbis* and *Segmentina*. Two hundred and eighty four samples from the VMM data set contained at least two individuals of at least two of the seven taxa. BBI was calculated for these samples when distinguishing the seven taxa and calculated again after summing the abundances of the seven taxa into one taxon, *Anisus*. For 34 samples (12.0%), summing the taxa resulted in a BBI decrease of one unit. The other samples were not affected.

Since there is no reason to assume that taxonomic modifications will not proceed in future, this problem can only be overcome by using a fixed list of taxa at all time (or, more correctly, a semi-fixed list; see further). The establishment of a common list of taxa was already recommended by Woodiwiss (1980). For the German saprobic index, a fixed taxon list is already in use (DIN, 1990).

Inconsistencies due to the introduction of exotic species

Adverse effects of invasive species on ecosystems have been discussed by several authors (e.g. Lodge, 1993; Cairns & Bidwell, 1996; Mack et al., 2000; Torchin et al., 2003). Invasion of exotic macroinvertebrate genera in Europe is increasing (e.g. Van den Brink et al., 1991; Bij de Vaate et al., 2002). These invasions cause controversy on the subject of index-based biological assessment, strongly related to the question whether or not a fixed taxa list is used. An important aspect of this controversy is the higher potential number of taxa present in monitoring samples due to these introductions, which may cause an increase in index number when using an index dependent on taxa richness. Though alpha diversity, expressed as number of taxa, may have risen, this will only be reflected in index calculation provided the new taxon is included in the list for index calculation. On the other hand, introduction of exotic species might as well cause a decrease of alpha diversity, which is masked due to a higher taxonomic identification level. For example, the invader *Dikerogammarus villosus* (Sowinsky, 1894) (Crustacea, Gammaridae) might outcompete a number of native gammarid species (e.g. Bij de Vaate et al., 2002), but this will not influence the results of the index calculation at family level of a given sample since Gammaridae are still present.

Table 4. Calculation of the **BBI** of two virtual samples

Species	Abundance	Tolerance class	Taxa according to the current state-of-the-art taxonomy (each at the applicable level)	Taxa according to the taxonomy as applied in De Pauw & Vannevel (1991) (each at the applicable level)
Panel A. Example resulting in a decreased **BBI** due to taxonomic changes				
Tubifex tubifex	100	6	Tubificidae	Tubificidae
Chironomus riparius	45	–	Chironomidae *non-thummi-plumosus*	Chironomidae *non-thummi-plumosus*
Erpobdella octoculata	4	–	*Erpobdella*	*Erpobdella*
Lymnaea stagnalis	5	4	*Lymnaea*	*Lymnaea*
Radix peregra	2	4	*Radix*	
Gerris lacustris	4	5	*Gerris*	*Gerris*
Aquarius najas	2	5	*Aquarius*	
Total number of taxa			7	5
Lowest tolerance class			4	4
Tolerance class frequency			2	1
BBI			5	4
Water quality class			III (yellow)	IV (orange)
Panel **B**. Example resulting in an increased **BBI** due to taxonomic changes				
Tubifex tubifex	100	6	Tubificidae	Tubificidae
Chironomus riparius	45	–	Chironomidae *non-thummi-plumosus*	Chironomidae *non-thummi-plumosus*
Erpobdella octoculata	4	–	*Erpobdella*	*Erpobdella*
Lymnaea stagnalis	1	4	(*Lymnaea*)	*Lymnaea*
Radix peregra	1	4	(*Radix*)	
Gerris lacustris	2	5	*Gerris*	*Gerris*
Sialis lutaria	10	–	*Sialis*	*Sialis*
Total number of taxa			5	6
Lowest tolerance class			5	4
Tolerance class frequency			1	2
BBI			3	5
Water quality class			IV (orange)	III (yellow)

The first and second column of both panels A and B list the species and their respective abundances, the third one the tolerance classes, the fourth one the taxa according to the current state-of-the-art taxonomy (each at the applicable level), and the fifth one the taxa according to the taxonomy as applied in De Pauw & Vannevel (1991) (each at the applicable level). At the bottom of the fourth and fifth column the **BBI** and the respective water quality class is indicated for both approaches.

Nguyen & De Pauw (2002) reported the invasion of the Asian clams *Corbicula fluminea* (Müller, 1774) and *Corbicula fluminalis* (Müller, 1774) (Mollusca, Corbiculidae) in the Belgian section of the river Meuse, and some of the connected canals in the early 1990s and the continuing colonisation of *Corbicula* species in Belgian watercourses. They could not establish a correlation between the clam density or proportion and the quality of the sediment. Since no tolerance class is defined for *Corbicula*, this may cause inconsistencies in BBI calculations due to a lack of consensus on how to deal with this phenomenon. The VMM encounters this genus more and more frequently in its biological samples. The question emerged whether or not this exotic genus should be included in BBI calculation, and if so, which tolerance class to use. A strict interpretation of the tolerance class as described by De Pauw & Vanhooren (1983) would lead to the inclusion of *Corbicula* in the standard list with a tolerance class of 4, being a non-sphaeriid mollusc, and thus being quite tolerant. By means of two calculation examples it is demonstrated that this may cause differences in index calculation (Table 5).

Table 5 (panels A and B) gives a list of taxa with their respective abundances. Then the BBI is calculated according to three different approaches. In the first approach, *Corbicula* is neglected, in the second it is included without tolerance class ('–') and in the third it is included with a tolerance class of 4. Note the difference between a tolerance class '–' and the absence of a tolerance class. With a '–' tolerance the taxon is only taken into account for taxon richness, while in the absence of a tolerance class the taxon is not included at all. The first example (Table 5, panel A) is a sample actually taken by the VMM on 6 May 1998 at a sampling site in the Albert Canal at Genk. In this case, the inclusion of *Corbicula* leads to an increase of the BBI from 6 to 7. The VMM reported the BBI of this sampling site as 7, and consequently this site met the basic water quality conditions (BBI = 7) thanks to *Corbicula*. In the second example (Table 5, panel B), a virtual sample, it is demonstrated that the three approaches can as well lead to three different BBI values.

Eighteen samples from the data set of the VMM contained *Corbicula* individuals. In twelve of these samples, at least two individuals were counted and hence *Corbicula* was included in the BBI calculation of these samples. For one sample (Table 5, panel A), the BBI was affected when *Corbicula* was discarded. The number of samples was however statistically insufficient and therefore conclusions on the probability of affecting the BBI could not yet be drawn. Nguyen & De Pauw (2002) found that including *Corbicula* species in the Biotic Sediment Index (BSI; De Pauw & Heylen, 2001), altered biological sediment quality classification in 52% of the cases.

In order to obtain a more reliable indication of the frequency of BBI alteration if an alien taxon would be discarded, the same calculation was performed for *Dreissena*, another alien bivalve that was already included in the taxa list of De Pauw & Vannevel (1991), with a tolerance class of 4. *Dreissena* is already present in Belgian waters for a longer time and consequently more data were available for comparing calculations. Four hundred and twenty one samples from the VMM data set contained at least two *Dreissena* individuals. The BBI was calculated for all samples and recalculated after exclusion of *Dreissena*. For 100 samples (23.8%), BBI values decreased when *Dreissena* was excluded. Ninety eight of these (23.3%) decreased with one BBI unit and two (0.5%) with two units.

Biodiversity loss that is not evident at the taxonomic level of the biotic index used, is a matter of bioconservation and not of biological assessment of water quality. Therefore the new genus should be included in the taxa list since it has become part of local biodiversity. A biotic index, *in casu* the BBI, is partly based on a rapid biodiversity survey (expressed as number of taxa) as an indicator of the water quality, not of the ecosystem stability. Furthermore, species that invaded our regions at earlier times were already included in water quality assessment and are nowadays commonly accepted. Therefore, it is recommended to include *Corbicula* in the standard taxa list, despite its potential harmful effects. To obtain insight in the adverse effects of the invasion of this genus, more detailed studies – at species level – are necessary.

The VMM has already added the genus *Corbicula* to its standard list for calculating the BBI, however without assigning a specific tolerance class to it ('–'). In this way, *Corbicula* only affects the BBI through the number of taxa and not

Table 5. Calculation of the BBI of a real (panel A) and a virtual (panel B) sample

Taxa	Abundance	Tolerance class without inclusion of *Corbicula*	Tolerance class if *Corbicula* is included without tolerance class	Tolerance class according to De Pauw & Vanhooren (1983) *sensu stricto*
Panel A. Sample taken by the VMM on 6 May 1998 at sampling site no. VMM-820000 in the Albert Canal at Genk				
Naididae	2	–	–	–
Tubificidae	11	6	6	6
Chironomidae *non-thummi-plumosus*	11	–	–	–
Helobdella	1			
Erpobdella	11	5	5	5
Gammaridae	11	4	4	4
Atyidae	11	–	–	–
Asellidae	1	–	–	–
Cambaridae	2	–	–	–
Bithynia	11	4	4	4
Ancylus	2	3	3	3
Dreissena	11	4	4	4
Sphaerium	11	5	5	5
Corbicula	2		–	4
Valvata	2	4	4	4
Physa	2	4	4	4
Pisidium	2	5	5	5
Ecnomidae	11	–	–	–
Total number of taxa		15	16	16
Lowest tolerance class		3	3	3
Tolerance class frequency		1	1	1
BBI		6	7	7
Water quality class		III (yellow)	II (green)	II (green)
Panel B. Virtual sample				
Tubificidae	100	6	6	6
Chironomidae *thummi-plumosus*	45	6	6	6
Asellidae	20	5	5	5
Erpobdella	4	5	5	5
Gerris	2	5	5	5
Corbicula	50		–	4
Total number of taxa		5	6	6
Lowest tolerance class		5	5	4
Tolerance class frequency		3	3	1
BBI		3	4	5
Water quality class		IV (orange)	IV (orange)	III (yellow)

The first column lists the taxa, the second one the abundances, the third one the tolerance classes if *Corbicula* is not included, the fourth one the tolerance classes if *Corbicula* is included without tolerance class ('–'), and the fifth one the tolerance classes according to De Pauw & Vanhooren (1983) *sensu stricto*.

through its tolerance class, which is also the case for e.g. the taxa of Plathelminthes and most Diptera.

A number of exotic species of Ponto-Caspian origin are invading European watercourses (e.g. Bij de Vaate et al., 2002). Many of these species such as *Dikerogammarus villosus*, belong to a taxon *(in casu* Gammaridae) that is already in the list, while others will have to be included in the list as new taxa, for the same reasons as *Corbicula*. Some of these are very likely to be encountered in Flemish watercourses in the near future. Anticipating this, two Ponto-Caspian taxa should already be added to the list: Ampharetidae (Polychaeta) and Janiridae (Crustacea).

The presence of *Hypania invalida* (Grube, 1860) (Polychaeta, Ampharetidae) was recently reported in the river Meuse (Vanden Bossche et al., 2001). Although not yet encountered in VMM samples, this may be expected in the near future, especially in the Flemish stretch of the river Meuse. Therefore, Polychaeta should be added as a new group, including one taxon, Ampharetidae, with tolerance class '–', the identification level being set at family (as for Oligochaeta). Another Ponto-Caspian invader, *Jaera istri* (Veuille, 1979) (Crustacea, Janiridae) has also recently been encountered in the river Meuse (Usseglio-Polatera & Beisel, 2003), although not collected in VMM samples so far. Consequently, the list of Crustacea should be extended with the family Janiridae, with tolerance class '–'.

List of taxa taken into consideration

There is indeed a growing need to ensure that the BBI-values remain comparable in future, which implies not altering the method itself, but rather clarifying the problems that emerge, to ensure its future application without being inconsistent with the past and current practice. Altering the method itself would imply making old and new applications incomparable; in other words, it would be a different index. The aim of this paper with regard to the BBI was to identify the problems that arose since 1991 and propose solutions to these problems.

Initially, a checklist by Vanhooren et al. (1982) was commonly used as a reference base for taxonomy of the systematic levels used in the BBI calculation. Some additional taxa were added later, e.g. due to the separation of the mollusc genus *Anisus* into *Anisus*, *Armiger*, *Bathyomphalus*, *Gyraulus*, *Hippeutis*, *Planorbis* and *Segmentina*.

Although the original description of the BBI method dates from 1983, the situation in 1991 was chosen as point of reference. At that moment, the aforementioned modifications were already established and commonly accepted. The situation in 1991 was chosen as point of reference for two reasons. The first reason is that at that time a large-scale monitoring network in Flanders was being initiated by the VMM, with the already cited modifications. The second reason is that the keys of the Pauw & Vannevel (1991) are nowadays widely used and accepted as standard reference for taxonomic identification levels with regard to the BBI.

In the previous paragraphs it has been shown that taxonomic modifications and alien invasions may both lead to biased BBI calculations. Although a change of one or two units in BBI (on a 0–10 scale) may seem insignificant, it is not. A small change in BBI may also lead to a change in the quality class (cf. Table 3). This may become (legally) crucial when this quality class boundary is also a quality standard, e.g. the boundary between the ecological quality classes 'good' and 'moderate', the target imposed by the European Water Framework Directive (EU, 2000). Moreover, a standardised assessment method should be unambiguously applicable and produce unbiased results at all times. This underpins the need for establishing a fixed taxa list. Because more exotic taxa can be expected to invade Belgian watercourses in the future, a fixed taxa list may need to be extended later with those taxa. Therefore a proposal for a fixed taxa list should be more likely called a semi-fixed list, leaving the possibility to add new taxa at a later time.

Table 6 is a proposal for a semi-fixed list to be used to calculate the BBI in order to eliminate the discussed calculation inconsistencies. This list contains 221 taxa and can be considered as a semi-fixed list, in the sense that the taxa already in the list cannot be altered (e.g. split up or lumped), but that the list may be extended with possible future invaders when necessary. The list is based on the taxa identification *sensu* De Pauw & Vannevel (1991) with the addition of the Polychaeta family

Table 6. Proposed semi-fixed taxa list of aquatic macroinvertebrates for calculating the BBI in order to avoid inconsistencies

Taxon	Tolerance class
Plathelmintes	
Bdellocephala	–
Crenobia	–
Dendrocoelum	–
Dugesia	–
Phagocata	–
Planaria	–
Polycelis	–
Polychaeta	
Ampharetidae	–
Oligochaeta	
Aelosomatidae	–
Branchiobdellidae	–
Enchytraeidae	–
Haplotaxidae	–
Lumbricidae	–
Lumbriculidae	–
Naididae	–
Tubificidae	6
Hirudinea	
Cystobranchus	5
Dina	5
Erpobdella	5
Glossiphonia	5
Haementeria	5
Haemopis	5
Helobdella	5
Hemiclepsis	5
Hirudo	5
Piscicola	5
Theromyzon	5
Trocheta	5
Mollusca	
Acroloxus	3
Ancylus	3
Anisus	4
Anodonta	4
Aplexa	4
Armiger	4
Bathyomphalus	4
Bithynia	4
Bythinella	4
Corbicula	–

Table 6. (Continued)

Taxon	Tolerance class
Dreissena	4
Ferrissia	3
Gyraulus	4
Hippeutis	4
Lithoglyphus	4
Lymnaea s.l.	4
Margaritifera	4
Marstoniopsis	4
Myxas	4
Physa s.l.	4
Pisidium	5
Planorbarius	4
Planorbis	4
Potamopyrgus	4
Pseudamnicola s.l.	4
Pseudanodonta	4
Segmentina	4
Sphaerium	5
Theodoxus	4
Unio	4
Valvata	4
Viviparus	4
Acari	
Hydracarina s.l.	–
Crustacea	
Argulidae	–
Asellidae	5
Astacidae	–
Atyidae	–
Cambaridae	–
Chirocephalidae	–
Corophiidae	–
Crangonyctidae	–
Gammaridae	4
Grapsidae	–
Janiridae	–
Leptestheriidae	–
Limnadiidae	–
Mysidae	–
Palaemonidae	–
Talitridae	–
Triopsidae	–
Ephemeroptera	
Baetis	3

Continued on p. 146

Table 6. (Continued)

Taxon	Tolerance class
Brachycercus	3
Caenis	3
Centroptilum	3
Cloeon	3
Ecdyonurus	1
Epeorus	1
Ephemera	3
Ephemerella	3
Ephoron	3
Habroleptoides	3
Habrophlebia	3
Heptagenia	1
Isonychia	3
Leptophlebia	3
Metreletus	3
Oligoneuriella	3
Paraleptophlebia	3.
Potamanthus	3
Procloeon	3
Rhitrogena	1
Siphlonurus	3
Odonata	
Aeshna	4
Anax	4
Brachytron	4
Calopteryx	4
Cercion	4
Ceriagrion	4
Coenagrion	4
Cordulegaster	4
Cordulia	4
Crocothemis	4
Enallagma	4
Epitheca	4
Erythromma	4
Gomphus	4
Ischnura	4
Lestes	4
Leucorrhinia	4
Libellula	4
Nehalennia	4
Onychogomphus	4
Ophiogomphus	4
Orthetrum	4
Oxygastra	4
Platycnemis	4

Table 6. (Continued)

Taxon	Tolerance class
Pyrrhosoma	4
Somatochlora	4
Sympecma	4
Sympetrum	4
Plecoptera	
Amphinemura	1
Brachyptera	1
Capnia	1
Chloroperla	1
Dinocras	1
Isogenus	1
Isoperla	1
Leuctra	1
Marthamea	1
Nemoura	1
Nemurella	1
Perla	1
Perlodes	1
Protonemura	1
Rhabdiopteryx	1
Taeniopteryx	1
Hemiptera	
Aphelocheirus	4
Arctocorisa	5
Callicorixa	5
Corixa	5
Cymatia	5
Gerris s.l.	5
Glaenocorisa	5
Hebrus	5
Hesperocorixa	5
Hydrometra	5
Llyocoris	5
Mesovelia	5
Micronecta	5
Microvelia	5
Naucoris	5
Nepa	5
Notonecta	5
Paracorixa	5
Plea	5
Ranatra	5
Sigara	5
Velia	5

Continued on p. 147

Table 6. (Continued)

Taxon	Tolerance class
Megaloptera	
Sialis	–
Coleoptera	
Dryopidae	–
Dytiscidae	–
Elminthidae	–
Gyrinidae	–
Haliplidae	–
Hydraenidae	–
Hydrophilidae	–
Hygrobiidae	–
Noteridae	–
Psephenidae	–
Scirtidae	–
Trichoptera	
Beraeidae	2
Brachycentridae	2
Ecnomidae	–
Glossosomatidae	2
Goeridae	2
Hydropsychidae	–
Hydroptilidae	2
Lepidostomatidae	2
Leptoceridae	2
Limnephilidae	2
Molannidae	2
Odontoceridae	2
Philopotamidae	–
Phryganeidae	2
Polycentropodidae	–
Psychomyidae	–
Rhyacophilidae	–
Sericostomatidae	2
Diptera	
Athericidae	–
Blephariceridae	–
Ceratopogonidae	–
Chaoboridae	–
Chironomidae *non-thummi-plumosus*	–
Chironomidae *thummi-plumosus*	6
Culicidae	–
Cylindrotomidae	–
Dixidae	–
Dolichopodidae	–

Table 6. (Continued)

Taxon	Tolerance class
Empididae	–
Ephydridae	–
Limoniidae	–
Muscidae	–
Psychodidae	–
Ptychopteridae	–
Rhagionidae	–
Scatophagidae	–
Sciomyzidae	–
Simuliidae	–
Stratiomyidae	–
Syrphidae-Eristalinae	7
Tabanidae	–
Thaumaleidae	–
Tipulidae	–

The first column lists the taxa, the second one the associated tolerance classes. *Lymnaea* s.l. = *Lymnaea* or *Stagnicola* or *Radix* or *Galba* ; *Physa* s.l. = *Physa* or *Physella*; *Pseudamnicola* s.l. = *Pseudamnicola* or *Mercuria* ; Hydracarina s.l. = Hydracarina or *Hydrozetes*; *Gerris* s.l. = *Gerris* or *Aquarius*.

Ampharetidae, the Mollusca genus *Corbicula* and the Crustacea family Janiridae. The notation 's.l.' (*sensu lato*) was added to those taxa that comprise one or more taxa in addition to the one actually mentioned. In the case of Hydracarina the notation s.l. already appeared on the original list of De Pauw & Vannevel (1991) because Hydracarina s.l. comprises *Hydrozetes* in addition to Hydracarina s.s. (*sensu stricto*). Because the Belgian Insitute of Normalisation has adopted the BBI as a standard method (IBN, 1984), it is recommended that its method description (NBN T92-402) be extended by including this new semi-fixed taxa list.

Taxa belonging to groups such as Bryozoa, Hydrozoa, Nemertea, Nematoda, Ostracoda and Porifera are not included in the new taxa list. Taxa from these groups are not frequently encountered in macroinvertebrate samples. These groups already did not appear on the original list in De Pauw & Vannevel (1991), and their addition would cause new inconsistencies between BBI calculations, since they may have been present in older samples. This problem does not arise with new, exotic taxa since they were not yet encountered in the older samples. For this reason, the mentioned groups of taxa were not added to the list.

Comparison of the tolerance classes of Table 6 with the indicator groups from Table 2 reveals some inconsistencies as well. *Acroloxus*, having a tolerance class 3, is not included in the appropriate column in Table 2. This is due to the fact that according to Vanhooren et al. (1982), *Acroloxus* belonged to the family Ancylidae, which is included in Table 2 among tolerance class 3. Since *Acroloxus* is now considered as belonging to a separate family (Acroloxidae), it should be included there as well. Furthermore, not only Sphaeriidae should be excluded from the Mollusca mentioned in tolerance class 4, but also *Corbicula*, Ancylidae and *Acroloxus*. All mentioned inconsistencies were corrected and indicated in bold in Table 2.

The proposal for future application of the BBI is therefore as follows:

(1) application of the taxa list from Table 6 with the associated tolerance classes;
(2) calculation of the index value based on all taxa of which more than one individual was found, using Table 2;
(3) determination of water quality class by means of Table 3.

Sampling macroinvertebrates and calculating the BBI is a rigorous task and should be performed with the highest possible care and precision. Along with the calculation method, many other sources of variability exist, such as seasonality (e.g. Hughes, 1978; Furse et al., 1984; Rosillon, 1989; Linke et al., 1999; Humphrey et al., 2000; Reece et al., 2001), operator (e.g. Humphrey et al., 2000) and sampling variation (e.g. Clarke, 2000; Clarke et al., 2002). Due to all these sources of variability, it is difficult to attain a high precision for the BBI. Nevertheless, these other sources of errors are an additional incentive for using a calculation method that is as rigorous as possible.

Conclusion

Lack of consensus on how to deal with taxonomic modifications and invasions of exotic species may lead to inconsistencies in biotic index calculation. This problem could be overcome by using a semi-fixed taxa list. A semi-fixed list of macroinvertebrate taxa including a tolerance class for each taxon is proposed in order to avoid inconsistencies in the calculation procedure of the BBI. This list is based on the taxa identification *sensu* De Pauw & Vannevel (1991) with the addition of the Polychaeta family Ampharetidae, the Mollusca genus *Corbicula* and the Crustacea family Janiridae. It is hoped for that this list may lead to a harmonisation of the BBI calculation practice so that the BBI values can still be compared unambiguously in the future.

Acknowledgements

The authors wish to thank the VMM, in particular S. De Smedt, A. De Winter, J. Mertens and T. Warmoes for stimulating discussions on the subject and for providing useful sampling data, and Dr. B. Goddeeris of the Royal Belgian Institute of Natural Sciences and an anonymous reviewer for their valuable comments.

References

Adriaenssens, V., F. Simons, L. T. H. Nguyen, B. Goddeeris, P. L. M. Goethals & N. De Pauw, 2004. Potential of bio-indication of chironomid communities for assessment of running water quality in Flanders (Belgium). Belgian Journal of Zoology 134: 15–24.

Bervoets, L ., B. Bruylants, P. Marquet, A. Vandelannoote & R. Verheyen, 1989. A proposal for modification of the Belgian Biotic Index method. Hydrobiologia 179: 223–228.

Bij de Vaate, A., K. Jazdzewski, H. A. M. Ketelaars, S. Gollasch & G. Van der Velde, 2002. Geographical patterns in range extension of Ponto-Caspian macroinvertebrate species in Europe. Canadian Journal of Fisheries and Aquatic Sciences 59: 1159–1174.

Bowman, M. F. & R. C. Bailey, 1997. Does taxonomic resolution affect the multivariate description of the structure of freshwater benthic macroinvertebrate communities? Canadian Journal of Fisheries and Aquatic Sciences 54: 1802–1807.

Cairns, J. Jr. & J. R. Bidwell, 1996. Discontinuities in technological and natural systems caused by exotic species. Biodiversity and Conservation 5: 1085–1094.

Clarke, R. T., 2000. Uncertainty in estimates of biological quality based on RIVPACS. In Wright, J. F., D. W. Sutcliffe & M. T. Furse (eds), Assessing the Biological Quality of Fresh Waters: RIVPACS and Other Techniques. Freshwater Biological Association, Ambleside, Cumbria, UK: 39–54.

Clarke, R. T., M. T. Furse, R. J. M. Gunn, J. M. Winder & J. F. Wright, 2002. Sampling variation in macroinvertebrate data and implications for river quality indices. Freshwater Biology 47: 1735–1751.

De Pauw, N. & G. Hawkes, 1993. Biological monitoring of river water quality. In Walley, W. J. & S. Judd (eds), River Water Quality Monitoring and Control. Aston University, Birmingham, UK: 87–111.

De Pauw, N. & S. Heylen, 2001. Biotic index for sediment quality assessment of watercourses in Flanders, Belgium. Aquatic Ecology 35: 121–133.

De Pauw, N. & G. Vanhooren, 1983. Method for biological quality assessment of watercourses in Belgium. Hydrobiologia 100: 153–168.

De Pauw, N. & R. Vannevel (eds), 1991. Macroinvertebrates and Water Quality (in Dutch). Stichting Leefmilieu, Antwerp, Belgium, 316 pp.

DIN, 1990. Biological–Ecological Analysis of Water (group M). Determination of the Saprobic Index (M 2). DIN 38 410, Part 2. Beuth Verlag GmbH, Berlin, Germany.

Dolédec, S., J. M. Olivier & B. Statzner, 2000. Accurate description of the abundance of taxa and their biological traits in stream invertebrate communities: effects of taxonomic and spatial resolution. Archiv für Hydrobiologie 148: 25–43.

Ellis, D., 1985. Taxonomic sufficiency in pollution assessment. Marine Pollution Bulletin 16: 459.

EU, 2000. Directive 2000/60/EC of the European Parliament and of the Council of 23 October 2000 establishing a framework for Community action in the field of water policy. Official Journal of the European Communities L 327 : 1–72.

Furse, M. T., D. Moss, J . F. Wright & D. Armitage, 1984. The influence of seasonal and taxonomic factors on the ordination and classification of running-water sites in Great Britain and on the prediction of their macro-invertebrate communities. Freshwater Biology 14: 257–280.

Gayraud, S., B. Statzner, P. Bady, A. Haybach, F. Schöll, P. Usseglio-Polatera & M. Bacchi, 2003. Invertebrate traits for the biomonitoring of large European rivers: an initial assessment of alternative metrics. Freshwater Biology 48: 2045–2064.

Ghetti, P. F., 1997. Manuale di applicazione Indice Biotico Esteso (I.B.E.). I macroinvertebrati nel controllo della qualità degli ambienti di aque correnti. Provincia Autonoma di Trento, Agenzia provinciale per la protezione degli ambienti, 222 pp.

Guérold, F., 2000. Influence of taxonomic determination level on several community indices. Water Research 34: 487–492.

Hughes, B. D., 1978. The influence of factors other than pollution on the value of Shannon's diversity index for benthic macro-invertebrates in streams. Water Research 12: 359–364.

Humphrey, C. L., A. W. Storey & L. Thurtell, 2000. AUSRIVAS: operator sample processing errors and temporal variability – implications for model sensitivity. In Wright, J. F., D. W. Sutcliffe & M. T. Furse (eds), Assessing the Biological Quality of Fresh Waters: RIVPACS and Other Techniques. Freshwater Biological Association, Ambleside, Cumbria, UK: 143–163.

IBN, 1984. Biological Quality of Watercourses. Determination of the Biotic Index Based on Aquatic Macroinvertebrates, NBN T92-402 (in Dutch and French). Belgian Institute for Normalisation, Brussels, Belgium, 11 pp.

Karr, J. R. & E. W. Chu, 1999. Restoring Life in Running Waters: Better Biological Monitoring. Island Press, Washington, DC, USA, 206 pp.

King, R. S. & C. J. Richardson, 2002. Evaluating subsampling approaches and macroinvertebrate taxonomic resolution for wetland bioassessment. Journal of the North American Benthological Society 21: 150–171.

Lenat, D. R. & V. H. Resh, 2001. Taxonomy and stream ecology – the benefits of genus-and species-level identifications. Journal of the North American Benthological Society 20: 287–298.

Linke, S., R. C. Bailey & J. Schwindt, 1999. Temporal variability of stream bioassessments using benthic macroinvertebrates. Freshwater Biology 42: 575–584.

Lodge, D. M., 1993. Biological invasions – lessons for ecology. Trends in Ecology & Evolution 8: 133–137.

Mack, R. N., D. Simberloff, W. M. Lonsdale, H. Evans, M. Clout & F. A. Bazzaz, 2000. Biotic invasions: causes, epidemiology, global consequences, and control. Ecological Applications 10: 689–710.

Nguyen, L . T. H. & N. De Pauw, 2002. The invasive *Corbicula* species (Bivalvia, Corbiculidae) and the sediment quality in Flanders, Belgium. Belgian Journal of Zoology 132: 41–48.

Olsgard, F., P. J. Somerfield & M. R. Carr, 1998. Relationships between taxonomic resolution, macrobenthic community patterns and disturbance. Marine Ecology Progress Series 172: 25–36.

Reece, P. F., T. B. Reynoldson, J. S. Richardson & D. M. Rosenberg, 2001. Implications of seasonal variation for biomonitoring with predictive models in the Fraser River catchment, British Columbia. Canadian Journal of Fisheries and Aquatic Sciences 58: 1411–1418.

Resh, V. H. & E. P. McElravy, 1993. Contemporary quantitative approaches to biomonitoring using benthic macroinvertebrates. In Rosenberg, D. M. & V. H. Resh (eds), Freshwater Biomonitoring and Benthic Macroinvertebrates. Chapman & Hall, New York, USA: 159–194.

Roach, A. C., A. R. Jones & A. Murray, 2001. Using benthic recruitment to assess the significance of contaminated sediments: the influence of taxonomic resolution. Environmental Pollution 112: 131–143.

Rosillon, D., 1989. The influence of abiotic factors and density-dependent mechanisms on between-year variations in a stream invertebrate community. Hydrobiologia 179: 25–38.

Schmidt-Kloiber, A. & R. C. Nijboer, 2004. The effect of taxonomic resolution on the assessment of ecological water quality classes. Hydrobiologia 516: 269–283.

Stubauer, I. & O. Moog, 2000. Taxonomic sufficiency versus need for information – comments based on Austrian experience in biological water quality monitoring. Verhandlungen der Internationale Vereinigung für Theoretische und Angewandte Limnologie 27: 2562–2566.

Torchin, M. E., K. D. Lafferty, A. P. Dobson, V. J. McKenzie & A. M. Kuris, 2003. Introduced species and their missing parasites. Nature 421: 628–630.

Tuffery, G. & J. Verneaux, 1968. Méthode de détermination de la qualité biologique des eaux courantes. Exploitation codifiée des inventaires de la faune du fond. Ministère de l'

Agriculture (France), Centre National d'Etudes techniques et de recherches technologiques pour l'agriculture, les forêts et l'équipement rural 'C.E.R.A.F.E.R.', Section Pêche et Pisciculture, 23 pp.

Usseglio-Polatera, P. & J. N. Beisel, 2003. International biomonitoring of the Meuse River: spatio-temporal analysis of benthic communities (1998–2001). Final report (in French). LBFE, Equipe de Démoécologie, Université de Metz, France, 107 pp. + appendices.

Vanden Bossche, J. -P., F. Chérot, E. Delooz, F. Grisez & G. Josens, 2001. First record of the Pontocaspian invader *Hypania invalida* (Grube, 1860) (Polychaeta: Ampharetidae) in the river Meuse (Belgium). Belgian Journal of Z oology 131: 183–185.

Vanhooren, G., A. Ovaere, H. De Schutter & C. Boelen, 1982. Coded list of aquatic macroinvertebrates (in Dutch and French). Instituut voor Hygiene en Epidemiologie, Brussels, Belgium, 58 pp. + appendix.

Van Den Brink, F. W. B., G. Van Der Velde & A. Bij de Vaate, 1991. Amphipod invasion on the Rhine. Nature 352: 576.

Verdonschot, P. F. M., 2000. Integrated ecological assessment methods as a basis for sustainable catchment management. Hydrobiologia 422/423: 389–412.

Warwick, R. M., 1988. The level of taxonomic discrimination required to detect pollution effects on marine benthic communities. Marine Pollution Bulletin 19: 259–268.

Williams, P. H. & K. J. Gaston, 1994. Measuring more of biodiversity: can higher-taxon richness predict wholesale species richness? Biological Conservation 67: 211–217.

Woodiwiss, F. S., 1964. The biological system of stream classification used by the Trent River Board. Chemistry and Industry 14: 443–447.

Woodiwiss, F. S., 1980. Biological Monitoring of Surface Water Quality. Summary Report. Commission of the European Communities, 45 pp.

Hydrobiologia (2005) 542:151–163
H. Segers & K. Martens (eds), Aquatic Biodiversity II
DOI 10.1007/s10750-004-7423-1

A large-scale stream benthic diatom database

Véronique Gosselain[1,*], Michel Coste[2], Stéphane Campeau[3], Luc Ector[4], Claude Fauville[1], François Delmas[2], Markus Knoflacher[5], Magdalena Licursi[1], Frédéric Rimet[4], Juliette Tison[2], Loïc Tudesque[4] & Jean-Pierre Descy[1]

[1]*Facultés Universitaires N-D de la paix, Department of Biology, URBO, rue de Bruxelles, 61, 5000 Namur, Belgium*
[2]*Cemagrel, Bordeaux (France)*
[3]*Université du Québec à Trois-Rivières, Section of Geography Trois-Rivières (Quebce, Canada)*
[4]*Centre de Recherche Public Gabriel Lippmann, CREBS, Luxembourg (Luxembourg)*
[5]*Austrian Research Center Seibersdorf, Vienna (Austria)*
(**Author for correspondence: E-mail: veronique.gosselain@fundp.ac.be*)

Key words: diatom taxonomy, environmental conditions, relational database, auto-ecology, diatom typology

Abstract

A relational database linking benthic diatom records, taxonomic nomenclature including synonyms, and corresponding environmental data has been built in MS Access. It allowed flexible and long-term use of a relatively important amount of data (~3000 records) gathered in the framework of the EC-funded PAEQANN project, gathering precise and documented information both about benthic diatoms and quantitative or semi-quantitative environmental data. Such a database has been shown to be a useful tool for the definition of benthic diatom typology at a multi-regional scale, the prediction of the impact of environmental characteristics on the structure of diatom communities, and additionally for a new insight on the auto-ecology of some taxa. This database could serve as a template for further work on diatoms and, after some implementation, on other freshwater communities. It could also be the basis for wider typology of stream diatoms, extended to other regions.

Introduction

In a context of environmental changes, there is an increasing need to organise information about biodiversity and community structure in natural or near-natural conditions, and to identify changes due to natural factors from those driven by changes from human activities. Benthic diatoms have long been recognised as excellent indicators of ecological status of water bodies (e.g. Descy, 1979; McCormick & Cairns, 1994; Prygiel et al., 1999). Therefore, diatoms have been used in water quality monitoring programs, in which there is, however, a demand from managers to simplify and reduce as much as possible identification level (e.g. Prygiel et al., 1996), to make the techniques accessible to non-specialists having received minimal, but adequate training.

Most studies on benthic diatoms have been carried out at a regional level, and only in the US variation of diatom composition along various gradients at a continental scale has been addressed (Pan et al., 1996, 2000; Potapova & Charles, 2002). Both for scientific and applied issues, structuring relevant and quantitative information about auto-ecology of diatom at a multi-regional level would certainly be valuable. This would allow, for instance, gathering in a single data matrix a large number of diatom records and corresponding environmental information from various regions, allowing statistical analysis and development of predictive models, and making information available for progress in ecological research.

Generalised databases have been developed for various purposes: paleo-environmental reconstruction (EDDI: Battarbee et al., 2000, 2001; DPDC: Sullivan & Charles, 1994), taxonomy (e.g. Kusber & Jahn, 2003; Index Nominum Algarum), collections (Alga Terra: Jahn et al., 2004; HANNA; The UCMP Collection Catalogue), images (e.g. ANSP; PID: O'Kelly & Littlejohn, 1994–2004; BGSU Algae Image Laboratory), and identification (OMNIDIA: Lecointe et al., 1993, 1999; Joynt & Wolfe, 1999).

In the framework of a European research program aiming at predicting aquatic communities in order to assess aquatic ecosystem quality and define river restoration objectives, we developed a diatom relational database at a European multi-regional scale. This database, built in MS Access, allowed flexible use and processing of a rather important amount of data (~3000 records). The choice to build a relational database instead of using spreadsheets arose when facing the two following methodological aspects, which were identified as needing particular attention: (1) the uneven quality of the environmental data within the database, and (2) the different nomenclature used for diatom data originating from different institutes and collected at different times, as well as the different taxonomical precision achieved by different institutes. General advantages of databases as compared to spreadsheets are, first, that information is partitioned over different tables, in order to be stored only once and not repeated for each record, and in a sequential format, avoiding empty cells in tables. This leads to a considerable reduction of disk space, in addition to reduction of errors. Second, queries allow an easy extraction of information, and are stored instead of the resulting tables. In addition, data can be stored along with meta-data about their origin and quality, and a link between biotic (diatom records) and environmental data can easily be established. Finally, in an Access database, all raw data can be stored together, which is not always possible in an Excel spreadsheet due to row and column number limitation. MS Access has been chosen as relatively easy to learn and use by non-IT-specialists.

The aim of this paper is to present the general structure of an Access database on stream diatoms and environmental conditions, which could serve as template for further applications in algal ecology, but also for other freshwater communities. A brief presentation of the actual dataset is given, as well as applications carried out. Limitations, and further possible and/or needed implementations and uses are discussed.

The database presented here was used as a scientific tool and it was not intended to put it on the Internet for public use. No user interface was built, neither to add nor to extract data. A lighter version was nevertheless created for use in the PAEQANN tool that was developed in the project, which is available for downloading on the Internet (http://aquaeco.ups-tlse.fr/).

Materials and methods

Benthic diatom records and corresponding environmental data under consideration have been gathered during the EC-funded PAEQANN project (EVK1-CT1999-00026). Part of the records and data were already available from previous studies carried out in several regions of Belgium, France, Luxembourg, and Austria. Another part was obtained by sampling new river sites, mostly located in regions which had not, or incompletely, been sampled in past studies. Diatom sampling, slide preparation, and counting under the microscope followed standard procedures (AFNOR, 2000; CEN 2002, 2004).

As diatom records originated from different laboratories and periods of time, an important harmonisation of the taxonomy had to be done. In addition, some slides were re-examined in order to take into account taxonomical updates, and some records counted with low level of discrimination between taxa were checked in order to distinguish morphologically close taxa with different ecology. Diatom identifications were based mainly on the Süßwasserflora von Mitteleuropa (Krammer & Lange-Bertalot, 1986, 1988, 1991a, b). Harmonisation of taxonomy and identification level was carried out at the scale of the entire database (see below).

Diatom records were all characterised by PSI index (Polluo-Sensitivity Index; Coste *in* Cemagref, 1982), as this index was used in further analyses. PSI is a water quality index, which is calculated from relative abundances of benthic diatoms collected in a given site. In the PSI system, a large number of stream diatoms have an indicator score, according to their sensitivity to pollution and ecological amplitude. PSI has been tested several times in different countries as, Finland, Germany, Poland, Portugal (Prygiel et al., 1999), and is usually considered as a reference method for water quality assessment using diatoms (Descy & Coste, 1991). It has been calculated using the OMNIDIA software (version 3.2.; Lecointe et al., 1993, 1999). When not available initially, records were re-encoded in OMNIDIA in order to generate the PSI. PSI was considered as an independent estimate of water quality, more reliable than physical and chemical water analysis, and was used to select records for defining reference conditions (see details and full discussion in Gosselain et al., in press).

Environmental data were collected at the time of sampling and/or provided by authorities in charge of monitoring and management of the sampled rivers. We estimated that the best expression of the water quality data to be considered when dealing with benthic diatom assemblages is a 3-month average of the measurements made at the sampling site. Whenever detailed data were available, environmental data were averaged over the 3-month period before sampling. However, the number of data taken into account varied greatly, depending on monitoring frequency. Due to practical issues, it was not possible however to store original values of environmental data in the database and process them through queries. Only averages were kept with information about the origin of data (see below).

The database was built using the MS Access 2000 software, after drawing a logical model.

The database structure

The database consists of 11 main tables and 13 dictionaries. Relationships have been created between and within tables to facilitate the organisation of information. In most of the tables, new IDs have been created to identify unequivocally the records. Introduction of possible duplicates was checked through appropriate queries. Referential integrity has been applied to relationships between tables to guarantee correct links between parent and child data.

The overall structure of the database, i.e. its logical model, is presented on Figure 1. Each table and each relationship have been defined in order to describe the most precisely as possible the content of the table or the nature of the relationship between two tables (or within a table), respectively (see example on Table 1). The full list of tables (entities) and fields (attributes) with their definition and description is available on the PAEQANN web site at http://aquaeco.ups-tlse.fr/Results/Data/DiatomsDatabaseAceess.htm. The main features of the diatom database are described hereafter.

Tables of the main path

The main path corresponds to a succession of tables from the river to the diatom counted, going through the site visited, the visit(s), i.e. the sampling occasions, the diatom samples taken on those occasions, the slides prepared from those samples, the diatom records obtained from microscope observation and counting, and the detail of diatom taxa counted.

Two different tables allowed defining the sampling position: SITE and STATION. The site corresponds to a certain area, homogenous for all environmental characteristics (water quality, habitat, . . .), which can contain several sampling points, the stations. The station is a precise sampling point, identified in a measurement network and/or precisely defined by geographic coordinates. The distinction was mainly needed as sampling points for diatoms might be slightly different than sampling points for water chemistry while both are perfectly compatible and could be used as corresponding data. In addition, a same station could be part of more than one measurement network. The station table was thus linked to the STATION_CODE table through a one-to-many relationship. Each station code ID was related to the dictionary of STATION NETWORK. Theoretically, at least some of the environmental variables of sites (see below) should have been related to the station instead, e.g. the geographical coordinates. Nevertheless, as the station concept was not taken into account when building the database and was added afterward, the rule was to describe the diatom sampling point as the 'site'. The whole

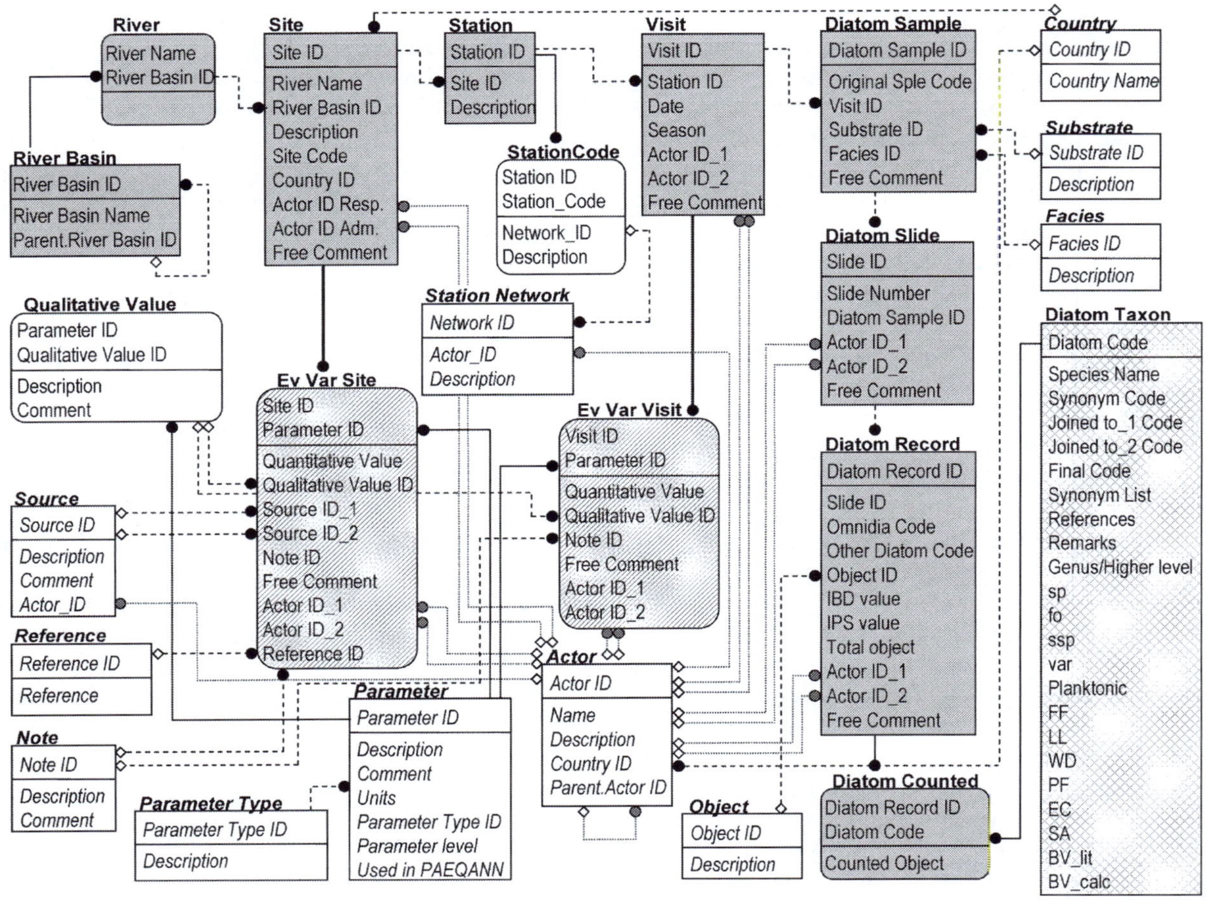

Figure 1. Logical model of the Diatom PAEQANN database. Legend: Ev_Var_Site: environmental variables related to site; Ev_Var_Visit: environmental variables related to a visit to the site; sp: indicates if the taxon is noted sp.; fo: form; ssp: sub-species; var: variety; Planktonic: indicates if the taxon is planktonic; FF: geometric shape; LL: length; WD: width; PF: depth; EC: fourth dimension; SA: surface area; BV_lit: biovolume as found in the literature; BV_calc: biovolume as calculated through the macro F_Algamica.

potential of the distinction between site and station has thus been by-passed in the present database, and could be easily implemented in further applications.

The Actor dictionary was linked to most of the tables. It contains name and contact information about all people and/or entities that were involved in a specific action in the project. This includes partners of the project, people actually involved in each step of the sample processing (visiting and collecting, mounting slides, counting diatoms), entities providing environmental data, administrative authorities from which depends a site or a sampling network.

Environmental data

Two different tables gathered the environmental data. One was linked to the site (Ev_Var_Site) and comprised environmental features of the site, i.e. characteristics that are not expected to change with time, provided that major physical changes (regulation, dredging, ...) are not made. The second table was linked to the visit (Ev_Var_Visit), containing environmental data that usually vary with time. Typically, environmental data associated with the visit were water quality data. Nevertheless, as the database comprised data ranging over more than 10 years, data as

Table 1. Example of (panel a) entity and (panel b) attribute definition in the Diatom PAEQANN database

Panel a

Entity name	Entity definition
ACTOR	An ACTOR is a person or an entity involved in a specific ACTION in the project
DIATOM TAXON	The DIATOM TAXON is the dictionary of all the species encountered in the PAEQANN project, that indicates synonyms and joined species, as well as dimensions and biovolumes, as found in the literature or calculated by the macro F-Algamica
NOTE	A NOTE is an information about the origin of the value, indicating whether the value has been calculated or estimated and whether it is an a unique value or an average
PARAMETER	A PARAMETER is a physical or chemical parameter
PARAMETER TYPE	A PARAMETER TYPE classifies parameter according to the type of information it gives, i.e. General, Stream or Water characteristics
VISIT	A VISIT made to the site at a specific date. Measurements and sampling(s) for further analyses and diatom counting are carried out. The visit always corresponds to the visit for diatom sampling.

Panel b

Attribute name	Attribute definition	Attribute required	Attribute entity name
Actor ID	An ACTOR ID is a code given to a person or an entity involved in a specific action in the project	No	ACTOR
		No	DIATOM RECORD
		No	EV VAR SITE
		No	EV VAR VISIT
		No	STATION NETWORK
		No	SOURCE
	ACTOR ID of the person or entity who carried out the diatom sampling	No	VISIT
Actor ID Adm	ACTOR ID Adm identifies the administrative entity which the site belongs to	No	SITE
Actor ID Resp	ACTOR ID Resp is a the code of the person in charge of the PAEQANN sub-database comprising this site	No	SITE
Diatom Code	A DIATOM CODE is a 4 letter code given to a diatom taxon, according to the OMNIDIA species list	No	DIATOM COUNTED
		Yes	DIATOM SPECIES LIST
Diatom Record ID	A DIATOM RECORD ID is a unique number of diatom record for the entire database	Yes	DIATOM COUNTED
		Yes	DIATOM RECORD
Parameter ID	A PARAMETER ID is a string abbreviation for a parameter, if necessary informing of the unit used when several units are possible in the database	Yes	EV VAR SITE
		Yes	EV VAR VISIT
		Yes	QUALITATIVE VALUE
		Yes	VISIT
		No	PARAMETER
		Yes	EV VAR VISIT

'presence of hydropower plant in the upstream 10 km' were encoded in the Ev_Var_Visit table.

The Parameter dictionary listed all parameters; the parameter_level field indicated the table in which data would be recorded,

Ev_Var_Site or Ev_Var_Visit. When entering new data in those tables, a control query was checking that they were entered in the right table. In addition, as environmental data could be quantitative or qualitative, the Qualitative_Value table listed all parameters and the qualitative values they can take; zero was used for quantitative or missing value. Additional information on data (origin, quality, ...) were given in the fields Note, Source, Reference, and Free Comments in one or both environmental tables. The first three fields were linked to the corresponding dictionaries (see Fig. 1). Information gathered in those fields was thus clearly defined and systematic; it could be used in a query or be sorted (see example in Table 2). In particular, the Note table contained information about the origin of values, indicating whether they had been calculated or estimated and whether it was a single value or a mean. In fact, as the more relevant environmental data chosen to be used with diatom data were means over 3 months, only values as close as possible as 3-month means were entered in the database.

Diatom tables and associated dictionaries

Diatom river samples, slides and records were placed in three different tables in order to allow multiple or sub-samples in each cases. In each

Table 2. Example of environmental data from the Ev_Var_Visit table, with information about their origin and quality

VISIT_ID	NH4 (mg/l N)	Note	Free comment	Date	Actor
2542	1.040	ASY	Tilleur, Sept. 27, 1979 & Sept. 20, 1980	15-sept-79	RW
2701	3.000	ASY	Sept. 1978	15-sept-79	RW
2685	7.500	ASY	Roselies, Sept. 18, 1978 & Oct. 19, 1978	15-sept-79	RW
2541	0.710	ASY	Ombret, Sept. 27, 1979 & Sept. 20, 1980	15-sept-78	RW
2532	0.265	ASY	Dinant, Sept. 27, 1979 & Sept. 20, 1980	15-sept-79	RW
2546	1.280	ASY	Cheratte, Sept. 27, 1979 & Sept. 20, 1980	15-sept-79	RW
379	0.070	ASY	1979	17-juin-80	AERMC
922	5.462	AY		10-sept-97	AERMC
921	0.056	AY		10-sept-97	AERMC
918	0.031	A2M		25-sept-92	AERM
917	0.023	A2M		02-juil-92	AERM
916	0.290	A2M		26-sept-92	AERM
923	0.150	AY		09-sept-97	AERMC
914	0.155	A2M		25-sept-92	AERM
926	0.079	AY		09-sept-97	AERMC
913	0.125	A2M		02-juil-92	AERM
912	0.405	A2M		26-sept-92	AERM
881	0.050	M1		25-juil-01	CMGRF
1031	0.350	A3M		26-août-97	AELB
1030	0.198	A3M		06-sept-96	AELB
1029	1.167	A3M		30-août-99	AELB
2734	0.120	A4F		01-juin-99	RW
2735	0.120	A4F		15-oct-99	RW
2764	0.020	A4F		24-sept-99	RW

Note description: ASY: average on available values of the same season of other years; AY: yearly average = average on every month available for the sampling year; A2M: average on 2 months, the sampling month and the previous one; A3M: average on the 3 months before and including diatom sampling; A4F: average on 4–5 weeks, up to 7 (rarely more) before and up to 10 days after the diatom sampling date (or sometimes up to the next month), corresponding to an average on 1–5 values + value measured on the field at the occasion of the diatom sampling; M1: measured once;
Actor description: AELB: Agence de l'Eau Loire-Bretagne; AERMC: Agence de l'Eau Rhône-Méditerrannée-Corse; CMGRF: Cemagref; RW: Région Wallonne.

table, a field has been allocated to the original code. This would allow an easy identification of the actual sample when needed. A unique code for the database was added. In the DIATOM_SAMPLE table, information of substrate and facies were noted. In fact, samples, while mainly collected on rocks, could also have been originating from other substrates (plants, sediments,...), which could have an impact on further analysis of assemblages. In addition, lotic and/or lentic facies were sampled. Again information was kept in order to allow further selection of cases through queries. As an example, in further analysis (e.g. Gosselain et al., 2003, in press), only slides from samples taken from rocks in lotic facies were considered.

The DIATOM_RECORD table provided general information about the record. It also indicated objects actually counted: single diatom valves, entire frustules [2 valves], or indifferently single valves and frustules. Additionally, PSI and IBD (Prygiel & Coste, 1999) indexes were given, for each record and when available, respectively. Actually, a more flexible and generalised system would be to create a dictionary of indication methods to which to refer. This would prevent loosing information from the originally available diatom records, when values for other indexes were available.

The DIATOM_COUNTED table was created to solve a one-to-many relationship from both DIATOM_RECORD and DIATOM_TAXON tables (resolution table).

Diatom taxon dictionary

One of the most important and useful operations carried out when building the database was the diatom dictionary. Each taxon was entered in the table using its name as in the initial record. The codes (DIATOM_CODE field) used to identify the taxa followed the codes defined in the OMNIDIA software 3.2. When this code was used originally, it was entered as recorded initially; this led to a few cases where a single taxon was identified by two different codes. The coding system consists of four letters that indicate genus (one letter), species and varieties (the 3 last letters). At the present stage, the SPECIES_NAME field contains both the taxon name and authorities, due to encoding in the original file. A second code field (SYNONYM_CODE) indicates taxonomic transfer to the code associated with the most current taxon name. This nomenclature mainly followed recent updates of diatom taxonomy (e.g. Round et al., 1990) compiled from recent journals like Diatom Research, Diatom Monographs, or taxonomic listings (Kusber & Metzeltin, 2001; Kusber & Jahn, 2003), as provided in the OMNIDIA software 3.2. For example (Fig. 2 & Table 3), *Achnanthes biasolettiana* Grunow var. *biasolettiana* Grunow in Cleve & Grunow, which has the code ABIA, has the associated SYNONYM_CODE ADBI, indicating that *Achnantidium biasolettianum* (Grunow in Cl &

DIATOM_CODE	SPECIES_NAME	SYNONYM	JOINED_TO_1	JOINED_TO_2	FINAL_CODE
AACU	Amphora acutiuscula Kutzing				☑
AAEQ	Amphora aequalis Krammer				☑
AAFF	Achnanthes affinis Grunow in Cleve & Grunow (Achnanthidium)	ADMF	ADMI	ADMI	☐
AAMB	Aulacoseira ambigua (Grunow) Simonsen				☑
AAMO	Achnanthes amoena Hustedt	KAMO			☐
AATG	Achnanthidium alteragracillima (Lange-Bertalot) Round & Bukhtiyarova				☑
AATO	Achnanthes atomus Hustedt				☑
AAUS	Achnanthes austriaca Hustedt	EUAU			☐
ABAH	Achnanthes bahusiensis (Grunow) Lange-Bertalot	ABHS			☐
ABHS	Astartiella bahusiensis (Grun.) Witkowski, Lange-Bertalot & Metzeltin				☑
ABIA	Achnanthes biasolettiana Grunow var. biasolettiana Grunow in Cleve & Grunow	ADBI			☐
ABIN	Achnanthes brevipes Agardh var.intermedia (Kutz.) Cleve		ABRE	ABRE	☐
ABIO	Achnanthes bioretii Germain	PBIO			☐
ABPA	Achnanthes brevipes Agardh var.parvula (Kutz.) Cleve	APAR			☐
ABRE	Achnanthes brevipes Agardh var. brevipes				☑
ABRY	Adlafia bryophila (Petersen) Moser Lange-Bertalot & Metzeltin				☑
ABSU	Achnanthes biasolettiana Grunow var. subatomus Lange-Bertalot	ADSU			☐
ABTH	Achnanthes biasolettiana Grun. var. thienemannii (Hustedt) Lange-Bertalot				☑
ACAC	Amphora coffeaeformis (Ag.) Kutzing var.acutiuscula (Kutzing) Rabenhorst		ACOF	ACOF	☐
ACAR	Achnanthes carissima Lange-Bertalot				☑
ACBO	Achnanthes clevei Grunow var. bottnica Cleve		KCLE	KCLE	☐
ACBR	Achnanthes conspicua A.Mayer var. brevistriata Hustedt	ACON			☐
ACEN	Achnantheiopsis engelbrechtii (Cholnoky) Lange-Bertalot				☑

Figure 2. Example of data records of the DIATOM_DICTIONARY table.

Table 3. Example of synonyms and joined taxa as extracted from the diatom dictionary of the Diatom PAEQANN database

F_CODE	F_SYN	DIATOM_CODE	SPECIES_NAME
ADBI	ADBI	ABIA	*Achnanthes biasolettiana* Grunow var. *biasolettiana* Grunow in Cleve & Grunow
		ADBI	*Achnanthidium biasolettianum* (Grunow in Cleve & Grunow) Round & Bukhtiyarova
	ADBT	ADBT	*Achnanthidium biasolettianum* (Grunow) Round & Bukhtiyarov fo. teratogene
ADMI	ADMF	AAFF	*Achnanthes affinis* Grunow in Cleve & Grunow
		ADMF	*Achnanthidium minutissima* (Kützing)Czarn. var. *affinis* (Grunow) Bukhtiyarova
		AMAF	*Achnanthes minutissima* Kützing var. *affinis* (Grunow) Lange-Bertalot
	ADMI	ADMI	*Achnanthidium minutissimum* (Kützing) Czarnecki
		AMIC	*Achnanthes microcephala* (Kützing) Grunow
		AMIN	*Achnanthes minutissima* Kützing var. *minutissima* Kützing
	ADMT	ADMT	*Achnanthidium minutissimum* (Kützing) Czarnecki fo.teratcgene
	ADSA	ADSA	*Achnanthidium saprophila* (Kobayasi & Mayama) Round & Bukhtiyarova
		AMSA	*Achnanthes minutissima* Kützing var. *saprophila* Kobayasi & Mayama
	AMJA	AMJA	*Achnanthes minutissima* Kützing var. *jackii* (Rabenhorst) Lange-Bertalot
		AMRO	*Achnanthes minutissima* Kützing var. *robusta* Hustedt
ADMS	ADMM	ADMM	*Adlafia minuscula* var. *muralis* (Grunow) Lange-Bertalot
		NMMU	*Navicula minuscula* Grunow var. *muralis* (Grunow) Lange-Bertalot
	ADMS	ADMS	*Adlafia minuscula* (Grunow) Lange-Bertalot
		NMIS	*Navicula minuscula* Grunow in Van Heurck
	CMNO	CMNO	*Craticula minusculoides* (Hustedt) Lange-Bertalot
		NMNO	*Navicula minusculoides* Hustedt
ADSU	ADSU	ABSU	*Achnanthes biasolettiana* Grunow var. *subatomus* Lange-Bertalot
		ADSU	*Achnanthidium subatomus* (Hustedt) Lange-Bertalot

DIATOM_CODE: code given to a taxon, following the Omnidia software 3.2.; F_SYN: code associated to the "final synonym", the most current name, to which the taxon is transferred; F-CODE: "final code" to which the taxon is associated, in the framework of the PAEQANN project, in order to harmonise level of identification at the scale of the whole database.

Grun.) Round & Bukhtiyarova is the current name for the diatom in question.

Identification levels were sometimes highly different between institutes, according to the purpose of the original countings. Consequently, both taxonomic and identification levels had to be harmonised prior to analysis. When allowed by the ecology of taxa, merging of distinct taxa was proposed in order to reach a single level of identification for the entire database. On the basis of expert knowledge, two levels of merging were proposed, in the fields (1) JOINED_TO_1, corresponding to the harmonised level for all but the Austrian data, and (2) JOINED_TO_2, corresponding to the more severe and common level at the scale of the entire database. In some cases, slides were re-examined in order to refine the taxonomy when dominant taxa of different ecology had been counted together initially. That was the case to make the distinction between *Achnanthes biasolettiana* Grunow var. *biasolettiana* Grunow in Cleve & Grunow and *Achnanthes biasolettiana* Grunow var. *subatomus* Lange-Bertalot, which were counted together by one of the laboratories. In some cases, nevertheless, recounting was not considered, when a taxon was rare, or when forms possibly corresponding to distinct taxa were not unanimously recognised. This was the case for *Achnanthidium saprophila* (Kobayasi et Mayama) Round & Bukhtiyarova that was joined to *Achnanthidium minutissimum* (Kütz.) Czarnecki (Table 3). *A. saprophila*, counted by one of the laboratories, however accounted for only 1.2% of all objects counted as *A. minutissima*, in 7.5% of the corresponding records, while true *A. minutissima* accounted for 97.9% of objects, in 85.9% of corresponding records. The 0.9% counted objects remaining were either *Achnanthes minutissima* Kützing var. *jackii* or *Achnanthidium minutissima* (Kützing) Czarn. var. *affinis*; they were present in 6.6% of the records. It is to be noted that decision about taxa to be joined should be reconsidered each time new sets of data would be added in the database.

Two fields gave the list of synonyms of the taxon, and the references and date of publication, respectively. A few additional fields allowed to indicate if the taxon was (1) a genus or a higher taxonomical level, (2) noted sp., (3) a form, (4) a sub-species, (5) a variety, (6) planktonic. Finally, the closest geometrical shape of the taxon was identified and values of linear dimensions, surface area and biovolumes were provided using published size data (Krammer & Lange-Bertalot, 1986–1991; OMNIDIA software 3.2, *op. cit.*), at least for taxa to be used in further analysis. The fields SA and BV_lit gave the surface area and biovolume as given in the literature, while the BV_calc field gave the biovolume as calculated through the F_ALGAMICA macro, following calculation provided in the counting program ALGAMICA (Gosselain & Hamilton, 2000; http://Algamica.ibelgique.com). In fact, diatom biovolumes spanned at least three orders of magnitude and, as long recognised by planktonologists, biovolume of an algal unit is directly related to its carbon biomass, as well as its nutrient uptake and growth rates. Therefore biovolume is particularly relevant from a functional and ecological point of view. However, despite their significance, biovolume and carbon biomass have not commonly been used in studies on benthic algae (see nevertheless Ghosh & Gaur, 1998; Sabater et al., 1998; Wargo & Holt, 1998; Mayer & Galatowitsch, 2001; Peterson et al., 2001; Gosselain et al., 2003).

Actual dataset

The database presented here contains 2847 diatom records associated with corresponding environmental variables, from 1472 sites and 696 rivers, covering 118 river basin systems and 4 countries (see http://aquaeco.ups-tlse.fr/Results/Data/Diatomsmain.htm for details). It comprised 59 variables in addition to geographic coordinates, among which 23 were actually used for the benthic diatom application (Table 4; http://aquaeco.ups-tlse.fr/Results/Data/DiatomEnvVar.htm, for the complete list of variables). As some water quality data were far from the ideal 3-month averages (Table 5), information about the values was helpful for further interpretation of results. The diatom dictionary presently contains a total of 1719 different codes and names, corresponding to 1255 different taxa.

Queries were run in order to retrieve data for further analysis, in particular to put together data constituting a data matrix. For diatoms, the queries allowed to carry out a first pre-treatment of raw data: selection of records comprising enough counted objects, selection of species by rejecting too rare taxa or taxa with too low frequency of occurrence.

In order to allow averaging diatom samples but only from same substrate and facies, a new ID had to be created from queries. This ID identified cases as used in further analyses, where a single record corresponded to the mean diatom record (practically one or rarely two) from a single visit, in a single facies and on a single substrate; mean values of PSI and IBD were calculated.

Discussion and conclusion

The Diatom PAEQANN database has been shown to be useful to tackle multiple practical issues both about diatom taxonomy, and multiple origins and references of related environmental data. The database was thus the primary tool that allowed further analyses at a multi-regional scale while keeping track of all original information. This was needed in a concern of reference and quality control of the data, e.g. allowing checking

Table 4. List of main environmental variables collected

Var.	Description (units)	Basic statistics/categories								
		Min.	Max.	Median	Mean	75%ClUp	75% ClLo	SD	Var	*n*
Quantitative variables										
ALT	Altitude (m)	1	2660	203	257	265	248	276	76152	1472
SLOPE	Slope (m km^{-1})	0.0	133.3	1.6	4.23	4.51	3.95	9.25	85.49	1472
DIST	Distance from source (km^{-1})	0.0	964.42	29.80	66.97	70.26	63.68	109.66	12025.80	1471
CAreaS	Catchment surface area up to the site (km^2)	0.0	115413	241	2299	2558.04	2039.98	8633.0	$7.4529\ 10^7$	1471
ALK_meq	Alkalinity (meq l^{-1})	0.03	12.84	2.20	2.65	2.70	2.61	1.92	3.69	2612
pH	Water pH	3.8	10.04	7.76	7.72	7.73	7.71	0.50	0.253	2755
COND_20	Conductivity at 20 °C (μS cm^{-1})	7.65	24500	383.33	493.50	513.66	473.33	919.88	846171	2755
TEMP	Water temperature (°C)	2.3	27.9	16.0	15.9	16.0	15.8	4.2	17.64	2755
DO	Dissolved oxygen (mg l^{-1})	0.10	26.45	9.35	9.14	9.19	9.09	2.26	5.12	2749
DOC	Dissolved organic carbon (mg $^{-1}$)	0	153.75	2.90	3.83	3.95	3.71	5.24	27.42	2701
NO_3	Nitrate (mg NO_3^--N l^{-1})	0	37.00	2.50	3.36	3.43	3.28	3.31	10.97	2742
NO_2	Nitrite (mg NO_2^--N l^{-1})	0	3.028	0.03	0.14	0.10	0.09	0.20	0.04	2740
NH_4	Ammonium (mg NH_4^+-N l^{-1})	0	35.93	0.08	0.65	0.74	0.56	2.38	5.66	2744
PO_4	Phosphate (mg PO_4^{3-}-P l^{-1})	0	14.03	0.07	0.26	0.29	0.23	0.78	0.61	2738
Semi-qualitative or qualitative variables		Categories								
Season	Season	SP = spring, SA = autumn, SW = winter; coded as 2 dummy variables								
Geol	Geology	'mudstone', 'limestone', 'sandstone', 'granitic', 'quaternary', 'mixed and other'; coded as 5 dummy variables								
Morph	River morphology	1 = natural, 2 = partly channelized, 3 = totally channelized								
Level	Water level	1 = lowest water levels, 2 = mid levels, 3 = flood levels								
Shad	Shading at the sampling site	1 = closed, 2 = mid, 3 = opened								
Hydropwr	Hydropower installation within 10 km upstream the sampling site	Yes or no								
RedFlow	Reduction of flow installation within 10 km upstream the sampling site	Yes or no								
Vel	Water velocity	1: <0.2 m s^{-1}, 2: 0.2–0.5 ms^{-1}, 3: >0.5 m s^{-1}								

Note: The zero (0) value for minima could either be actual zero or mean 'below detection limit.'

Table 5. Summary of data available for water quality in the PAEQANN database, according to the NOTE given to the data

Average on 3 months	731	27%
Other average	1186.4	43%
Single value	562.56	21%
Estimated value	256.67	9%
Without note	14	1%
	2741.3	100%

Numbers given here are mean on all parameters.

outliers data in analysis (Table 5). However, due to practical issues, it was not possible, at this stage, to gather and process through queries original values of water quality data. This is now arising as the main weakness of the actual dataset.

The long-term use of the database has been guaranteed by some choices about its structure. The reference to dictionaries of parameters instead of limited lists of parameters included into environmental data tables, while requiring more complex queries to retrieve data, allows the introduction of values for new parameters in the future. In fact, they can be added without limitation in corresponding dictionaries. Information has been split into numerous tables in order to anticipate as most as possible the different concrete cases that could arise (duplicate samples, slides, counts, ...). In order to avoid the creation of a new ID for cases through queries (see above), a table should be included between the Visit and Diatom_Sample tables, to first define the characteristics of the sample, in term of facies and substrate.

Further work on this database should deal with the development of an easy update procedure of the diatom dictionary and a more flexible taxonomic system. In addition, taxon names and authorities should be split in two different fields. Finally, the distinction between site and station, and related environmental data, should be fully implemented.

Data extracted from the database, analysed through artificial neural networks, allowed the definition of a typology of benthic diatom for near-natural conditions at a European multi-regional scale (Gosselain et al., in press), and analysis of diatom records originating from both undisturbed and disturbed conditions, providing a fresh insight about the changes of diatom assemblages along disturbed ecological gradients (unpublished). One of the objectives was to design a tool for prediction for water quality management (http://aquaeco.ups-tlse.fr/). Those analyses also offered new insight on the auto-ecology of some diatom taxa (Gosselain et al., 2003, in press). Establishing a correspondence between biotic and environmental data also allowed prediction of diatom assemblages from environmental conditions as well as the identification of main environmental conditions driving the occurrence of specific biotypes (Gosselain et al., in press). Similar application was also carried out at a regional level comprising relatively diverse environmental conditions but few cases for which both diatoms and corresponding environmental records were available (<100). This was possible due to the existence of the multi-regional database, providing extra cases for similar environmental conditions (Darchambeau et al., submitted).

In auto-ecological studies, taxonomical revisions are a common problem hampering the use of 'old' ecological and ecophysiological data. Therefore, a database designed for storing ecological records has to include precise and harmonized taxonomy and possibilities for updating, along with data on environmental conditions (Gosselain et al., in press; Darchambeau et al., submitted). At a time of high concern about assessment of ecological status of surface water bodies and identification of reference conditions for the various freshwater biota (Wallin et al., 2003), the development of databases gathering precise and documented information about aquatic communities, and corresponding high quality environmental data, becomes of prime interest. We suggest that scientists involved in ecology of freshwater communities should pay more attention to such problems, in order to save relevant ecological information in well-structured databases.

Acknowledgements

This work has been carried out in the framework of the PAEQANN project [standing for 'Predicting Aquatic Ecosystem Quality using Artificial Neural Networks: impact of environmental characteristics on the structure of aquatic communities (algae, benthic and fish fauna)'], supported by the European Commission under the Fifth Framework Programme and contributing to the implementation of the Key Action 'Sustainable Management and Quality of Water' within the Energy, Environment and Sustainable Development [Contract no: EVK1-CT1999-00026]. The authors wish to thank French Water Agencies and Water Authorities, the Ministry of Environment of Luxembourg, the Ministry of the Wallonia Region (Belgium), and the Upper Austrian Water Authority, for providing physical and chemical data, diatom data, and for supporting a great number of ecological and monitoring studies that allowed gathering this enormous amount of data. VG also warmly thanks A. Audet and K. Teymouri, who introduced her to the world of databases and accompanied her first steps in the PAEQANN database. The authors finally thank three anonymous reviewers for their comments and suggestions that helped improving an earlier version of the manuscript.

References

AFNOR, 2000. Qualité de l'Eau. Détermination de l'indice biologique diatomées (IBD) – Norme NF T90-354, 63 pp.

ANSP Algae Image Database from the Phycology Section, Patrick Center for Environmental Research, The Academy of Natural Sciences at http://diatom.acnatsci.org/AlgaeImage/

BGSU Algae Image Laboratory: http://www.bgsu.edu/departments/biology/facilities/ algae/html/Image_Archive.html

Battarbee, R. W., S. Juggins, F. Gasse, N. J. Anderson, H. Bennion & N. G. Cameron, 2000. European Diatom Database (EDDI). An Information System for Palaeoenvironmental Reconstruction. European Climate Science Conference, Vienna City Hall, Vienna, Austria, 19–23 October, 1998: 1–10.

Battarbee, R. W., S. Juggins, F. Gasse, N. J. Anderson, H. Bennion, N. G. Cameron, D. B. Ryves, C. Pailles, F. Chali & N. Telford, 2001. European Diatom Database (EDDI). An Information System for Palaeoenvironmental Reconstruction. ECRC Research Report, 81, 210 pp.

Cemagref, 1982. Etude des méthodes biologiques d'appréciation quantitative de la qualité des eaux. Rapport Q.E. Lyon A.F. Bassin Rhône-Méditérannée-Corse, 218 pp.

CEN, 2002. Water quality – Guidance standard for the routine sampling and pretreatment of benthic diatoms from rivers, prEN13946, Final draft, 14 pp.

CEN, 2004. Water quality – Guidance standard for the identification, enumeration and interpretation of benthic diatom samples from running waters, EN14407: 2004, Final draft.

Darchambeau, F., V. Gosselain, C. Fauville & J.-P. Descy, in prep. Definition of regional reference conditions for diatoms based on a multi-regional typology. To be submitted to Freshwater Biology.

Descy, J.-P., 1979. A new approach to water quality estimation using diatoms. Nova Hedwigia, 64: 305–323.

Descy, J.-P. & M. Coste, 1991. A test of methods for assessing water quality based on diatoms. Verhandlungen der Internationalen Vereinigung für theoretische und angewandte Limnologie, 24: 2112–2116.

Ghosh, M. & J. P. Gaur, 1998. Current velocity and the establishment of stream algal periphyton communities. Aquatic Botany 60: 1–10.

Gosselain, V. & P. Hamilton, 2000. Algamica: revisions to a key-based computerized counting program for free-living, attached, and benthic algae. Hydrobiologia 438: 139–142.

Gosselain, V., C. Fauville, S. Campeau, M. Gevrey & J.-P. Descy, 2003. Typology and prediction of diatom assemblages in rivers: building of database and first predictive model. In Symoens, J.-J. & K. Wouters (eds), Biological Evaluation and Monitoring of Surface Water Quality. National Committee of Biological Sciences and National Committee SCOPE, Brussels: 45–57.

Gosselain, V., S. Campeau, M. Gevrey, M. Coste, L. Ector, Y. S. Park, S. Lek & J.-P. Descy. Diatom typology of reference situations at a large multi-regional scale: combined results of multivariate analysis and SOM. In Lek, S., M. Scardi, P. Verdonschot, Y. S. Park & J.-P. Descy (eds), Modelling Community Structure in Freshwater Ecosystems, Springer-Verlag, in press.

HANNA Database, California Academy of Sciences Diatom Collection, http://www.calacademy.org/research/diatoms/index.html#collection

Index Nominum Algarum, University Herbarium, University of California, Berkeley. Compiled by Paul Silva. Available online at http://ucjeps.berkeley.edu/INA.html

Jahn, R., W.-H. Kusber, L. K. Medlin, R. M. Crawford, D. Lazarus, T. Friedl, D. Hepperle, B. Beszteri, K. Hamann, F. Hinz, S. Strieben, V., Huck, J., Kasten, A. Jobst & K. Glück, 2004. Taxonomic, molecular and ecological information on diatoms: the information system AlgaTerra. In Poulin, M. (ed.), Seventeenth International Diatom Symposium 2002. Biopress, Bristol: 121–128. [AlgaTerra Homepage at www.algaterra.net]

Joynt, E. H. III & A. P. Wolfe, 1999. An image database for diatom identification and nomenclature. Journal of Paleolimnology 22: 109–114.

Kusber, W.-H. & D. Metzeltin, 2001. Checklist of new diatom combinations, replaced names and validations published by Horst Lange-Bertalot until the year 2000 and additional validations. In Jahn, R., J. P. Kociolek, A. Witkowski & P. Compère (eds), Studies on Diatoms, Lange-Bertalot – Festschrift, Gantner, Ruggell: 585–633.

Kusber, W.-H. & R. Jahn, 2003. Annotated list of diatom names by Horst Lange-Bertalot and co-workers – Version 3.0. [http://www.algaterra.org/Names_Version3_0.pdf] pdf-file, Version 3.0, (23 June 2003).

Krammer, K. & H. Lange-Bertalot, 1986. Bacillariophyceae. 1. Teil: Naviculaceae. In Ettl, H., J. Gerloff, H. Heynig & D. Mollenhauer (eds), Süsswasserflora von Mitteleuropa, Band 2/1. Gustav Fischer Verlag, Stuttgart, New York: 876 pp.

Krammer, K. & H. Lange-Bertalot, 1988. Bacillariophyceae. 2. Teil: Bacillariaceae, Epithemiaceae, Surirellaceae. In Ettl, H., J. Gerloff, H. Heynig & D. Mollenhauer (eds), Süsswasserflora von Mitteleuropa, Band 2/2. VEB Gustav Fischer Verlag, Jena: 596 pp.

Krammer, K. & H. Lange-Bertalot, 1991a. Bacillariophyceae. 3. Teil: Centrales, Fragilariaceae, Eunotiaceae. In Ettl, H., J. Gerloff, H. Heynig & D. Mollenhauer (eds), Süsswasserflora von Mitteleuropa, Band 2/3. Gustav Fischer Verlag, Stuttgart, Jena: 576 pp.

Krammer, K. & H. Lange-Bertalot, 1991b. Bacillariophyceae. 4. Teil: Achnanthaceae, Kritische Ergänzungen zu Navicula (Lineolatae) und Gomphonema, Gesamtliteraturverzeichnis Teil 1-4. In Ettl, H., G. Gärtner, J. Gerloff, H. Heynig & D. Mollenhauer (eds), Süsswasserflora von Mitteleuropa, Band 2/4. Gustav Fischer Verlag, Stuttgart, Jena: 437 pp.

Lecointe, C., M. Coste & J. Prygiel, 1993. 'OMNIDIA' software for taxonomy, calculation of diatom indices and inventories management. Hydrobiologia 269/270: 509–513.

Lecointe, C., M. Coste, J. Prygiel & L. Ector, 1999. Le logiciel OMNIDIA version 2, une puissante base de données pour les inventaires de diatomées et pour le calcul des indices diatomiques européens. Cryptogamie Algolologie 20: 132–134.

Mayer, P. M. & S. M. Galatowitsch, 2001. Assessing ecosystem integrity of restored prairie wetlands from species production–diversity relationships. Hydrobiologia 443: 177–185.

McCormick, P. V. & J. Cairns, 1994. Algal as indicators of environmental change. Journal of Applied Phycology 6: 509–526.

O'Kelly, C. J. & T. Littlejohn, 1994–2004. PID: Protist Image Database. Distribution: http://megasun.bch. umontreal.ca/protists/protists.html

Pan, Y., R. J. Stevenson, B. H. Hill, A. T. Herlihy & G. Collins, 1996. Using diatoms as indicators of ecological conditions in lotic systems: a regional assessment. Journal of North American Benthological Society 15: 481–495.

Pan, Y., R. J. Stevenson, B. H. Hill & A. T. Herlihy, 2000. Ecoregions and benthic diatom assemblages in Mid-Atlantic Highlands streams, USA. Journal of North American Benthological Society 19: 518–540.

Peterson, C. G., H. M. Valett & C. N. Dahm, 2001. Shifts in habitat templates for lotic microalgae linked to interannual variation in snowmelt intensity. Limnology and Oceanography 46 : 858–870.

Potapova, M. G. & D. F. Charles, 2002. Benthic diatoms in USA rivers: distributions along spatial and environmental gradients. Journal of Biogeography 29: 167–187.

Prygiel, J. & M. Coste, 1999. Progress in the use of diatoms for monitoring rivers in France. In Prygiel, J., B. A. Whitton & J. Bukowska (eds), Use of Algae for Monitoring Rivers III. Agence de l'Eau Artois-Picardie, Douai: 165–179.

Prygiel, J., L. Leveque & R. Iserentant, 1996. L'IDP: Un nouvel Indice Diatomique Pratique pour l'évaluation de la qualité des eaux en réseau de surveillance. Revue des Sciences de l'Eau, 9 : 97–113.

Prygiel, J., B. A. Whitton & J. Bukowska, 1999. Use of Algae for Monitoring Rivers III. Agence de l'Eau Artois-Picardie, Douai.

Round, F. E., R. M. Crawford & D. G. Mann, 1990. The Diatoms. Biology and Morphology of the Genera. Cambridge University Press, Cambridge, 747 pp.

Sabater, S., S. V. Gregory & J. R. Sedell, 1998. Community dynamics and metabolism of benthic algae colonizing wood and rock substrata in a forest stream. Journal of Phycology 34: 561–567.

Sullivan, T. J. & D. F. Charles, 1994. The feasibility and utility of a paleolimnology/ paleoclimate data cooperative for North America. Journal of Paleolimnology 10: 265–273.

The UCMP Collection Catalogue : http://www.ucmp.berkeley.edu/collections/micro.html

Wallin, M., T. Wiederholm & R. K. Johnson, 2003. Guidance on establishing reference conditions and ecological status class boundaries for inland surface waters, final draft, version 7.0, produced by CIS working group 2.3. – REFCOND, 5 March 2003. 93 pp.

Wargo, M. J. & J. R. Holt, 1998. Determination of stream reaches in a ridge and valley creek using diatom periphyton communities. Journal of Freshwater Ecology 13: 447–456.

Hydrobiologia (2005) 542:165–190
H. Segers & K. Martens (eds), Aquatic Biodiversity II
DOI 10.1007/s10750-004-0756-y

A review on the present status and management of mangrove wetland habitat resources in Bangladesh with emphasis on mangrove fisheries and aquaculture

Md. Shahidul Islam[1,*] & Md. Abdul Wahab[2]
[1]*Division of Applied Biosciences, Faculty of Fisheries, Graduate School of Agriculture, Kyoto University, Kyoto 606-8502, Japan*
[2]*Department of Fisheries Management, Faculty of Fisheries, Bangladesh Agricultural University, Mymensingh 2202, Bangladesh*
*(*Author for correspondence: E-mail:msi@kais.kyoto-u.ac.jp)*

Key words: mangrove, Sundarban, shrimp farming, fisheries, Bangladesh

Abstract

The mangrove forest of Bangladesh, the largest continuous mangrove bulk, is one of the most important features of the coastal area of the country. The existence of the mangrove has increased the values of other coastal and marine resources such as the coastal and marine fisheries by increasing productivity and supporting a wide biological diversity. The deltaic mangrove of Bangladesh is ecologically different from the other, mostly nondeltaic mangroves of the world and is unique also in its floral and faunal assemblage; therefore, a number of endangered plants and animals that are extinct from other parts of the world, are existing in Bangladesh mangrove. However, the mangrove has been under intensive pressure of exploitation for the last few decades which, in addition to direct clearance and conversion have placed the mangrove under extreme threat. Shrimp farming is the most destructive form of resource use the mangrove has been converted to, which contributed significantly to mangrove destruction with a corresponding loss of biological resources. Concerns have been raised among the ecologists, biologists, managers and policy makers since the early 1990s; deliberate destruction of mangrove and unplanned development of coastal aquaculture particularly shrimp aquaculture have been put under extreme criticism and the sustainability has been questioned. The present status of the mangrove resources including mangrove fisheries and aquaculture and management practices have been reviewed in this paper; impacts of different forms of human interventions and resource use have also been discussed. It is suggested that the management options and the policy aspects should be critically reviewed and amended accordingly; beneficiaries and stakeholders at all levels of resource exploitation must take part and contribute to conservation and management. An immediate need for mangrove conservation has been identified.

Introduction

Mangrove forest is one of the primary features of the coastlines throughout the tropics and subtropics of the world and, locally known as 'Sundarbans', mangroves in Bangladesh is one of the major important coastal habitats and a major component of the vegetative cover in the coastal region. Substantial areas of mangrove forests are found in Bangladesh, Cambodia, India, Indonesia, Malaysia, Myanmar, New Guinea, Pakistan, Philippines, Thailand, Vietnam, and in the West India (Fig. 1). The Sundarbans mangrove forest is located in the Gangetic delta (Ganges-Brahmaputra-Meghna) of India and Bangladesh, is the largest single chunk of continuous mangrove forest in the world (Gittings & Akonda, 1982; Chaudhury & Naithani, 1985; Hussain & Acharya, 1994), covering an area of 10 000 km^2 of which 62% fall within the territory of Bangladesh, covering an area of 6500 km^2 (FAO, 2003) (Fig. 1). At one time, mangrove forest in Bangladesh was

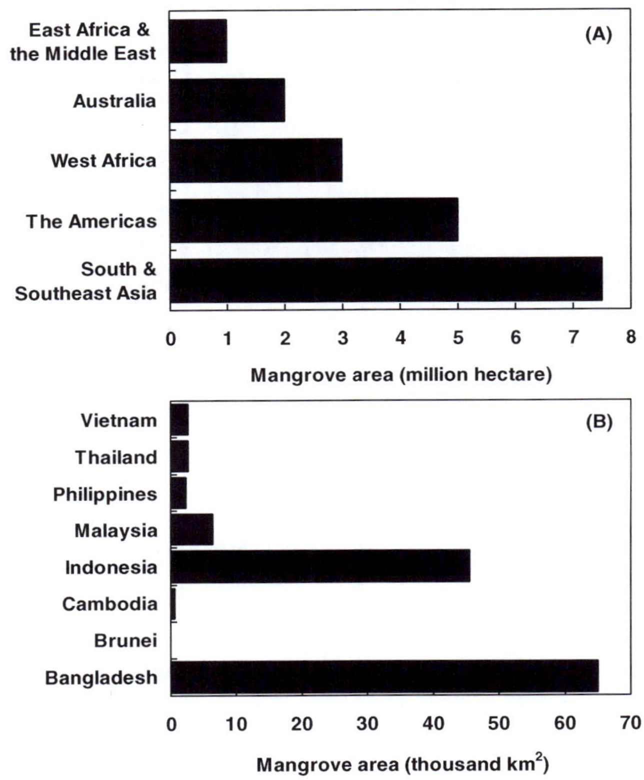

Figure 1. Relative proportions of mangrove coverage. (A) Global distribution of mangrove forests with the South and Southeast Asia having the largest mangrove area (Kathiresan & Bingham, 2001). (B) Distribution of Asian mangrove shcwing the largest area in Bangladesh (FAO, 2003).

distributed along much of the coastal belt of the southwestern region, but is now confined to the Sundarban reserve forest and offshore islands (Fig. 2). In the eighteenth century, before the process of land reclamation began, the total area of the forests including the Indian part was 16 700 km^2. Being the source of a variety of renewable resources, mangroves are playing significant role in the local coastal economy and livelihood of the people, in the national economic development and also in the global environmental balance. Baran & Hambrey (1998) described the major features of the estuarine environment including the dependence of coastal resources on estuarine and mangrove environment and also highlighted why mangroves are a particular case of an estuarine environment. The ecological significance of mangrove forests has now been universally recognized. Forests in Bangladesh cover an approximate area of 577 000 hectares (Hussain & Acharya, 1994) which stand for about 17% of the total area of the country (Das & Siddiqi, 1985), of which 401 600 hectares are land forests and the remaining 175 600 hectares are temporarily or permanently inundated by water in the form of rivers, canals or tidal floodlands. The mangrove is one of the three major forest types of Bangladesh and one of the most important features of the coastal areas comprising about 50% of the productive forests of Bangladesh (FAO, 1981, 1993).

Mangroves in Bangladesh have two major regional distributions. The first and the largest of which is the southwest region covering greater Khulna District, namely, the Sundarbans mangrove forest and the other is the southeast region covering greater Chittagong District, namely, the Chakaria Sundarbans. The mangrove forest is very rich in biotic diversity supporting around 330

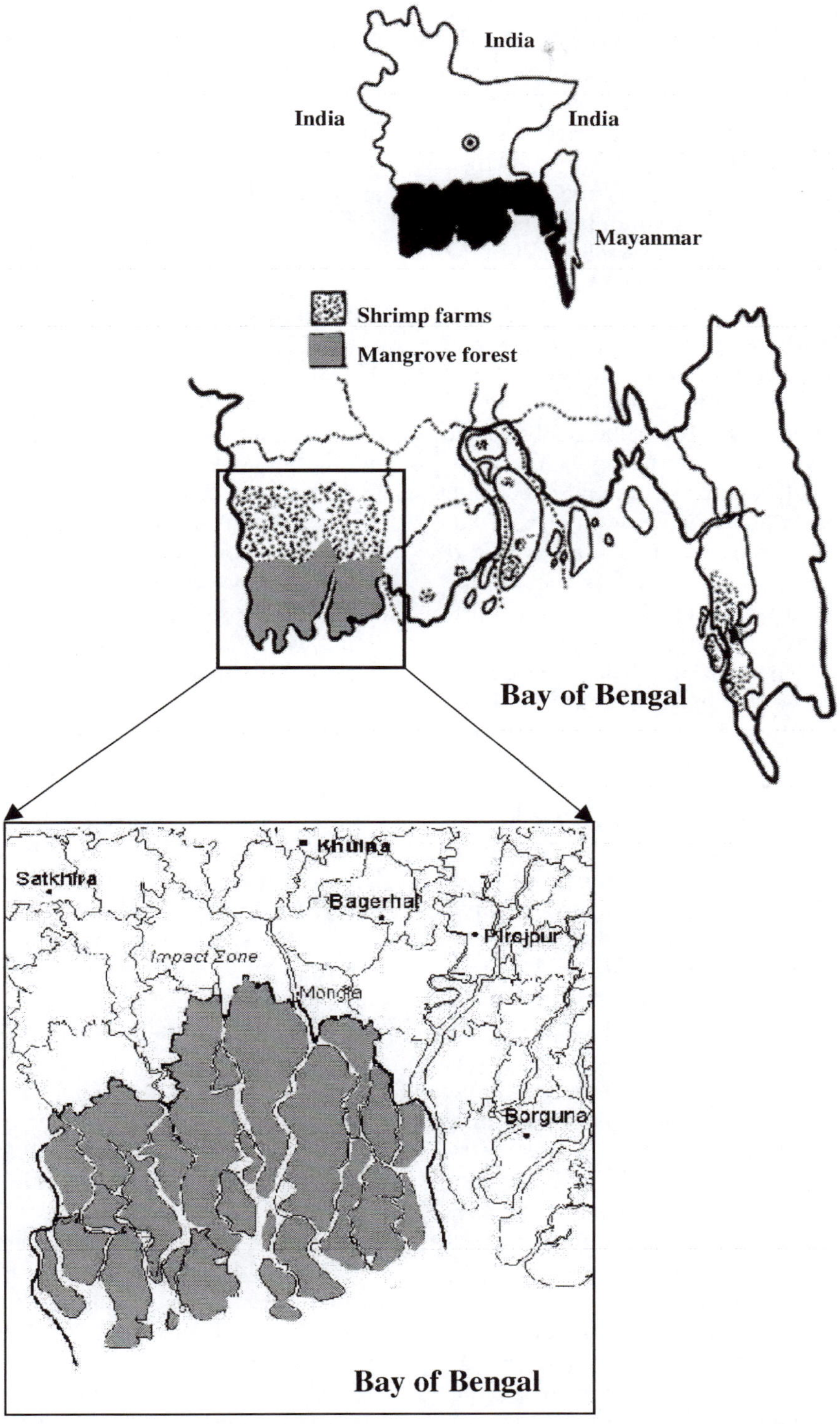

Figure 2. Map showing the mangrove forest and associated coastal areas of the Bay of Bengal, Bangladesh.

species of plants, as many as 400 species of fishes, about 40 species of shrimps and other crustaceans, 35 species of reptiles, over 270 species of birds, and 42 species of mammals (Hussain & Acharya, 1994). Besides, the Sundarbans is the largest single remaining habitat of the famous Royal Bengal Tiger (*Panthera tigris*) and estuarine crocodile (*Crocodylus porosus*). A large number of threatened and endangered species are still found in the Sundarbans mangrove areas, which are not available in other parts of the country.

The Sundarbans mangrove forest is used as the basis for various commercial and subsistence activities. About 10 million people of the coastal regions are dependent directly or indirectly on the mangrove for a variety of purposes such as agriculture, fishing, cattle rearing, human settlement, collection of housing materials, fuelwood and human foods and employment opportunities in forestry practices. Along with the subsistence fishing by numerous small fishermen, large-scale export-oriented aquaculture industries have been developed in the coastal areas of Bangladesh for the last few decades, using the mangrove as the basic productive unit. Thus, the mangrove is playing a very significant role in uplifting the coastal regional as well as the national economy.

Although mangroves are exploited for human interests from time immemorial, conversion of mangroves for different land use purposes is almost a new phenomenon. With increase in population pressure, the dependence of man on forests for various purposes has been intensified. In recent years, the exploitation and conversion of the forestlands have reached such a level that irreversible changes in the natural balance of the ecosystem have taken place, resulting in serious loss of sustenance. Among others, shrimp farming is one of the major and perhaps the most detrimental activity the mangroves have been converted to. Shrimp culture by removing mangroves in Bangladesh expanded so rapidly during the past two decades that concerns on the possible detrimental effects became obvious. Hopefully, plans and policies are currently being formulated and implemented with a view to conserve and protect the mangrove resources and to ensure sustainable resource use and management.

This paper reviews the present status of the mangrove resources and the main forms of coastal resource use with special emphasis on mangrove fisheries and aquaculture. The paper also discusses existing management practices and policy aspects, legislative and institutional issues as well as the impacts of various form of resource use and, finally, suggests options for better management and probable modes of rehabilitation of the exploited resources.

Physical characteristics

The Sundarbans is part of the world's largest delta (100 000 km^2) formed from sediments deposited by three great rivers, the Ganges, Brahmaputra and Meghna, which converge on the Bengal Basin (Seidensticker & Hai, 1983). The land is moulded by tidal action, resulting in a distinctive physiography. The soil is a silty clay loam with alternate layers of clay, silt and sand. The surface is clay except on the seaward side of islands in the coastal limits, where sandy beaches occur. The pH averages 8.0 (Christensen, 1984). The Sundarbans mangrove forest is intersected by an elaborate network of rivers, channels and creeks; innumerable small canals drain the land at the ebb. These waterways are of varying width and length and occupy a substantial part of the forest. Of the total area of the Bangladesh Sundarbans almost 62% is land and the rest water (Christensen, 1984). Generally, the rivers flow north to south and are connected with a number of side channels that connect two rivers and facilitate exchange of waters. The larger rivers, passing through the forest join together, forming estuaries at the confluence where they meet near the sea. The surface water flow in the rivers of the Sundarbans mainly comes from the Ganges-Padma River Systems through Garai-Madhumati Rivers and from the lower Meghna through the Swarupkati-Kocha Rivers. The other source of surface water flow in the rivers is the runoff from the local catchment area.

The Sundarbans mangrove forest is located south of the Tropic of Cancer and at the northwest of the Bay of Bengal (21° 30′–22° 30 ′N, 89° 12′–90° 18′ E). The four main seasons are pre-monsoon (March–May), monsoon (June–September), post-monsoon (October–November) and the dry winter season (December–February). Each of the seasons has its distinct characteristic features which influence mangrove ecology. The area inundated by tidal water increases and, due to increasing maritime influence, the salinity of river water also reaches a maximum during the post-monsoon season because of its characteristic southern winds, high temperature and higher evaporation rates with occasional heavy thunderstorms. The monsoon brings high rainfall and cloudy cover resulting in a rise in water levels and increased sediment loads in upper catchment areas. Very low or no tidal inundation resulting in a dry, exposed period is characteristic of the winter season.

Rainfall is heavy and humidity high (80%) due to the proximity of the Bay of Bengal; about 80% of the rainfall is in the monsoon. Mean annual rainfall varies from about 1800 mm at the north of the Sundarbans to 2790 mm on the coast. There is a 6-month dry season during which evapotranspiration exceeds precipitation. Conditions are most saline in February–April, the depletion of soil moisture being coupled with reduced freshwater flow from upstream. Temperatures rise from daily minima of 2–4 °C in winter to a maximum of about 43 °C in March and may exceed 32 °C in the monsoon. Storms are common in May and October–November and may develop into cyclones, usually accompanied by tidal waves of up to 7.5 m high (Seidensticker & Hai, 1983). Climatic data for Khulna are summarized by Christensen (1984); annual changes in temperature, rainfall, humidity and evaporation have been shown in Figure 3.

The Sundarbans is a low-lying deltaic swamp drained by a network of small creeks, which flow into the main watercourses on the ebb-flow of the tide. These are, in some instances, eroded by the tide to form saucer-shaped depressions in which there may be standing water. For about seven months of the year, from November to May, potential evaporation exceeds precipitation by nearly 500 mm. For more than half of the year, therefore, the vegetation is heavily dependent on soil–water, which is recharged largely by tidal intrusion and the reduced overflow. During this period of precipitation deficit, soil salinity temporarily increases. The forest receives large volumes of freshwater from inland rivers and of saline waters from the twice-daily tidal inundation from the sea, creating in the mangrove forests an open

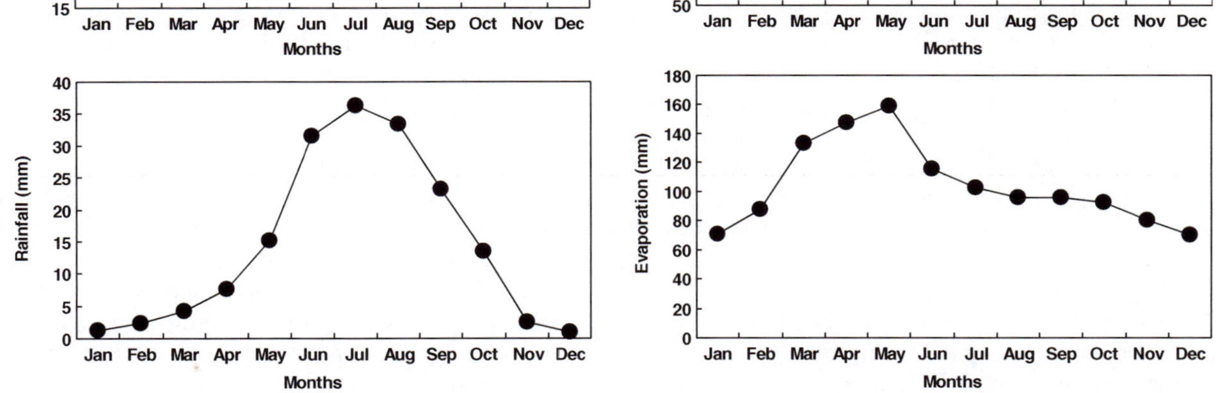

Figure 3. Seasonal patterns in some of the meteorological parameters of the Sundarban mangrove forest of Bangladesh (Islam, 2003).

ecological system. The freshwater charged with alluvium that contains plant nutrients and which, together with the salinity of the tidal water, is a major factor affecting the forest ecosystem. The relative contributions from the two sources of water are not constant throughout the year. The bulk of the freshwater reaches the forest in the monsoon season. During the dry season the freshwater flow decreases progressively, resulting in a prolonged saline condition which persists until the freshwater flow increases again. Mangroves are vegetated tidelands and function as a nutrient producer, water purifier, nutrient and sediment trap, storm barrier, shore stabilizer, aesthetic attraction, energy storage unit, transportation networks nursery and feeding grounds of many major coastal and ocean fisheries.

Mangrove flora

The study of the floral diversity of the Sundarbans is limited to a group of commercially important species and few others. The forest supports a diverse plant genetic stock including timber species, wild relatives of cultivated plants and a number of endemic species. While most of the mangroves in other parts of the world are characterized by members of the Rhizophoraceae, Avicenniaceae, and LagancuIariaceae, the mangroves of Bangladesh are dominated by the Streculiaceae and Euphorbiaceae. The mangroves of the Sundarbans are unique when compared to non-deltaic coastal mangrove forest; the reason for this difference is the large freshwater influence in the north-eastern part and the elevated level of the ground surface. Sunduri (*Heritiera fomes*) and Gewa (*Excoecaria agallocha*) are the dominant species, occurring in suitable localities along the coast. Unlike former times, Rhizophoraceae are now of minor importance. Inventory of the Sundarbans shows that there has been a change in favour of mixed *Heritiera fomes* and *Excoecaria agallocha* at the expense of pure *Heritiera fomes* forest. A further important change which the forest has undergone is a reduction in canopy density. In 1959, 78% of the forest had a canopy closure of 75% or more; now only 65% of the forest has canopy closure of 70% or more. Current inventory also shows that approximately 70% of *Heritiera fomes* stems are moderately or severely affected by topdying.

Seidensticker and Hai (1983) report a total of 334 plant species, representing 245 genera. The natural mangroves are halophytic, two-storied woodland, seldom exceeding 10 m in height with the notable exception of Sunduri (*Heritiera fomes*) and Keora (*Sonneratia apetala*) and Bean (*Avicennia officinalis*). Stem diameters are generally less than 30–40 cm at breast height although few species attain diameter up to about 1 m. Among the dicotyledonous plants, the tree species are represented by 30 genera under 22 families. Rhizophoraceae is represented by all the known four genera and at least six species. Avicenniaceae is represented by three species, Meliaceae by three species, Combretaceae by two species and Sonneratiaceae by two species. Among the tree species, *E. agallocha* and *H. fomes* are found as the most common species. The shrubs and scandent shrubs are represented by 12 species belonging to 11 genera under 7 families. The forest is also abundant in climbers and epiphytes. At least 11 species belonging to 6 families have so far been identified. The monocotolydenous herbs are represented by Graminae, Palmae, and Pendanaceae. Besides, the rooted plants, the swamp forest species also include the epiphytic parasitic flora that live in the rooted trees and do not come into contact with forest floor. The epiphytes include *Hoya parasitica*, *Dischidia numularia*, 13 species belonging to the family Orchidaceae, 7 epiphytic ferns, *Lycopodium* and *Psilotum*. Like the epiphytes, woody species parasitic on tree crowns are also found to some extent. One of the most important non-wood forest products is *Nypa fruticans* leaf.

Several reports suggest that a number of plant species in the mangrove forest are endangered to some extents. The changing composition of the forest is a clear manifestation of the uniqueness of its ecological condition and the management regime. *Heritiera fomes* is not an endangered species but the dominance of the species as a forest type is threatened and reflects the changing health of the ecosystem, particularly due to reduced supply of freshwater. The decline in growth of many freshwater loving plants such as *Nypa fruticans* has also been noted. Previous reports mentioned at least four common species of the genus *Bruguiera* but only one species, *B. parviflora* is found in the Sundarbans today. *Lumnitzera racemosa* and *Barringtonia* spp. are also very limited in the present forest composition. Except for the commercially important species the remaining species are subjected to unregulated felling practices. Some of these species, particularly *Cynometra*, *Amoora cuculata* and *Rhizophora* spp. are threatened due to such unregulated felling.

Mangrove fauna

The mangrove forest provides a unique ecosystem and extensive habitats for a wide variety of faunal species. The Sundarbans is a very high quality wildlife conservation area of regional and international importance. Many authors have studied and given lists of the wildlife assemblages of the Sundarbans (Hendrichs, 1975; Mukherjee, 1975; Khan, 1982; Hussain et al., 1983; Sanyal, 1983, 1987; Khan, 1986, 1987; Seidensticker & Hai, 1983). The international importance of the Sundarbans as an essentially long-term wildlife habitat has also been documented (Gittins & Akonda, 1982; Seidensticker, 1986, 1987). Several studies (Seidenstiker & Hai, 1983; Blower, 1985a, b; Khan, 1986) reported that the mangroves of Bangladesh support at least 120 species of commercially important fishes, 290 species of birds, 42 species of mammals, 35 reptiles, and 8 species of amphibians. The mangroves also support a large number of species, which are now extinct elsewhere in the country. These include 2 amphibians, 14 reptiles, 25 birds, and 5 mammals (Sarker & Sarker, 1986; Sarker, 1993).

The Sundarbans is the only remaining habitat in the lower Bengal Basin for a variety of faunal species. The presence of 49 mammal species has been documented. Of these, no less than five spectacular species, namely Javan rhinoceros *Rhinoceros sondaicus*, water buffalo *Bubalus bubalis*, swamp deer *Cervus duvauceli*, gaur *Bos frontalis* and probably hog deer *Axis porcinus* have become locally extirpated since the beginning of this century (Salter, 1984). The Sundarbans of Bangladesh and India support one of the largest populations of tiger *Panthera tigris*, Spotted deer *Cervus axis*, and wild boar *Sus scrofa*, smooth-coated otter *Lutra perspicillata*, three species of wild cat, *Felis bengalensis*, *F. chaus* and *F. viverrina*, and Ganges River dolphin *Platanista gangetica*, which occurs in some of the larger waterways. Species accounts and a check-list are given by Salter (1984).

A total of 315 species of bird have been recorded (Hussain & Acharya, 1994), including about 95 species of waterfowl (Scott, 1989) and 38 species of raptors (Sarker, 1985a, b). Among the many which may be readily seen by the visitor are no less than nine species of kingfisher, including brown-winged and stork-billed kingfishers, *Pelargopsis amauropterus* and *P. capensis*, respectively; the magnificent white-bellied sea-eagle *Haliaeetus leucogaster* which, at a density of one individual per 53.1 km of waterways (Sarker, 1985a), is quite common; also the much rarer grey-headed fish eagle *Ichthyophaga ichthyaetus*, Pallas's fish-eagle *Haliaeetus leucoryphus* and several other raptors. Herons, egrets, storks, sandpipers, whimbrel, curlew and numerous other waders are to be seen along the muddy banks and on the chars or sandbanks which become exposed during the dry season. There are many species of gulls and terns, especially along the coast and the larger waterways. Apart from those species particularly associated with the sea and wetlands, there is also a considerable variety of forest birds such as woodpeckers, barbets, shrikes, drongos, mynahs, minivets, babblers and many others (Salter, 1984). Scott (1989) gives further details of the avifauna.

Some 53 reptile species and eight of amphibians have been recorded (Hussain & Acharya, 1994). Of these mugger *Crocodylus palustris* is now extinct, probably as a result of past over-exploitation. Estuarine crocodile *C. porosus* still survives but its numbers have been greatly depleted through hunting and trapping for skins. There are also three species of monitor, *Varanus bengalensis*, *V. flavescens* and *V. salvator*, and Indian *python Python molurus*. Four species of marine turtle have been recorded from the area, olive ridley *Lepidochelys olivacea* being the most abundant. Green turtle *Chelonia mydas* is rare due to excessive fishing, while loggerhead *Caretta caretta* and hawksbill *Eretmochelys imbricata* are not common although there have been some reported on the beaches (Hussain & Acharya, 1994). River terrapin *Batagur baska* is also present. The eighteen recorded snake species include king cobra *Ophiophagus hannah* and spectacled cobra *Naja naja*, three vipers and six sea-snakes (Salter, 1984).

Mangrove fish and related fauna

Plankton

Algal flora is one of the most poorly studied components of the mangrove ecosystem. Algae can be seen attached to the pneumatophores, stilt roots, tree trunks and on the mud floor. Mangrove algae can also be found in some areas forming red, brown or green mats on intertidal mud flats. Due to muddy shorelines and low saline conditions, typical euryhaline benthic marine algae are negligible in the region. Although not complete, so far, 34 species of phytoplankton have been listed (Islam, 1973; Hussain & Acharya, 1994). Belonging to the families Volvocaceae, Palmelaceae, Ulvaceae, Cladephoraceae, Zygnemaceae, and Codiaceae, the Division Chlorophyta is represented by 9 species under 6 genera. Division Cyanophyta is represented by 16 species under 9 genera belonging to the families Chroococcaceae, Oscillatoriaceae, Nostocaceae, and Rivulariaceae. Bacillariophyta is represented by 16 genera and 35 species and the Euglenophyta is represented by 2 genera consisting of 2 species. It was observed that most of the members of Bacillariophyta and Cyanophyta occur in abundance in the lower littoral zone. The distribution and abundance of algae in the mangrove areas are greatly influenced by the ecological characteristics, particularly the mixing of saline and freshwater resulting in salinity fluctuation and the silt carried by rivers.

The complex mangrove ecosystems are highly suitable to support very complex food chains and for organisms of all trophic levels and niches to occur. The mangroves of Bangladesh are extraordinarily rich in zooplankton production. Thirteen major taxa of zooplankton namely, Copepods, Amphipodes, Mysids, Acetes, Chaetognaths, Polychaetes, Lucifers, Hydromedusae, Shrimp larvae, Fin fish larvae, Crab larvae, Squilla, and Horseshoe crab larvae were found to occur in this region. Mahmood (1985) recorded 23 species of ichthyoplankton dominated by *Glossogobius giuris*, *Odontablyopsus rubiundus*, *Mugil* spp. etc.

Fish fauna

The fishery resources are one of the primary concerns in the Sundarbans mangrove ecosystem. The fishery resources are characterized by a very high degree of endemism and diversity. The number of fish species available in the Sundarbans and their taxonomic status have been described among others by Rahman (1975, 1989) with varying degrees of accuracy. According to Hussain and Acharya (1994) the mangrove waters support 53 species of pelagic fish belonging to 27 families and 124 species under 49 families of demersal fish. However, this number is expected to be as high as 400 species using the mangrove swamps in various ways. Diadromous species such as *Pangasius pangasius*, *Hilsa ilisha*, *Lates calcarifer* are common in the low saline zone. The moderately saltwater zones are dominated by *Hilsa ilisha*, *Pomadasys hasta*, Polynemus spp., *Coilia* spp., *Johnius* spp. etc. The highly saline zones are likely to support species such as *Harpodon nehereus*, *Trichiurus savala*, *Setipinna* spp., *Pampus* spp., *Sardinella* spp., *Salar* spp. etc. Considerable seasonal changes also occur in the quality and quantity of fish fauna in the region.

Over 120 species of fish are reported to be commonly caught by commercial fishermen in the Sundarbans (Seidensticker & Hai, 1983). According to Mukherjee (1975) only brackish water species and marine forms are found in the Indian Sundarbans, freshwater species being totally absent. This may be assumed to apply also to the Bangladesh Sundarbans, except possibly in the eastern portion where there is freshwater in Baleswar River. Mention should also be made of mud-skippers or gobies which occur in large numbers and are a characteristic feature of mangrove swamps.

Non-fish organisms

Besides the large number of fish species, the mangroves also support many nonpiscine species of very high degree of commercial value. These include 24 species of shrimps belonging to 5 families, 7 crab species under 3 families, 2 species of gastropods, 6 species of Pelecypods, 8 species of lobsters and 3 species of turtles (Hussain & Acharya, 1994; Blower, 1985a, b). Being a transitional zone between freshwater and seawater the mangrove serves as a seasonal habitat for many freshwater and marine species and, as such, is richer than either of the two in terms of its biological genetic resources. Many species, both freshwater and marine, are dependent on the mangroves for their nursery areas.

Crustacea account for by far the largest proportion of animal biomass, with an estimated 40 million kilograms of fiddler crabs and 100 million kilograms of mud crabs (Hendrichs, 1975). The nutrient-rich waters of the Sundarbans also yield a considerable harvest of shrimps, prawns and lobsters. The area supports a varied insect population including large numbers of honey bees, honey and beeswax being among the economically important products. The insect life of the Sundarbans has been little studied.

Mangrove forest and fisheries

There are two basic types of capture fisheries in the mangrove areas, namely, inshore fishery and offshore fishery. Usually small boats fishing in relatively shallow waters, ranging in depths from 2 to 8 m constitute the inshore fishery. Various types of gears are used by artisanal and commercial fishermen and about one-third of annual fish production in the mangrove areas is attributable to inshore fishery. The offshore fishery covers estuarine and coastal waters. Among different types of fishing gears used by the fishermen, the most common ones are 'behundi jal' (set bag net), 'ilisha jal' (hilsa gill net), and long line.

In addition to commercial fishing, another important fishing activity that has gained considerable attention throughout the past two decades is 'shrimp fry fishery', the collection of shrimp fry from mangrove estuaries and nearshore waters of the Bay of Bengal. Until recently, wild post-larvae (PL) of tiger prawn (Penaeus monodon) were the only source of fry for stocking the coastal brackish water ponds of Bangladesh. Following increasing demand, intense and widespread fishing of fry is being practiced in the estuaries and creeks along the coast. The total shrimp fry caught during the peak period between mid-February and mid-March in three coastal districts, namely, Satkhira, Khulna, and Cox's Bazar was estimated to be 75 000 post-larvae in 1990 which was increased as much as 1453 million in 1993 (Chantarasri, 1994). Over 100 000 fishermen were reported to engage in fry collection.

The importance of mangrove ecosystems to coastal fisheries has been described among others by Pauly (1985), Twilley et al. (1996), Baran & Hambrey (1998) and Kathiresan & Bingham (2001). Yanez-Arancibia et al. (1985), in the Gulf of Mexico, have shown a clear positive correlation between commercial finfish catches and the total area of mangroves. Similarly, positive correlations of mangrove were reported by De Graaf & Xuan (1997) with finfish catches, by Martosubroto & Naamin (1977) with annual catch of prawns, by Paw & Chua (1989) with penaeid shrimp catch and by Staples et al. (1985) with the annual catch of banana prawn. Pauly & Ingles (1986) showed that most of the variance of the MSY of penaeids could be explained by a combination of area of mangrove habitats and latitude. This implies that a certain minimum mangrove area is necessary for a high production, a point also noted by Pauly & Ingles (1986) who suggest that the impact of destruction of a mangrove area might be greater if this area is small and residual.

Conversion of mangrove into aquaculture

Many areas of coastal districts such as Khulna, Barisal, Patuakhali, Bagerhat, Chittagong, and Cox's Bazar were once ornamented with dense mangrove vegetation, but, in the course of time, mangroves over large areas have been cleared and converted to other land uses. In the southeast part in the delta of Matamuhuri River, one of the oldest mangrove forests of the subcontinent, the 'Chakaria Sundarbans' had an area of 18 200 ha which was dramatically reduced to 5446 ha as consequence of human intervention such as salt production and shrimp cultivation in the last two decades. The Chakaria Sundarbans typical of the southwestern mangrove forests in terms of growth, with unique dense vegetation of what is locally known as 'Chulia Kanta' (*Dalbergia spinosa*) and 'Nunia Kanta' (*Aegialitis rotundifolia*). Now, virtually no compact mangrove forest is found in these areas except some sporadic vegetation.

The ecology of coastal areas and the climatic condition of Bangladesh are extremely suitable for shrimp culture with a very low production costs. Shrimp culture is mostly practiced in low-lying tidal flats within Bangladesh Water Development

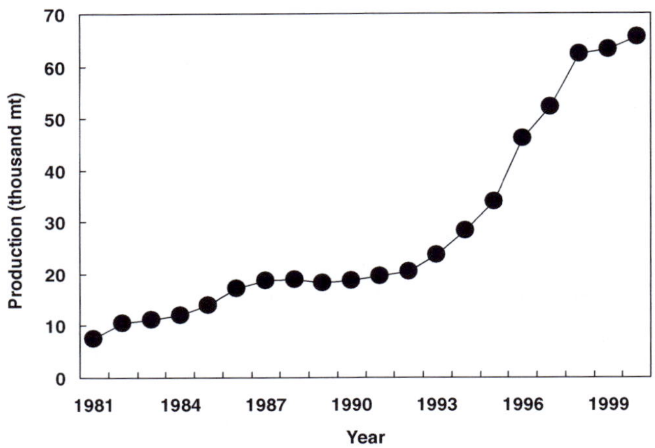

Figure 4. Showing a steady rise in the production of cultured shrimp over the last two decades, indicating an expansion in shrimp farming areas (DOF, 2001).

Board (BWDB) polders. The polders were originally constructed to prevent the land from brackish water inundation and to use the reclaimed land for agriculture.

Shrimp culture in Bangladesh started to expand in 1970s and, as a source of immediate economic gain, shrimp farming has expanded steadily over the last two decades (Fig. 4). According to survey result of 1982–1983, the area under shrimp culture was 52 000 ha, which has increased to about 0.11 million ha by 1995. About 75% of this land is located in Khulna, Bagerhat and Satkhira districts in the south and the rest in the Cox's Bazar district in the southeastern region of the country (DOF, 1999). From the early eighties, the government of Bangladesh has been endeavoring to improve the traditional culture practices.

The development shrimp farming took place in a speculating and traditional manner. Because of unscientific management, water quality problems are common in almost all farms in each year, which are further compounded by higher stocking density which results from the intention of getting more production. However, at the end, very poor production with respect to the total input is experienced by most producers. Frequent occurrence of disease and mortality of shrimps is another consequence of poor water quality. Shrimp culture techniques existing at present in Bangladesh and its environmental and socio-economic have been described by many authors (Karim & Aftabuzzaman, 1997; Deb, 1998; Islam, 2003).

Along with the shrimps that are introduced by trapping during watering of the ghers, a number of finfish species are also introduced and raised with shrimps. At the end of the culture period, large quantities of different species of finfishes are harvested. The major fish types are *Mystus*, *Wallago*, *Pangasius*, *Glossogobius*, *Liza* etc. At present, commercial culture of several species of finfishes and one crab species is being carried out on a small-scale. The major species cultured are *Lates calcarifer*, *Liza cascasia*, *L. ologolepis*, and *L. tade* among the finfishes and *Scylla serrata* (the mud crab) among the crab species. However, development of suitable culture technologies for these species and some others is under trial.

Utilization of mangrove vegetation

Heritiera fomes is the principal timber species in Bangladesh. This plant along with some others, namely, *Xylocarpus mekongensis* and *Bruguiera sexangula* are highly priced construction material and are exported to distant markets. *Ceriops decandra*, *Phoenix paludosa*, *Sonneratia apetala*, and *Avicennia officinalis*, are extensively used as poles for construction of the hunts. *Nypa fruticans*, *Typhia elephantina* and *Saccharum cylindricum* are used for thatching and roofing. Among these, the leaves of *Nypa fruticans* (Nipa) are used extensively throughout the country. The timbers are used among other purposes for boat building and bridge construction. Various household items and agricultural implement are made from mangrove materials. Some products like mats, baskets etc. are sold in distant markets.

The Sundarbans mangrove forest is the largest source of timber, firewood, fuelwood and other forest products. *E. agallocha* is used as raw materials in the paper and match industry. Annual coupes are laid out each year and pulpwood is extracted under a selection system where all trees above 20 cm diameter at breast height are extracted. The two major fuel wood species in the mangroves are *Heritiera fomes* and *Ceriops decandra*. The trimmed parts of these two species, originated during harvesting of timber are separated for use as fuel. Some healthy trees below the exploitable diameter are also removed to use as fuel during thinning of congested and dense stands. However most of the fuelwood harvested commercially from the mangrove is transported to the towns.

The mangroves are the main source of honey and wax. The Sundarbans possesses numerous species of trees that produce abundant nectar and pollen and these floral resource are used by the giant honey bee (*Apis dorsata*). An estimated amount of 200 metric tones of honey and 55 metric tonnes of wax are collected annually from the forest (Das & Siddiqui, 1985). Local people are seasonally engaged in honey collection. The value of mangrove-marketed products can be expressed in economic terms. However, the free services like protection, accretion and land stabilization provided by the mangroves are difficult to quantify in economic terms.

Impacts of mangrove exploitations

Environmental impacts

Destruction of the mangrove forests has been occurring for a very long time and the total mangrove forest in Bangladesh today is less than half the size it was two centuries ago. Many authors have reported that removal of mangrove for shrimp farming is responsible for rapid and irreversible environmental changes, some of which are so severe that the shrimp culture itself is under threat. Most shrimp farming in Southeast Asia has occurred or still takes place on reclaimed mangrove forests.

Although there are still many of the more traditional style aquaculture farms in production, the dominant trend during the past decade has seen most farms built as either semi-intensive or intensive operations. Unfortunately, a number of climatic, environment and monetary factors make the shrimp farmers follow the principle of adapting short-cut ways for profit making and, therefore, farmers have been moving to even greater levels of intensification and, as the extensive style has given way to intensification and industrialization, the degree and extent of environmental damage intensifies too.

Modern shrimp farming is clearly unsustainable ecologically because its operational requirements vastly exceed the carrying capacity of surrounding ecosystems. The environmental pressures from industrialized shrimp farms have impacts well beyond the boundaries of the immediate site itself. For instance, studies have shown that a one hectare (2.5 acres) semi-intensive shrimp culture system producing about 4000 kg of shrimp annually requires the productive and assimilative capacity of between 38 and 189 hectares of natural ecosystem per year. Higher intensity farming operations require even greater levels of support from the surrounding environment. Such systems are extremely inefficient from an ecological energy standpoint, using approximately 295 Joules of ecological work in order to produce just one Joule of edible shrimp (Mahmood, 1986; Larrson et al., 1994; Mahmood et al., 1994; Deb, 1998).

The shrimp culture industry is mostly dependent on wild sources for fry. Fry catching has become such a profitable business that thousands of poor landless and unemployed coastal people are engaged along the coastline for catching tiger shrimp fry from the nearshore waters of the Bay of Bengal. The catch includes different types of marine, estuarine and freshwater shrimps and prawns, crabs and mollusks, and a wide variety finfish and zooplankton. A large number of young shrimps and fishes are destroyed before attaining biologically sustainable size due to intensive fishing pressure. The overexploitation has been reported to result in decreasing the availability of shrimp fry from year to year and posing disastrous impacts on artisanal and commercial fisheries

(Deb, 1998). Besides, most of the semi-intensive shrimp farms depend on fishmeal-based pelleted feeds which will lead to even greater fishing pressure globally as demand for fishmeal increases with increase in area and intensity of shrimp farming. This contributes to the already unsustainable fishing pressure that is pushing many fisheries in the developing coastal countries into quite serious declines.

It is reported that the target species, the tiger shrimp, P. monodon constitutes only a very small portion of the total catch. Consequently, huge mortality and loss of other species have been reported for every single P. monodon post-larvae collected from the wild. Hoq et al. (2001) reported that 12–551 post-larvae of other shrimps, 5–152 finfish larvae and 26–1636 macrozooplankton are wasted during the collection of a single P. monodon PL in major rivers of the mangrove forest in Bangladesh. Karim (1986) and Kamal (2000) suggested that *P. monodon* post-larvae constitute at most 1% of the total collection in Bangladesh, amounting 1–3 billion/year. Therefore, the remaining 99% would count about 99–297 billion. Recent report shows that more than 100 000 collectors were engaged in this activity with a total loss of about 180 000 million other aquatic organisms in Bangladesh (Kamal, 2000). Clearly, the shrimp fry fishery posed serious impacts on the regional biodiversity and aquatic community structure not only by directly killing a vast number of the fish and shrimp species but also by reducing the food availability of other organisms such as aquatic birds, reptiles linked through the food web.

The sustainability of the practice of harvesting wild PL has been questioned, in particular as the aquaculture industry has expanded and the demand for post-larvae has increased over the past two decades. Begum & Nazmul (2002) stated that the overall survival from the egg stage to recruitment is highly density dependent; they suggested that if density-dependence occurs after the PL stage, then the harvesting of PL would have little effect on recruitment and if density-dependence occurs before the PL stage, then PL harvesting would have a direct and proportional effect on recruitment and fishery yield. Based on the comparison of growth rate and corresponding economic value of the PL between those left in the wild (to be harvested by capture fisheries) and those collected for aquaculture rearing (6.0 g and 0.01 US$/PL for those left in the wild vs. 0.38 g and 0.0025 US$/PL collected for aquaculture rearing), they concluded that PL harvesting is not intrinsically irrational or wasteful. However, this conclusion was not based on practical time-series data on the stock size, recruitment pattern and the level at which wild shrimp fry can be exploited without affecting the original wild stock. The authors also ignored the ecological importance of the fishery as well as the loss of other fishery resources associated with shrimp fry collection. Such comment is, therefore, under extreme criticism while shrimp fry fishery itself is in a steadily declining trend. Hoq et al. (2001) reported that the abundance of P. monodon PL was significantly reduced from 1992 to 1999 and the shrimp fry fishery is responsible for declining other important fisheries such as Hilsa fishery and crab fishery in coastal Bangladesh.

Mangrove removal is associated with coastal erosion, change in sedimentation pattern and shoreline configuration. Massive destruction of mangroves in the southeastern part (the Chakaria Sundarbans) of Bangladesh since 1982 in the name of 'improvement of shrimp culture technology' was patronized by the 'Asian Development Bank' (ADB) and the World Bank (Deb, 1998). Huge areas of mangrove forests were cleared for shrimp farming. Rapid deforestation has led to increase in both the discharge rate and sediment load of the river resulting in increasing siltation of the distributary channels of the delta.

Massive clearance of large areas of mangroves has resulted in reduction of natural production of fish and shrimp larvae that use mangrove for feeding, breeding, and nursery grounds. Therefore, at present, shrimp seeds are not sufficiently available for stocking and this situation, compounded by unavailability of enough hatchery produced seed has resulted empty and abandoned farms. According to FAO/NACA (1995), because of largescale destruction of mangroves for shrimp culture, the natural breeding grounds have been lost to a great extent. The clearing of mangroves resulted in an increased vulnerability to cyclones and tidal waves. The conversion of mangrove forests would not appear to be economic if the potential yields were compared with the combined yields of the forests and traditional fisheries, which are now both lost.

Mangrove soils developed from seawater sediments contain high sulfides, which occur in the form of iron sulfides (FeS and FeS_2). Under conditions of severe acidity, solubility of aluminium, iron, and manganese increases and these may cause phytotoxicity, unavailability of nutrients and reduction in soil fertility. Deb (1998) reported the effects of high concentrations of aluminium and high acidity on shrimp farms constructed through mangrove cut. Potential acidity causes severe stress and deaths of fish and shrimps have been reported due to gill clogging by ferric hydroxide resulting from pyritic soils through complex ecological processes. Because of the growing toxicity, a number of agricultural and aquacultural farms are left abandoned at present.

Mangrove forests have been playing a very important role as a protecting wall against the devastating cyclones and tidal surges by deflecting and reducing energy. It has been reported that the southeastern coastal zone of Bangladesh with fewer mangroves had been more vulnerable than that of southwestern zone with more mangroves. Inundation of large areas with water for shrimp farming has led to intrusion of salts in the surrounding soils resulting in salinization of soil and altering soil chemistry. Long time inundation of lands does not allow other cropping and prevents free nitrogen fixation and mineralization and soil fertility reduces within few years (Rahman, 1993). Continuous inundation also posed tremendous impacts on the underground water of the coastal districts of Bangladesh.

Underground water is used to reduce the salinity of water at the early stage of culture period. Intensive culture systems also need water exchange at a rate of 10–50% of the pond volume each day. This water demand places a tremendous burden on local ground water supplies, rapidly depleting local freshwater resources. Huge amounts of water are used for this purpose in each cycle of culture resulting in lowering in underground water table and consequently drying of wells. In the coastal areas of the country, loss in crop production, freshwater crisis, loss of fodder as consequences of salinity intrusion are common.

It has been reported that only 17% of the feeds supplied to the shrimp farms become harvestable tissues and the rest (83%) is lost and find way into the water in various forms of wastes (Primavera, 1994; Deb, 1998). As a result, the waters discharged from the shrimp farms are extraordinarily high in organic loads from dissolved and particulate organic matters in the form of detritus and unused feeds, metabolites, various forms of toxic and hazardous chemicals, fertilizers, feces, microbes etc. The volume of wastes occurs at a very high amount when the ponds are dried for complete harvesting of the crop. The nutrients and organic loads in shrimp pond effluents are potential sources of detrimental effects on the receiving waters such as reduction in dissolved oxygen, production of hazardous gases and off-odors through microbial decomposition, hypernutrification and eutrophication, higher rate of sedimentation resulting in altering primary and secondary productivity, changes in benthic community structure etc. The wastewater is discharged directly to the surrounding waters.

The higher intensification requires regular fertilization and supplemental feeding to stimulate higher growth. In addition, a wide variety of other chemicals such as herbicides, pesticides, lime, antibiotics, preservatives, disinfectants are used in shrimp farms. It is supposed that some of these chemicals particularly the antibiotics and some persistent ones might lead to the proliferation of resistant pathogens complicating further disease treatment and due to the residual effects of the persistent chemicals, they get deposited through various food chains in different animals even in man.

Facing scarcity of wild shrimp seeds, substantial amounts of shrimp fry were imported from other countries. There is evidence that shrimp pathogens have been widely disseminated through introduction and transfer of shrimp fry. The 'Infectious Hypodermal Hematopoietic Necrosis Virus (IHHNV)' is suspected of being spread from its natural range on the Pacific Coast of Latin America to the Middle East and Asia through transshipment of infected shrimps (Deb, 1998). The 'Monodon Baculovirus' (MBV) disease is also suspected to spread from the same source (Lightner et al., 1989). The serious outbreak of disease in the shrimp ponds of Bangladesh since mid 1990s hints to the danger of importing shrimp fry.

Socio-economic impacts

Although shrimp industry tries to promote itself as a boon to the local economies, it is largely only a relatively few investors that have received the lion's share of benefits, while large portions of society, particularly the rural poor, have become disenfranchised and marginalized into severely degraded environments. Repeated claims have revealed that the ownership and the tenurial processes of shrimp culture ponds are handled by the socially advantaged groups of people who are sufficiently powerful and have access to the necessary capitals and credits and these groups usually gain the benefits (Deb, 1998). It is very interesting and surprising that despite the involvement of internationally reputed organizations like ADB and World Bank in the policy making and implementation, the original objectives of various national and international efforts for improvement of shrimp culture are not met in most instances. The land use patterns, and leasing and sub-leasing systems in the coastal regions of Bangladesh have resulted in a chain of tenurial hierarchy. Competition between traditional fishermen and family (peasant) farmers, and entrepreneurial shrimp farmers for the use of good quality land, coastal fisheries habitats like mangroves, and other critical areas along the coastline increases.

The transformation of multiple-use coastal resource system into a privately owned single-purpose use deprives the coastal communities of their traditional resource use rights who have been using the resource in question for generations. This has also been reported from many other tropical countries (Balley, 1988). The expropriation further aggravates the livelihood of the vast majority of desperately poor coastal communities, which, can be quoted after Deb (1998) as 'As observed in the aquatic trophic structure, predatory "big fishes" of the society engulf the benefits of "small" social creatures'. The traditional patron-client relationship among the local rich and poor in the form of inter-linkages of various exchanges is being gradually reputed to such an extent that the poor are either migrating out of the area, or forced to accept a demeaning life style.

It is said that shrimp play a significant role in providing animal protein to the people. But, in fact, this seems to be not true. The coastal people have never had the opportunity to taste the tiger shrimp. Most of the coastal people do not have the buying capacity of the costly shrimps, which are processed and exported abroad. However, low quality and diseased shrimps discarded from the processing plants are sometimes sold to the local markets.

Modern shrimp farming is capital rather than labor intensive and, therefore, results in labor utility discrimination and provides limited employment opportunities for coastal residents and most of those are typically poorly paid seasonal and non-skilled jobs, offering no long-term job security. It has been reported that shrimp farming displaces more jobs than it creates. Again, the leaseholders from outside generally do not trust the local laborers and rather hire workforce from their own localities. The unemployment problem being very acute in the coastal areas and such bias against local laborers is a source of discontent among them. As a result, people are forced either to migrate to the urban areas compounding the already complex environmental and social conditions of the urban areas or to resort to destructive fishing or improper methods of husbandry in order to extract livelihoods from the coastal areas.

In the coastal areas, the poor are becoming economically more and more vulnerable due to shrimp culture. Livestock and trees are important private resources which save the poor in overcoming vulnerabilities. In the shrimp farming areas, common grazing grounds have been converted to shrimp ghers (large ponds) and owners do not allow the cattle to pass over the dikes, thus the grazing areas are restricted (the results of such restrictions on the livestock and plant will be discussed in the next section). The depressions and canals, earlier used by the poor fishermen for their livelihoods have been turned into shrimp ponds. In the remaining canals virtually no fish are available due to discharge of the shrimp farm effluents containing toxic chemicals and hazardous substances and to overfishing by the fry catchers. Fishing by cast nets around shrimp culture areas has virtually been banned for security reasons of the shrimp farms. All these have resulted in extremely damaging effects on local food security.

Economic and social marginalization is not the end of the sufferings of the ill-fated poor. The coastal paddy farmers very frequently fall in victim in the hand of shrimp farm owners. There are growing conflicts among the shrimp farm owners and paddy cultivators in the coastal areas, which frequently end up through tragic bloodshedding incidents and even deaths. The ultimate consequence of such conflicts further marginalizes the victims and the local administrative bodies have been reported to favor the powerful owners. There are also enough incidents of murders through clashes among the farm owners themselves on the issues of territorial disputes and occupying lands forcefully. The shrimp farm owners hire professional terrorists to fight with local peoples when disputes reach climax.

The biodiversity impacts of mangrove utilization are profound on the wild fish stock. It has been reported that about 12 species of fish are considered as endangered or threatened due to their gradual disappearance as a result of change or conversion of habitats (FAO/NACA, 1995). Although precise information regarding the effects of import of shrimp fry from other countries on the native genetic stock are not available, there are many evidences that introduced genetic material often has a lower reproductive successes. The data on the reduction in plant and animal biodiversity and production in the coastal residential areas due to shrimp culture are very frustrating as well as alarming. It has been shown that during 1986–1989, total production of cows and buffalos have been declined by 47% and yield by 25%; mortality of chicken is very high due to higher salinity resulted from salinity intrusion; poultry production declined by 36% in shrimp culture areas (Deb, 1998). Poultry farming have become difficult due to reduced natural food and degrading economy of the rural people which does not permit them to buy feeds for the poultry. Productions of other green crops along with paddy have been reduced to a great extent. Significant reduction in the fruit trees such as coconut, mango, jackfruit, bettlenut etc. have been reported from both Cox's Bazar and Khulna regions. Most of the fewer trees still remaining are almost fruitless even in the season (Deb, 1998).

Worldwide, rural community people affected by the encroachment of shrimp farmers have started striking back at aquaculture owners. Shrimp farming is now meeting with significant resistance by local communities to its further expansion. Protests have centered on issues of pollution, takeover of lands, access to water resources, destruction of mangroves, and the deterioration of surrounding soil and water quality. Sometimes violent clashes occur, some ending in beatings and even death for some protesters. In one incident in Bangladesh two villagers lost their lives, one of them killed by a bomb attack arranged by shrimp-farm owners. Similar protests have been reported in other shrimp producing countries such as Ecuador, India, and Honduras.

Habitat restoration and conservation

The Sundarbans mangrove forest of Bangladesh is most likely to be changed directly or indirectly by human activities and a priority question in conservation planning is what to expect from the Sunderbans in the future. It is very difficult or almost impossible to rehabilitate a mangrove forest to its original state once it is destroyed severely or completely. This is because the changes that the ecosystem undergoes and the conditions under which the natural mangrove vegetation flourish are hardly likely to restore. However, like some other countries, some efforts and strategies have been made in Bangladesh and, hopefully, many authors have reported the Sundarbans as the most important area left for the long-term survival of many forms of wildlife because the area is large enough to support a large effective population size (Seidensticker, 1986, 1987; Tilson & Seal, 1987).

Although mangroves are exploited for a diverse array of purposes, the practice of afforestation in the mangrove areas is almost recent. Information regarding the nursery raising and planting techniques for mangrove species still are not adequate and complete (Siddiqi & Khan, 1996). However, to some extent, afforestation with mangrove species have been standardized and are implemented with varying degrees of success in sites which have not been too badly degraded (Siddiqi et al., 1993).

The program of afforestation in Bangladesh was initiated in 1966, and up to 1990 and an area of 0.12 million hectares had been brought under plantations which have substantially increased the total area of mangrove during the early 1990s (Fig. 5). Primarily, the plantation was restricted to the commercially important species. Virtually, all the commercially important species have been

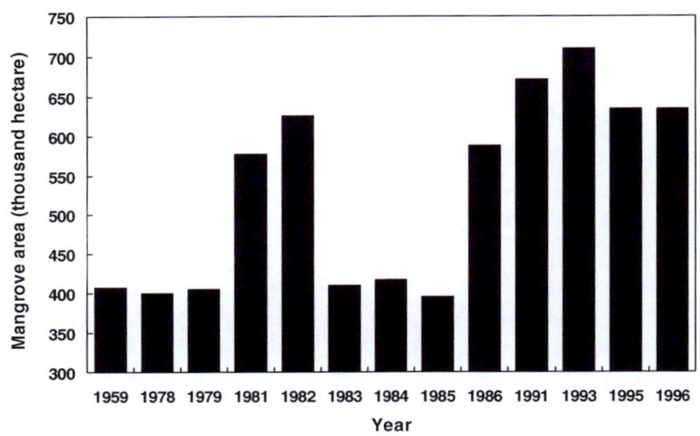

Figure 5. Changes in the total areas of mangrove in Bangladesh over the last few decades.

tried in the coastal belt of Bangladesh, but, among them, only two species, namely, *Sonneratia apetala* and *Avicennia officinalis* showed better performance constituting about 90% of the plantations (Siddiqi et al., 1993). Consequently, being the major planting species, more information on the nursery techniques on these two species are available in comparison to the other species. At present, trials with all the commercially important plant species are being carried out with the objectives of accelerating the process of siltation and stabilization of soil, creating forest shelterbelts to protect life and property of inland from tidal bores, creating an urgently needed resource to add to the national wealth, creating job opportunities for the rural communities, and creating an environment for wildlife, fishes and other estuarine and marine fauna.

Species used in coastal afforestation include *Sonneratia apetala*, *Avicennia officinalis*, *Rhizophora gymnorhiza* and *Casuarina equisetifolia* (MOEF, 1996, 1998; Haque, 1999). For plantation in the mangrove areas, species that produce long tapering propagules are usually desirable because they are more likely to propagate easily and flourish naturally. But, plantations with such species have been difficult, particularly in cases where soils have become badly desiccated and hard; similarly, in areas where pan formation has taken place below the soil surface, plantations with species which have long roots have been difficult (Hussain & Acharya, 1994). However, different types of approaches and their alternatives are employed to overcome these difficulties.

Annual planting continuously increased and reached a peak of 22 800 ha in 1985, of which coastal plantations were about 10 000 ha (MOEF 1993a). Recent inventories and estimates generally note that 20–30% of all plantations established during the last 30 years no longer exist. Officially, the reported total plantation area in the country in 1990 was 332 000 ha of which 113 000 ha were in coastal regions (MOEF, 1993b).

The Forestry Master Plan (1993) suggested two scenarios. The first one proposed an annual planting target of about 18 000 ha during 1993–2002 and 21 000 ha during 2003–2012 (MOEF, 1993b). However, use of quality planting material, site preparation and post-establishment maintenance has not been given adequate attention. Due to budgetary and legal constraints adequate protection of plantations from fire, grazing, illegal removal and encroachment has not been provided (MOEF, 1993b). The second scenario emphasizes on the development of wildlife sanctuaries; consequently, three areas of the Sundarbans Forest Division have been declared wildlife sanctuaries. However, the total area of these sanctuaries is not sufficient to provide long-term protection to the wildlife of the forest.

Planning and policy aspects

The first management plan for the Sundarbans was formulated in 1893 for a period of 6 years. The Sundarbans was declared a reserve forest in 1875–1876, under the forest act of 1855. The Pakistan portion of the Sundarbans became under Forest Division after the partition of India in 1947. The Sundarbans is now administered by the Bangladesh Forest Department under the Ministry of Environment and Forest. To regulate the activities within the forest and to ensure sustainable resource use from the forests, several policies have been formulated, amended, modified and improved over years.

After the liberation the first National Forestry Policy of the Government of Bangladesh was formulated in July, 1979. Since then, efforts have been made to orient the policies to meet current need, particularly as they relate to the rapid depletion of forestry resources owing to numerous socio-economic factors. As part of this, the Government has formulated another and the latest National Forestry Policy (1994) for a period of 20 years. The objectives and the major policies have been summarized below (Hussain & Acharya, 1994; ADB/UNDP, 1995; FAO/UNDP, 1998).

The national forestry policy

To meet the basic needs of the present and future generations and also to ensure greater contribution of the forestry sector in the economic development, about 20% of the total area of the country will be afforested under this policy by taking up various afforestation programs. Side by side, fallow lands (lands not useful for the purpose of agriculture) hinter lands and in other possible areas, Government sponsored afforestation programs will be implemented. Moreover, appropriate measures will be taken to encourage afforestation on private land and technical support and services regarding the production of forest crops have to be ensured. The major policies are as follows:

- By creating employment opportunities, strengthening the rural and national economy, the scope for poverty alleviation and trees and forest based rural development sectors will be extended and consolidated.
- Biodiversity of the existing degraded forests will be enriched by conserving the remaining natural habitat of birds and other animals.
- Agricultural sector will be strengthened by extending assistance to the sectors related with forest development, especially by conserving the land and the water resources.
- National responsibilities and commitments will be fulfilled by implementing various international efforts and government ratified agreements relating to global warming, desertification and control of trade and commerce of wild birds and animals.
- Through the participation of the local people, illegal occupation of the forest lands, illegal tree felling and hunting of the wild animals will be prevented.
- Effective use and utilization of the forest goods at various stages of processing will be encouraged, and Implementation of the afforestation programs, on both public and private lands will be provided with encouragement and assistance.
- Forest shall be carefully preserved and scientifically managed.
- Attempts will be made to bring about 20% of the country's land under the afforestation programs.
- Private initiatives will be encouraged to implement the programs.
- Attempts will be made to increase the amount of protected area by 10% of the reserved forest land by the year 2015.
- Multiple use of forest, water and fish of the Sundarbans through management will be ensured keeping the bio-environment of the area intact.
- Keeping in view the ecology, the management of the forest lands will be brought under profit oriented business.
- Inaccessible areas as slopes of the hills, fragile watersheds, swamps etc. will be identified and kept as protected forests. Modern technology shall be employed for extraction and utilization of the forest produce.
- Emphasis will be given to the forest based industries to ensure effective utilization of the forest raw materials and profit oriented management systems under the free market economy;
- Rules and procedures regarding forest produce in the country will be simplified and made up-to-date.
- Reserved forest cannot be used for non forest purposes without the permission of the Head of the Government.

- Women will be encouraged to participate in homestead and farm forestry, and participatory afforestation programs.
- Fund from different donors will be used to promote private forestry organization and such programs like training, technical and financial support will be imparted at an increasing rate.
- Timber resource to be increased by establishing large scale plantation.
- Ecotourism, related to forest and wildlife, is recognized as forestry related activity, which will be promoted taking into consideration the carrying capacity of the nature.
- Forest Department will be strengthened in order to achieve the goals and objectives.
- Research, education and training will be organized to meet the scientific, technological and administrative needs of the country.
- Laws, rules and regulations relating to the forestry sector will be amended and, if necessary, new laws and rules will be promulgated in consonance with goals and objectives of the National Policy.

Legislation and regulation

The broad principles for forest management in Bangladesh are encapsulated in the Forest Act 1927 (amended in 1989). This Act consolidates the laws relating to forests, the transit of forest produces and the duty leviable on timber and other forest produce. It also makes provision for the government to declare any forest or wasteland as reserved forests, village forests or protected forests. Reserved forests prohibit a variety of activities, including agricultural clearing, cattle grazing, fuelwood collection and harvesting of most forest products. Protected forests provide a lesser degree of regulation in that these activities may be carried out under license. Rights to government forests may also be assigned to communities as village forests. More recently, the Forest Law of 1993 has been enacted to provide a legal basis for implementing forest management practices envisaged under the Forestry Sector Master Plan. A variety of specific forestry, environment and land acts have been formulated, having substantial impacts on forest management. These include: the State Acquisition and Tenancy (SAT) Act, 1950; the Acquisition of Wasteland Act, 1950; the Bangladesh Private Forest Act (PFA), 1959; and the Bangladesh Wildlife (Preservation) (Amendment) Act, 1974.

The Government of Bangladesh launched a Master Plan for Forestry Development in 1990 in an effort to counter ongoing deforestation and forest degradation. The objectives of the Master Plan are to optimize the contribution of forest resources for environmental stability and economic and social development. The Plan covers a 20 year period 1993–2013, and has been incorporated into the country's Fifth National Five Year Plan (1995–2000). As a first step in implementing the Master Plan a new national forest policy was promulgated in October 1994. Three sets of programs were formulated under this policy: people-oriented programs, production directed programs, and institutional strengthening programs. In 1994, a National Environmental Management Plan was also prepared. The main forest-related provisions of this plan are for participatory management of plantations and natural forests.

Forest act

The Sundarbans mangrove forest was declared as 'Reserve Forest' in 1875–1876 under the first Forest Act of the then British India. This act, virtually, transferred the ownership and all the rights of the forest to the government through the Department of Forest and restricted privileges of the public in the forest. The act empowered the Forest Department to permit or prohibit all activities and also made the department responsible for formulating and implementing management plans. The act contains 86 sections of which 25 are concerned with reservation, 10 with protection, 1 with village forests, 6 with transit of forest produce, 7 with drift timber, 24 with penalties and procedures, and the remaining 13 with miscellanea. The important features of this act with respect to mangrove are as follows:

- All rights or claims over the forest have been settled at the time of reservation of the Sundarbans and the grant of any new right to individuals or communities is prohibited.

- Any activity within the Reserve Forests of Sundarbans, unless permitted by the Forest Department, is prohibited.
- Any kind of violation in the Sundarbans Reserve Forest may result in court cases with fine or rigorous imprisonment, or both.
- Any offense may be compounded by the Forest Department upon realization of the fine.
- Without any warrant, the Forest Official can arrest anyone for violation of the Forest Act.
- The Forest Department is also empowered to regulate the watercourses within the Sundarbans.
- The Forest Department is also empowered to regulate the rivers with regard to transit to and transportation of timber and other forest produce.

Wildlife act

In order to strengthen the power of the Forest Department with regard to conservation of wildlife, the Wildlife Act was promulgated in 1973. The act has 48 sections, 18 of which relate to procedures regarding the disposal of offenses and claims, 2 to framing of rules, 7 to export, import and possession and the remaining 21 relate to various other aspects of wildlife. The major features of the act are:

- The forest officials are empowered to prosecute violators for emphatically protecting the wildlife.
- Hunting of wildlife is restricted only to certain species.
- Hunting and trapping of any of the threatened or endangered species is prohibited.
- Possession of specified wildlife materials requires certificates from the Forest Department.
- Upon realization of fine, some of the offenses can be compounded by higher officials.
- Man-eating tigers cannot be killed without permission from the Chief Warden.
- Forest Officials can arrest anyone for violating the provisions of the act.

Brick burning act

Fuelwood was once used extensively for manufacturing bricks. The Brick Burning Act was promulgated in 1991 to regulate the use of fuelwood in brick manufacturing plants. Heritiera fomes and Ceriops decandra are considered high quality fuelwood for brick burning and were in high demand for this purpose. At present, a limited quantity of fuelwood is allowed but use of coal fuel in brick burning is strongly recommended.

Sundarbans forest transit rules

The Forest Transit Rules formulated in 1906 and revised in 1969 are in action till now. There are 15 rules 6 of which are concerned with details of working, 7 with permit issuance, 2 with licensing of boats used in transportation of the forest produce. The important features of the rules are:

- Prohibition of cutting or converting timber without a pass.
- Prohibition of transit of forest produces without a valid transit document.
- Any permit issued for collection of any forest produce must contain specific information about what type of produce to be collected, where to be collected, and when the collection would have been completed.
- The boat used in the transportation of the forest produce must have a valid Boat License Certificate (BLC).
- Any forest produce on transit from the forest will pass through one of the 16 notified Forest Revenue Stations located on the banks of the rivers on the boundary of the forest.

Felling and cutting rules

These rules focused mainly on those purposes for which trees are most likely to be collected such as for use as firewood and on those species which are most likely to be destroyed such as Ceriops decandra and Nypa fruticans. Besides the rules and regulations, the government has made several agreements with the industries, which are dependent on the forest for raw materials. The industries are newsprint mills, match factory, hard board mills etc. which require huge quantities of *Excoecaria agallocha*, Heritiera fomes, and *Cynometra ramiflora*. Upon the agreement, the government has decided to reduce the annual supply of these woods, but on a regular basis, which, on one hand, ensures regular supply of the raw materials to the industries, and help conserving the resources on the other.

Institutional issues

In view to maintain the ecological balance and to develop the resource for sustainable utilization, the Government of Bangladesh has formulated and established different management policies and action plans. At the same time, various non-government organizations (NGOs) and private organizations have engaged themselves in research, management and development processes. The Government also provides institutional, infrastructural and legal support to encourage participation of such organizations. Moreover, International organizations such as CIDA, AIDAB, USAID, JSPS, UNDP, UNESCO, FAO, ADB, and World Bank have sponsored research programs on mangrove ecosystems and various development activities.

The FAO/SIDA Bay of Bengal Program (BOBP), First Aquaculture Development Project (ADB), Shrimp Culture Project (IDA), Second Aquaculture Development Project (ADB), Third Fisheries Project (World Bank) have all contributed along with the national policy making and funding in the improvement of shrimp culture in Bangladesh. The currently operating Fourth Fisheries Project funded by the World Bank also has its component for promotion of shrimp aquaculture.

Status and trends in forest management

Scientific forest management and conservation in Bangladesh has a history of more than a century. Forest management in the country naturally encompassed the various silvicultural and management techniques developed under British colonial era. The strictly ecological approach to forest management has foundered in more recent years in the face of rapidly mounting population pressures and poverty-driven demands for additional agricultural lands. Consequently, the focus of forest management has moved quite rapidly toward a community-based approach. This approach recognizes that effective forest management in conditions of intense population pressure requires community participation in planning and implementation.

The Sundarbans mangrove forest has a distinct forest management history. The area was mapped as early as 1764, soon after proprietary rights were obtained by the East India Company in 1757. The first Forest Management Division with jurisdiction over the Sundarbans was established in 1869 and the first management plan was introduced in 1892.

Current management objectives

The Master Plan for Forestry Development stipulates that forest resources should be managed in an ecologically sustainable manner to ensure species diversification and continuous improvement of environmental conditions. State-owned forests, comprising natural forests, hill plantations and Sal (*Shorea robusta*) plantations are used to produce forest products, except for reserved forests, which are set aside to conserve soil and water resources, and for the maintenance of biodiversity. Commercial management of forests is expected to be profit oriented. The Master Plan aims at rehabilitating denuded or encroached forest areas using participatory approaches. The Master Plan recommends a strategy incorporating six key points for forest management:

- Enhancing environmental preservation and conservation;
- Introducing rational forest land use;
- Increasing public participation and benefits from resource management;
- Expanding the resource base;
- Improving management practices; and
- Undertaking efficient resource allocation.
- Forest management plans.

The shift toward participatory forestry means that new plans, with significantly altered management prescriptions, will be required. This planning process is, however, very much in its infancy. Nonetheless, efforts to involve stakeholders in forest planning are being made in Bangladesh. Workshops and seminars have been arranged at divisional levels to enable 'grass roots' views to be expressed and to share local experiences in forest planning. Forests managed for production purposes comprise around 550 000 hectares, conservation forests constitute 220 000 hectares, while forests for coastal protection, most notably the Sundarbans area, cover 350 000 hectares. The Sundarbans Reserved Forest is the focus of several conservation efforts, including an FAO-executed project, which helped draw up an integrated resource management plan for the area. The planwas published in 1998 and has objectives: to produce a plan for integrated resource management designed to enhance the supply of wood and non-wood products; to conserve and manage aquatic and terrestrial wildlife resources; to study the potential for mobilizing and assisting people's participation in income- and employment generating activities in the area (with particular focus on disadvantaged groups); to develop the tourism and recreational potential of Bangladesh's forests; and to enhance the protective role of forests against cyclones, soil erosion and tidal surges. In mangrove forests the principal silvicultural system is 'Selection cum Improvement'. This system involves selection of individual trees and small coupes for commercial felling, based on specified diameter limits. Diseased and undesirable species are also felled, to promote growth of more valuable species.

Mangrove management problems

The problems associated with management of the mangrove resources are many and varied. These problems are closely related to the rapid destruction of the mangroves, conversion of mangroves to aquaculture and other forms of land uses and other related impacts. In defining the management problems of mangroves it is better to identify the root causes of the problems rather than to look at the most dramatic forms of alteration of the mangrove system alone. Identification of the problems associated with the mangrove must also take into account the social, economic and other related events particularly when the development issues are concerned. Some of the major problems concerning mangrove resources management and development are listed below.

The management agencies are formed on the basis of economic sectors giving major consideration on economic benefit from the resources. The complex biophysical mechanisms occurring in the ecosystems and their relations with the management are rarely considered. Consequently, persons with sufficient knowledge on the ecological basis of mangrove management and development are ignored from the overall management systems.

Mangroves are very complex ecosystems formed by interactions between land, water, flora and fauna. But the country's management policies are such that, in most cases, management policies are formulated separately for each individual unit by the authorities concerned. In most instances, there is no coordination between different sectors, and their policies are conflicting. The need for a quite different management strategy for mangrove forests rather than managing separately each unit has not yet been realized. As a result, effective exploitation, management and conservation of mangrove resources are far away.

Many of the mangrove development policies cannot be successfully implemented due to various reasons. Corruption at different levels of management systems, illegal felling of trees, poaching of woods are some of the major constraints in successful implementation of development projects. There are frequent claims that the law enforcing agencies and the management bodies themselves are sometimes engaged with the illegal actions. If this situation prevails, virtually, no policy, rule or regulation could contribute to the development of the resources.

Although the mangrove lands are under the ownership of the state the accessibility of the general people is not always regulated properly. People even engage themselves with agricultural activities within the restricted areas. Implementation of development policies is hindered to a great extent by shortage of manpower for official management, including monitoring, evaluation, supervision etc. As in other Asian nations, there is a general shortage of resource management expertise in proportion to the scale and rate of development related pressures in Bangladesh.

Although the management history of the mangrove is quite old, yet, management strategies based on logical and scientific basis have not been developed. Sufficient research efforts have not yet been paid to find out effective management policies. Although the forest management system of the Sundarbans has focused on the sustained production of forest products the management systems has not taken into account the mechanisms to maintain best achievable ecosystem function (Clark, 1974).

Poor management policies, weak organizational and legal efforts and, above all, the population pressure have subjected the mangrove forests to massive destruction. Destruction of mangrove resources in Bangladesh is so high that irreversible long-term changes have taken place in the ecosystem affecting the balance and sustenance of the system. There is an annual loss of some 1500 ha of natural forest which is more than compensated for by an annual plantation establishment of about 18 000 ha. For the sectoral economic interest, traditional patterns of use of forest by the local people have been restricted to a great extent, which, appeared as threat to the poor rural economy of the country.

Any sector without a reliable guideline can be treated as a boat without a helm. The need for a basic guideline for mangrove resources is now fully recognized to overcome the problems of unplanned destruction. Most of the people in the rural areas of Bangladesh are poor and illiterate. They do not have the knowledge of any biophysical or socio-economic event of the mangroves and are not aware of even their own legal rights. As a result, very often they are deprived and, when they get chance, they exploit the resources in an irrational way.

Best practice recommendations

The principal challenges for Bangladesh arise from intense population pressures on meagre and stressed forest resources. Nearly half the country's population is living below the poverty line and there is, consequently, severe conflict between environmental protection and enhancement objectives on one hand, and economic development and poverty eradication objectives on the other. The overriding imperative is to balance social, economic and environmental goals to avoid ecological disaster. In this context, the key issues for forestry are to bring forests under environmentally sound management, while also enhancing their economic productivity.

Problems regarding the exploitation, management and development of the mangroves in Bangladesh are many and varied as well as complex. To ensure sustainable resource use and development of the mangrove resource, a well-coordinated and complete management plan is the first need for the mangrove of Bangladesh. The plan should be formulated on the basis of a clear understanding of the mangrove systems including the factors affecting it and the impacts of different forms of resource use on the system rather than considering the forest as merely a collection of trees. The plan should also consider the social and economic conditions of the people directly or indirectly associated with the mangrove forests. Sustainable development requires, among others, an integrated decision making process, a political system that secures effective citizen participation, an economic system that generates surpluses on a sustained basis, a social system that provides solution to the tensions arising from disharmonious development, a production system that respects the obligation to preserve the ecological base, and an administrative system that is flexible and has the capacity for self-correction (Deb, 1998).

Regular and up-to-date data of the forest and other related resources should be available of the authority. For this, a continuous and effective program of resource assessment of the mangrove should be formulated and implemented. Intensive and exhaustive research should be undertaken to study the basic functional systems of the mangrove ecology and the factors affecting the system. This will help finding out all possible way of mangrove utilization as well as developing models for optimizing and integrating multiple use of resources without causing any temporary or permanent damage to the system. The research should be conducted on the basis of a multi-disciplinary approach, ensuring participation of all the related disciplines.

A national institute for multi-disciplinary research on mangroves should be established. This institute, along with research programs, may take the responsibilities for formulation of management plans and conservation strategies. Bangladesh is also suitable for establishment of a Regional Mangrove Research Institute. Establishment of such institute could be helpful in disseminating and exchanging research findings amongst the regional countries as well as sharing benefits from the experience of other countries. Intensive studies should be conducted on the environmental impact of different forms of resource use, which must take into consideration the socio-economics of the mangroves.

Conversion of mangroves to aquaculture or any other land use should proceed with extreme caution and must be followed by the environmental and socio-economic impact assessment, considering the probable damage to the resources because of such conversions. All conversion programs should be carefully evaluated both ecologically and socio-economically. For promotion of recreational and tourism activity in the mangrove areas, 'Mangrove national Parks' and 'Virgin Jungle Reserves' can be established, which can help developing public awareness. Experience can be shared from other mangrove-dominated countries like Malaysia or Thailand.

A 'National Mangrove Committee' should be established which will be responsible for all aspects of mangrove utilization, conservation, research and management. To facilitate the development of an integrated management plan, there should have regular communication and exchange of data among groups, which actively support sustainable use of mangroves.

For the fisheries and aquaculture sector, there should have clear understanding of the fisheries implication of the mangrove forests. Studies should be conducted on the biology and life history as well as on the potential benefits that could be obtained from the mangrove fisheries resources without affecting the long-term sustenance. Emphasis should be given on technical and financial assistance, which stimulates more efficient use of existing resources rather than that, which encourages further destruction and degradation. The experts should look forward to stimulate the development of such kinds of aquaculture in mangrove, which are compatible with the mangrove forests or to the probable alternative sites outside the mangrove areas for establishment of aquaculture industries.

The population pressure and the economic situation of the country is such that there will be a continuing pressure on the forest resources of the country in the form of settlement, farming activities, cattle grazing etc. which could lead to serious consequences. The experts and authorities should, therefore, look forward to alternative sources and patterns of land use.

In Bangladesh, a realistic national policy for shrimp cultivation is yet to be formulated. The Government should formulate a complete and compatible guideline for brackish water aquaculture particularly for shrimp culture which must provide specification on all aspects of shrimp farming activities including the types and amounts of drugs and chemicals that can be used in the farms, the effluent discharge, import of shrimp fry, catching of fry from the wild stock etc. Such a policy should ensure regular and sufficient institutional supervision and monitoring of the operational procedures, application of the impact assessment, and implementation of the regulatory measures. The guideline should necessarily include an appropriate land use pattern to resolve the socio-economic conflicts. The Government should strictly regulate the import of the banned poisonous 'dirty dozen' and other chemicals of the 'red group' from other countries through back doors.

For large-scale production of shrimp seeds, it is extremely important to establish commercially viable hatcheries to support the shrimp industries of Bangladesh. This will play a great role in enhancing production by reducing the problems originated from import of fry as well as to conserve the biodiversity by reducing dependence on the wild stock for fry. By the side of the government, the non-government organizations (NGOs) and the private sectors could play a significant role in this process. The government should provide infrastructural and legal support to the NGOs and private sectors to encourage hatchery establishment. Development of nursery and larval rearing techniques and facilities for regular monitoring of water quality are also of considerable importance. It is also important to ensure regular supply of quality shrimp feeds at low cost.

Destruction of mangroves in the name of shrimp farming must be banned. To protect the resources from destruction, participatory approaches including the local people could be effective. To help promote the development of the brackish water aquaculture, bank financing should be extended to the private entrepreneurs.

Conclusion

The beneficial influence of mangrove forests is not limited to the immediate geographical area in which they grow. They exert balancing ecological influences that are felt far beyond their immediate boundaries. Although exploitation of mangroves for a long time and large-scale conversion of mangroves to aquaculture and other forms of resources have caused serious consequences, it is not impossible to rehabilitate the resources if proper attention is paid and effective management programs is formulated and practiced. A number of reports suggest that the mangrove Sundarbans of Bangladesh has been under a simple, easy-to-implement management regime which was installed more than a hundred years ago. The management regime has ensured sustainability and if the management practices are followed properly, this will ensure maintenance of the stands in stable condition for an indefinite period of time.

Bangladesh is as rich in terms of its natural resources as it is poor in terms of its inadequate resource utilization and mobilization of its large human population to productive and development purposes. Therefore, Bangladesh has to give a new look to its natural resources particularly to the coastal resources of how they can be used for sustainable social and economic development of the country.

In old days, there was very poor knowledge on mangroves, and the forests were exploited without any scientific background, not considering the future implication of the exploitation. However, in recent days, knowledge of different aspects of the mangrove resources has increased greatly and mangrove resource conservation, management and development have been paid considerable national and international attention. It has been fully realized and recognized in recent times that we are dependent to a great extent on the mangrove forests and they must be conserved for our own interests.

Destruction of mangrove for coastal farming is now a global problem that has created considerable debate worldwide and need proper attention and immediate action. The numbers of shrimp producing and consuming nations are very few. These relatively few countries should provide a clear field of action on specific demands. The governments of these producing and consuming countries have a fundamental responsibility to enact legislation and enforce regulations to protect coastal and marine environments from destructive aquaculture practices. The multilateral development agencies such as the World Bank and the UN Food and Agriculture Organization, along with various national foreign aid programs, which have promoted and/or helped finance the destructive shrimp farming industry must review and amend their policies in this field, and redirect their support into rectifying the damage done.

References

ADB/UNDP, 1995. Development perspectives of the Forestry Sector Master Plan. Bangladesh. Ministry of Environment and Forest. Government of the Peoples' Republic of Bangladesh.

Balley, C., 1988. The social consequences of tropical shrimp mariculture development. Ocean Shore Management 11: 31–44.

Baran, E. & Hambrey, J., 1998. Mangrove conservation and coastal management in Southeast Asia: what impact on fishery resources? Marine Pollution Bulletin 37(8–12): 431–440.

Begum, A. & Nazmul, S. M., 2002. Social aspects of Shrimp aquaculture in Bangladesh. Report prepared under the World Bank, NACA, WWF and FAO Consortium Program on Shrimp Farming and the Environment. Work in Progress for Public Discussion. Published by the Consortium.

Blower, J., 1985a. Sundarbans Forest Inventory Project, Bangladesh. Wildlife conservation in the Sundarbans. Project Report 151. Overseas Development Administration, Land Resources Development Centre, Surbiton, UK, 39 pp.

Blower, J. H., 1985b. Sundarbans forest inventory project. Bangladesh Wildlife Conservation in Sundarbans. Overseas Development Administration (ODA), England.

Chaffey, D. R. & Sandom, J. H., 1985. Sundarbans Forest Inventory Project, Bangladesh: a glossary of vernacular plant named and a field key to the trees. Overseas Development Administration, England. 23 pp.

Chantarasri, S., 1994. Integrated resource development of the Sundarbans Reserved Forest. Fisheries Resources Management for the Sundarbans Reserved Forest. UNDP/FAO. BGD/84/056. Khulna, Bangladesh.

Chaudhury, A. B. & Naithani, H. B., 1985. A comprehensive survey of tropical mangrove forests of Sundarbans and Andamans. Part-I. International Book Distributors, Dehra Dun, India, 41 pp.

Christensen, B., 1984. Ecological aspects of the Sundarbans. FO: TCP/BGD/2309 (Mf). FAO, Rome, 42 pp.

Clark, J., 1974. Coastal Ecosystems. The Conservation Foundation, Washington, DC.

Das, S. & Siddiqi, N. A., 1985. The mangroves and mangrove forests of Bangladesh. UNDP/FAO Project BGD/79/017. 142 pp.

De Graaf, G. J. & Xuan, T. T., 1997. Shrimp farming and natural fisheries in the southern provinces of Vietnam. Coastal Wetlands Protection and Development Project, Ho Chi Minh City, Vietnam.

Deb, A. K., 1998. Fake blue revolution: environmental and socio-economic impacts of shrimp culture in the coastal

areas of Bangladesh. Ocean and Coastal Management 41: 63–88.
DOF, 2001. Fishery Statistical Yearbook of Bangladesh. Department of Fisheries, Dhaka, Bangladesh.
DOF, 1999. Special publication on fish fortnight (in Bengali). Department of Fisheries, Dhaka, Bangladesh.
FAO, 1981. Forest Resources of Tropical Asia, Tropical Forest Resources Assessment Project, Technical Report 3, UN 32/6.1301-78-04, FAO, Rome, 475 pp.
FAO, 1993. An Overview Document, The Existing Situation on Forestation and Future Requirements for Improved Productivity of Man-made Forests in the Member Countries of FORTIP (Forest Tree Improvement Project). In UNDP/FAO Regional Project, Field Document No. 10 RAS/91/004. Rome.
FAO, 2003. FAO global mangrove area estimate. Food and Agricultural Organization of the United Nations. Rome.
FAO/NACA, 1995. Regional Study and Workshop on the Environmental Assessment and Management of Aquaculture Development (TCP/RAS/2253). NACA Environment and Aquaculture Development Series No. 1. Network of Aquaculture Centers in Asia-Pacific, Bangkok, Thailand.
FAO/UNDP, 1998. Integrated Resource Management Plan for the Sundarbans Reserved Forest (vols. 1–3): UNDP/FAO project BGD/84/056: Integrated Resource Development of the Sundarbans Reserved Forest.
Gittins, S. P., & Akonda, A. W., 1982. What survives in Bangladesh? Oryx 16: 275–281.
Haque, M. A., 1999. Site, Technology and Productivity of Teak plantations in Bangladesh, Paper presented to Regional Seminar on this subject at Chiang Mai, Thailand 1999.
Hendrichs, H., 1975. The status of the tiger *Panthera tigris* (Linne, 1758) in the Sundarbans mangrove forest (Bay of Bengal). Saugetierkundliche Mitteilungen 23: 161–199.
Hoq, M. E., Islam, M. N., Kamal, M. & Wahab, M. A., 2001. Abundance and seasonal distribution of *Penaeus monodon* postlarvae in the Sundarbans mangrove, Bangladesh. Hydrobiologia 457: 97–104.
Hussain, K. Z. Sarker, S. U. & Rahman, M. M., 1983. Summer birds of the Sundarbans' 'Nilkamal Sanctuary', Bangladesh. Bangladesh Journal of Zoology 11(1): 48–51.
Hussain, K. Z. & Acharya, G. (eds.), 1994. Mangroves of the Sundarbans. Volume two: Bangladesh. IUCN, Bangkok, Thailand.
Islam, A. K. M. N., 1973. The algal flora of the Sundarbans mangrove forests, Bangladesh. Bangladesh Journal of Botany 2(2): 11–36.
Islam, M. S., 2003. Perspectives of the coastal and marine fisheries of the Bay of Bengal, Bangladesh, Ocean and Coastal Management 46(8): 763–796.
Kamal, M., 2000. Assistance to Fisheries Research Institute – A Report Prepared for the 'Assistance to Fisheries Research Institute'. Consultancy report on Marine Fisheries Resource Management. BGD/89/012. FRIGOB/UNDP/FAO.
Karim, M. & Aftabuzzaman, A., 1997. Brackish and marine water aquaculture: potential, constraints and management needs for sustainable development. In: Report of the National workshop on fisheries resources development and management in Bangladesh. BOBP/REP/74. Bay of Bengal Program. 91, St. Mary's Road. Chennai, India.
Karim, M., 1986. Brackishwater aquaculture in Bangladesh: a review. Report of FAO/UNDP TA-project BGD 83-010. IDA/WB agricultural research II project report no 12.
Kathiresan, K. & Bingham, B. L., 2001. Biology of mangroves and mangrove ecosystems. Advances in Marine Biology 40: 81–251.
Khan, M. A. R., 1982. Wildlife of Bangladesh, a checklist. University of Dhaka, Dhaka.
Khan, M. A. R., 1986. Wildlife in Bangladesh mangrove ecosystem. Journal of the Bombay Natural History Society 83: 32–48.
Larrson, J., Folke, C. & Kautsky, N., 1994. Ecological limitations and appropriation of ecosystem support by shrimp farming in Colombia. Environmental Management 18(5): 663–676.
Lightner, D. V., Redman, R. M., Bell, T. A. & Thurman, R. B., 1989. Geographic dispersion of the viruses IHHN, MBV and HPV as a consequence of transfers and introductions of penaeid shrimp to new regions for aquaculture purposes. Journal of Shellfish Research 7: 554–555.
Mahmood, N., 1985. On Fishery significance of the mangroves of Bangladesh. Paper presented at the Workshop on 'Coastal Aquaculture and Environmental Management' held during 25–28 April, 1995 at Cox's Bazar, organized and sponsored by Institute of Marine Science, Chittagong University (IMS, CU).
Mahmood, N., 1986. Effect of shrimp farming and other impacts on mangroves of Bangladesh. Paper presented at the 3rd session of the IPFC working party of experts on inland Fisheries, Bangkok, Thailand, June 19–27, 1986.
Mahmood, N., Chowdhury, N. J. U., Hossain, M. M., Haider, S. M. B. & Chowdhury, S. R., 1994. An environmental Assessment of the Bay of Bengal region. Swedish Centre for Coastal Development & Management of Aquatic Resources (SWEDMAR). Bay of Bengal Program. BOBP/Rep/67; 75–129.
Martosubroto, P. & Naamin, N., 1977. Relationship between tidal forests (mangroves) and commercial shrimp production in Indonesia. Marine Research in Indonesia 18: 81–86.
MOEF, 1993a. Forestry Master Plan, Statistical data, Government of Bangladesh, Ministry of Environment of Forests, Asian Development Bank (TA No. 1355-BAN).
MOEF, 1993b. Forestry Master Plan, Main Plan 1993-2013, Volume I, Government of Bangladesh, Ministry of Environment of Forests, Asian Development Bank (TA No. 1355-BAN).
MOEF, 1996. Questionnaire reply on the Evaluation of Forest Plantation by the Ministry of Environment and Forests, Government of Bangladesh through FAO Representative dated 4 July 1996.
MOEF, 1998. Final Report: Forest Inventory of the Coastal Afforestation Divisions. By Revilla, J. A. V., Ahmad, I. U. and Hussain and Acharya M. Dhaka. Ministry of Environment and Forests.
Mukherjee, A. K., 1975. The Sundarbans of India and its biota. Journal of the Bombay Natural History Society 72: 1–20.
Pauly, D., 1985. Ecology of coastal and estuarine fishes in southeast Asia: a Philippine case study. In Yanez-Arancibia (ed.), Fish Community Ecology in Estuaries and Coastal

Lagoons, towards an Ecosystem Integration. UNAM Press, Mexico: 499–535.

Pauly, D. & Ingles, J., 1986. The relationship between shrimp yields and intertidal vegetation (mangrove) areas: a reassessment. In IOC/FAO Workshop on Recruitment in Tropical Coastal Demersal Communities. submitted. Ciudad de Carman, Mexico, 21–25 April 1986. IOC, Unesco Paris: 227–284.

Paw, J. N. & Chua, T. E., 1989. An assessment of the ecological and economic impact of mangrove conversion in Southeast Asia. Marine Pollution Bulletin 20(7): 335–343.

Primavera, J. H., 1994. Environmental and socio-economic effects of shrimp farming: the Philippine experience. INFOFISH International.

Rahman, A., 1993. The impact of shrimp culture on the coastal environment. Grassroots – An Alternative Development Journal Published by Association of Development Agencies in Bangladesh (ADAB) 2(8): 5–17.

Rahman, A. K. A., 1975. A checklist of the freshwater bony fishes of Bangladesh. Bull. Freshw. Fish. Res. Sta. Chandpur. 1: 1–18.

Rahman, A. K. A., 1989. Freshwater fishes of Bangladesh. The Zoological Society of Bangladesh.

Salter, R. E., 1984. Integrated development of the Sundarbans, Bangladesh: status and utilization of wildlife. FO: TCP/BGD/2309(MF). Report No. W/R0034. FAO, Rome, 59 pp.

Sanyal, P., 1983. Mangrove tiger land: the Sundarbans of India. Tigerpaper 10(3): 1–4.

Sanyal, P., 1987. Managing the man-eaters in the Sundarbans Tiger Reserve of India – a case study. In Tilson, R. L. & Seal, U. S. (eds), Tigers of the World: The Biology, Biopolitics, Management, and Conservation of an Endangered Species. Noyes Publications, Park Ridge, NJ: 427–434.

Sarker, M. N., 1993. Marine Fishery Resources of Bangladesh (in Bengali). Bangla Academy Science Newsletter: 113–157.

Sarker, S. U., 1985a. Ecological observation on the endangered whitebellied sea eagle *Haliaeetus leucogaster* (Gmelin) in the Sundarbans, Bangladesh. In: Symposium on endangered marine animals and marine parks, Vol. 4. Endangered and/or vulnerable other marine invertebrates and vertebrates. Paper No. 58. Marine Biological Association of India, Cochin.

Sarker, S. U., 1985b. Density, productivity and biomass of raptorial birds of the Sundarbans, Bangladesh. Proceedings of SAARC Seminar on Biomass Production, 15 April 1985, Dhaka: 84–92.

Sarker, S. U. & Sarker, N. J., 1986. Status and distribution of birds of the Sundarbans, Bangladesh. The Journal of Noami 3: 19–33.

Scott, D. A. (ed.), 1989. A Directory of Asian Wetlands. IUCN, Gland, Switzerland and Cambridge, UK. 1,181 pp.

Seidensticker, J. & Hai, M. A., 1983. The Sundarbans wildlife management plan: conservation in the Bangladesh coastal zone. International Union for the Conservation of Nature and Natural Resources, Gland, 120 pp.

Seidensticker, J., 1986. Large carnivores and the consequences of habitat insularization: ecology and conservation of tigers in Indonesia and Bangladesh. In Miller, S. D. & Everett, D. D. (eds), Cats of the World: Biology, Conservation, and Management. National Wildlife Federation, Washington, D.C.

Seidensticker, J., 1987. Managing tigers in the Sundarbans: experience and opportunity. In Tilson, R. L. & Seal, U. S. (eds), Tigers of The World: The Biology, Biopolitics, Management, and Conservation of an Endangered Species. Noyes Publications, Park Ridge, NJ: 416–426.

Siddiqi, N. A. & Khan, M. A. S., 1996. Planting techniques for mangroves on new accretions in the coastal areas of Bangladesh. In Field, C. D. (ed.), Restoration of Mangrove Ecosystems, International Society for Mangrove Ecosystems, Okinawa, Japan: 143–159.

Siddiqi, N. A., Islam, M. R., Khan, M. A. M., & Shahidullah, M., 1993. Mangrove nurseries in Bangladesh. Mangrove Ecosystems Occasional Papers 1. ISME, Okinawa: 14 pp.

Staples, D. J., Vance, D. J. & Heales, D. S., 1985. Habitat requirements of juvenile penaeid prawns and their relationship to offshore fisheries. In Second Australian National Prawn Seminar, Kooralbyn (Australia), 22 October 1984 eds. Rothlisberg: 47–54.

Tilson, R. L. & Seal, U. S., 1987. Preface. In Tilson, R. L. & Seal, U. S. (eds), Tigers of the World: the Biology, Biopolitics, Management, and Conservation of an Endangered Species. Noyes Publications, Park Ridge, NJ: vii–viii.

Twilley, R. R., Snedaker, S. C., Yanez-Arancibia, A. & Medina, E., 1996. Biodiversity and ecosystem processes in tropical estuaries: perspective of mangrove ecosystems. In Mooney et al. (eds), Functional Roles of Biodiversity: A Global Perspective. Wiley, New York: 327–370.

Yanez-Arancibia, A., Soberon-Chavez, G. & Sanchez-Gil, P., 1985. Ecology of control mechanisms of natural fish production in the coastal zone. In Yanez-Arancibia (ed.), Fish Community Ecology in Estuaries and Coastal Lagoons, towards an Ecosystem Integration. UNAM Press, Mexico: 571–595.

Hydrobiologia (2005) 542:191–202
H. Segers & K. Martens (eds), Aquatic Biodiversity II
DOI 10.1007/s10750-004-6947-8

Primary Research Paper

Coexistence of two similar copepod species, *Eudiaptomus gracilis* and *E. graciloides*: the role of differential predator avoidance

C.D. Jamieson
Department of Statistics, University of Wyoming, P.O. Box 3332, Laramie, Wyoming 82071-3332, USA
E-mail: colleenjamieson@lycos.com

Key words: DVM, fish predation, lekking, diaptomid, selection

Abstract

Habitat choice in relation to environmental factors of two coexisting calanoid copepod species, *Eudiaptomus gracilis* and *E. graciloides*, was studied in a mesotrophic lake and in large indoor mesocosms. Both species and sexes showed pronounced diel vertical migration (DVM) in the field. In 12 m deep mesocosms with free ranging fish DVM was observed and species increased day depth over time. No changes were observed in copepod day depth over time in experiments with fish kairomone. It is hypothesized that fish kairomone acts as an early warning system to copepods which respond by moving deeper, but only as far as the thermocline. For full DVM, a nearby mechanical stimulus is necessary. Thus, as fish go deeper to feed, copepods retreat. The response of copepods to fish predation, in the presence of low and high numbers of *Daphnia,* shows that copepods effectively use *Daphnia* as living shields to avoid predation. The two species adopt different vertical migration strategies depending on whether there are high or low numbers of *Daphnia* present. A dominant feature of mesocosm experiments was the night time aggregating (lekking) of *E. gracilis* males at the surface. When the spring and autumn percentages of risk takers in the epilimnion were compared, *E. gracilis*, particularly males, suffered the greatest cost.

Introduction

Hutchinson (1961) in his discussion of the 'paradox of the plankton' pointed out that changing environmental conditions could cause shifts in competitive ability that promoted species persistence, despite resource limitation and competition. Thus, very similar species that have identical requirements and similar capabilities can coexist due to fluctuating advantage. Temporal variability in resource supply, temperature and predation, can lead to such shifts in the competitive advantage of different species. For example, the onset of fish predation in spring shifts the competitive advantage among zooplankters and promotes summer biodiversity. Typically, the spring bloom of *Daphnia* is suppressed by fish predation (Sommer et al., 1986) and a number of smaller species are found in the plankton during summer. Small size and differential responses to fish predation may allow some of these species to coexist.

The two freshwater copepod species *Eudiaptomus gracilis* and *E. graciloides* are widely distributed over all Europe with *E. gracilis* predominating in the west (*E. graciloides* being absent from the British Isles) and south, and *E. graciloides* being more abundant in the east (Kiefer & Fryer, 1978). In areas where both species occur, lakes are mostly inhabited by only one of the two species, but several cases have been described where both species coexist in one lake, for example, in Northern Germany (Hofmann, 1979) and Sweden (Nauwerck, 1980). Nauwerck reported between 2 and 12% of the lakes in different regions of Sweden harboured both species. He related their distribution to recolonization after glaciation with *E. graciloides* being the faster colonizer, but *E. gracilis* the better competitor. Although *E. graciloides* is the only species in temporary water bodies (Kiefer & Fryer, 1978), and *E. gracilis* seems to be more abundant in eutrophic lakes (Hofmann, 1979), there is still no lake characteristic known to support the coexistence of the two species.

E. gracilis and *E. graciloides* are morphologically very similar. Although their body size depends on the nutritional situation, *E. gracilis* is slightly larger and carries larger clutches when the two species coexist (Santer et al., 2000). The main difference between the two species relates to their life cycle. *E. gracilis* produces subitaneous eggs exclusively and does not enter diapause. On the other hand *E. graciloides* produce subitaneous eggs and resting eggs, and adults diapause in winter (mid November–January (Kiefer & Fryer, 1978; Pasternak & Arashkevich, 1999; Santer et al., 2000).

Due to their morphological and ecological similarity, the two *Eudiaptomus* species have been considered to be competitors. Nauwerck (1980) assumed that *E. gracilis* having larger clutches should replace *E. graciloides*. Santer et al. (2000), found consistently larger clutches and lower food (POC) thresholds for egg production by *E. gracilis* compared to *E. graciloides* in the Schöhsee, which supports the idea of *E. gracilis* being the better competitor. However, the existence of pure populations of either species, and of lakes where both species have coexisted for extended periods of time in the same area, suggests that neither historical reasons nor competitive exclusion provide a good explanation for local distribution patterns.

Differential susceptibility to predation may structure the copepod community. Unfortunately, there is no comparative study on fish predation effects on *Eudiaptomus*, but within one species, Svensson (1997) demonstrated the preference of young roach for larger specimens and egg-bearing females of *E. gracilis*. As *E. gracilis* is larger than *E. graciloides* and has larger clutches it is likely to be more prone to fish predation. Moreover, the production of resting eggs stored in the egg bank may be a mechanism to withstand periods of severe fish predation by *E. graciloides*. Thus, fish predation may balance the competitive advantage of *E. gracilis* over *E. graciloides*.

If the two species are differentially susceptible to fish predation, they can be expected to differ in their diel vertical migration behaviour and vertical habitat choice (Lampert, 1993). I tested two hypotheses; firstly, that different responses to fish predation reduce spatial overlap of the coexisting copepod populations, and secondly, that disproportionate fish predation may balance the competitive advantage of *E. gracilis* over *E. graciloides*. I studied the vertical distributions and diel vertical migrations in the lake to check for species-specific responses. I then tested in manipulative experiments if the distributions are in fact, a response to the presence of planktivorous fish. Finally, to detect selection by fish, I compared the proportion of risk-takers in the epilimnion in spring and in autumn in the presence of fish kairomone in a second set of manipulative experiments. Both sets of experiments were performed in large indoor mesocosms.

Methods

Field study

Vertical distributions of *E. gracilis* and *E. graciloides* were monitored in the Schöhsee, a mesotrophic lake in front of the Max Planck Institute of Limnology at Plön, Northern Germany. Lake characteristics are given in Santer et al. (2000). The dominant planktivorous fish in the Schöhsee are young perch (*Perca fluvialitis*) and roach (*Rutilus rutilus*) during late spring and summer. For this study, I considered only adult copepods, as the discrimination of nauplii and copepodites of the two species in field samples is impossible. Zooplankton, temperatures, and oxygen were sampled, and a secchi transparency reading was taken at a single station from 31 May to 29 August 2001 at 1400 h, at either 2-weekly intervals or more frequently when night time samples were taken during the summer months. Temperature and oxygen concentrations were measured with a WTW probe at 1 m intervals. Three net hauls per sample were taken with a conical closing net (opening diameter 9 cm, mesh size 100 μm) at 2-m intervals from the surface to near the bottom (18 m), starting at 1400 h on each sampling day. Samples were concentrated and preserved with alcohol or with 2% formalin in the laboratory within 2 h of capture. Twice a month (within a few days of each other), starting at 1400 and 2400 h respectively, day and night time zooplankton samples were taken to provide two replicate day/night zooplankton series each for June, July and August 2001. The total numbers of male and female *E. gracilis* and *E. graciloides* in the combined samples were counted under a dissecting microscope (×25–40). The proportions of male and female *E. gracilis* and *E. graciloides* as a percentage of the total from the surface to the bottom in each 2 m vertical haul were then calculated.

Mesocosm experiments

Experiments were conducted in a system of two stainless steel indoor columns (height 11.5 m, width 0.9 m), which were identical in every respect to the Plankton Towers described by Lampert & Loose (1992). The towers were filled with 10 μm-filtered water from the mesotrophic Schöhsee. Temperature in the towers was regulated to give a stable temperature distribution of 20 °C in the epilimnion and 10 °C in the hypolimnion with a thermocline between 2 and 3 m. Light was provided by a light source (Osram Metallogen HMI 1200W/GS) with a light:dark cycle of 14:10. The light intensity at the water surface was approximately 400 μE m^{-2} s^{-1}.

Cryptomonas sp. from the Max Planck Institute culture collection grown in 10 l batch cultures in WC medium (Guilllard & Lorenzen, 1972) served as food during the experiment. Algae were added to the towers before the addition of zooplankton and regularly throughout the experiment to give a POC concentration in the epilimnion of 0.43 ± 0.23 mg C l^{-1}. However, microscopic examination of the samples showed that this was not the only alga present in the towers. The volumes of algae present were determined with a particle counter between 5 and 25 μm (CASY, Schärfe Systems). Biovolumes were converted to carbon contents according to a previously established relation. Literature suggests that one carbon:volume relation can be used for taxonomically diverse protist plankton excluding diatoms (Rocha & Duncan, 1985).

Zooplankton to inoculate the towers was collected from the Schöhsee with a 250-μm mesh net (diameter 0.75 m). Four net hauls from 18 m to the surface were put in 10 l plastic containers and immediately taken back to the laboratory. They were initially filtered through a wide mesh net to remove large *Daphnia* and *Chaoborus*. Then samples were diluted with 10-μm filtered lake water in two 150 l rain barrels and bubbled with air for 3 h to reduce the number of small *Daphnia*. Any *Daphnia* caught in the surface film were removed with a fine mesh scoop net. The zooplankton was thoroughly mixed in the barrels and equal amounts were added to each tower.

The vertical distribution of zooplankton and algae was monitored by sampling through a selection of the 23 ports arranged at distances of approximately 50 cm at each tower. Small samples for the determination of algal biovolume in the particle counter were withdrawn through a syringe needle. Zooplankon was sampled by glass traps (for details see Lampert & Loose, 1992) filtering a volume of 48 l. Zooplankton samples were preserved with alcohol or 2% formalin. The number of male and female *E. gracilis* and *E. graciloides* in each sample were counted under a dissecting microscope (×25–40).

Two sets of experiments were carried out. The first set compared copepod behaviour in the presence and absence of free-ranging fish. The second series consisted of two experiments and compared the behaviour of spring and autumn (2001) copepod populations, at the start and after the field-sampling period. Both experiments were performed under identical conditions with fish kairomone.

Free-ranging fish

Although this experiment was conducted as a follow on experiment after the autumn kairomone experiment with the addition of more freshly caught zooplankton (as described above), it is more informative to discuss it first. After the addition of zooplankton, 22 small young-of-the-year perch (*Perca fluviatilis*) were added at night to one tower with the other tower acting as a control (no fish). Starting with the third night after the addition of fish, copepods were sampled daily at 10 sampling ports (0.1, 1.6, 2.5, 3.5, 4.6, 5.6, 6.5, 8.5, 9.5 and 10.4 m depth) during the day and night. Algal concentrations were adjusted by daily additions of *Cryptomonas* sp. at the surface 4–5 h before night time sampling. However, in the no fish tower the *Daphnia* population exploded which resulted in the algal concentration being lower in this tower than in the fish tower after 24 h. After 8 days, 22 young-of-the-year perch fish were released into the no fish tower, and another 8-day series of samples were taken during the day but not at night. Night time samples were so consistent in the first series it seemed unnecessary to take further samples.

In the tower that had fish from the beginning, the *Daphnia* could not multiply and were quickly suppressed to a low level. This tower was the low *Daphnia* treatment (mean density 1.4 *Daphnia* l^{-1}), and the no fish tower became the high *Daphnia* treatment (mean density 7 *Daphnia* l^{-1}) in the second half of the experiment after the addition of fish. At the completion of the experiment, the fish were released back into the lake they were captured from.

Seasonal effects

To assess the effects of selection by fish in a 'common garden setting' the proportions of risk-taking copepods in the epilimnion were determined in spring and in autumn. Copepods were provided with a chemical clue (kairomone) that has been shown to elicit diel vertical migration in *Daphnia* (Loose, 1993). To signal to copepods the presence of fish without direct predation effects, six small (< 8 cm) carp *Cyprinus carpio* were held in a cage and fed to produce the kairomone. This cage was placed in an aquarium next to one of the towers and an empty cage (control) was put in to an aquarium next to the other tower. The water was circulated from the aquaria to the epilimnion of the towers (surface −2.5 m) with pumps at a rate of 100 L h^{-1}. The response of copepods to fish kairomone was very similar to their response to fish on the first day of the free-ranging fish experiment. Tower conditions were maintained as similar as possible between seasons; for example, there were no significant differences in seston concentration or temperature profiles during the spring and autumn series. However, zooplankton composition was similar but not identical between seasons.

The towers were sampled at 8 depths (0.1, 1.2, 1.6, 2.5, 3.5, 4.6, 5.6 and 6.5 m) at least 48 h after the addition of zooplankton. Each tower was sampled twice in a 24-h period 12 h apart: 3 h after dark and 5 h after light. The basic sampling design was to take a series of three day/night samples from both towers on alternate days and then swap the fish cage to the aquarium at the other tower. After another 48 h a new series of day/night samples was started. Food levels were adjusted on the days in between successive sampling. Although the same population of copepods was used for the whole sampling period the individual day/night samples are considered independent replicates. This was justified as the vertical distributions changed between day and night such that two complete re-distributions took place between samplings. The spring experiment consisted of six replicates, but only four replicates could be used in the autumn experiment due to problems with the mass cultures of *Cryptomonas*.

Statistics

All vertical distributions were converted to percentages at each depth. Percentage data were summarized using principal component analysis based on the variance–covariance matrix. This is in contrast to the majority of principal component analyses in the biological literature (e.g. community analyses of species by sites, morphometric data), which are carried out using the correlation matrix. For percentage data all variables are on the same scale, and use of the variance–covariance matrix means depths that have the most variation in percentages have the heaviest weights in the eigenvectors. Use of the correlation matrix is preferable when variables need to be standardized (i.e. mean zero and variance one) so that different scales of measurement (e.g total body length compared to antennal length) do not influence the analysis. In both procedures the first principal component represents a weighted combination of variables to account for the maximum possible amount of variance. The choice of matrix to use in a principal component analysis is an option in most statistical packages.

The first principal component from the analyses of the field data and the mesocosm experimental data produced consistent results. The upper depths (with positive weights in the eigenvector) were contrasted with lower depths (negative weights in the eigenvector) (Table 1). Thus, the lower the value of the first principal

Table 1. Eigenvectors for the first principal component summarizing distributions in the Schöhsee and the free-ranging fish experiment

Schöhsee		Free-ranging fish experiment		
Mean depth (m)	Eigenvector	Actual depth (m)	Day eigenvector	Night eigenvector
1	0.56	0.1	0.20	0.99
3	0.38	1.5	0.26	−0.44
5	0.30	2.5	0.58	−0.72
7	0.09	3.5	0.23	−0.62
9	−0.15	4.5	0.04	−0.64
11	−0.15	5.5	−0.07	−0.40
13	−0.12	6.5	−0.09	−0.26
15	−0.30	8.5	−0.18	−0.33
17	−0.50	9.5	−0.45	−0.32
		10.5	−0.51	−0.33
%	62.9	%	51.6	75.8

%, variation accounted for by pc1.

component (pc1), including negative values, the deeper the copepods. The contrasting signs in the eigenvector indicate that when numbers are high in the upper part of the water column they are low in the lower part of the water column and *vice versa.* Subsequent principal components are not discussed as the eigenvectors were not as easily interpreted and were not consistent between field and experimental analyses.

For the field data the principal component analysis was followed by an analysis of variance on the principal component scores to detect significant differences between day and night. Only significant interactions and lower order terms necessary to maintain hierarchy were included in ANOVA models. The day/night series were analysed with Time (day or night) Month, Species and Sex as effects.

For the free-ranging fish experiment the day and night distributions were summarized in separate principal component analyses as the model was too complex to interpret with all factors included, and the daytime responses were clearly different from night time responses. Equipment failure on day 4 and at 6.5 m in the control tower meant that this day and depth were excluded in the calculation of the principal components. A generalized linear model was fitted to the daytime principal component scores with Day, Fish, Species, Sex and interactions of these factors as effects as the design was unbalanced. There were three levels for fish: fish with low numbers of *Daphnia*, fish with high numbers of *Daphnia*, and no fish. A separate analysis of variance was carried out on night time data.

All procedures were carried out with the programme NCSS (Hintz, 1998), except the analysis of variance on the daytime data from the experiment with free-ranging fish, where the model was fitted using the programme Genstat (Lawes Agricultural Trust, 1998). Eigenvectors from the principal component analyses are shown in Table 1.

For the seasonal experiments with kairomone, the percentage of copepods at the surface and 1 m during the day were compared with an analysis of variance with Season (spring or autumn), Fish (present or absent), Species (*E. gracilis* or *E. graciloides*) and Sex (female or male) as effects. This was followed by a *post hoc* test (Tukey's HSD) to distinguish between the means. Data were $\log(x + 1)$ transformed to satisfy homogeneity of variance assumptions.

Results

Field observations

The Schöhsee was thermally stratified during the three summer months (June to August), the thermocline being permanently between 7 and 8 m depth. Oxygen started to decline near the bottom

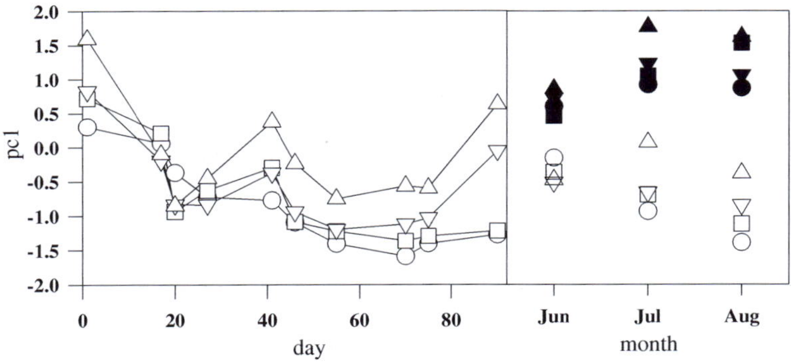

Figure 1. First panel: Principal component scores (pc1) summarising vertical distributions during the day over the summer (2001). Second panel: Principal component scores (pc1) summarising vertical distributions during the day (open symbols) and at night (closed symbols) in the Schöhsee over the summer. (*E. gracilis* females, circles; *E. gracilis* males, squares; *E. gracilioides* females, downward triangles; *E. gracilis* males, upward triangles).

in mid-July, and oxygen depletion progressed until values close to zero were found near the bottom at the end of August. Secchi transparency declined steadily from 6.3 m on 1 June to 2.0 m on 29 August. The Schöhsee regularly shows increasing turbidity due to calcite precipitation during the summer, which was the case during July. However, microscopic observations of water samples showed that the change in secchi transparency in August was mainly due to increased algal abundance. The Schöhsee is a mesotrophic lake with summer levels around 4 μg chlorophyll a l^{-1} and 0.4 mg l^{-1} POC (Müller-Navarra, 1996). Similar chlorophyll levels were measured in June/July 2001 (K. Jürgens, personal. communication).

In June both copepod species were found around 9 m where oxygen over saturation provided evidence for a deepwater chlorophyll maximum as has been frequently observed in the lake. During July, *E. gracilis* and *E. graciloides* went continuously deeper, but *E. graciloides* never went as deep as *E. graciloides*. In August, both species responded to decreasing water transparency (lower secchi disc readings) and moved up in the water column with the start of the autumn bloom. *E. graciloides* responded to the changing conditions to a greater extent than *E. gracilis* (Fig. 1).

In the principal component analysis of vertical distributions in the Schohsee, the first principal component contrasts the epilimnion (1–7 m) with the hypolimnion (9–17 m) and explains 63% of the variance (Table 1). The ANOVA of the day/night principal component scores detects significant effects of Time, Species, Sex and the Month × - Time interaction (Table 2). Both species performed diel vertical migrations so there were pronounced differences in vertical distributions between day and night (Fig. 1). During the day, both females and males of *E. gracilis* usually dwelled deeper in the water column than *E. graciloides*. The significant Sex effect confirms the observation of males staying higher in the water than females in June and July (Fig. 3), and the Month × Time interaction reflects the seasonal trend in day depth.

Free-ranging fish experiment

Both species exhibited increasing day depth over time in both low and high *Daphnia* treatments but not in the no fish treatment (Fig. 2 and Table 3). This increasing day depth in the presence of free-ranging fish suggests a continuing accumulation in deep water, but this is misleading

Table 2. Results of ANOVA on the first principal component of *Eudiaptomus* species day and night distributions (2 days per month) in the Schöhsee during summer 2001

Effect	DF	MS	F	p
Month	2	0.19	1.52	0.231
Time	1	33.75	265.60	<0.001
Species	1	1.13	8.88	<0.001
Sex	1	0.97	7.65	0.005
Month × Time	2	1.40	10.95	0.009
Error	40	0.13		

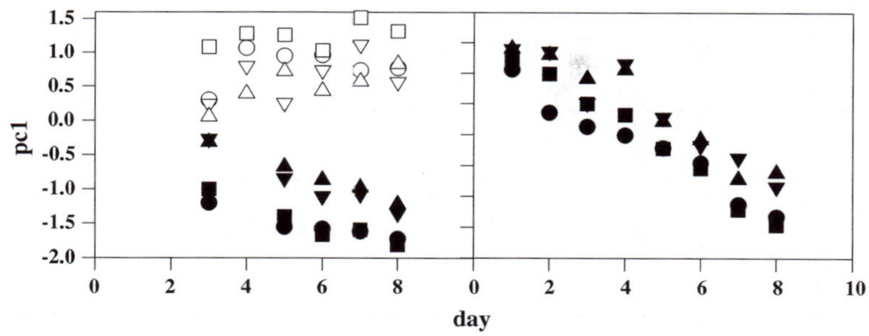

Figure 2. Daytime principal component scores (pc1) summarising vertical distributions in no fish (open symbols), low and high *Daphnia* over time (closed symbols). (*E. gracilis* females, circles; *E. gracilis* males, squares; *E. gracilioides* females, downward triangles; *E. gracilis* males, upward triangles).

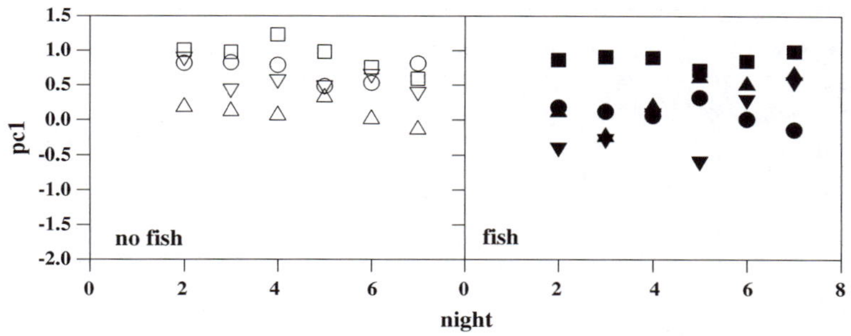

Figure 3. Night time principal component scores (pc1) summarising vertical distributions in no fish (open symbols) and low *Daphnia* (closed symbols) over time. (*E. gracilis* females, circles; *E. gracilis* males, squares; *E. gracilioides* females, downward triangles; *E. gracilis* males, upward triangles).

as the distribution changed completely every night and was re-establishes everyday. *E. gracilis* the larger species went deeper earlier than *E. graciloides* the smaller species, and both species went deeper earlier with low numbers of *Daphnia* than with high numbers of *Daphnia* (Table 3). In contrast, a constant daytime depth was found for *Daphnia* in the same experiment. Daily observations on the fish through the tower portholes revealed that at first fish remained exclusively in the epilimnion, but then they began feeding forays into progressively deeper waters during the following days. Forays into deeper water began earlier in the low *Daphnia* treatment than in the high *Daphnia* treatment. What is clearly visible in the raw results (Table 3) is that each day more copepods seek refuge below the *Daphnia* layer.

When copepods distributions were summarized with a principal component analysis (Table 1), daily depth changes can be seen as a decreasing pc1 over time in fish treatments, and in contrast a stable pc1 over time in the no fish treatment (Fig. 2). For daytime distributions, the first principal component explains 51.6% of the variance and contrasts proportions at the thermocline (2.5 m) with the deepest layers (Table 4). *E. gracilis* was deeper in both fish treatments (high and low *Daphnia*), but *E. graciloides* was deeper in the no fish treatment. In low *Daphnia*, copepods went deeper earlier (day 3 mean pc1 = −0.698 ± 0.237) than in high *Daphnia* (day 3 mean pc1 = 0.514 ± 0.160).

Results of the model fitted to pc1 from daytime distributions with Day, Fish, Species and Sex as main effects are shown (Table 5). Day and the Day × Fish interaction are significant showing that the copepods go significantly deeper in fish treatments over the time of the experiment. Species and the Species × Fish interaction are also significant, indicating the species differ in their response to fish. There is also significant Species × Sex and

Table 3. Actual counts from samples taken during the free-ranging fish experiment from day 3 after the addition of fish

Depth (m)	Low *Daphnia*						High *Daphnia*					
Day:	3	4	5	6	7	8	3	4	5	6	7	8
E. gracilis												
0.0	0	0	1	0	0	0	2	1	0	0	0	0
1.5	0	0	0	0	0	0	1	1	0	0	0	1
2.5	3	4	0	0	0	0	15	9	2	0	0	0
3.5	12	3	1	0	0	0	28	26	14	7	2	0
4.5	14	16	2	3	1	0	31	24	20	18	4	2
5.5	17	11	9	5	2	1	23	16	32	37	4	6
6.5	7	9	7	7	8	1	14	19	18	26	14	4
7.5	10	16	4	6	2	3	4	9	15	22	19	14
8.5	33	52	29	20	26	10	1	5	2	16	5	9
9.5	68	*	65	53	43	47	8	7	14	18	30	40
10.5	52	48	66	96	90	79	3	3	3	9	14	19
E. graciloides												
0.0	1	0	0	1	1	0	5	0	0	0	0	0
1.5	0	0	0	0	0	0	1	0	0	0	0	0
2.5	3	0	1	0	0	0	17	25	4	0	0	0
3.5	19	6	3	0	0	0	17	24	24	22	3	2
4.5	12	12	18	7	4	0	8	12	11	30	9	10
5.5	15	14	18	19	18	4	5	6	8	26	19	13
6.5	2	11	8	12	13	10	2	1	3	10	13	12
7.5	7	6	5	6	8	7	2	2	1	8	6	4
8.5	6	11	17	19	8	18	0	1	1	3	4	6
9.5	12	*	20	24	21	27	4	0	5	9	8	12
10.5	12	19	18	22	19	19	4	2	4	7	6	10

Daphnia depth is distinguished with bold font.

Fish × Species × Sex interactions, which show that females go deeper than males particularly for *E. gracilis* (Fig. 2).

Night time distributions were strikingly different to daytime distributions (Fig. 3). Copepods

Table 4. Results of ANOVA on the first principal component (pc1) for night time distributions of *Eudiaptomus* species in the free-ranging fish experiment

Effect	DF	MS	*F*	*p*
Fish	1	22.46	14.74	<0.001
Species	1	5.66	28.94	<0.001
Sex	1	4.54	23.19	<0.001
Fish × Species	1	3.07	15.66	<0.001
Fish × Sex	1	2.66	13.57	<0.001
Error	40	0.20		

Table 5. Results of the model fitted to the first principal component (pc1) for daytime distributions of *Eudiaptomus* species in the free ranging fish experiment.

Effect	DF	MS	*F*	*p*
Day	1	12.84	291.57	<0.001
Fish	2	22.09	501.56	<0.001
Species	1	1.24	28.08	<0.001
Sex	1	0.23	5.25	0.025
Day × Fish	2	4.36	99.07	<0.001
Fish × Species	2	2.27	51.52	<0.001
Fish × Sex	2	0.01	0.32	0.727
Species × Sex	1	0.22	4.95	0.030
Fish × Species × Sex	2	0.15	3.47	0.038
Error	61	0.04		

migrated into the epilimnion at night, but there were no systematic changes in depth distributions over time. *E. gracilis* migrated into the surface waters at night with both sexes migrating higher in the no-fish than in the fish treatment. *E. gracilis* males aggregated extremely close to the surface, and the largest part of the male population was above 1m depth every night. *E. graciloides* showed less pronounced migration behaviour. Both sexes migrated upwards at night, but they were usually deeper than *E. gracilis.*

The first principal component (pc1) for night time distributions contrasts the proportions in the surface layer with the remaining water column and explains 75.8% of the variation in the data (Table 1). The ANOVA on pc1 shows that all main effects (Fish, Species and Sex) are significant. Also, both two-way interactions with fish (Fish × Species, Fish × Sex) are significant. These results confirm the visual impression (Fig. 3) of the two species and sexes reacting differently to the presence of fish at night when predation risk is low.

The night time surface-aggregating behaviour *of E. gracilis* males was a dominant feature of experiments. They aggregated at the surface every night in the presence and absence of fish and fish kairomone (season experiment). In fact, 70–80% of the copepods found at the surface at night were *E. gracilis* males (Fig. 4). In the absence of fish this represented on average 73% of the male *E. gracilis* population.

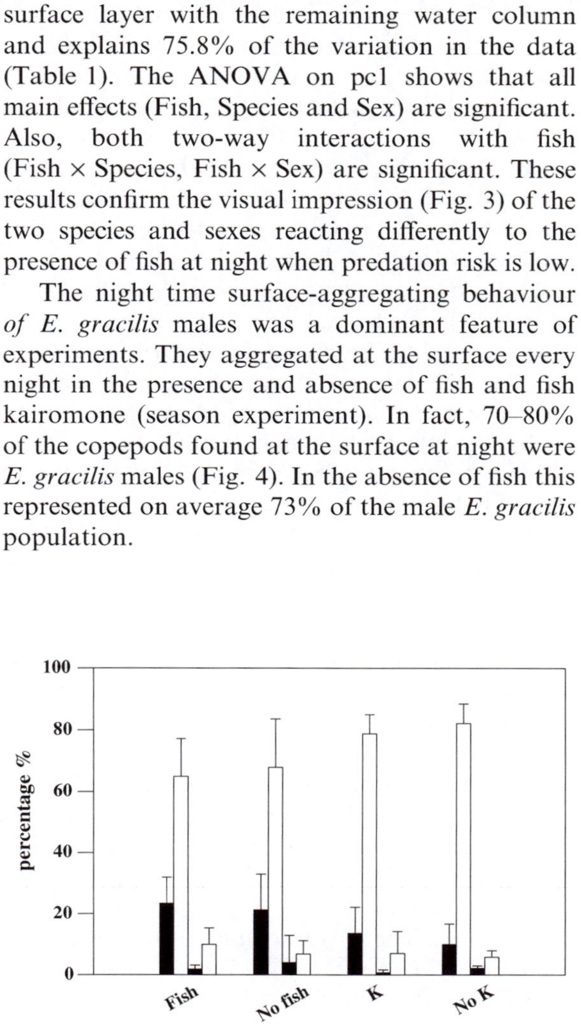

Figure 4. Night time percentage composition at the surface. (*E. gracilis* females, circles; *E. gracilis* males, squares; *E. graciloides* females, downward triangles; *E. gracilis* males, upward triangles).

Seasonal effects (kairomones)

A comparison was made between spring and autumn populations in the percentage of risk-takers in the epilimnion in the presence of fish kairomone (Fig. 5). A significantly smaller percentage of *E. gracilis* males and females were found at the surface in the autumn than in the spring. There were also fewer *E. gracilis* males at 1m in the autumn, but not females. There were no differences between spring and autumn *E. graciloides* populations (Fig. 5).

Discussion

In the free ranging fish experiments and in the lake *E. gracilis* went deeper during the day than *E. graciloides*, and females went deeper than males. The differences in day depth and migration behaviour are consistent with the predator avoidance hypothesis for DVM. Visually hunting planktivorous fish select large, conspicuous prey items. *E. gracilis*, being larger than *E. graciloides* is under greater predator threat, and females carrying egg sacs are more conspicuous than males. *E. gracilis* showed clearer day–night differences than *E. graciloides*, which was due to both a deeper daytime depth and a shallower night-time depth.

In the Schöhsee both species increased their migration amplitude during summer, but this trend halted when visibility decreased. Ringleberg et al. (1991) also found that *E. gracilis* increased the amplitude of DVM in Lake Maarsseveen in the first half of the summer, and then it decreased gradually in the second half of the summer. They associated amplitude changes with the presence of high concentrations of chlorophyll-a, and fish behaviour. At the end of August, a considerable part of the *E. gracilis* population in the Schöhsee was found under low oxygen conditions during the day but *E. graciloides* being smaller and less visible aggregated higher in the water column and avoided the hypoxic conditions. The fact that *E. gracilis* accepted the costs of living under low oxygen conditions suggests that a strong factor forced them down.

In the tower experiments *E. gracilis* went deeper during the day than *E. graciloides* to avoid fish predation, but at night *E. graciloides* was usually

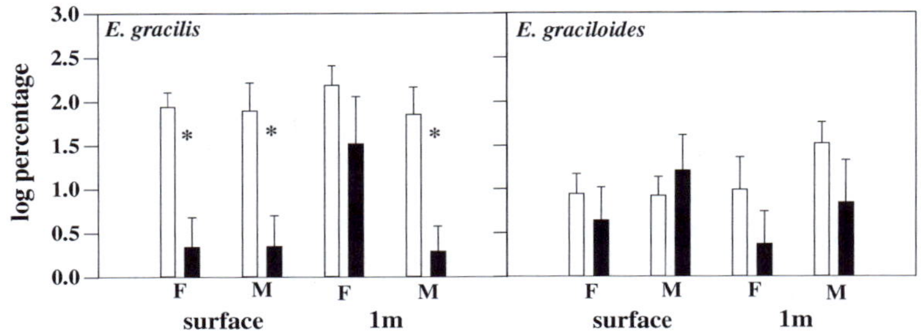

Figure 5. Log percentage ($x + 1$) at the surface and 1 m during spring (empty bars) and autumn (closed bars). (Female, F; male, M; * significantly different $p < 0.01$).

deeper. Both species went a little deeper with fish kairomone but required the physical presence of fish nearby for full migration. Consequently, daytime depths of copepods increased over time as fish went deeper to search for prey. The increasing day depth does not result from a permanent displacement of the populations, but from increasing migration amplitudes. Bollens & Frost (1989) found a fish mimic was necessary as a mechanical stimulus to induce vertical migration in *Acartia hudsonica. Daphnia* also go deeper with free-ranging fish present than with fish kairomone, but in contrast to *E. gracilis* and *E. graciloides* they adopt a constant depth over time if all else is equal (De Meester et al., 1995).

No trend over time was found for night time distributions. Both species migrated into the epilimnion and *E. gracilis* ascended higher in the water column than *E. graciloides*. Male *E. gracilis* reached very high densities at the surface at night which resembles lekking behaviour. In the no-fish treatment and in the kairomone experiment this night surface aggregation behaviour was extreme. *E. gracilis* males aggregated immediately under the surface and they did this every night. With free-ranging fish not as many were found at the surface at night, but they were still higher in the water column than the other groups, and they were the dominant group at the surface. As this behaviour has been demonstrated in three different experiments, it seems to be a general phenomenon. Field data collected as part of this study did not reflect this phenomenon, as samples were collected over 2 m. However, on two separate occasions the immediate surface layers of the Schöhsee were sampled, and on both occasions over 50% of the calanoid copepods in the sample were *E. gracilis* males. Svenson (1997) described the horizontal and vertical distributions of male and female *E. gracilis* in a 6.3 m deep lake in Sweden and found that males aggregated at the surface and in the littoral zone of the lake. If this surface aggregating behaviour is associated with reproduction it may indicate differences between the species in the duration and/or frequency of mating. Mating in copepods has been associated with an increased risk of predation by fish (Maier et al., 2000), thus it may be advantageous to mate at night rather than during the day. Byron et al. (1993) hypothesized that male copepod aggregations could operate in the same way as aerial swarms of insects (Höglund & Alatoto, 1995). Although it cannot be confirmed that these aggregations have a social function, I did find that virtually every male carried a spermatophore. Hayward (1981) also found males carrying spermatophores were more abundant in surface waters at night. These similar observations may indicate that this strategy could be relatively common among marine and freshwater copepods. Lonsdale et al. (1998), in their review on copepod pheromones, suggest that species-specific swarms may increase mating success.

Males were often late to descend in the morning, and they could be seen cruising around at the surface of the plankton towers 2–3 h after light (pers. obs.). This is extremely risky behaviour and the predation theory for DVM would predict a heavy cost. There were significantly lower percentages of *E. gracilis* males and females at the surface in the autumn than in spring, and of *E. gracilis* males at 1 m, but no differences were found for *E. graciloides*.

Lack of replication arguments could be produced to counter the validity of these results. However, the daily downward shift did occur in both fish treatments, which does establish a trend, and the extent of this trend was constant for each treatment, that is, we can fit straight lines to the principal components with high R^2 values (88.9–99.4%). Furthermore, supporting evidence comes from the fact that the same trend was found in field data.

It is of interest to compare the vertical migration strategies of daphniids and copepods. The onset of fish predation in the spring significantly reduces the number of *Daphnia* in the water column and copepods usually dominate the plankton during the summer (Sommer et al., 1986). This would suggest that copepods are smarter than daphniids at avoiding fish predation. Copepods rely on a nearby mechanical stimulus to establish full DVM therefore they can maximize the time spent in food-rich warm waters by going progressively deeper. They can afford to do this because fish selectively prey upon *Daphnia* (Lampert & Sommer, 1997). Thus, *Daphnia* effectively become living shields for the copepods. Finally, when predation pressure is too intense, the copepods migrated below the edible wall of migrating *Daphnia*. In contrast, *Daphnia* adapt daytime depth to their body size, light intensity, kairomone strength, fish presence and food availability, but not apparently to numbers of competing zooplankters. More experiments are necessary to unravel the complexity of interspecific relationships among zooplankters.

The results of these experiments are all logically equivalent with a hypothesis of fish predation changing the competitive advantage between the two species. The larger species goes deeper than the small species, and females deeper than males. There is evidence for selection acting on *E. gracilis* the larger species, but not on *E. graciloides* the smaller species. The group taking the greatest risks, *E. gracilis* males suffered the greatest cost. To otherwise explain these results would require a number of ad hoc hypotheses. What is surprising is that night time depth selection behaviour was also critical. Copepods stayed deeper in the water column at night as well as during the day in the presence of fish. I found no evidence of significant selection acting on the smaller cautious *E. graciloides* which staying deeper at night than *E. gracilis*. Zaret & Suffern (1976) found the planktivore, *Melaniris chagresi* fed almost exclusively on *Diaptomus* on calm moonlight or starlight nights but not in other weather. In these conditions *E. gracilis* would be under greater predation threat than *E. graciloides*.

Finally, I conclude that different responses to fish predation reduce the spatial overlap of *E. gracilis* and *E. graciloides* during the day. At night, spatial overlap is also reduced by species-specific reproductive behaviour.

Acknowledgements

By request of the funding organization, these experiments are to remain unacknowledged. Thanks are extended to the graduate students who helped with field works.

References

Bollens, S. M. & B. W. Frost, 1989. Predator-induced diel vertical migration in a planktonic copepod. Journal of Plankton Research 11: 1047–1065.

Byron, E. R., P. T. Whitman & C. R. Goldman, 1983. Observation of copepod swarms in Lake Tahoe. Limnology and Oceanography 28: 378–382.

De Meester, L., L. J. Weider & R. Tollrian, 1995. Alternative antipredator defences and genetic polmorphism in a pelagic predator-prey system. Nature 378: 483–485.

Guillard, R. R. L. & C. J. Lorenza, 1972. Yellow-green algae with chlorophyllidae c. Journal of Phycology 8: 10–14.

Hayward, T. L., 1981. Mating and depth distribution of an oceanic copepod. Limnology and Oceanography. 26: 374–377.

Hintz, J., 1998. NCSS 2000. 329 North 1000 East, Kaysville, Utah 84037, USA.

Hofmann, W., 1979. Characteristics of syntopic populations of *Eudiaptomus gracilis* (Sars) and *E. graciloides* (Lilljeborg) in three lakes of different trophic levels. Archive fur Hydrobiologie 86: 1–12.

Höglund, J. & R. V. Alatoto, 1995. Leks. Princeton University Press.

Kiefer, F. & G. Fryer, 1978. Zooplankton of Inland Waters. Inland waters Monographs on Limnology and Related Fields. Vol. 26. Part 2. e. Schweizerbatsche Verlagsbuchhandlung: Stuttgart, West Germany, 380 pp.

Lampert, W., 1993. Ultimate causes of diel vertical migration. Functional Ecology 3: 21–27.

Lampert, W. & C. J. Loose, 1992. Plankton towers: bridging the gap between laboratory and field experiments. Archive-fur Hydrobiologie 126: 53–66.

Lampert, W. & U. Sommer, 1997. Limnoecology: The Ecology of Lakes and Streams. Oxford University Press, New York, Oxford. 382 pp.

Lawes Agricultural Trust (Rothamsted Experimental Station), 1998. Genstat Release 4.1 for Windows.

Lonsdale, D. J., M. A. Frey & T. W. Snell, 1998. The role of chemical signals in copepod reproduction. Journal of Marine Systems 15: 1–12.

Loose, C. J., 1993. *Daphnia* diel vertical migration behavior: response to vertebrate predator abundance. Archive fur Hydrobiology 39: 29–36.

Nauwerck, 1980. The distribution of the diaptomidae in Northern Sweden. Archive fur Hydrobiologie 89: 247–264.

Pasternak, A. F. & E. G. Arashkevich, 1999. Resting stages in the life cycle of *Eudiaptomus graciloides* (Lill.) (Copepoda: Calanoida) in Lake Glubokoe. Journal of Plankton Research 21: 309–325.

Ringleberg, J., B. J. G. Flik, D. Lindenaar & K. Royackers, 1991. Diel vertical migration of *Eudiaptomus gracilis* during a short summer period. Hydrobiolosy Bulletin 25: 77–84.

Rocha, O. & A. Duncan, 1985. The relationship between cell carbon and cell volume in freshwater algal species used in zooplanktonic studies. Journal of Plankton Research 7: 279–294.

Santer, B., E. Blohm-Sievers, C. E. Cárceres & N. G. Jr. Hairston, 2000. Life history variation in the coexisting freshwater copepods *Eudiaptomus gracilis* and *Eudiaptomus graciloides*. Archive Hydrobiologie 149: 441–458.

Sommer, U., Z. M. Gliwicz, W. Lampert & A. Duncan, 1986. The PEG-model of seasonal succession of planktonic events in fresh waters. Archive Hydrobiologie 106: 433–471.

Svensson, J.-E., 1997. Fish predation on *Eudiaptomus gracilis* in relation to clutch size, body size, and sex: a field experiment. Hydrobiologia 344: 155–161.

Zaret, T. M. & J. S. Suffern, 1976. Vertical migration in zooplankton as a predator avoidance mechanism. Limnology and Oceanography 21: 804–813.

Hydrobiologia (2005) 542:203–220
H. Segers & K. Martens (eds), Aquatic Biodiversity II
DOI 10.1007/s10750-004-8930-9

Native and exotic Amphipoda and other Peracarida in the River Meuse: new assemblages emerge from a fast changing fauna

Guy Josens[1,*], Abraham Bij de Vaate[2], Philippe Usseglio-Polatera[3], Roger Cammaerts[1,4], Frédéric Chérot[1,4], Frédéric Grisez[1,4], Pierre Verboonen[1] & Jean-Pierre Vanden Bossche[1,4]
[1]*Université Libre de Bruxelles, Service de systématique et d'écologie animales, av. Roosevelt, 50, cp 160/13, B-1050 Bruxelles*
[2]*Institute for Inland Water Management & Waste Water Treatment, Lelystad, The Netherlands*
[3]*Université de Metz, LBFE, équipe de Démoécologie, av. Général Delestraint, F-57070 Metz*
[4]*Centre de Recherche de la Nature, des Forêts et du Bois, DGRNE, Ministère de la Région wallonne, Avenue Maréchal Juin, 23, B-5030 Gembloux*
*(*Author for correspondence: e-mail: gjosens@ulb.ac.be)*

Key words: aquatic biodiversity, alien species, invasive species, invasibility, community dynamics, *Dikerogammarus villosus*, *Chelicorophium curvispinum*

Abstract

Samples issued from intensive sampling in the Netherlands (1992–2001) and from extensive sampling carried out in the context of international campaigns (1998, 2000 and 2001) were revisited. Additional samples from artificial substrates (1992–2003) and other techniques (various periods) were analysed. The combined data provide a global and dynamic view on the Peracarida community of the River Meuse, with the focus on the Amphipoda. Among the recent exotic species found, *Crangonyx pseudogracilis* is regressing, *Dikerogammarus haemobaphes* is restricted to the Condroz course of the river, *Gammarus tigrinus* is restricted to the lowlands and seems to regress, *Jaera istri* is restricted to the 'tidal' Meuse, *Chelicorophium curvispinum* is still migrating upstream into the Lorraine course without any strong impact on the other amphipod species. After a rapid expansion *Dikerogammarus villosus* has continued its upstream invasion between 1998 and 2002 at a rate of 30–40 km per year, but no further progression was noticed in 2003. Locally and temporarily the native species (*Gammarus fossarum* and *G. pulex*) and naturalized species (*G. roeseli* and *Echinogammarus berilloni*) *may* have been excluded by the most recent invaders (mainly *D. villosus*), but none of the native and naturalized species has disappeared completely. Therefore, the number of amphipod species found in the River Meuse has increased. Moreover, the native and naturalized species keep on dominating the tributaries from which the recent invaders seem to be excluded. A changing Peracarida community structure is observed along the course of the River Meuse: four native or naturalized species inhabit the upstream (Lorraine) course, three invasive species dominate in the middle reach (Ardenne-Condroz zone), one exotic species is housed in the Border Meuse and three or four invasive species dominate the assemblages in the lowlands.

Introduction

Introductions of exotic species dramatically increased in the past decades. Rivers and canals are considered among the most invadable systems, especially if they have been subjected to anthropogenic deterioration followed by environmental rehabilitation (Van der Velde et al., 2002). In the case of the macro-crustaceans it is supposed that the assemblages in the Western European rivers were rather stable until the mid 1960s (Pinkster et al., 1992) and that most introductions occurred after 1993 (Van der Velde et al., 2002, Wouters, 2002).

One of the most puzzling questions about the introduction of exotic species is the unpredictable character of any particular invasion (Kornberg & Williamson, 1987) as well as the unpredictable invasibility of any particular community (Levine & D'Antonio, 1999). Some statistical predictions, however, can be made when numerous introductions occur, e.g. the so-called tens rule. This rule holds that one of 10 imported species appears in the wild, one of 10 of such introduced species becomes established and one of 10 of the established species becomes a pest, but this rule suffers many exceptions as pointed out by its authors (Williamson & Fitter, 1996). For some aquatic groups this rule is indeed confirmed (Van der Velde et al., 2002).

Many papers on invasions deal with rather restricted ecosystems regardless of what is going on elsewhere. In this paper, as in Usseglio-Polatera & Beisel (2002b), we intend to analyse the situation for the whole course of the River Meuse, taking profit from the results of national and international surveys. The River Meuse flows through three countries: France (km 0–489), Belgium (km 489–633) and the Netherlands (km 685–943). From km 633–685 the Border Meuse marks the frontier between Belgium and the Netherlands. As most large European rivers, it has been changed strongly by human activities. The alterations (dams, weirs, locks,...) to improve navigation in the river started in the Netherlands in 1825 and in Belgium in 1850 (Vereerstraeten, 1971).

For geo-morphological reasons it is convenient to divide the River Meuse into four main zones: the Lorraine course (km 0–405), the Ardenne-Condroz course (km 405–632), the Border Meuse course (km 633–685) and the lowland course (sand Meuse and tidal Meuse) (km 685–943). In this study on the peracarid communities, however, the Meuse was subdivided into six sections that are defined further in the paper.

From Pagny-sur-Meuse (km 127) to Sedan (km 350) some navigation occurs in the canal de l'Est, running parallel to the river, with some short common reaches. From Sedan to Borgharen (km 632) the river has been deepened and the current speed is slowed down by 58 weirs and locks in France, nine in Belgium and one in the Netherlands. From Namur to Liège, the river is heavily navigated and the banks were petrified in order to withstand the heavy ship's backwash. From Borgharen (km 632) to Maasbracht (km 685) navigation takes place in the parallel Juliana Canal and not in the river itself, the so-called Border Meuse. From Maasbracht to the sea six weirs and locks make the river again commercially and intensively navigated. Its banks were strengthened with blocks.

The nowadays quality of the water can be inferred from regular surveys made on request of the 'Commission Internationale de la Meuse/Internationale Maascommissie', formerly the 'Commission Internationale pour la Protection de la Meuse/Internationale Commissie voor de Bescherming van de Maas' (C.I.M.-I.M.C., 2003, C.I.P.M.-I.C.B.M., 2001, 2002a, b).

The headwaters of the river, surveyed at Goncourt (km 46) and Brixey (km 87) are eutrophicated and contain high level of organic matter; dissolved oxygen can be in deficit. Downstream, from Saint Mihiel (km 177) until Tailfer (km 526), the chemical indicators show a noticeable reduction of the organic load but eutrophication does not decrease much. A local mercury contamination is detected at Ham-sur-Meuse (km 472). From Liège (km 599) to Eijsden (km 620) various indicators (oxygen deficit, organic matter, eutrophication, several heavy metals and anthropogenic bacteria) show multipollution. In the subsequent Border Meuse, surveyed at Lanaken (km 633) and Kinrooi (km 679) the water still contains relatively high levels of various heavy metals, organic mater and is eutrophicated. The load of organic matter decreases at Belfeld (km 716) but it increases again at Keizersveer (km 884). The eutrophication remains rather high. However, it is no longer the case at the mouth (Haringvlietdam, km 943) where the waters from the rivers Rhine and Meuse have mixed. Other quality indicators (pH, temperature, chloride and sodium concentrations) fluctuate in favourable ranges. The calcium concentration is naturally high in the headwaters and decreases downstream.

Materials and methods

Sampling methods

The present paper presents a meta-analysis and a synthesis of data from various sources: 459 samples ('historical' samples and samples from the tributaries not included) were taken into consideration. No selection on unity in the sampling methods was made. Responding to different objectives, the following methods were used.

(a) Handnet sampling (25 samples): bank sampling of various substrates with a 500 μm mesh size handnet. This method was performed for the Direction générale des Ressources naturelles et de l'Environnement (Ministère de la Région wallonne) in Wallonia in 1991, 1995 and 1998 (Vanden Bossche et al., 1999, Vanden Bossche, 2002, this paper).
(b) Artificial substrates of the C.I.P.M.-I.C.B.M. sampling campaign (117 samples), of the I.B.G.A. type: strong 10-mm mesh nylon netbags are filled with 5 l of 4- to 8-cm stones and a rope about 1.2 m long and left submerged for 28 days (Gay-environnement, 1997). This method was applied in 1998, 2000 and 2001 in France, Belgium and the Netherlands.
(c) Larger artificial substrates used by the R.I.Z.A. (256 samples): stainless steel cages of eight litres are filled with 2-cm glass marbles (De Pauw et al., 1994, Ketelaars & Frantzen, 1995, R.I.Z.A., 1998). This method was applied 12 or 14 times per year from 1992 to 2001 at Borgharen and Grave, the Netherlands.
(d) Small artificial substrates (61 samples) specifically designed for sampling Amphipods in this study: polypropylene nets with 1-cm mesh size are filled with 1 l of calcareous gravel (size: 5.2 ± 3.4 cm^3: av. $\pm$ st. dev.). The substrates were left in the water for 2–3 weeks and were used in 2002 in France and Wallonia and in 2003 all along the river.

Samples provided by artificial substrates can be considered as reproducible (De Pauw et al., 1986) and thus provide semi-quantitative results. However, artificial substrates of different sizes should not be compared with each other. Unless otherwise stated, the data used in the figures are restricted to those provided by one kind of artificial substrate at a time.

Extra samples were collected in some tributaries, acting as possible reservoirs of species, in the last few kilometres before their confluence with the River Meuse: mainly in the Rivers Semois, Viroin, Lesse and Berwinne. Other extra, 'historical', samples from the collection of the Natural Science Institute of Belgium are mentioned (Wouters, personal communication).

For the sake of clarity the zone or section limits and the sampling points are mentioned with a kilometric reference counted from the source at Pouilly-en-Bassigny (km 0) to the mouth, closed by the Haringvlietdam (km 943). These distances refer to old national systems that were combined, but since the river has been straightened at many places, the actual distances are probably shorter (Fig. 1).

Peracarid species found in the River Meuse

Amphipoda
Corophiidae
Chelicorophium curvispinum (Sars, 1895)
Crangonyctidae
Crangonyx pseudogracilis, Bousfield, 1958
Gammaridae
Echinogammarus berilloni (Catta, 1878),
Dikerogammarus villosus (Sowinsky, 1874),
Dikerogammarus haemobaphes (Eichwald, 1841)
Gammarus fossarum, Koch, 1835
Gammarus pulex (L., 1758)
Gammarus roeseli (Gervais, 1835)
Gammarus tigrinus, Sexton, 1939
Talitridae
Orchestia cavimana, Heller, 1865

Isopoda
Asellidae
Asellus aquaticus, L., 1758
Proasellus meridianus (Racovitza, 1919)
Proasellus coxalis (Dollfus, 1892)
Janiridae
Jaera istri, Veuille, 1979

Mysidacea
Mysidae
Hemimysis anomala, Sars, 1907
Limnomysis benedeni, Czerniavsky, 1882

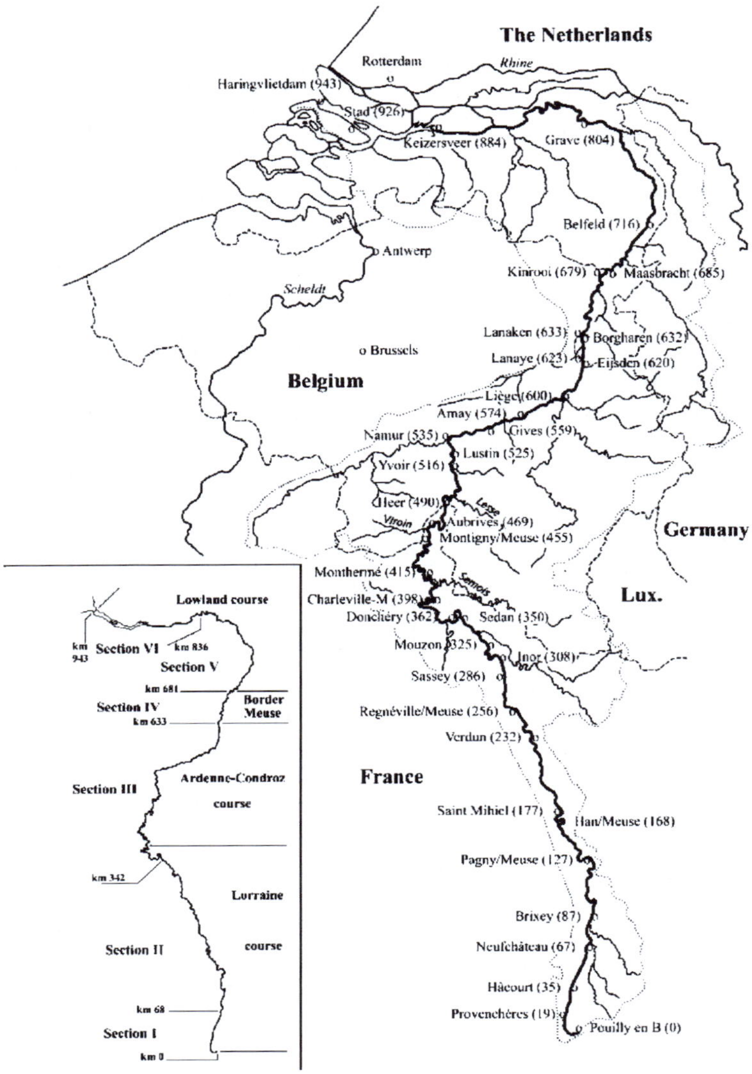

Figure 1. Map of the River Meuse with most of the locations mentioned in text. The numbers in brackets refer to the kilometric distances from the source. The zones and sections are figured in the insert.

Multivariate analysis

Prior to perform a correspondence analysis (CA) the data set was worked out. Two locations (Borgharen and Grave), with 12 or 14 samples per year during ten years, contributed for 55% to the whole data set. In order to avoid an imbalance with the other locations, these data were reduced to annual averages. Twelve samples, which did not contain any amphipods or isopods (and would be positioned at the origin of the axes of the factorial planes), have been discarded. The four geo-morphological zones of the Meuse previously defined were reorganized in six sections based on their peracarid community (see further in the text). Finally the log transformed data of the remaining matrix (215 samples × 11 taxa) were processed by correspondence analysis with the software ADE 4.0 (Thioulouse et al., 1997).

Other statistical analyses were made on log transformed numbers of individuals per substrate (or ind. subs.$^{-1}$) with the software STATISTICA 6.

Results

Analytical results

Amphipoda

Gammarus fossarum
G. fossarum was present in the artificial substrates at Brixey (km 87) where this species was rather abundant in 1998 and 2000 (29 and 16 ind. subs.$^{-1}$, respectively) but absent in 2001. It was again present at Hâcourt (km 35) in 2002 and at Savigny (km 91) in 2003 (Fig. 2).

G. fossarum was mentioned at several places in 1980–84 between Aubrives (km 469) and Rivière (km 522), always just downstream the old weirs, in turbulent water (Meurisse-Genin et al., 1987), at Tailfer and Hastière in 1989 (Frantzen, 1991) and at Hastière (km 494) in 1991 (Ketelaars, 1993). Some individuals were caught at Grave (km 804) in 1996.

It was present in samples from the River Viroin and was the dominant gammarid in 2001 in the River Semois (tributaries of the Meuse at km 465 and 415, respectively).

Gammarus pulex
G. pulex was present in the artificial substrates from Brixey (km 87) or Hâcourt (km 35) in 2002, to various places from Heer (km 490) in 2000, to Kinrooi (km 679) in 1998 (Fig. 2). In 2001 one isolated specimen was also collected at Lanaken (km 633). In the Dutch Meuse it was found in small numbers at Borgharen (km 632) in 1994 and 1995 and in larger numbers at Grave (km 804) from 1992 to 1996. It vanished there after the arrival of *D. villosus*, but some isolated specimens were caught again in 1999 and 2001 (Fig. 6).

In the Belgian Meuse historical samples showed the presence of *G. pulex* in 1934 at Namèche (km 545) and in 1943–1945 at Hastière (km 494), Hermeton (km 492) and Chertal (km 609) (Wouters, personal communication). In 1980–1984 it was collected at 13 places from Aubrives (km 469) to Amay (km 574) (Meurisse-Genin et al., 1987). It is mentioned at Hastière (km 494) in 1984–1993 by Ketelaars & Frantzen (1995). In our samples the species was dominant in 1991 (80–100% of the gammarids) at four localities from Heer (km 490) to Lustin (km 525). In 1995 it was still dominant at 10 localities sampled between Heer (km 490) and Petit Lanaye (km 624). In 1998 it was still present at eight localities and dominant at two of them (Heer, km 490 and Lustin, km 525). However, in 2000 it had vanished from the Belgian Meuse except from Heer (Vanden Bossche, 2002, this paper). In 2001 and 2002 it was locally present again in small numbers.

G. pulex was fairly common in 2002 in the samples from the Rivers Semois, Viroin and Lesse (tributaries of the Meuse at km 415, 465 and 505, respectively).

Gammarus roeseli
G. roeseli, originating from the Balkan area (Jazdzewski, 1980), can be considered as a naturalized species (it is a former exotic species, well established since a long period and in balance with its environment). It was present in the artificial substrates from Brixey (km 87) to Heer (km 490) (Fig. 2). In the Dutch Meuse it was consistently present at Grave (km 804) in small numbers from 1992 to 2001 (Fig. 6).

In the Belgian Meuse, historical samples (1943–1945) showed the presence of G. *roeseli* at Hastière (km 494) and Hermeton (km 492) (Wouters, personal communication). In 1980–1984 it was mentioned at four places from Chooz (km 480) to Bas-Oha (km 562) (Meurisse-Genin et al., 1987) and at Hastière (km 494) in 1987 (Frantzen (1991). From 1991 to 1998 the species was sporadically found at Heer (km 490) and Yvoir (km 516) and more regularly at Anseremme (km 504). After the invasion by *D. villosus*, it seemed to have vanished from the Ardenne-Condroz part of the river, but some specimens were caught again at Anseremme (km 504) in 2002.

It occurs in the River Sambre and in some canals connected with the River Meuse.

Gammarus tigrinus
G. tigrinus originates from North America. It was introduced in the Netherlands via Great Britain, probably in 1960 (Pinkster et al., 1992), it invaded the River Meuse at the latest in 1983 and became very abundant in 1991 at Keizersveer (Ketelaars, 1993, Ketelaars & Frantzen, 1995). It is still present from Belfeld (km 716) to Keizersveer (km 884) (Fig. 2).

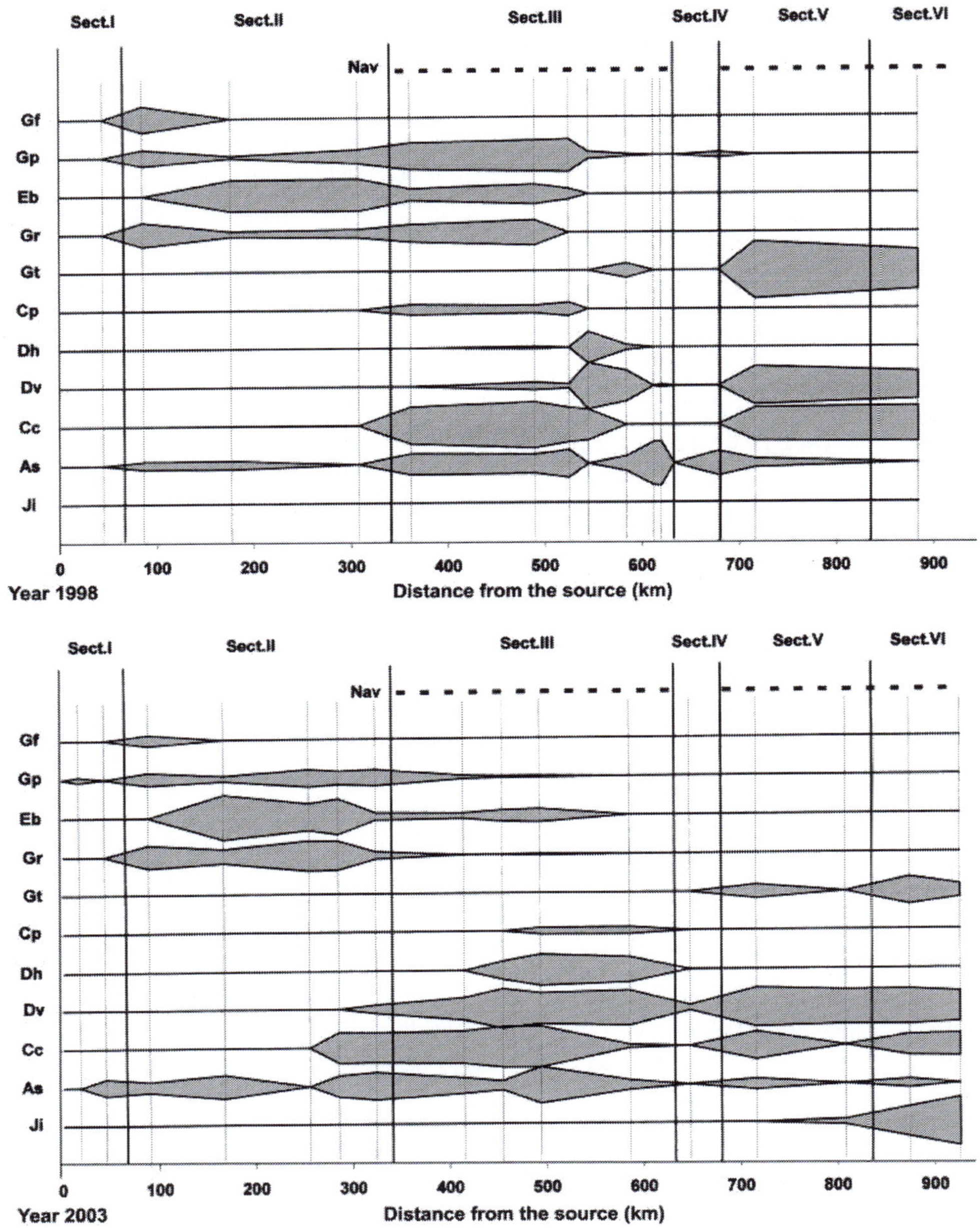

Figure 2. Longitudinal distribution of the Amphipoda and Isopoda in the River Meuse, above: in 1998 (C.I.P.M.-I.C.B.M. substrates, but at km 632 and 804: larger substrates) and below: in 2003 (small artificial substrates). Vertical solid lines = section limits, vertical dotted lines = sampling points. Horizontal dotted line (nav) = navigated zone. Other horizontal figures = species densities, the height at each sampling point is proportional to the log of the mean number of individuals per artificial substrate. As = Asellidae (*Asellus aquaticus* + *Proasellus meridianus* + *Proasellus coxalis*), Cc = *Chelicorophium curvispinum*, Cp = *Crangonyx pseudogracilis*, Dh = *Dikerogammarus haemobaphes*, Dv = *Dikerogammarus villosus*, Eb = *Echinogammarus berilloni*, Gf = *Gammarus fossarum*, Gp = *Gammarus pulex*, Gr = *Gammarus roeseli*, Gt = *Gammarus tigrinus* and Ji = *Jaera istri*.

At Grave (km 804) it was the dominant gammarid species from 1992 to 1996 and declined, but did not vanish, after the arrival of *D. villosus* (Fig. 6).

It was collected at Borgharen (km 632) in small numbers in 1992 and from 1994 to 1996, at Petit Lanaye (km 624) in 1995 and at Chokier (km 584) in 1998.

Echinogammarus berilloni

E. berilloni, originating from South-West France, can be considered as a naturalized species. It was present in the artificial substrates depending on years from Saint-Mihiel (km 177) or Han-sur-Meuse (km 168) to various places from Heer (km 490) to Anseremme (km 504). In 2001 some isolated specimens were also collected at Gives (km 559). It builds up densities that seem to complement those of *G. pulex* and *G. roeseli* (Fig. 2).

E. berilloni is present in large proportions (between 15 and 75% of the gammarids) in the Lorraine Meuse. However, it is absent from the headwaters of the river.

In the Belgian Meuse, historical samples showed the presence of *E. berilloni* in 1933 at Dave (km 529) and in 1943–1945 at Hastière (km 494) and Hermeton (km 492) (Wouters, 2002). In 1980–1984 it was common from Aubrives (km 469) to Amay (km 574) (Meurisse-Genin et al., 1987), it was also common in 1989 at Tailfer (km 526) and occurred at Hastière (km 494) in the period 1984–1993 (Ketelaars & Frantzen, 1995). This species was absent from our samples collected in 1991 from Heer (km 490) to Lustin (km 525) but present in the same localities in 1995 and 1998. In the Border Meuse, it was reported until the mid sixties (Pinkster et al., 1992).

E. berilloni is the dominant amphipod in the lower courses of the River Viroin (Schmit & Josens, 2004) and the Rivers Lesse and Berwinne (tributaries of the Meuse at km 465, 505 and 618, respectively), but seems to be absent from the lower course of the River Semois.

Dikerogammarus haemobaphes

The Ponto-Caspian *D. haemobaphes* invaded the Rhine River basin through the Main-Danube Canal after its opening in 1992 (Schleuter et al., 1994). From then, it expanded massively in Germany (Tittizer et al., 2000; Müller, 2001). Despite the intensive sampling in the Dutch Meuse it was collected only once, at Grave (km 804), in 2000.

This species is quoted here for the first time from Belgium. It appeared in the Meuse in 1998 (concurrently with *D. villosus*), in small numbers at four localities between Namèche (km 545) and Petit Lanaye (km 624). One isolated specimen was also caught the same year at Heer (km 490). In 2000 and the following years it became common in the whole Belgian Meuse, including the blind arms (Vieille Meuse and Nouvelle Gravière) at Lanaye (km 623), but in its upstream expansion it hardly crossed the French-Belgian border (in 2002 and 2003 it was found in small numbers at Montigny-sur-Meuse, km 455) and it remained absent from the Border Meuse (Fig. 2).

Dikerogammarus villosus

The Ponto-Caspian *D. villosus* invaded the Netherlands in 1994 (Bij de Vaate & Klink, 1995) through the Main-Danube Canal and the River Rhine (Tittizer et al., 2000) and was first caught in the River Meuse in 1996 at Grave (km 804) (Fig. 6). It reached Borgharen (km 632) in 1997 and in 1998 it was present at 12 localities between Heer (km 490) and Keizersveer (km 884), but it was absent from the Border Meuse (Usseglio-Polatera & Beisel, 2002a, Vanden Bossche, 2002). In the subsequent years, its invasion extended both into the Lorraine course and into the Border Meuse (Fig. 2).

The ongoing upstream invasion by *D. villosus* in the Meuse can be characterised by (a) a front of adults in small numbers (generally one or two individuals per artificial substrate) (b) about 30–60 km downstream a much higher density (> 50 individuals per artificial substrate) including adults and juveniles and (c) an upstream progression at a speed of 30–40 km per year between 1998 and 2002. The front was at Heer (km 490) in 1998, at Ham-sur-Meuse (km 472) in 2000 and at Mouzon (km 325) in 2002. However, in 2003 it had not moved, it was still found at Mouzon but not upstream (Fig. 3).

Crangonyx pseudogracilis

C. pseudogracilis originates from North America. It appeared in northern Netherlands in 1979 (Pinkster & Platvoet, 1983), but it was never collected in the Dutch Meuse and was first caught in the Belgian Meuse at Heer (km 490) in 1991 (Vanden Bossche, this paper). In 1995 it was present from Anseremme (km 504) to Namur (km

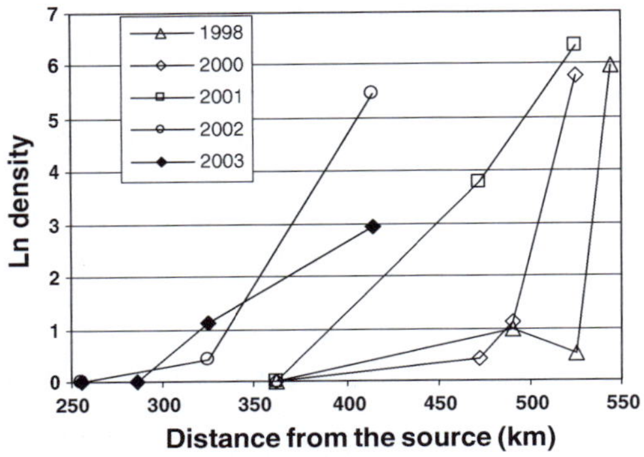

Figure 3. The ongoing upstream invasion *by Dikerogammarus villosus* (natural log of the mean number of individuals per artificial substrate) in the Meuse from 1998 to 2002. For the sake of clarity only the three or four most upstream locations where *D. villosus* was present have been figured. The data of 1998, 2000 and 2001 were acquired with C.I.P.M.-I.C.B.M. substrates, those of 2002 and 2003 with small artificial substrates.

535) and was common from Yvoir (km 516) to Loyers (km 541). In 1998 it was still present in small numbers from Donchéry (km 362) to Lustin (km 525) (Fig. 2) and had become abundant in a blind arm (Vieille Meuse) at Lanaye (km 623). In 2000 it seemed to be vanishing: few specimens were found at Donchéry (km 362) and Chokier (km 584) and one isolated specimen was caught in the Border Meuse at Kinrooi (km 679). In 2001 it was present again from Lustin (km 525) to Lanaye (km 621). In 2001 and 2002 it was no longer found in the French Meuse but continued to be abundant in the blind arms at Lanaye (km 623).

C. pseudogracilis was reported to be abundant in some gravel pits, liable to flooding, that were dug along the Border Meuse (Klink & de la Haye, 2000).

Chelicorophium curvispinum

The Ponto-Caspian C. *curvispinum* was first recorded in the Belgian Meuse near Huy (km 566) in 1981 and had become a dominant species of the benthic fauna at Champalle (km 515) in 1986. In the same year it had extended its expansion until Montigny-sur-Meuse (km 455) (Wouters, 1985, d'Udekem d'Acoz & Stroot, 1988) and was abundant at Tailfer (km 526) in 1989 (Ketelaars & Frantzen, 1995) and at Hastière (km 494) in the period 1984–1993 (Ketelaars & Frantzen, 1995). In France it is still migrating upstream: it was present in the artificial substrates at Donchéry (km 362) in 1998, Inor (km 308) in 2000 and 2001 and Sassey (km 286) in 2003. Its average migration speed is therefore about 15 km per year. In 1987 the species invaded the Dutch part of the River Rhine from the German part of that river (Van den Brink et al., 1993) and subsequently the Dutch Meuse (Van der Velde et al., 1998). Presently it can be found from the lower Lorraine stretch until the mouth of the river with a gap in the Border Meuse (Fig. 2).

At Grave (km 804) the Dutch survey showed that its density increased rather steadily from 1992 to 2001 with a temporary decrease in 1997, at the onset of *D. villosus*. At Borgharen (km 632) some scattered individuals were collected from 1993 to 2001 (Fig. 6).

Some specimens were found in the River Lesse in 2002 at about 5 km from the confluence with the Meuse (km 505) and repeatedly caught, equally in small numbers, in the River Viroin in 2002 and 2003 at about 5 km from the confluence with the Meuse (km 465).

Orchestia cavimana

O. cavimana lives along the river banks. It is therefore rarely sampled with artificial substrates and is not included into the figures. It is locally common along the river, under stones or dead plant debris. It was introduced in the early 20th century and was mentioned at Jambes (km 533) in 1983 and along the Zuid-Willemsvaart Canal in 1980 (Wouters, 2002). In 2002 and 2003 special attention was paid at this species and it was found at Montigny-sur-Meuse (km 455), Givet (km 489), Waulsort (km 498), Gives (km 559), Lanaye (km 621), Maasmechelen (km 648), Belfeld (km 716), Balgoij (km 808) and Stad aan 't Haringvliet (km 926).

Isopoda

Asellus aquaticus

A. aquaticus is present in small numbers in the Lorraine and Ardenne courses of the Meuse and becomes more common downstream the French-Belgian border. It reaches high densities downstream from Liège and is often the dominant peracarid at Eijsden (km 620), in the still water of the blind arms at Lanaye (km 623) and at Borgharen (km 632). It is present in the Border Meuse and in the lowlands: at Grave (km 804) its density decreased in 1997 when *D. villosus* became dominant (Fig. 6). It is thus the most ubiquitous peracarid in the Meuse (Fig. 2).

In 1980–1984 it was common at 17 locations from Aubrives (km 469) to Lixhe (km 618) (Meurisse-Genin et al., 1987) and in 1984–1993 at four locations from Hastière (km 494) to Borgharen (km 632) but rare at Keizersveer (km 884) (Ketelaars & Frantzen, 1995, Ketelaars, 1993).

Proasellus meridianus

P. meridianus is an exotic not invasive species originating from southern Europe (Tittizer et al., 2000). It was regularly collected, but always in small numbers, from Brixey (km 87) to Gives (km 559) and, in 2003, from Goncourt (km 46) to Flémalle (km 586). It was also sporadically collected at Eijsden (km 620) and Keizersveer (km 884). In the intensive surveys at Borgharen (km 632) it was caught sometimes from 1994 to 1997 and at Grave (km 804) from 1992 to 1995.

In 1980–1984 it was present at 11 locations from Aubrives (km 469) to Amay (km 574) (Meurisse-Genin et al., 1987) and in 1984–1993 occasionally at Hastière (km 494) and Keizersveer (km 884) (Ketelaars & Frantzen, 1995).

Proasellus coxalis

A. coxalis is another exotic not invasive species originating from southern Europe (Tittizer et al., 2000). It was collected once in the intensive surveys at Borgharen (km 632) in 1996 and several times at Grave (km 804) from 1992 to 1996.

Jaera istri

The Ponto-Caspian *J. istri* invaded the River Rhine through the Main-Danube Canal and reached the German-Dutch border in 1997 (Tittizer et al., 2000). It was first found in the River Meuse at Grave (km 804) in 1999 where it became abundant in 2001 (Fig. 6). In 2003 it was collected from Balgoij (km 808) to Stad aan 't Haringvliet (km 926). More upstream it was collected only once at Belfeld (km 716) in 2000 (Fig. 2).

Mysidacea

Hemimysis anomala

H. anomala is not correctly sampled with artificial substrates. The species is therefore not included into the figures. This Ponto-Caspian species was first discovered in 1997, near Keizersveer, in a

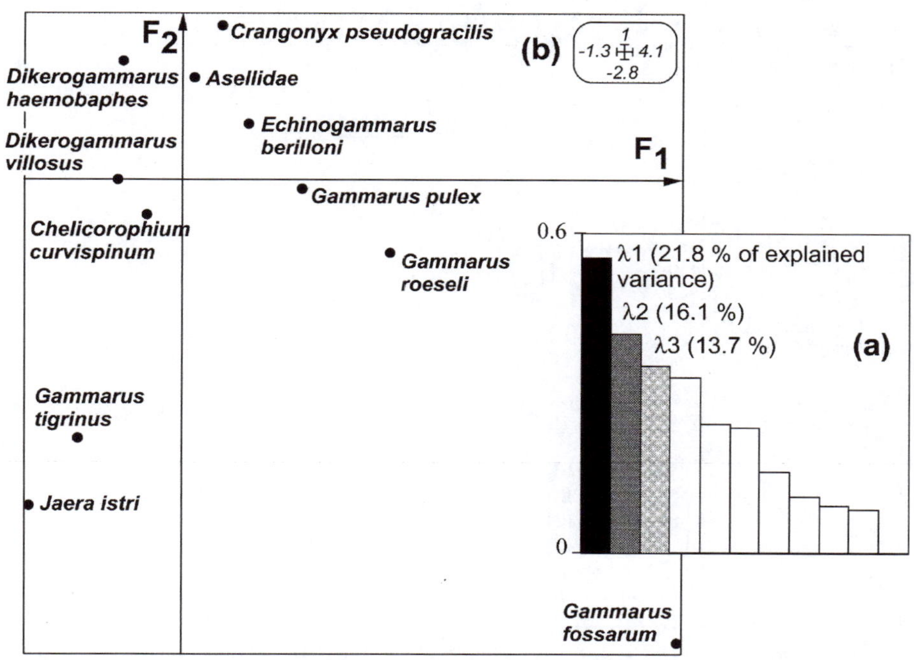

Figure 4. Ordination of species by Correspondence Analysis, (a). Histogram of eigenvalues, (b). Distribution of species (= solid circles) on the F1–F2 factorial plane.

storage reservoir fed with Meuse water. The first record in the river Meuse itself (Keizersveer, km 884) dates from 1998 (Ketelaars et al., 1999). It was also collected at Grave (km 804) in 1999. In Belgium it was found at Gives (km 559), Lustin (km 525) and Heer (km 490) in 2000 and in the blind arms at Lanaye (km 623) in 2001 (Vanden Bossche, 2002). All these findings in the River Meuse are unique.

Limnomysis benedeni

L. benedeni is not correctly sampled with artificial substrates and is therefore not included into the figures. This Ponto-Caspian species was first discovered in the Biesbosch, Dutch Meuse basin, in 1998 (Ketelaars et al., 1999). In 2000 and 2001 it was collected in the River Meuse itself at Keizersveer (km 884) and Belfeld (km 716).

Multivariate analysis

Results of the correspondence analysis are shown in Figures 4 and 5. The two first factors explain 21.8% and 16.1% of the variance, respectively. They are both linked with the longitudinal gradient of the river and therefore generate an arch (Guttman effect). The factorial plane F1–F2 will be interpreted in terms of amphipod and isopod communities (see discussion).

The third factor (not shown) is linked with the temporal change of the fauna and explains 13.7% of the total variance.

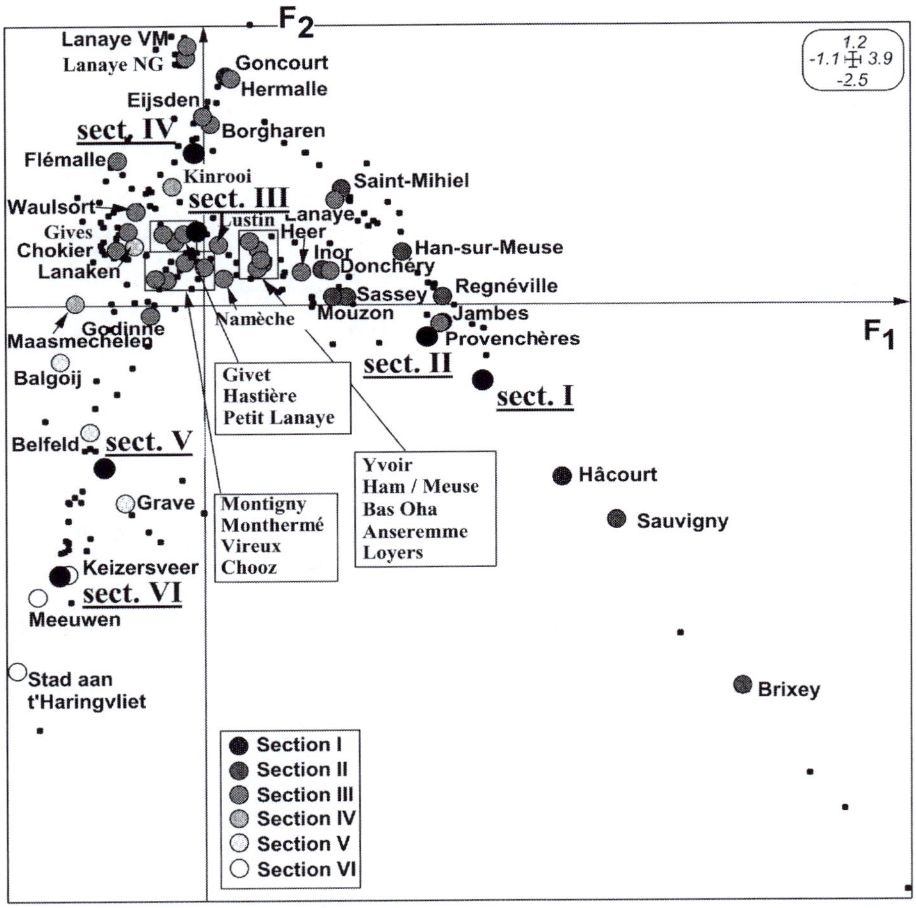

Figure 5. Ordination of samples, sites and sections by Correspondence Analysis. Distribution of samples (=small solid squares) on the F1–F2 factorial plane of the analysis. Each site (solid circle) was positioned at the weighted average of the sample positions representing this site. Each section (black solid circle) was positioned at the weighted average of the sample positions representing this section.

Discussion

Few information is available on the former composition of the Amphipoda and Isopoda communities of the River Meuse. However, from the samples collected during the second half of the 20th century until the early eighties, it can be inferred that the community of the Ardenne-Condroz course of the river consisted of four Amphipoda, *G.pulex*, *G. roeseli*, *G. fossarum* (the last species located just downstream the old weirs), *E. berilloni* and two Isopoda, *A. aquaticus* and *P. meridianus*, (Meurisse-Genin et al., 1987, Vanden Bossche, 2002, Wouters, 2002 and personal communication). In the lowland courses the Meuse housed mainly *G. pulex* and *E. berilloni* (Pinkster et al., 1992) and typical estuarine species as *Gammarus duebeni* Liljeborg, 1851 and *G. zaddachi* (Sexton, 1912) which were present in the Haringvliet (Den Hartog, 1964; Pinkster et al., 1992).

New communities have developed in the River Meuse as a consequence of successive introductions: *C. curvispinum* in 1981 (Wouters, 1985, d'Udekem d'Acoz & Stroot, 1988) and 1987 (Van der Velde et al., 1998), *O. cavimana* in 1983 or earlier (Wouters, personal communication), *G. tigrinus* in 1983 (Frantzen, 1991, Ketelaars & Frantzen, 1995), *C. pseudogracilis* in 1991 (Vanden Bossche, this paper), *D. villosus* in 1996 (this paper) and *D. haemobaphes* in 1998 (this paper).

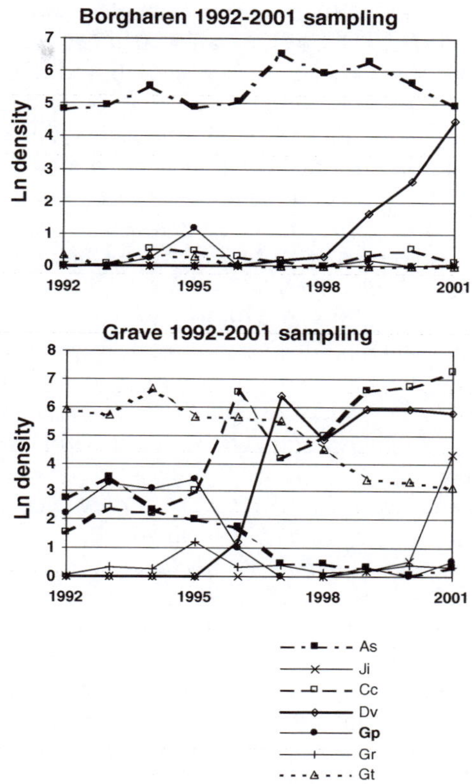

Figure 6. Temporal changes of the Amphipoda and Isopoda in the River Meuse, above: at Borgharen (km 632) and below: at Grave (km 804). Every annual point is the natural log of the average of 12 or 14 samples collected mostly from May to October with large artificial substrates. As = Asellidae (*Asellus aquaticus* + *Proasellus meridianus* + *Proasellus coxalis*), Cc = *Chelicorophium curvispinum*, Dv = *Dikerogammarus villosus*, Gp = *Gammarus pulex*, Gr = *Gammarus roeseli*, Gt = *Gammarus tigrinus* and Ji = *Jaera istri*.

The community of the head of the river (km 0–67) (section I)

Section I (km 0–67, 12 samples) (FI > 0 in the first factorial plane, Figs. 4 and 5) is characterized by an assemblage of three native species (*G. fossarum*, *G. pulex* and *A. aquaticus*) and one naturalized species (*G. roeseli*) with relatively low densities: some scattered individuals were caught at Provenchères (km 19), Hâcourt (km 35) and Goncourt (km 46). Six out of the 12 samples that did not contain any amphipods or isopods (discarded from the factorial analysis) came from this section. This rarity is thought to be linked with (a) a deepened riverbed and a very slow current, favouring the sedimentation of organic and locally anoxic mud, (b) the erosion of the surrounding clayey soils resulting in a clogged bottom substrate, (c) consequently the scarcity of suitable hiding places despite the presence of some aquatic vegetation (mainly helophytes) and (d) the fish abundance (Usseglio-Polatera, personal observations).

The native-naturalized community in the upper Lorraine course (km 68–341) (section II)

Section II (km 68–341, 35 samples) (FI > 0 in the first factorial plane, Figs 4 and 5) is characterized by an assemblage of native and naturalized species: *G. fossarum*, *G. pulex*, *G. roeseli*, *E. berilloni*, *A. aquaticus* and *P. meridianus*, making a community that seems mature and rather stable: (a) from 1998 to 2001 every C.I.P.M.-I.C.B.M. artificial substrate (except one) retrieved between 50 and 120 individuals ($n = 21$ samples) and (b) in 2002 and 2003 every smaller artificial substrate (except three) retrieved between 70 and 270 individuals ($n = 14$ samples). In the community, as suggested by Figure 2, (a) a significant negative correlation exists between *E. berilloni* and *G. roeseli* and (b) a significant positive one exists between *G. pulex* and *G. roeseli*.

This section of the River Meuse has not been modified for navigation. We observed that it offers a variety of biotopes with locally shallow and fast flowing water and diverse aquatic vegetation. At Regnéville (km 256), for example, the algae *Enteromorpha intestinalis* and *Chara vulgaris* and the moss *Fontinalis antipyretica* coexist with a variety of spermaphytes: *Myriophyllum spicatum*, *Ceratophyllum demersum*, *Elodea* sp., *Potamogeton pectinatus*, *P. perfoliatus*, *Lemna minor*, *L. trisulca*, *Nuphar lutea*, *Butomus umbellatus*, *Sagitaria sagittifolia* and *Sparganium demersum*.

The most downstream portion of this community was invaded by *C. curvispinum* without any apparent damage: the presence of 50 or more individuals per substrate at Sassey (km 286), Inor (km 308) or Mouzon (km 325) has not reduced the density of the native-naturalized community. *C. curvispinum* should not compete with the gammarids and asellids since it fills a totally different niche (Van der Velde et al., 1998). However, this community is downstream threatened by *D. villosus*, which reached Mouzon (km 325) in 2002. It is worth to notice that in 2003 *D. villosus* did not move further upstream (see section III). Perhaps the species has reached its uppermost extension in the River Meuse.

Despite the impression of a stable community it is worth to mention that significant variations (ANOVA, $p < 0.05$) between years were recorded for *G. fossarum* (from 0 to 29 ind. subs.$^{-1}$) and for *G. pulex* (from 8 to 24 ind. subs.$^{-1}$) in localities that were not subjected to any recent invasion. On the other hand in the same locations *E. berilloni* exhibited little variation between years (from 25 to 29 ind. subs.$^{-1}$).

The invasive community of the lower Lorraine and Ardenne-Condroz courses (km 342–632) (section III)

Section III (km 342–632, 88 samples) (F2 > 0 in the first factorial plane, Figs. 4 and 5) associates native or naturalized species (*G. pulex*, *E. berilloni* and *A. aquaticus*) and recent dominating invaders (*C. pseudogracilis*, *D. haemobaphes*, *D. villosus* and *C. curvispinum*). Two of them (*C. pseudogracilis* and *D. haemobaphes*) that reach the most positive values on the axis F2 are thus restricted to this section.

The amphipod community in the Ardenne-Condroz course was in 2002–2003 dominated by recent invaders: (a) by *D. villosus* and *C. curvispinum* from Monthermé (km 415) until Givet (km 489), (b) by *D. villosus*, *D. haemobaphes* and *C. curvispinum* from Waulsort (km 498) to Gives (km 559) and (c) by *D. haemobaphes* and *C. pseudogracilis* in the blind arms of Lanaye (km 623). Those invaders were probably transported by ship in ballast water or among fouling organisms attached to the hull. Most of them may have been released near Liège which is the second European inland harbour and all along the section, which is submitted to an upstream decreasing gradient of navigation (Vanden Bossche et al., 1999).

A transition zone extends in the lower Lorraine and upper Ardenne-Condroz courses (approximately km 350–500) where native and exotic species are struggling. This unstable community reached its highest species richness with nine peracarid species (of which five introduced) in samples taken at Heer (km 490) in 2000 and at Hastière (km 494) in 2003. Most of the samples of the transition zone contained six or seven species versus the presence of three or four in the native-naturalized community.

The success of *C. curvispinum* in the River Rhine was claimed to be linked with relatively high chloride concentration (Van den Brink et al., 1993). However, Harris & Bayliss (1990) showed its acclimation capacity toward lower salt concentrations. This may have been achieved in the Meuse where the annual average chloride concentration, i.e. 0.41 mMol l^{-1} from 1999 to 2002 at the French-Belgian border (C.I.M.-I.M.C., 2003, C.I.P.M.-I.C.B.M., 2001, 2002a, b), is lower than the threshold initially proposed for this species (Bayliss & Harris, 1990). It is even lower in the Rivers Viroin and Lesse where *C. curvispinum* is also present.

It was also claimed that *C. curvispinum* had filled an 'empty niche' (Den Hartog et al., 1992). Actually it was the first tubiculous amphipod that colonised the Rivers Rhine and Meuse (Van der Velde et al., 1998) and thus filled a niche that never existed there before. It is an active filter feeder (Hynes, 1970) that benefits from the high phytoplankton contents of large rivers (Van der Velde et al., 1998). Therefore the high phytoplankton content of the Lorraine Meuse should favour its continuing upstream migration in the Meuse, but the lesser phytoplankton content of the tributaries will probably hamper its invasion into those streams.

C. curvispinum is a rather fast migrant: d'Udekem d'Acoz & Stroot (1988) estimated its upstream migration speed at about 17 km per year. From those observations in 1986 to our first data, in 1998, this species migrated from Montigny-sur-Meuse (km 455) to Donchéry (km 362), i.e. at a speed of about 8 km per year. During the present study it reached Sassey (km 286) in 2003, its speed thus had again increased at about 15 km per year. Since upstream migration was regular we suggest that it moved upstream actively, i.e. in addition to ship transportation.

The behaviour of the most aggressive invader, *D. villosus*, can be characterized by (a) the high speed of upstream range extension: 30–40 km per year, i.e. about 100 m per day. As section III is entirely navigable, this high upstream speed may be favoured by ship transportation. However, migration occurs regularly and proceeds without 'gaps'. It might thus also be an active migration in addition to ship transport. The structure of the Meuse cut by weirs and locks into reaches of slow flowing water may obviously facilitate this migration, (b) Its high and fast conquering capacity: once the first adults have appeared in the samples the species becomes dominant already one year later. This invasive efficiency can be related with some of its biological traits such as its intra-guild predation behaviour (Dick & Platvoet, 2000) and its ability to colonize a wide range of substratum types (Devin et al., 2003). (c) Once established this invasive gammarid often builds up higher densities (200–500 ind. subs.$^{-1}$) than the previous native-naturalized community (50–120 ind. subs.$^{-1}$). The densities that *D. villosus* achieves are so high that they can hardly be supported by its predatory behaviour alone, therefore we suggest that it is also able to exploit the available food resources more efficiently than the native gammarid community.

The annual changes in the communities in section III suggest that *D. villosus* eliminates firstly *G. roeseli* and secondly *G. pulex*. *E. berilloni* seems to withstand this competition better at least at some locations. This is consistent with the dynamics observed in the annual cycle of the gammarids at Montigny-sur-Meuse (km 455) (Josens, unpublished).

Despite its migration dynamics, *D. villosus* has not yet been found in the Rivers Semois, Viroin and Lesse even at a short distance from their confluence with the Meuse. There does not seem to be any physical barrier preventing the upstream migration. Those tributaries are more natural, not navigated, flow faster, contain less dissolved salts, are cooler and poorer in phytoplankton than the Meuse, but it is too early to state which factor(s) could explain their non invasibility. Those tributaries therefore keep on sheltering the native species that could be able to reinvade the Meuse in case of a decreasing aggressiveness of *D. villosus*.

It is worth to mention the amazing extension of *D. haemobaphes* which is almost restricted to the Belgian Meuse. This invading species has found there favourable conditions that seem to be rather far from those occurring in Lake Balaton, where it lives among floating *Myriophyllum spicatum* (Musko, 1993). As it was pointed out by Usseglio-Polatera & Beisel (2002b) locations from km 362–525 display close physico-chemical characteristics but faunal organization vary deeply from the French to the Belgian Meuse, the frontier being located at km 490. The river is navigable for boats and barges of 300 and 1350 tons upstream and downstream from Givet (km 489), respectively. Indeed some relevant habitat parameters, such as the heterogeneity of current velocity, depth and substrates and the extent of impoundment and embankment for navigation change definitely at Givet and may be decisive in explaining *D. haemobaphes* upstream migration restriction.

The Border Meuse community (km 633–680) (section IV)

Section IV (km 633–680, 149 samples) ($F1 < 0$ and $F2 > 0$ in the first factorial plane, Figs 4 and 5) is an impoverished stretch dominated by *A. aquaticus* and *D. villosus*. Five out of the 12 samples that did not contain any amphipods or isopods (discarded from the factorial analysis) came from this section. Between the Condroz and the lowland courses, the Border Meuse used to be a 'no-Amphipod stretch' in the period 1981–1990. According to Bij de Vaate (1995) a permanent high concentration of cadmium was responsible for this former amphipod absence. Although the cadmium concentration has been reduced substantially since 1992 (at <0.5 μg l^{-1}: Volz et al., 2002), this section still suffers multipollution and its gammarid community is very slowly recovering. In 1998 some specimens of *G. pulex* were caught at Kinrooi (km 679). The first *D. villosus* were collected in the Border Meuse in 2000 at Lanaken (km 633) and Kinrooi (km 679) and catches were a little more abundant in 2001 at Lanaken. *D. villosus* was still the only gammarid present at Maasmechelen (km 648) in 2003 in small numbers.

Very nearby, at Borgharen (km 632) the intensive 1992–2001 sampling suggests that the conditions were getting better for the gammarids there than in the Border Meuse: *D. villosus* was collected first in 1997 and its density steadily increased (Fig. 6). However, it did not increase as fast as in the invaded locations of section III. Moreover a dramatic drop of asellid densities can be mentioned in 1998 from Borgharen (km 632) to the Border Meuse, one km downstream (Fig. 2).

Although the region of Liège (km 588–610) has been incriminated (Volz et al., 2002), it is worth to point out the complex situation downstream this industrial area. At this point the River Meuse undergoes a unique feature: it feeds three canals (Albert, Juliana and Zuid-Willemsvaart), leaving, when discharge is low, a minimal residual discharge of 10 m^3 s^{-1} through the dam of Borgharen in the Meuse itself (treaty of January 17, 1995, between Flanders and the Netherlands). There is no navigation on the Border Meuse. However, in the Border Meuse, the volume, current and depth fluctuate dramatically in short periods of time. In this area the amphipod communities show a complex pattern. On the one hand the samples taken in section III of the main channel of the river [at Hermalle-sous-Argenteau (km 612) in 1998, at Lixhe (km 618) in 2000 and at Eijsden (km 620) in 1998, 2000 and 2001] all contained either no amphipods or low densities (0–5 ind. subs.$^{-1}$). On the other hand the samples collected in the blind arms near Lanaye (km 623) in 1998, 2001 and 2002 did contain high densities (90–190 ind. subs.$^{-1}$) of *C. pseudogracilis* and both *Dikerogammarus* species (mainly *D. haemobaphes*). These blind arms, however, are widely connected to the main channel and thus similarly subjected to the same chemical characteristics. Moreover the samples collected at Borgharen (km 632) in still water contained only few amphipods in 1998 and 1999 but increasing densities in 2000 and 2001 (Fig. 6), whereas the densities stayed low at Lanaken (km 633) in the same years. Therefore a pollution effect alone cannot explain the absence or rarity of amphipods. It is striking that they are present in still waters and lacking in either strongly waved water (by navigation) or swiftly flowing water with huge level changes (in the Border Meuse).

Hynes (1954) established that water movement increases the osmoregulation capacity of gammarids through a better uptake of soluble ions: this could also be true for pollutants. It can be thought that animals that are undermined by multipollution can less easily withstand flowing or turbulent water and this should even be worse if the water level unpredictably changes as it is the case just downstream the dam of Borgharen (km 632).

Moreover, more downstream, in section V, the amphipods regain very high densities e.g. at Belfeld (km 716), Grave (km 804) and Keizersveer (km 884) despite the fact that cadmium bio-availability seems to stay high: the zebra mussel bioaccumulates as much Cd at Keizersveer as at Eijsden (km 620) (Maas, 2001).

The sand Meuse community (km 681–835) (section V)

Section V (km 681–835, 140 samples) (F1 < 0 and F2 < 0 in the first factorial plane, Figs 4 and 5) houses only recent invaders (*D. villosus*, *C. curvispinum* and *G. tigrinus*).

This section is heavily navigated and the banks have been strengthened with blocks. The phytoplankton is less abundant than in the upstream sections and the vegetation is dominated by filamentous green algae (personal observations).

The 10-year intensive sampling at Grave (km 804) shows a contrasted situation before (1992–1995) and after (1997–2001) the establishment of *D. villosus* (Fig. 6). In 1992–1995 the community was dominated by *G. tigrinus* and *G. pulex* and the

Asellidae were moderately abundant. *G. roeseli* was present but rare and *C. curvispinum* increased slowly. *D. villosus* was collected first in May 1996 and became already dominant in 1997. From 1997 to 2001 the Asellidae and native gammarids almost vanished and *G. tigrinus* was reduced drastically. However, *C. curvispinum* kept on increasing and *J. istri* became moderately abundant in 2001.

The population dynamics followed a little different pattern at Belfeld (km 716): *D. villosus* became dominant in 2000 and *G. tigrinus* regressed. In contrast to Grave, *C. curvispinum* regressed and *J. istri* did not achieve its settlement. It is indeed known that *D. villosus* preys upon other macroinvertebrates such as *G. tigrinus* and *C. curvispinum* (Dick & Platvoet, 2000). The results from Belfeld could therefore be expected, but those from Grave show that another outcome is possible.

The 'tidal'Meuse community (km 836–943) (section VI)

Section VI (km 836–943, 10 samples) (Fl < 0 and F2 < 0 in the first factorial plane, Figs 4 and 5) is rather similar to section V and also houses only recent invaders (*C. curvispinum*, *G. tigrinus* and *J. istri*). *D. villosus* here became dominant with some delay.

The 'tidal' Meuse, downstream the Lith dam, has no longer been subjected to tides since the closure of the Haringvliet in 1970. Therefore the typical estuarine species *Gammarus duebeni* and *Gammarus zaddachi* disappeared from the Haringvliet (Pinkster et al., 1992). Although few differences could be expected from section V, the 'tidal' Meuse is considered apart because it is directly and strongly influenced by the Waal, one of the major distributaries of the River Rhine.

Two species were very abundant in the 1996 samples from Keizersveer (km 884): *G. tigrinus* and *C. curvispinum*. *D. villosus* was first found in 1998, but it might have been present already in 1997. The subsequent increase of its density was rather slow and it became dominant only in 2001. Its aggressiveness towards the other species seems to be milder since *G. tigrinus* and *C. curvispinum* hardly regressed.

The samples collected in 2003 at Meeuwen (km 874) and Stad aan 't Haringvliet (km 926) are very similarly dominated by *D. villosus* and *C. curvispinum*. An extra species, *J. istri*, was the most abundant in one of the samples collected at Stad, in a muddy place, which contrasts with its reputation of a 'strictly rheophil' organism in the River Danube (Tittizer & Banning, 2000).

Invasibility of the River Meuse

Obviously the different sections of the Meuse are neither equally likely to be invaded nor invadable by the same way. Sections VI and V can be invaded easily either by ship transportation, by active migration or simply by drifting from the River Rhine (via connecting canals in the case of section V). Section III can be invaded only by ship transportation as section IV, till now, does not offer favourable survival conditions for most peracarids and should constitute a barrier against active migration. Exotic species have indeed been introduced into Section III mostly between Namur and Liège where heavy navigation occurs. From Namur the upstream migration of exotic species progresses regularly, which suggests that, besides transport by ship, it may be due to active migration. Section II is not navigable, but navigation takes place in the canal de l'Est running parallel to the river with some short common reaches. Therefore if some ship transportation occurs then the invaders should easily reach Pagny-sur-Meuse (km 127), the uppermost navigable point of the Meuse. The non navigable section I, as well as the tributaries, can be invaded only through an active spread of the invaders and this has not yet occurred. However, in some cases canoeing activities might also facilitate the transportation of exotic species (e.g. in the River Lesse).

The invasibility of the River Meuse seems to be lower than of the River Elbe. In the latter, *D. villosus* was probably introduced near Magdeburg in 1998 (Tittizer et al., 2000) and spread in the same year over more than 200 km. In 1999 the species occurred over a stretch of 500 km and by 2001 it had invaded the whole German course of the Elbe, which is totally navigable, except the tidal section of the river (Krieg, 2002).

Among the predictions made for invasive fishes (Moyle & Light, 1996) some could apply to the peracarid fauna of the Meuse: (a) the most successful invaders are those adapted to the local hydrologic regime, (b) a much wider range of species can invade systems with higher levels of human disturbance and (c) a much wider range of species can invade systems with assemblages or organisms that have been temporarily disrupted or depleted. In its native region *E. berilloni* inhabits the middle reaches of streams (Pinkster, 1993), which predisposed this species to invade the tributaries of the Meuse and the Meuse itself but not its headwaters. On the other hand *D. villosus* and other recent invaders do not enter into the non navigable tributaries of the Meuse probably because the hydrologic regimes of their native water bodies are those of navigable or canalised streams, which are totally different from the shallow and rather fast flowing tributaries of the Meuse.

Moyle & Light (1996) and Tittizer (1996) pointed out the fact that extinctions following invasions in rivers and canals are rare. In case of the Meuse this applies only if the whole river is taken into consideration. At a more local scale, the native species have actually been driven to extinction.

Conclusions

In general, the peracarid species richness of the River Meuse has increased because of the arrival of exotic species, but locally the native species have been driven to extinction. The success of the invasive species in each section of the river depends merely on the local hydrological conditions: most of the recent invaders originating from large water bodies have taken advantage of the impoundment of the river in its sections III, V and VI.

The Meuse basin shows a *gradient* of invasibility linked with the gradient of anthropogenic impacts. It exhibits a rather high level of resistance against invasions in its upper course as well as in its tributaries that have not been impounded and that act as reservoirs for the native species.

Acknowledgements

We want to thank the A.E.R.M. (Agence de l'Eau Rhin-Meuse) and the research consultancy GREBE for the loaning of the C.I.P.M.-I.C.B.M. samples collected in France, the CRNFB-MRW (Centre de Recherche de la Nature, des Forêts et du Bois, DGRNE, Ministère de la Région wallonne) for the loaning of the C.I.P.M.-I.C.B.M. samples collected in Wallonia, the VMM (Vlaams Milieu Maatschappij) for the loaning of the C.I.P.M.-I.C.B.M. samples collected in Flanders, the R.I.Z.A. (Institute for Inland Water Management & Waste Water Treatment), that provided us the results of the C.I.P.M.-I.C.B.M. samples collected in the Netherlands and their ten-year surveys at Borgharen and Grave, the Netherlands.

We are grateful to Karel Wouters for the information he provided about the collections of the IRSNB-KBIN and to two anonymous referees whose comments contributed to improve the manuscript.

References

Bayliss, D. & R. R. Harris, 1990. Chloride ion regulation in the freshwater amphipod *Corophium curvispinum* and acclimatory effects of external Cl^-. Journal of Comparative Physiology B 158: 81–90.

Bij de Vaate, A., 1995. Macroinvertebrate communities in the Grensmaas stretch of the River Meuse: 1980–1990. Journal of Freshwater Ecology 10: 75–82.

Bij de Vaate, A. & A. G. Klink, 1995. *Dikerogammarus villosus* Sowinsky (Crustacea: Gammaridea) a new immigrant in the Dutch part of the lower Rhine. Lauterbornia 20: 51–54.

C.I.M.-I.M.C. (Commission Internationale de la Meuse – Internationale Maascommissie), 2003. Résultats du réseau de mesures homogène – Resultaten van het homogeen meetnet 2002, 99 pp. Palais des Congrès, Liège or http://www.meuse-maas.be.

C.I.P.M.-I.C.B.M. (Commission Internationale pour la Protection de la Meuse – Internationale Commissie voor de Bescherming van de Maas), 1997. La qualité de la Meuse en 1994. De kwaliteit van de Maas in 1994, 114 pp.

C.I.P.M.-I.C.B.M., 2001. Résultats du réseau de mesures homogène – Resultaten van het homogeen meetnet 1999, 159 pp. Palais des Congrès, Liège or http://www.meuse-maas.be.

C.I.P.M.-I.C.B.M., 2002a. Résultats du réseau de mesures homogène – Resultaten van het homogeen meetnet 2000, 139 pp. Palais des Congrès, Liège or http://www.meuse-maas.be.

C.I.P.M.-I.C.B.M., 2002b. Résultats du réseau de mesures homogène – Resultaten van het homogeen meetnet 2001, 92 pp. Palais des Congrès, Liège or http://www.meuse-maas.be.

Den Hartog, C., 1964. The amphipods of the deltaic region of the Rivers Rhine, Meuse and Scheldt in relation to the hydrography of the area; part III. The Gammaridae. Netherlands Journal of Sea Research 2: 407–457.

Den Hartog, C., F. W. B. van den Brink & G. van der Velde, 1992. Why was the invasion of the River Rhine by *Corophium curvispinum* and *Corbicula* species so successful? Journal of Natural History 26: 1121–1129.

De Pauw, N., V. Lambert, A. van Kenfoven & A. bij de Vaate, 1994. Comparison of two artificial substrate samplers for macroinvertebrates in biological monitoring of large and deep rivers and canals in Belgium and the Netherlands. Journal of Environmental Monitoring & Assessment 30: 25–47.

De Pauw, N., D. Roels & A. P. Fontoura, 1986. Use of artificial substrates for standardized sampling of macroinvertebrates in the assessment of water quality by the Belgian Biotic Index. Hydrobiologia 133: 237–258.

Devin, S., C. Piscart, J. N. Beisel & J. C. Moreteau, 2003. Ecological traits of the amphipod invader *Dikerogammarus villosus* on the mesohabitat scale. Archiv für Hydrobiologie 158: 43–56.

Dick, J. T. A. & D. Platvoet, 2000. Invading predatory crustacean *Dikerogammarus villosus* eliminates both native and exotic species. Proceedings of the royal Society of London, Series B: Biological Sciences 267: 977–983.

d'Udekem d'Acoz, C. & P. Stroot, 1988. Note sur l'expansion de *Corophium curvispinum* Sars, 1895 en Meuse (Crustacea, Amphipoda: Corophiidae). Annales de la Société royale zoologique de Belgique 118: 171–175.

Frantzen, N., 1991. De kwaliteit van Rijn- en Maaswater in de periode 1983–1989. Beoordeling met behulp van macro-evertebraten. RIWA, Amsterdam.

Gay Environnement, 1997. Indice biologique global adapté aux grands cours d'eau et aux rivières profondes (I.B.G.A.). Protocole expérimental. Agence de l'Eau Rhône-Méditerranée-Corse, 45 pp.

Harris, R. R. & D. Bayliss, 1990. Osmoregulation in *Corophium curvispinum* (Crustacea: Amphipoda), a recent coloniser of freshwater. III. Evidence for adaptive change in sodium regulation. Journal of comparative Physiology B 160: 85–92.

Hynes, H. B. N., 1954. The ecology *of Gammarus duebeni* Lilljeborg and its occurrence in fresh water in western Britain. Journal of Animal Ecology 23: 38–84.

Hynes H. B. N., 1970. The ecology of running waters. Liverpool University Press, 541 pp.

Jazdzewski, K., 1980. Range extension of some Gammaridean species in European inland Waters caused by human activity. Crutaceana Suppl. 6: 84–107.

Ketelaars, H. A. M., 1993. Die Makro-Invertebratenbesiedlung von künstlichem Substrat in den Flüssen Maas und Rhein im Jahre 1991. Wasser Abwasser 134: 533–541.

Ketelaars, H. A. M. & N. M. L. H. F. Frantzen, 1995. One decade of benthic macroinvertebrate biomonitoring in the River Meuse. Netherlands Journal of Aquatic Ecology 29: 121–133 with addendum.

Ketelaars, H. A. M., F. E. Lambregts-van de Clundert, C. J. Carpentier, A. J. Wagenvoort & W. Hoogenboezem, 1999. Ecological effects of the mass occurrence of the Ponto-Caspian invader *Hemimysis anomala* G.O. Sars, 1907 (Crustacea: Mysidacea), in a freshwater storage reservoir in the Netherlands, with notes on its autecology and new records. Hydrobiologia 394: 233–248.

Klink, A. G. & M. A. A. de la Haye, 2000. Inventarisatie van macrofauna in de Limburgse Maasplassen. Reports of the project 'Ecological rehabilitation of the River Meuse' RIZA report 34–2000: 73 pp.

Kornberg, H., & M. H. Williamson (eds), 1987. Quantitative Aspects of the Ecology of Biological Invasions. The royal Society, London.

Krieg, H. -J., 2002. Biomonitoring der Amphipodenfauna in der Oberen, Mittleren und Unteren Elbe. ARGE ELBE (Arbeitsgemeinschaft für die Reinhaltung der Elbe). http://www.arge-elbe/wge/Download/Dtexte.html.

Levine, C. M. & C. M. D'Antonio, 1999. Elton revisited: a review of evidence linking diversity and invasion. Oikos 87: 15–26.

Lonsdale, W. M., 1999. Global patterns of plant invasions and the concept of invasibility. Ecology 80: 1522–1536.

Maas, H., 2001. Ecotoxicology. In Liefveld, W.M., K. van Looy & K. H. Prins (eds), Biological monitoring of national fresh waters, the River Meuse 1996. RIZA report 2002.022: 23–24.

Meurisse-Genin, M., A. Reydams-Detollenaere, Ph. Stroot & J. -C. Micha, 1987. Les macroinvertébrés benthiques de la Meuse belge: bilan de cinq années de recherches (1980 à 1984). Archiv für Hydrobiologie 109: 67–88.

Moyle, P. B. & T. Light, 1996. Fish invasions in California: do abiotic factors determine success? Ecology 77: 1666–1670.

Müller, J., 2001. Taxonomic status and genetic structure of *Dikerogammarus*-species invading the River Rhine, http://perdrix.biologie.uni-mainz.de/gammarus/Dikerogammarus.

Musko, I. B., 1993. The life history *of Dikerogammarus haemobaphes* (Eichw.) (Crustacea, Amphipoda) living on macrophytes in lake Balaton (Hungary). Archiv für Hydrobiologie 127: 227–238.

Pinkster, S., 1993. A revision of the genus *Echinogammarus* (Stebbing, 1899) with some notes on related genera (Crustacea, Amphipoda). Memorie del Museo di storia naturale (II° serie) Scienze della vitta (a Biologia) 10: 9–183.

Pinkster, S. & D. Platvoet, 1983. Further observations on the distribution and biology of two alien amphipods, *Gammarus tigrinus* Sexton, 1939 and *Crangonyx pseudogracilis* Bousfield, 1958 in the Netherlands (Crustacea, Amphipoda). Bulletin zoölogisch Museum Universiteit van Amsterdam 9: 153–164.

Pinkster, S., M. Scheepmaker, D. Platvoet & N. Broodbakker, 1992. Drastic changes in the amphipod fauna (Crustacea) of Dutch inland waters during the last 25 years. Bijdragen tot de Dierkunde 61: 193–204.

Rajagopal, S., G. van der Velde, B. G. P. Paffen & A. bij de Vaate, 1998. Growth and production of *Corophium curvispinum* G.O. Sars, 1895 (Amphipoda) an invader in the lower Rhine. Proceedings of the fourth international Crustacean Congress 1998: 457–472.

R.I.Z.A. (Rijksinstituut voor Integraal Zoetwaterbeheer en Afvalwaterbehandeling), 1998. Proposal for the harmonisation of macroinvertebrate monitoring in the River Meuse, 15 pp.

Schmit, O. & G. Josens, 2004. Preliminary study of the scars borne by Gammaridae (Amphipoda, Crustacea). Belgian Journal of Zoology 134: 75–78.

Schleuter, M., A. Schleuter, S. Potel & M. Banning, 1994. *Dikerogammarus haemobaphes* (Eichwald, 1841) (Gammaridae) aus der Donau erreicht über den Main-Donau-Kanal den Main. Lauterbornia 21: 155–159.

Thioulouse, J., S. Dolédec, D. Chessel & J. M. Olivier, 1997. ADE software: multivariate analysis and graphical display of environmental data, Software per l'Ambiente, Guariso G. & A. Rizzoli (eds), Bologna: 57–62.

Tittizer, T., 1996. Vorkommen und Ausbreitung aquatischer Neozoen (Makrozoobenthos) in den Bundeswasserstrassen. In Gebhardt, H., R. Kinzelbach & S. Schmidt-Fischer (eds), Gebietsfremde Tierarten – Auswirkungen auf einheimische Arten, Lebensgemeinschaften und Biotope – Situationsanalyse: 49–86.

Tittizer, T. & M. Banning, 2000. Biological assessment in the Danube catchment area: indications of shifts in species composition induced by human activities. European Water Management 3: 35–45.

Tittizer, T., F. Schöll, M. Banning, A. Haybach & M. Schleuter, 2000. Aquatische Neozoen im Makrozoobenthos der Binnenwasserstrassen Deutschlands. Lauterbornia 39: 1–72.

Usseglio-Polatera, Ph. & J. N. Beisel, 2002a. Traitement des résultats du monitoring international de la faune de macroinvertébrés de la Meuse, 108 pp. Rapport C.I.P.M.-I.C.B.M., Palais des Congrès, Liège.

Usseglio-Polatera, Ph. & J. N. Beisel, 2002b. Longitudinal changes in macroinvertebrate assemblages in the Meuse River: anthropogenic effects versus natural change. River Research and Applications 18: 197–211.

Vanden Bossche, J. -P., E. Delooz & F. Grisez, 1999. Macroinvertébrés benthiques et qualité biologique de la Meuse wallonne en 1998. Centre de Recherche de la Nature, des Forêts et du Bois. D.G.R.N.E. Ministère de la Région wallonne. Gembloux, 60 pp.

Vanden Bossche, J. -P., 2002. First records and fast spread of five new (1995–2000) alien species in the River Meuse in Belgium: *Hypania invalida*, *Corbicula fluminea*, *Hemimysis anomala*, *Dikerogammarus villosus* and *Crangonyx pseudogracilis*. Bulletin de l'Institut royal des Sciences naturelles de Belgique Biologic, 72(Suppl.): 73–78.

Van den Brink, F. W. B., G. van der Velde & A. bij de Vaate, 1993. Ecological aspects, explosive range extension and impact of a mass invader, *Corophium curvispinum* Sars, 1895 (Crustacea, Amphipoda) in the lower Rhine (the Netherlands). Oecologia 93: 224–232.

Van der Velde, G., I. Nagelkerken, S. Rajagopal & A. bij de Vaate, 2002. Invasions by alien species in inland freshwater bodies in Western Europe: the Rhine delta. In Leppäkoski, E., S. Gollasch & S. Olenin (eds), Invasive aquatic species of Europe. Distribution, impacts and management. Kluwer Academic Publ., Dordrecht: 360–372.

Van der Velde, G., S. Rajagopal, F. W. B. van den Brink, B. Kelleher, B. G. P. Paffen, A. J. Kempers & A. bij de Vaate, 1998. Ecological impact of an exotic invasion in the River Rhine. In Nienhuis, P. H., R. S. E. W. Leuven & A. M. J. Ragas (eds), New concepts for sustainable management of river basins, 159–169. Backhuys Publ., Leiden.

Vereertsraeten, J., 1971. Le bassin de la Meuse, étude de géographie hydrologique. Revue belge de Géographie 94: 1–339.

Volz, J., H. Ketelaars & A. Wagenvoort, 2002. Cinquante années d'évolution de la qualité de l'eau de la Meuse – un aperçu. H_2O 3: 21–26.

Williamson, M. & A. Fitter, 1996. The varying success of invaders. Ecology 77: 1661–1666.

Wouters, K. A., 1985. *Corophium curvispinum* Sars, 1895 (Amphipoda) in the River Meuse, Belgium. Crustaceana 48: 218–220.

Wouters, K., 2002. On the distribution of alien non-marine and estuarine macro-crustaceans in Belgium. Bulletin de l'Institut royal des Sciences naturelles de Belgique, Biologie 72 (Suppl).: 119–129.

Hydrobiologia (2005) 542:221–233
H. Segers & K. Martens (eds), Aquatic Biodiversity II
DOI 10.1007/s10750-004-1389-x

Phylogeography and speciation in the *Pseudocrenilabrus philander* species complex in Zambian Rivers

Cyprian Katongo[1], Stephan Koblmüller[2], Nina Duftner[2], Lawrence Makasa[3] & Christian Sturmbauer[2,*]
[1]*Department of Biological Sciences, University of Zambia at Lusaka, Zambia*
[2]*Department of Zoology, University of Graz, Universitätsplatz 2, 8010 Graz, Austria*
[3]*Department of Fisheries at Mpulungu, Ministry of Agriculture and Cooperatives, Zambia*
*(*Author for correspondence: Tel.: +43-316-380-5595, Fax: +43-316-380-9875,*
E-mail: christian.sturmbauer@uni-graz.at)

Key words: mtDNA sequences, geographic speciation, phylogeography

Abstract

Haplochromine cichlids form the most species-rich lineage of cichlid fishes that both colonized almost all river systems in Africa and radiated to species flocks in several East African lakes. The enormous diversity of lakes is contrasted by a relatively poor albeit biogeographically clearly structured species diversity in rivers. The present study analyzed the genetic structure and phylogeographic history of species and populations of the genus *Pseudocrenilabrus* in Zambian rivers that span two major African drainage systems, the Congo- and the Zambezi-system. The mtDNA phylogeny identifies four major lineages, three of which occur in the Congo-system and one in the Zambezi system. Two of the Congo-clades (Lake Mweru and Lunzua River) comprise distinct albeit yet undescribed species, while the fish of the third Congo-drainage clade (Chambeshi River and Bangweulu swamps), together with the fish of the Zambezi clade (Zambezi and Kafue River) are assigned to *Pseudocrenilabrus philander*. Concerning the intraspecific genetic diversity observed in the sampled rivers, most populations are highly uniform in comparison to lacustrine haplochromines, suggesting severe founder effects and/or bottlenecking during their history. Two bursts of diversification are reflected in the structure of the linearized tree. The first locates at about 3.9% mean sequence divergence and points to an almost simultaneous colonization of the sampled river systems. Subsequent regional diversification (with about 1% mean sequence divergence) occurred contemporaneously within the Kafue River and the Zambezi River. The clear-cut genetic biogeographic structure points to the dominance of geographic speciation in this lineage of riverine cichlid fishes, contrasting the importance of *in situ* diversification observed in lake cichlids.

Introduction

With an estimated number of 2400 species (Snoeks, 2001) cichlid fishes represent the most species-rich family of vertebrates. More than 1500 species live in the Great East African Lakes alone (Turner et al., 2001). The cichlid species flocks of these lakes are the most spectacular examples of adaptive radiation and explosive speciation within a single group of organisms (Kosswig, 1947; Fryer & Iles, 1972; Greenwood, 1984). Numerous species of haplochromine cichlids occur in North–Central–East and Southern African rivers, of which many are still undescribed. The relationships of riverine cichlids are poorly known to date, but their diversity seems to follow a biogeographic pattern, in that distinct genera are restricted to particular biogeographic regions or river systems

(Bell-Cross, 1966, 1975; Roberts, 1975; Greenwood, 1984; Meyer et al., 1991; Stiassny, 1991; Skelton, 1994; De Vos & Seegers, 1998; Salzburger et al., 2002; Seehausen, 2002; Verheyen et al., 2003). Only a few genera (*Tilapia*, *Oreochromis*, *Thoracochromis* and *Pseudocrenilabrus*) are known to have distributions spanning many river systems and ichthyogeographic regions. The genus *Pseudocrenilabrus* occurs from the northernmost edge of Egypt down to South Africa.

Morphologically, the genus *Pseudocrenilabrus* is characterized by its small body size, an upper pharyngeal apophysis of the *Haplochromis*-type, a moderately protractile upper jaw, bicuspid and/or conical teeth in the outer row and tricuspid teeth in the inner rows of both jaws, ctenoid scales, an interrupted single-pored lateral line and a rounded caudal fin as autapomorphy. The males lack egg spots in the anal fin, displaying a distinct red or orange spot on the distal tip of the anal fin. They are maternal mouthbrooders with up to 120 fry per clutch. The genus comprises three species currently considered as valid, even if the actual taxonomic situation is considerably more complex (Greenwood, 1989). One species, *Pseudocrenilabrus multicolor*, consists of two highly distinct subspecies, one colonizing the Nile system and the second the Lake Victoria region. *Pseudocrenilabrus nicholsi* is only found in the Republic of Congo, and the third nominal species, *Pseudocrenilabrus philander* comprises three sub-species plus a complex of distinct geographically separated populations pointing to a considerable sub-structuring of this species in southern Africa: *P. philander luebberti* in the area of Otavi in SW-Africa, *P. philander dispersus* from SW-Africa to Mozambique, and *P. philander philander* in SE and southern Africa (Skelton, 1991). In Zambia, the target region of this study, only *P. philander philander* has been reported to date. In terms of phylogenetic placement of the genus among other haplochromine lineages, a single molecular phylogenetic study exists that included *P. multicolor* (Salzburger et al., 2002). According to the study of Salzburger et al. (2002), the diversification of haplochromine cichlids was connected to the "primary lacustrine radiation" of Lake Tanganyika cichlids, comprising both riverine haplochromine lineages and endemic Tanganyikan lineages, summarized as the 'H-lineage' (Nishida 1991; 1997, redefined by Salzburger et al., 2002). Interestingly, haplochromine cichlids are likely to be paraphyletic with respect to several endemic Tanganyikan lineages. *Pseudocrenilabrus* was placed as a member of the H-lineage in a clade containing both riverine taxa and the members of the endemic Tanganyikan tribe Tropheini. De Villiers et al. (1992) showed by means of analysis of restriction enzyme variation that allopatric populations of *P. philander* exhibit only a low degree of genetic variation despite their sometimes clear morphological distinctness.

The present study aims to place the genus Pseudocrenilabrus in a representative taxonomic framework of African haplochromines and focuses on the phylogeographic structuring of *P. philander philander* in Zambian river systems. Therefore, we used a molecular approach to study the pathway of diversification of this lineage of cichlid fishes in the context of Southern Africa's geological and hydrographic history. We sequenced three mitochondrial gene segments of nine populations in six Zambian rivers that span two major African drainage systems, to derive a mtDNA phylogeny.

Material and methods

In total we analyzed 43 individuals of the genus Pseudocrenilabrus plus 23 species from other African cichlid lineages. DNA sequences were obtained for three mitochondrial gene segments [1047 bp of the NADH2-gene (26 individuals), 402 bp of cytochrome *b* (26 individuals); and 358 bp of the control region (43 individuals)]. Detailed information about the sampled individuals is given in Table 1 and Figure 1. Voucher specimens are available from the authors. DNA-extraction, PCR-amplification, and automatic DNA-sequencing were performed according to standard methods (see Salzburger et al., 2002; Koblmüller et al., 2004) using published primers for all gene segments (Kocher et al., 1989, 1995; Meyer et al., 1994). DNA sequences were aligned using Clustal X (Thompson et al., 1997) and alignments were further improved by eye in the case of the control region. Phylogenetic analyses were performed in two steps. The first step

Table 1. Characterization of the studied species, with sampling locality, drainage system, and GenBank accession numbers

Species	Locality	Drainage[a]	GeneBank accession number		
			NADH 2	cytochrome *b*	control region
Eretmodus cyanostictus	Lake Tanganyika	LT	AF398220[b]	AF428155[b]	–
Tanganicodus irsacae	Lake Tanganyika	LT	AF398219[b]	Z21778[b]	–
Spathodus erythrodon	Lake Tanganyika	LT	AF398218[b]	AF428156[b]	–
Altolamprologus compressiceps	Lake Tanganyika	LT	AF398229[b]	AF428163[b]	–
Neolamprologus brichardi	Lake Tanganyika	LT	AF398227[b]	Z29997[b]	–
Lamprologus mocquardi	Congo River	CO	AF398225[b]	Z29995[b]	–
Cyphotilapia frontosa	Lake Tanganyika	LT	U07247[b]	AF428169[b]	–
Triglachromis otostigma	Lake Tanganyika	LT	AF398217[b]	Z30004[b]	–
Limnochromis auritus	Lake Tanganyika	LT	AF398216[b]	Z21775[b]	–
Xenotilapia sima	Lake Tanganyika	LT	U07270[b]	Z21772[b]	–
Ophthalmotilapia ventralis	Lake Tanganyika	LT	U07257[b]	Z21771[b]	–
Callochromis macrops	Lake Tanganyika	LT	U07242[b]	Z21760[b]	–
Perissodus microlepis	Lake Tanganyika	LT	AF398222[b]	AF428168[b]	–
Plecodus straeleni	Lake Tanganyika	LT	AF398221[b]	Z21777[b]	–
Cyprichromis leptosoma	Lake Tanganyika	LT	AF398224[b]	AF428154[b]	–
Paracyprichromis brieni	Lake Tanganyika	LT	AF398223[b]	Z21776[b]	–
Astatotilapia burtoni	Lake Tanganyika	LT	AF317266[b]	Z21773[b]	–
Gnathochromis pfefferi	Lake Tanganyika	LT	U07248[b]	AF428166[b]	–
Tropheus moorii	Lake Tanganyika	LT	U07267[b]	Z12037[b]	–
Lobochilotes labiatus	Lake Tanganyika	LT	U07254[b]	AF428170[b]	–
Astatoreochromis alluaudi	?	LV	AF398234[b]	AF428157[b]	–
Orthochromis malagaraziensis	Malagarazi River	LT	AF398232[b]	AF428161[b]	–
Orthochromis polyacanthus[c]	?	CO/LU	AF398231[b]	AF428159[b]	–
Pseudocrenilabrus multicolor	? (ornamental fish trade)	NI	AY602992	AY600141	AY602995
Pseudocrenilabrus nicholsi	? (ornamental fish trade)	CO	AY602994	AY600143	AY602996
Pseudocrenilabrus philander	Mukula Stream–Bangweulu	LU	–	–	AY612141
					AY612143
					AY612145
					AY612160
					AY612161
Pseudocrenilabrus philander	Mukula Stream–Bangweulu	LU	–	–	AY612142
					AY612144
Pseudocrenilabrus philander	Chambeshi–Chambeshi River	LU	–	–	AY612137
					AY612138
					AY612139
					AY612155
					AY612156
Pseudocrenilabrus sp. *nov.* 'yellow'	Lunzua River	LT	–	–	AY612140
					AY612146
					AY612147
					AY612148
					AY612149

Table 1. (Continued)

Species	Locality	Drainage[a]	GeneBank accession number		
			NADH 2	cytochrome *b*	control region
Pseudocrenilabrus sp. *nov*. 'blue'	Lunzua River	LT	–	–	AY612150
					AY612151
					AY612152
					AY612153
					AY612154
					AY615797
Pseudocrenilabrus sp. *nov*. 'orange'	Mwatishi River–Lake Mweru	LU	–	–	AY612133
					AY612134
					AY612135
					AY612136
Pseudocrenilabrus philander	Lake Ithezi-Thezi	KF	–	–	AY612158
Pseudocrenilabrus philander	Lake Ithezi-Thezi	KF	–	–	AY612159
Pseudocrenilabrus philander	Lukanga Swamps	KF	–	–	AY612124
Pseudocrenilabrus philander	Lukanga Swamps	KF	–	–	AY612125
Pseudocrenilabrus philander	Lukanga Swamps	KF	–	–	AY612126
Pseudocrenilabrus philander	Lake Kariba	MZR	–	–	AY612127
Pseudocrenilabrus philander	Lake Kariba	MZR	–	–	AY612128
Pseudocrenilabrus philander	Lake Kariba	MZR	–	–	AY612129
Pseudocrenilabrus philander	Lake Kariba	MZR	–	–	AY612130
Pseudocrenilabrus philander	near Marromeu	LZR	–	–	AY612157
Pseudocrenilabrus philander	Kabala	UZR	AY602993	AY600142	AY612131
Pseudocrenilabrus philander	Kabala	UZR	–	–	AY612132
Pseudocrenilabrus philander	Kabala	UZR	–	–	AY612162
					AY612163

Note: Species names were assigned according to fishbase (http://www.fishbase.org). For *Pseudocrenilabrus* only distinct haplotypes per locality are listed with separate accession numbers for each individual.

[a] Drainage systems to which sampling sites belong to: CO, Congo basin; LT, Lake Tanganyika; LU, Luapula River; LV, Lake Victoria; KF, Kafue River; LZR, lower Zambezi River; MZR, middle Zambezi River; UZR, upper Zambezi River; NI, Nile basin.

[b] GenBank accession numbers of sequences published elsewhere (Sturmbauer & Meyer, 1992; Sturmbauer & Meyer, 1993; Sturmbauer et al., 1994; Kocher et al., 1995; Salzburger et al., 2002).

[c] There is a taxonomic problem with *Orthochromis polyacanthus*. This is reviewed by Greenwood & Kullander (1994), who state that the distribution of this species is limited to Lake Mweru. Due to the unknown origin of the sample and the earlier supposed synonymy of *O. polyacanthus* with *O. stormsi*, the sample used might be a different species of the genus *Orthochromis*.

addressed the phylogenetic placement of the genus *Pseudocrenilabrus* among African cichlids and was based on a combined data set of cytochrome *b* and NADH 2, defining three representatives of the Tanganyikan cichlid tribe Eretmodini as the outgroup, justified by Salzburger et al. (2002). The second step analyzed the phylogenetic structure among nine Zambian populations of *Pseudocrenilabrus* using *P. multicolor* and *P. nicholsi* as outgroups. Phylogenetic trees were constructed using three alternative algorithms, maximum parsimony (MP), neighbor-joining (NJ) and maximum likelihood (ML) using the computer-program PAUP* 4.0b2a (Swofford, 2000).

In MP transversion mutations were weighted over transition mutations, according to the observed frequency in regions of the same degree of variation, as derived from a sliding-window

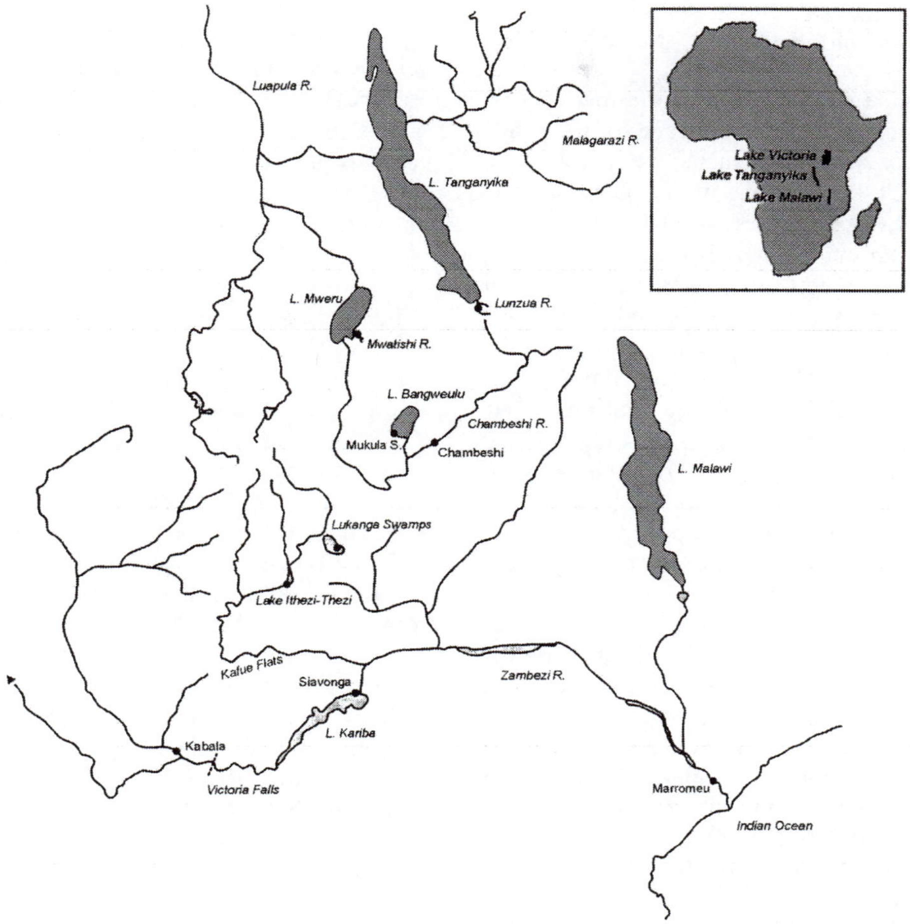

Figure 1. Map of the Zambian River systems, showing the sampling sites.

analysis (Sturmbauer & Meyer, 1992). Specifically, weightings were set to 2:1 in regions of more than 10% of genetic variation, and 3:2 in regions of less than 10% genetic variation. For MP of the combined data set of cytochrome *b* and NADH 2 we applied the same weighting scheme as Salzburger et al. (2002). The appropriate model of molecular evolution for the maximum likelihood analysis was evaluated by the likelihood ratio test implemented in the computer program Modeltest 3.0 (Posada & Crandall, 1998) to be used for NJ and ML. This test justified the use of the TrN model of molecular evolution (Tamura & Nei, 1993) for the combined data set of cytochrome *b* and NADH 2 and the HKY model (Hasegawa et al., 1985) for the control region. For the combined data set we applied the estimated proportion of invariable sites (0.45), the gamma shape parameter (0.95), and the base frequencies (A: 0.27; C: 0.34; G: 0.11; T: 0.28). For the control region we applied the base frequencies (A: 0.36; C: 0.19; G: 0.14; T: 0.32) and the transition/transversion ratio (2.24). We assessed the robustness of the resulting topologies by applying standard measures of confidence (bootstrap and quartet-puzzling frequencies) with 1000 pseudoreplicates (bootstrap) and 25 000 random quartets (quartet-puzzling), and in the case of maximum likelihood under the fast stepwise addition option in PAUP* 4.0b2a (Swofford, 2000).

To visualize the relatedness and the degree of genetic diversity within and among populations of *Pseudocrenilabrus* we constructed a minimum spanning tree, based upon the NJ- and an unweighted parsimony trees of step two of our analysis.

To determine the sequence of the major cladogenic events in the Zambian lineages of the genus *Pseudocrenilabrus*, we constructed a linearized tree based on a 358 bp segment of the D-Loop performing the two-cluster test implemented in the computer program LINTRE (Takezaki et al., 1995). First, rate constancy was tested for all internal nodes in the topology based on the sequences of the control region using the branch length test of LINTRE. In our case, the test was performed applying the HKY model of molecular evolution. No rate heterogeneity was detected at a high significance level ($p_\alpha < 0.01$) so that none of the taxa had to be excluded from further analyses. Then, a tree for the given topology was constructed under the assumption of rate constancy, which is termed a linearized tree. Since no accurate geology-based dating for the rate of base substitution of the control region is available to date for African cichlids we refrain from absolute age estimates and compare average observed levels of genetic divergences to those observed in other lineages. Our pairwise calculations of mean sequence divergence were based upon the Kimura-2-parameter model (K2P; Kimura, 1980), since this model was also used by Sturmbauer et al. (2001) and Baric et al. (2003).

Results

Phylogenetic placement of the genus Pseudocrenilabrus

Maximum parsimony resulted in two most parsimonious trees [2637 steps; CI excluding uninformative sites, 0.46; retention index (RI), 0.60; and rescaled consistency index (RC), 0.36; trees not shown]. The ML-tree is depicted in Figure 2. The following relationships were consistently found by all three algorithms: the non-mouthbrooding Lamprologine cichlids occupied the most ancestral split followed by a series of short-branched splits of seven lineages, corresponding to the radiation of the H-lineage (Limnochromini, Ectodini, Cyphotilapiini, Cyprichromini, Perissodini, Haplochromini I and Haplochromini II/Tropheini; see also Salzburger et al., 2002). The branching order among these seven lineages varied with respect to the tree building algorithm. *Pseudocrenilabrus* was unambiguously placed as sister group to the Congo River species *Orthochromis polyacanthus* and as ancestral branch within the Haplochromini II/Tropheini-lineage.

Phylogeographic structure among Zambian Pseudocrenilabrus populations

Phylogenetic analysis identified four major lineages, three of which occur in the Congo-system and one in the Zambezi system. Maximum parsimony resulted in two most parsimonious trees [268 steps; CI excluding uninformative sites, 0.75; retention index (RI), 0.96; and rescaled consistency index (RC), 0.77; trees not shown]. The strict consensus tree of the two MP trees, the NJ- and the ML-tree are depicted in Figure 3. MP and ML resulted in almost identical topologies and placed the populations of Chambeshi River and Mukula Stream in the Bangweulu swamps (both Luapula System) as most ancestral split, followed by the Lunzua River population (Lake Tanganyika – Congo River drainage), followed by the population of the Mwatishi River estuary in Lake Mweru (Luapula System), followed by a more complex clade of populations of the Zambezi River drainage. In this clade the samples from Kafue River were placed ancestral to the population samples from three localities of the Zambezi River. In NJ, the clades of the Lunzua population and the Lake Mweru population changed their relative position.

The minimum spanning tree (Fig. 4), derived from one of the two MP trees clearly reflects the four distinct lineages: the Chambeshi River and Lake Bangweulu individuals, the Lunzua River individuals, the Lake Mweru specimens, and finally the Kafue- and Zambezi-River samples. The Chambeshi-Bangweulu clade is separated from the Lunzua River population by 11–12 mutations, from the Lake Mweru population by 19–20 mutations, and from the Kafue- and Zambezi-River samples by 14–24 mutations. The Lunzua River population is separated from the Lake Mweru population by 12 mutations and from the Kafue- and Zambezi-River samples by 8–16 mutations. Finally, the Lake Mweru population is separated from the Kafue- and Zambezi-River samples by 10–18 mutations. Concerning the

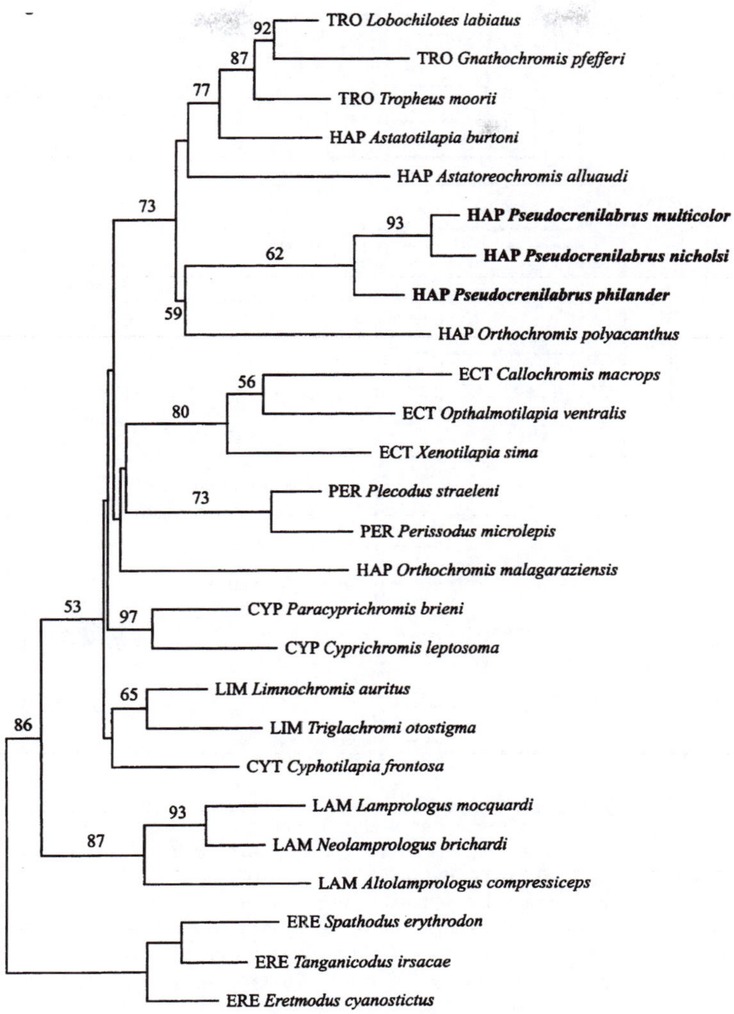

Figure 2. Phylogenetic placement of the genus *Pseudocrenilabrus*. Maximum likelihood tree comprising 26 species of East African cichlids, representing 9 distinct tribes [ERE, Eretmodini; LAM, Lamprologini; CYT, Cyphotilapiini; LIM, Limnochromini; CYP, Cyprichromini; HAP, Haplochromini; Per, Perissodini; ECT, Ectodini; TRO, Tropheini; assignment to tribes follows Salzburger et al. (2002) and Takahashi (2003)], based upon 1047 bp of the NADH 2 gene and 402 bp of the cytochrome *b* gene, using the substitution model TrN + I + Γ (Tamura & Nei, 1993). *Eretmodus cyanostictus*, *Tanganicodus irsacae* and *Spathodus erythrodon* were used as outgroup taxa. Quartet puzzling values (Strimmer & Von Haeseler, 1996) larger than 50 are shown above the branches.

intra-lineage diversity, three of the four lineages are highly uniform and contain identical haplotypes only, except for the Bangweulu-population in which two individuals differ from the others by one mutation, albeit five of the seven sampled Bangweulu-individuals share the same haplotype with the specimens from the Chambeshi River that were sampled about 200 km apart. Only the lineage from the Kafue- and Zambezi-River shows genetic and geographic structure in that the individuals from Kafue River are sister to those from the Zambezi River populations. Within these populations, a greater genetic diversity is observed, but no subdivision between the populations above and below the Victoria Falls is detected.

The linearized tree (Fig. 5) points to two diversification events, the first (3.9% K2P-distance ± 1.2%) concerning an almost contemporary colonization of all sampled river systems, and the second (1.0% K2P-distance ± 0.6%) a further

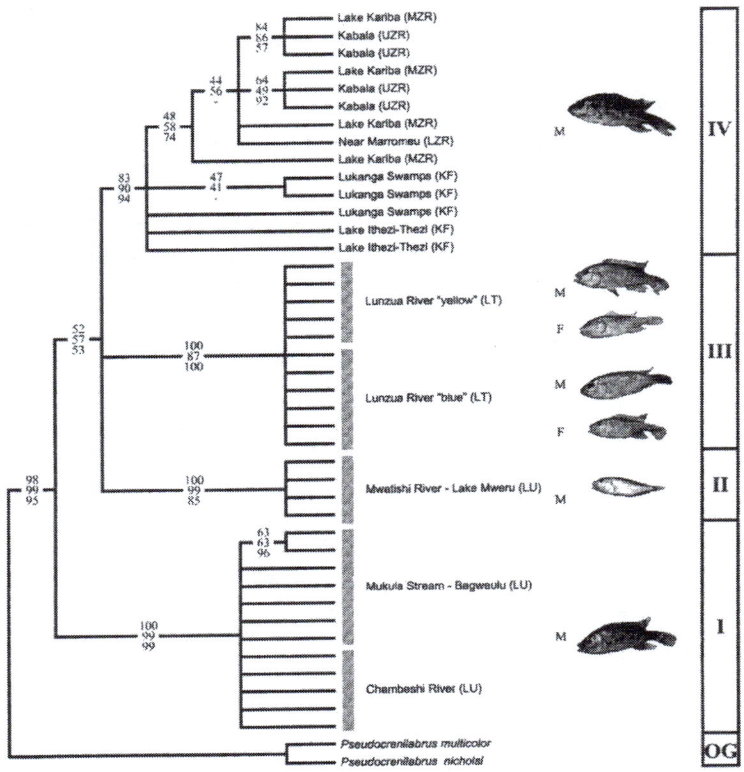

Figure 3. Strict consensus tree of two MP trees [268 steps; CI excluding uninformative sites, 0.75; retention index (RI), 0.96; rescaled consistency index (RC), 0.77], the NJ and the ML tree [substitution model HKY (Hasegawa et al., 1985] of 41 taxa of the *Pseudocrenilabrus philander* species complex from nine distinct localities, plus the two outgroup taxa *Pseudocrenilabrus multicolor* and *P. nicholsi*, based upon 358 bp of the most variable part of the control region. Bootstrap values obtained from neighbor joining are shown above the branches, while numbers in the middle represent parsimony bootstrap values. Quartet puzzling values are depicted below the branches. Only bootstrap and quartet puzzling values larger than 40 are shown. Abbreviations in parentheses refer to major drainage systems (see Table 1). Roman numerals refer to the four distinct clades within the Zambian *Pseudocrenilabrus*: I, Chambeshi–Bangweulu clade; II, Lake Mweru clade; III, Lunzua clade; IV, Kafue–Zambezi clade. Pictures of the corresponding fishes are shown beside the clades. For the Lunzua clade both distinct color morphs, blue and yellow, are depicted (M, male; F, female).

regional diversification within the Kafue–Zambezi clade. In fact, the conflicting branching order among the Lunzua- and Chambeshi populations among MP-ML versus NJ can be interpreted by the linearized tree as a contemporaneous diversification.

Discussion

Our analysis indicates a sister group relationship of the genus *Pseudocrenilabrus* to *Orthochromis polyacanthus* from the Congo River. Such a close relationship is corroborated morphologically by the lack of egg-spots on the anal fin in both genera. Concerning the intrageneric relationships of *P. philander*, *P. nicholsi* and *P. multicolor*, *P. philander* forms the most ancestral split (6.4% K2P-distance ± 1.3%), placing *P. nicholsi* and *P. multicolor* as sister taxa (4.7% K2P-distance), pointing to an origin of the genus in the Congo River, followed by a south- and northward range expansion. Also, the closest non-mouthbrooding sister group of the H-lineage, the lamprologine cichlids, are restricted to Lake Tanganyika, the Malagarazi and the Congo River. Salzburger et al. (2002) suggested a connection of the radiation of riverine haplochromines to the primary lacustrine radiation of the H-lineage, thus constraining the origin of haplochromines to the early Congo-Tanganyika basin (see also Lévêque, 1997). The level of genetic divergence between *Pseudocrenila-*

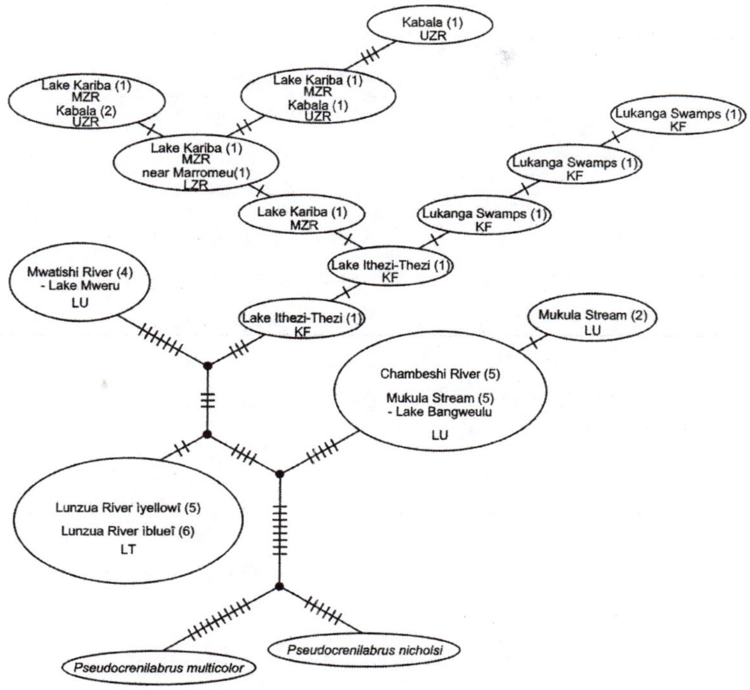

Figure 4. Minimum spanning tree of 14 haplotypes of the *Pseudocrenilabrus philander* species complex based on 358 bp of the most variable part of the mitochondrial control region (outgroup taxa: *Pseudocrenilabrus multicolor*, *P. nicholsi*). The topology corresponds to one of the most parsimonious trees that was most similar to the NJ tree. Each crossbar indicates one base substitution. •, branching points for which no intermediate haplotype was found. Abbreviations refer to major drainage systems (see Table 1). Numbers in parantheses refer to the number of identical haplotypes.

brus philander and the other two species of *Pseudocrenilabrus* (6.4% K2P-distance ± 1.3%) is highly similar to that observed in the endemic Tanganyikan genus *Tropheus* (6.1% K2P-distance ± 1.4%, Baric et al., 2003), suggesting a similar evolutionary age for diversification and spread of these riverine haplochromines and this lineage of endemic lake cichlids, given that the rates of molecular evolution are roughly the same. The colonization of North-, Central-, East- and Southern Africa by transgression of present day water sheds must be interpreted in the light of the complex interactions of a series of geological and palaeoclimatological changes (Roberts, 1975; Skelton, 1994; reviewed in Lévêque, 1997).

Phylogeographic structure among Zambian Pseudocrenilabrus populations

The evolutionary relationships of the surveyed populations show a clear phylogeographic pattern in that all water bodies contain a genetically and sometimes also morphologically distinct entity. This is in contrast to the study of de Villiers et al. (1992) who found identical or almost identical restriction enzyme patterns in distant populations in Zimbabwe, Namibia and South Africa. The individuals sampled in Lunzua River are strikingly different from all remaining populations. Lunzua River drains into Lake Tanganyika, which in turn is connected to the Congo River via the Lukuga, but its upper reaches are separated from the lake by high water falls. Below the falls, only *Astatotilapia burtoni* was caught which was assigned to the 'modern haplochromines' (Salzburger et al., 2002). Interestingly, the *Pseudocrenilabrus* population in the upper Lunzua River comprises two distinct color morphs, one being bluish on the head and the body sides and the second one being yellowish (see photographs in Fig. 3). This distinction is also clearly visible in females: blue females do not have a yellowish-orange anal fin

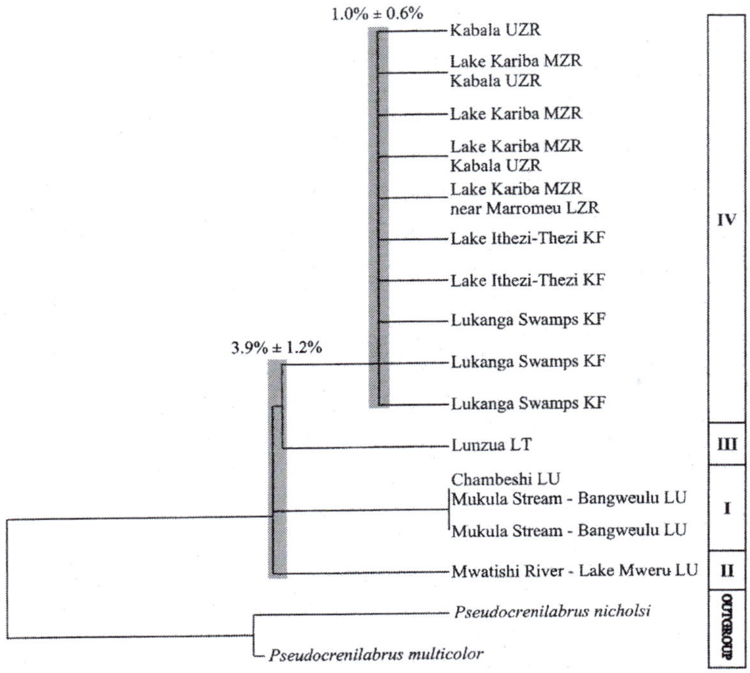

Figure 5. Linearized tree based on a 358 bp segment of the most variable part of the mitochondrial control region. The linearized tree was compiled with the computer program LINTRE (Takezaki et al., 1995) after performing a branch length test (Takezaki et al., 1995) to test for differences in base substitution rates, using the substitution model HKY (Hasegawa et al., 1985). Gray bars refer to major diversification events. The observed mean sequence divergences using the substitution model K2P (Kimura, 1980) are depicted for the corresponding diversification events. Abbreviations in parentheses refer to major drainage systems (see Table 1). Roman numerals refer to the four distinct clades within the Zambian *Pseudocrenilabrus*: I, Chambeshi–Bangweulu clade; II, Lake Mweru clade; III, Lunzua clade; IV, Kafue–Zambezi clade.

while yellow females do. However, in terms of mtDNA, they all share a single haplotype in the control region ($n = 11$), so that nuclear data need to be carried out to further analyze their distinctness. Moreover, it became evident that the Lake Mweru population is clearly distinct from the Chambeshi–Bangweulu population, not only in terms of genetics, but also in terms of morphology, and is currently being described as a new species (Jos Snoeks, personal communication), which will make *P. philander* paraphyletic. This distinctness of the Lake Mweru population reflects the separation of the upper Luapula River system in two faunal sub-regions by the Johnston- and further upstream the Mumbatuta Falls, which are likely to represent important dispersal barriers (Jackson, 1961). In contrast to the specimens in Lake Mweru, the individuals from Chambeshi and Bangweulu are morphologically highly similar to *Pseudocrenilabrus philander philander* from the Zambezi drainage, even if they display similar genetic distances (about 4% K2P-distance) to the fish from the Zambezi system as the two other Congo-drainage clades, as is also evident from the linearized tree analysis (Fig. 5). Within the clade of the Zambezi drainage, a clear geographic substructure became evident: The samples from Kafue River occupy the most ancestral branches, while all individuals from the Zambezi River form a clade. Interestingly, no genetic substructure seems to exist among the Zambezi River individuals that were sampled at three localities above and below the Victoria Falls, representing the three distinct sections of the Zambezi, each having distinctive geomorphic characters (Wellington, 1955). Given that the Victoria Falls are a geologically ancient barrier, this finding is highly surprising. Since the upper reaches of the Kafue River system are in close geographical vicinity to the upper reaches of the Luapula River system (Fig. 1), we hypothesize that the Kafue River was first colonized by *Pseudocrenilabrus* via an ancient connection to the Luapula system, according to the model of drainage evolution for southern Africa proposed by Skelton (1994) which suggests a series of opportunities for faunal exchange between the southwest, southeast and central African river systems during the tertiary. In general, the fish fauna of the upper Zambezi River and the Kafue River shows clear links to the Congo (Jackson, 1961). The occurrence of common fish species in the Kafue and Congo drainage requires an earlier link between these river systems. Moore & Larkin (2001) interpreted the abrupt change in course of the Chambeshi River, which at present forms a headwater of the Luapula system, as a capture elbow, implying that this river originally formed the headwaters of the Kafue River. Given the impossibility for *Pseudocrenilabrus* to get up the Victoria Falls, the upper reaches of the Zambezi River must have been colonized first, by an ancient connection to the Kafue River which might have been closed by a river capture event in the late Pleistocene (Thomas & Shaw, 1991; Moore & Larkin, 2001). Since the upper reaches of the Zambezi River and its tributaries are in close geographical vicinity to the Kafue River (Fig. 1), this scenario seems to be likely, also supported by the fact that the fish fauna of the Kafue River shows considerable similarity to the upper Zambezi River despite its present connection to the middle Zambezi River (Jackson, 1961; Skelton, 1994). This implies that the middle and lower Zambezi may have been colonized by *Pseudocrenilabrus* from two directions, via the Kafue and by individuals "jumping" down the Victoria Falls.

Diversification and speciation in Pseudocrenilabrus

The genetic diversity observed within the sampled populations is strikingly uniform in all except for the Zambezi drainage populations. These low levels of genetic diversity contrast all data of lacustrine cichlid species in the three Great East African Lakes (Moran & Kornfield, 1993; Moran et al., 1994; Verheyen et al., 1996; Sturmbauer et al., 1997; Nagl et al., 1998; Albertson et al., 1999; Rüber et al., 1999; Baric et al., 2003; Verheyen et al., 2003), suggesting severe founder effects during colonization. In addition, *Pseudocrenilabrus* seems to avoid water current of the main river beds and predominately occurs in quiet side arms and swamps. Many individuals are found in temporarily flooded areas during rainy season, so that seasonal bottlenecking might also contribute to the striking genetic uniformity of most sampled populations. The relatively greater degree of genetic variation in the Zambezi system (1.0% K2P-distance ± 0.6%) may be due to a larger and more stable effective population size in Kafue, and due to a bi-directional colonization of the middle and lower Zambezi, via the Kafue River and via the upper Zambezi. These hypotheses can be tested further when data from additional cichlid species of the Kafue–Zambezi drainage are available.

The observed species diversity of the genus *Pseudocrenilabrus* is low in comparison to other riverine cichlid genera in Zambia (Skelton 1991). Each of the rivers seems to be inhabited by one species only. In the upper Lunzua River we only found *Tilapia cf. sparrmanii* in addition to *Pseudocrenilabrus*. In all other rivers, several other species of the genera *Serranochromis*, *Sargochromis*, *Pharyngochromis*, *Tilapia* and *Oreochromis* were found. Thus, *Pseudocrenilabrus* is likely to be well adapted to one particular niche but seems not competitive enough against other cichlids to undergo speciation by niche segregation. Speciation in *Pseudocrenilabrus* is rather likely to be driven by geographic separation.

Acknowledgements

We wish to thank C. Kapasa, H. Phiri and the team at the Department of Fisheries, Ministry of Agriculture and Cooperatives, and L. Mumba of the University of Zambia for their support during fieldwork. We are further indebted to O. Seehausen for providing tissue samples and K. Sefc for valuable comments on the manuscript. S. K., N. D. and C.S. were supported by the Austrian Science Foundation (grant P15239), C. K. by the OEAD, Austrian Ministry of Foreign Affairs and also by a fellowship from the Royal Museum for Central Africa, Belgium. S.K. and N.D. were also supported by the University of Graz.

References

Albertson, R. C., J. A. Markert, P. D. Danley & T. D. Kocher, 1999. Phylogeny of a rapidly evolving clade: the cichlid fishes of lake Malawi, East Africa. Proceedings of the Natural Academy of Science, USA 96: 5107–5110.

Baric, S., W. Salzburger & C. Sturmbauer, 2003. Phylogeography and evolution of the Tanganyikan cichlid genus *Tropheus* based upon mitochondrial DNA sequences. Journal of Molecular Evolution 56: 54–68.

Bell-Cross, G., 1966. The distribution of fishes in Central Africa. Fish Research Bulletin of Zambia 4: 3–15.

Bell-Cross, G., 1975. A revision of certain *Haplochromis* species of Central Africa. Occasional Papers of the National Museums and Monuments of Rhodesia 7: 405–464.

De Villiers, D. L., E. H. Harley & A. J. Ribbink, 1992. Mitochondrial DNA restriction enzyme variation in allopatric populations of *Pseudocrenilabrus philander*. South African Journal of Science 88: 96–99.

De Vos, L. & L. Seegers, 1998. Seven new *Orthochromis* species (Teleostei: Cichlidae) from the Malagarasi, Luiche and Rugufu basins (Lake Tanganyika drainage), with notes on their reproductive biology. Ichthyological Explorations of Freshwaters 9(4): 371–420.

Fryer, G. & T. D. Iles, 1972. The Cichlid Fishes of the Great Lakes of Afrika: Their Biology and Evolution, TFH Publications, Neptune City. 641 pp.

Greenwood, P. H., 1984. African cichlid fishes and evolutionary theories. In Echelle, A. A. & I. Kornfield (eds), Evolution of Fish Species Flocks. University of Maine at Orono Press, Orono: 141–154.

Greenwood, P. H., 1989. The taxonomic status and phylogenetic relationships of *Pseudocrenilabrus* Fowler (Teleostei, Cichlidae). Ichthyological Bulletins of the J.L.B. Smith Institute of Ichthyology, 16 pp.

Greenwood, P. H. & S. O. Kullander, 1994. A taxonomic review and redescription of *Tilapia polyacanthus* and *T. stormsi* (Teleostei: Cichlidae), with descriptions of two new *Schwetzochromis* species from the Upper Zaire River drainage. Ichthyological Explorations of Freshwaters 5(2): 161–180.

Hasegawa, M., H. Kishino & T. Jano, 1985. Dating of the human-ape splitting by a molecular clock of mitochondrial DNA. Journal of Molecular Evolution 22: 160–174.

Jackson, P. B. N., 1961. The Fishes of Northern Rhodesia. A Check List of Indigenous Species. Government Printer, Lusaka, Zambia.

Kimura, M., 1980. A simple method for estimating evolutionary rate of base substitutions through comparative studies of nucleotide sequences. Journal of Molecular Evolution 16: 111–120.

Koblmüller, S., W. Salzburger & C. Sturmbauer, 2004. Evolutionary relationships in the sand-dwelling cichlid lineage of Lake Tanganyika suggest multiple colonization of rocky habitats and convergent origin of biparental mouthbrooding. Journal of Molecular Evolution 58: 79–96.

Kocher, T. D., W. K. Thomas, A. Meyer, S. V. Edwards, S. Pääbo, F. X. Villablanca & A. C. Wilson, 1989. Dynamics of mitochondrial DNA evolution in animals: amplification and sequencing with conserved primers. Proceedings of the National Academy of Sciences, USA 86: 6196–6200.

Kocher, T. D., J. A. Conroy, K. R. McKaye, J. R. Stauffer & S. F. Lockwood, 1995. Evolution of NADH dehydrogenase subunit 2 in East African cichlid fishes. Molecular Phylogenetics and Evolution 4: 420–432.

Kosswig, C., 1947. Selective mating as a factor for speciation in cichlid fish of East African lakes. Nature 159: 604.

Lévêque, C., 1997. Biodiversity Dynamics and Conservation: The Freshwater Fish of Tropical Africa. Cambridge University Press, Cambridge, New York.

Meyer, A., T. D. Kocher & A. C. Wilson, 1991. African fishes. Nature 350: 467–468.

Meyer, A., J. M. Morrissey & M. Schartl, 1994. Recurrent origin of sexually selected trait in *Xiphophorus* fishes inferred from a molecular phylogeny. Nature 368: 539–541.

Moore, A. E. & P. A. Larkin, 2001. Drainage evolution in south-central Africa since the breakup of Gondwana. South African Journal of Geology 104: 47–68.

Moran, P. & I. Kornfield, 1993. Retention of an ancestral polymorphism in the mbuna species flock (Teleostei: Cichlidae) of Lake Malawi. Molecular Biology and Evolution 10(5): 1015–1029.

Moran, P., I. Kornfield & P. N. Reinthal, 1994. Molecular systematics and radiation of haplochromine cichlids (Teleostei: Perciformes) of Lake Malawi. Copeia 1994(2): 274–288.

Nagl, S., H. Tichy, W. E. Mayer, N. Takahata & J. Klein, 1998. Persitence of neutral polymorphisms in Lake Victoria cichlid fish. Proceedings of the Natural Academy of Science, USA 95: 14238–14243.

Nishida, M., 1991. Lake Tanganyika as an evolutionary reservoir of old lineages of East African fishes: inferences from allozyme data. Experientia 47: 974–979.

Nishida, M., 1997. Phylogenetic relationships and the evolution of Tanganyika cichlids: a molecular perspective. In Kawanabe, H., M. Hori & M. Nagoshi (eds), Fish communities in Lake Tanganyika. Kyoto University Press, Kyoto, Japan: 3–23.

Posada, D. & K. A. Crandall, 1998. Modeltest: testing the model of DNA substitution. Bioinformatics 14: 817–818.

Roberts, T. R., 1975. Geographical distribution of African freshwater fishes. Zoological Journal of the Linnean Society 57: 249–319.

Rüber, L., E. Verheyen & A. Meyer, 1999. Replicated evolution of trophic specializations in an endemic cichlid lineage from Lake Tanganyika. Proceedings of the Natural Academy of Science, USA 96: 10230–10235.

Salzburger, W., A. Meyer, S. Baric, E. Verheyen & C. Sturmbauer, 2002. Phylogeny of the Lake Tanganyika cichlid species flock and its relationships to the Central and East African haplochromine cichlis fish faunas. Systematic Biology 51: 113–135.

Seehausen, O., 2002. Patterns in fish radiation are compatible with Pleistocene desiccation of Lake Victoria and 14 600 year history for its cichlid species flock. Proceedings of the Royal Society, London B 269: 491–497.

Skelton, P. H., 1991. *Pseudocrenilabrus*. In Daget, J., J. P. Gosse, G. G. Teugels & D. F. E. T. van den Audenaerde (eds), Check-list of freshwater fishes of Africa (CLOFFA).

ISNB, Brussels; MRAC, Tervuren; and ORSTOM, Paris 4: 394–398.

Skelton, P. H., 1994. Diversity and distribution of freshwater fishes in East and Southern Africa. Annales du Musée de la République de l'Afrique Centrale, Zoologique 275: 95–131.

Snoeks, J., 2001. Cichlid diversity, speciation and systematics: examples from the Great African lakes. Journal of Aquariculture & Aquatic Sciences 9: 150–166.

Stiassny, M. L. J., 1991. Phylogenetic interrelationships of the family Cichlidae: an overview. In Keenleyside, M. H. A. (ed.), Cichlid Fishes. Behaviour, Ecology and Evolution. Chapman & Hall, London: 1–35.

Strimmer, K. & A. Von Haeseler, 1996. Quartet puzzling: a quartet maximum-likelihood method for reconstructing tree topologies. Molecular Biology and Evolution 7: 964–969.

Sturmbauer, C. & A. Meyer, 1992. Genetic divergence, speciation, and morphological stasis in a lineage of African cichlid fishes. Nature 358: 578–581.

Sturmbauer, C. & A. Meyer, 1993. Mitochondrial phylogeny of the endemic mouthbrooding lineages of cichlid fishes of Lake Tanganyika, East Africa. Molecular Biology and Evolution 10: 751–768.

Sturmbauer, C., E. Verheyen & A. Meyer, 1994. Mitochondrial phylogeny of the Lamprologini, the major substrate spawning lineage of cichlid fishes from Lake Tanganyika in eastern Africa. Molecular Biology and Evolution 11: 691–703.

Sturmbauer, C., E. Verheyen, L. Rüber & A. Meyer, 1997. Phylogeographic patterns in populations of cichlid fishes from rocky habitats in Lake Tanganyika. In Stepien, C. A. & T. D. Kocher (eds), Molecular Systematics of Fishes. Academic Press, San Diego: 97–111.

Sturmbauer, C., S. Baric, W. Salzburger, L. Rüber & E. Verheyen, 2001. Lake level fluctuations synchronize genetic divergence of cichlid fishes in African lakes. Molecular Biology and Evolution 18: 144–154.

Swofford, D. L., 2000. PAUP*: Phylogenetic Analysis Using Parsimony (and other methods), Beta Version 4.0. Sinauer, Sunderland, MA.

Takahashi, T., 2003. Systematics of Tanganyikan cichlid fishes (Teleostei: Perciformes). Ichthyological Research 50: 367–382.

Takezaki, N., A. Rzhetsky & M. Nei, 1995. Phylogenetic test of the molecular clock and linearized trees. Molecular Biology and Evolution 12: 823–833.

Tamura, K. & M. Nei, 1993. Estimation of the number of nucleotide substitutions in the control region of mitochondrial DNA in humans and chimpanzees. Molecular Biology and Evolution 10(3): 512–526.

Thomas, D. S. G. & P. A. Shaw, 1991. The Kalahari Environment. Cambridge University Press, UK. 284 pp.

Thompson, J. D., D. G. Higgins & T. J. Gibson, 1997. CLUSTAL X windows interface: flexible strategies for multiple sequence alignment aided by quality analysis tools. Nucleic Acids Research 24: 4876–4882.

Turner, G. F., O. Seehausen, K. E. Knight, C. J. Allender & R. L. Robinson, 2001. How many species of cichlid fishes are there in African lakes? Molecular Ecology 10: 793–806.

Verheyen, E., L. Rüber, J. Snoeks & A. Meyer, 1996. Mitochondrial phylogeny of rockdwelling cichlid fishes reveals evolutionary influence of historical lake level fluctuations of Lake Tanganyika, Africa. Philosophical Transactions of the Royal Society, London B 351: 797–805.

Verheyen, E., W. Salzburger, J. Snoeks & A. Meyer, 2003. Origin of the cichlid fishes from Lake Victoria, East Africa. Science 300: 325–329.

Wellington, J. H., 1955. Southern Africa – A Geographic Study. Vol. 1, Physical Geography. Cambridge University Press, UK, 528 pp.

Hydrobiologia (2005) 542:235–247
H. Segers & K. Martens (eds), Aquatic Biodiversity II
DOI 10.1007/s10750-004-8332-z

Short term spatial and temporal variation of phytoplankton in a shallow tropical oligotrophic reservoir, southeast Brazil

Maria Rosélia Marques Lopes[1], Carlos E. de M. Bicudo[2] & M. Carla Ferragut[2]
[1]*Environmental Sciences Department, Universidade Federal do Acre, BR-364, km 4, 69915-900, Rio Branco, AC, Brazil*
[2]*Ecology Section, Instituto de Botânica, Caixa postal 4005, 01061-970, São Paulo, SP, Brazil*

Key words: phytoplankton, atelomixis, shallow oligotrophic reservoir, Brazil

Abstract

Locally called IAG pond, system is a small, oligotrophic, shallow, urban reservoir located in the Parque Estadual das Fontes do Ipiranga Biological Reserve, south of São Paulo municipality, southeastern Brazil. Study was carried out in two phases to cover the two climatic periods (dry and rainy) that characterize the area. Daily samples were collected on seven consecutive days (dry = 20–26 August 1996; rainy = 22–28 January 1997), three times a day (07:00 h, 13:00 h, and 19:00 h), following the water column vertical profile at reservoir's deepest site (Z_{max} = 4.7 m). The phytoplankton's community structure's diurnal variation was compared with environmental variables (thermal structure of the water column, pH, CO_2, dissolved oxygen, total phosphorus, and NH_4). Analysis of variance (ANOVA) showed significant variation among sampling hours and days but not among depths for most species, thus revealing a tendency towards a homogeneous vertical distribution of phytoplankton in the reservoir. Canonical Correspondence Analysis (CCA) showed a significant relationship between distribution of phytoplankton species and environmental variables used for ordination. Water temperature was the environmental variable that best explained distribution of sample units. Atelomixis was the only possible explanation for how a non-motile alga as, for example, the chlorococcal *Oocystis lacustris* Chodat and the diatom *Cyclotella stelligera* Cleve & Grunow, can be so widespread and abundant in the IAG pond during the stratified period.

Introduction

Many biotic and abiotic factors govern spatial and temporal distribution of phytoplankton in continental aquatic systems making their joint explanation a difficult task (Tilzer, 1973). Lacustrine ecosystems' limnetic regions' energy fluxes and nutrient cycling patterns, associated with the water's thermal stratification allow algae to maintain vertical distribution (Hutchinson, 1967; Wetzel, 1993). Vertical mixing, on the contrary, distributes phytoplankton in the water column and transfers nutrients from the much richer layers to the poorer ones. Consequently, stratification and mixing control the two most important factors in phytoplankton ecology: light and nutrient availability (Tilzer & Goldman, 1978). Studies carried out in tropical systems addressing to the role of stratification and mixing on the phytoplankton community structure at short time intervals are extremely scarce. Emphasis should be given to those of Barbosa & Padisák (2002) and Barbosa (2003) that focused on the atelomixis.

Thus, investigation of the structure and function of phytoplankton communities are of utmost importance for studies of the lacustrine ecosystems' dynamics. Among the most relevant aspects of such studies are those related to population abundance and diversity, spatial and temporal distribution, as well as those referring to the ecological factors directly related with the determination of primary productivity levels of such communities (Huszar et al., 1990).

In Brazil, studies focusing on biotic and abiotic variable analysis, aimed at determining which variables are responsible for variation of a system's phytoplankton community, are still incipient. Present study reports the spatial and temporal variation of phytoplankton in a shallow tropical oligotrophic reservoir, as well as up to what extent the main environmental factors are responsible for the system's phytoplankton variability.

Material and methods

Locally called a 'pond', the studied system is, in fact, a small, shallow, oligotrophic reservoir located in the Parque Estadual das Fontes do Ipiranga Biological Reserve, south of São Paulo municipality (23°38′08″S-23°40′18″S and 46°36′48″ W-46°38′00″W), southeastern Brazil. According to Bicudo et al. (2002), reservoir's maximum depth is 4.7 m, maximum length 311.5 m, maximum width 45.5 m, and volume 76 653 m^3.

Samplings were performed during 7 consecutive days at a single station (maximum depth 4.7 m) both during the dry (20–26 August 1996) and rainy (22–28 January 1997) periods. Abiotic and biotic variables studied were collected from the entire water column at three different times (07:00, 13:00, and 19:00 h). Measurements of temperature (°C) were obtained by using a portable thermistor at each 10 cm of the water column vertical profile. Temperature isolines graphs were prepared with the aid of the Surfer 6.01 package for Windows. During the rainy period, presence of thermocline defined the different strata of the water column in which samples were collected. At 19:00 h of day 22 there was no sampling due to very severe weather conditions. Presence of thermocline was defined previously to each gathering by taking the water column vertical thermal profile (measurements at each 10 cm), and when a thermal gradient with minimum differences of 0.5 °C m^{-1} was observed. Mixing zone, i.e. the layer subjected to turbulent mixing was considered the layer delimited by the water body surface and that in which the density gradient surpassed 2 x 10^{-2} kg m^{-3} m^{-1} (Reynolds, 1984).

Quantification of phytoplankton was done using Utermöhl's (1958) sedimentation chambers and following recommendations in Lund et al. (1958). Phytoplankton population density was calculated by counting all individual specimens present in every other field. As many vertical and horizontal transects were counted totalling at least 100 individuals of the most frequent species according to the species rarefaction curve, i.e. 10 consecutive fields without the appearance of new individuals. Consequently, according to Lund et al. (1958) error is less than 20% with a confidence coefficient of 95% for species with 100 individual specimens counted. Every cell, coenobium, colony, or filament was considered one individual. To define abundant and/or dominant species, Lobo & Leighton (1986) criteria were followed.

For statistical analysis of data, all taxa totaling over 0.5% of total phytoplankton community density, which corresponded to the abundant species, were considered. Aiming at establishing the degree of significance of values obtained for the variables and the different sampling depths and hours in each sampling day, Variance Analysis (ANOVA) with a level of significance of about 5% for each sampling day was performed using Systat 5.0 package for Windows.

Canonic Correspondence Analysis (CCA) was used to ordinate dry and rainy periods to determine the main environment variables responsible for the major variability trends of the system's phytoplankton community. Statistical package PC-ORD 3.11 for Windows was used and all matrices were transformed by ranging with the help of WinMat package prepared by Prof. G.J. Shepherd, of the Botany Department of the State University of Campinas.

To run the analysis, each species with more than 0.5% of total phytoplankton density was considered. Altogether all such species contributed over 90% of total density. Environment variables were selected on the basis of the Principal Components Analysis (PCA) using Pearson and Kendhal correlation matrix. Variables considered were only those showing high correlation ($r > 0.500$) with ordination axes 1 and 2. All combined variables were eliminated from analysis since they would promote information redundancy (Hall & Smoll, 1992) and analysis distortion (Pielou, 1984). According to Ter-Braak & Prentice (1988), CCA is much more robust when just a few environment variables are needed to identify the species distribution. Consequently, 22 biotic and 6 abiotic variables were used in present CCA.

To test the significance level of the first two canonic axes, Monte Carlo test was used and data variability was explained by using the canonic coefficient (Ter-Braak, 1986). Furthermore, Pearson and Kendhal correlation coefficient (*r*) resulting from the relation between ordination values was used, as well as individual variables for the ordination construction (McCune & Mefford, 1997).

Results

Based on physical and chemical characteristics (Table 1), IAG reservoir is classified as an oligotrophic tropical system. Despite the reservoir's shallowness (Z_{max} = 4.7 m), it was possible to detect thermal stratification during the two periods studied. Stratifications were temporary during the dry period with night circulation (Fig. 1), and long-lasting during the rainy period with no evidence of night circulation, except for the 07:00 h of day 28 January 1997 (Fig. 2). Depth of mixing zone was less during the rainy period, indicating greater water column stability than in the dry period.

One hundred and twenty taxa were presently identified in the reservoir. In regards to number of taxa, Chlorophyceae was the best represented class with 39, immediately followed by the Bacillariophyceae with 26, Euglenophyceae with 15, Cyanophyceae with 13, Zygnemaphyceae with 10, Cryptophyceae and Chrysophyceae with 6 each, Dinophyceae and Oedogoniophyceae with 2 each, and Xanthophyceae with just a single taxon. Number of taxa identified per sample unit varied between 15 and 29 during the dry period and 19 and 36 during the rainy one.

IAG reservoir phytoplankton community showed 17 abundant species defining 9 describing

Table 1. Mean and variation range of physical and chemical variables of the IAG reservoir during the study period

Variable	Dry period	Rainy period
Temperature (°C)	16.0 (14.5–19.5)	20.6 (19.9–24.9)
Secchi disk depth (m)	2.1 (1.9–2.4)	1.3 (0.9–1.7)
Dissolved oxygen (mg l^{-1})	10.0 (9.5–10.9)	5.3 (4.9–5.4)
Turbidity (NTU)	4.3 (3.7–5.0)	12.9 (4.9–22.0)
Z_{mix} (m)	1.2 (0.5–2.5)	1.0 (0.5–2.0)
Z_{eu}/Z_{mix}	7.8 (2.4–14.4)	3.1 (2.4–9.6)
Z_{mix}/Z_{max}	0.3 (0.1–0.5)	0.2 (0.1–0.4)
Light attenuation coefficient	0.8 (0.7–0.90	1.3 (1.0–1.9)
Conductivity (μS cm^{-1})	33.1 (32.1–34.3)	38.8 (34.7–42.1)
PH	6.3 (6.2–6.4)	5.7 (5.5–6.0)
Free CO_2 (mg l^{-1})	4.1 (2.2–8.0)	15.0 (8.6–26.2)
$Si(OH)_4$ (mg l^{-1})	0.9 (0.6–1.5)	0.7 (0.6–0.9)
N-NO_3 (μM)	2.7 (1.3–6.5)	4.2 (2.3–7.3)
N-NH_4 (μM)	1.1 (0.2–2.6)	4.9 (3.6–7.1)
Total nitrogen (μM)	8.2 (4.7–13–5)	11.3 (8.3–15.2)
Total phosphorus (μM)	0.2 (0.1–0.2)	0.8 (0.4–1.2)
NT/PT	54.8 (21.3–99.1)	10.3 (5.2–41.2)
Alkalinity (mEq l^{-1})	0.07 (0.06–0.09)	0.09 (0.08–0.2)
P-PO_4^{3-} (*)		
N-$N0_2$ (*)		

(*) <0.11 μM (undetectable by methods).

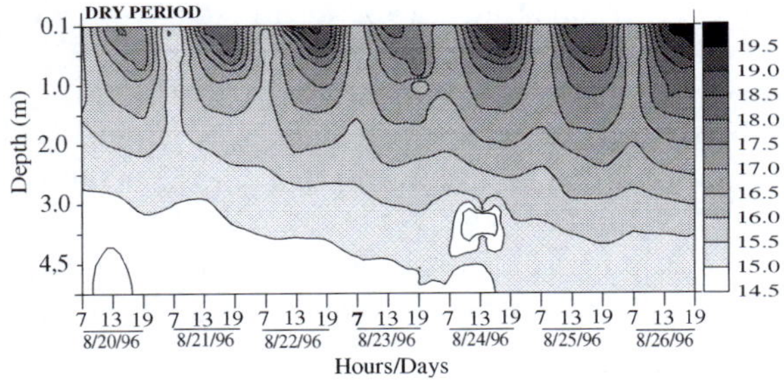

Figure 1. Depth–time diagram of water temperature (°C) recorded for dry period (August 1996).

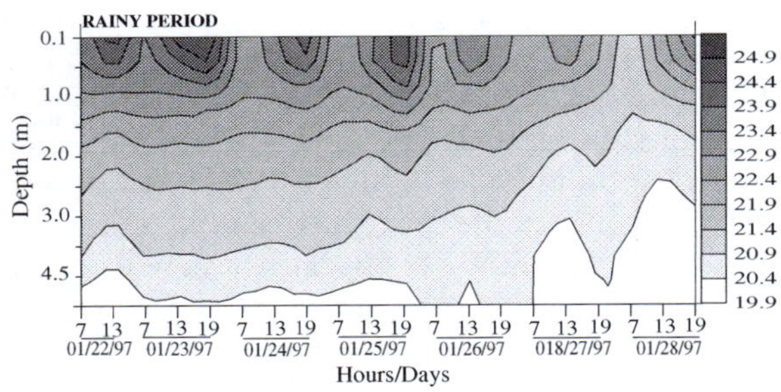

Figure 2. Depth–time diagram of water temperature (°C) recorded for rainy period (January 1997).

ones (contribution >3% of total density), 5 (84.4%) of which during the dry period and 6 (80.2%) during the rainy one. From the latter, three were common to both climatic periods (*Chlamydomonas planctogloea* Skuja, *Oocystis lacustris* Chodat, and *Chlamydomonas* sp.) and the following ones stood out in terms of relative abundance during both periods: *Chlamydomonas planctogloea* and *Oocystis lacustris* (nannoplanktonic), and *Synechococcus* aff. *nidulans* (picoplanktonic). *Chlamydomonas planctogloea*, a nannoplanktonic flagellate, was the only species that contributed with a high percentage to the phytoplankton total density in both periods. Of the 17 species occurring in both periods of present study, 76.4% are nannoplanktonic and 58.8% Chlorophyceae (Tables 2, 3).

During the dry period, vertical and temporal distribution variation of densities of the five species considered representative of the community were significantly different only among sampling days (Table 4), causing an indefinite pattern of vertical distribution with a tendency towards species homogeneity in the system. *Chlamydomonas* sp., species with the greatest density value, was detected at the reservoir bottom, whereas *Oocystis lacustris* occurred at the surface layers, except for the very last sampling day (26 August 1996). *Chlamydomonas planctogloea* dominated during the dry period's first 2 days and maintained its greatest abundance during the whole period, except for the very last day (26 August 1996) when *Oocystis lacustris* overtook in density by about 25% (Fig. 3).

During the rainy period, vertical and temporal distribution of representative species differed significantly among sampling days, but not among sampling hours and depths. This caused a homogeneous vertical distribution pattern of the species in the reservoir, except for *Chlamydomonas* sp. and *Oocystis lacustris*, whose variations were significant only among depths and, consequently, showed a tendency towards heterogeneous vertical distribution (Table 5). *Chlamydomonas planctogloea* (nannoplanktonic) and *Synechococcus* aff. *nidulans* (Pringsheim) Komárek (picoplanktonic) presented a homogeneous vertical distribution pattern, with detection of *Synechococcus* aff. *nidulans* in the reservoir only after the second sampling day. *Monoraphidium arcuatum* (Koršikov) Hindák and *Monoraphidium tortile* (West & West) Komárková-Legnerová did not show a definite distribution pattern, i.e. just the opposite of *Chlamydomonas* sp. whose vertical distribution was very well defined, with its greatest concentration at the bottom of reservoir, and *Oocystis lacustris* at the surface layers (Fig. 4). Noticeable temporal variation of *Synechococcus* aff. *nidulans* and *Chlamydomonas planctogloea* was due to the enormous growth of the first species and the decrease of the latter over the sampling days. *Synechococcus* aff. *nidulans* surpassed *Chlamydomonas planctogloea* by about 28% during the very first day of the present study (Fig. 4).

Main environmental variables responsible for the greatest phytoplankton community variability were identified with CCA that simultaneously represents environmental (abiotic) and biological variables in a bidimensional space. Monte Carlo permutation test showed that axes 1 and 2 were statistically significant ($p \leq 0.05$), indicating the impossibility for hazardous events. In other words, results strictly represent the existing relationships between environment variables and phytoplankton

Table 2. Abundant species and respective percentage of contribution at the IAG reservoir during the period 20–26/August/1996 (dry)

Class	Fraction	Habit	Species	Association
Chlorophyceae (86.9%)	Nanoplankton	Flagellate (50.9%)	*Chlamydomonas planctogloea* (43.8%)	X2
			Chlamydomonas sp. (5.4%)	X2
		Colonial (32%)	*Oocystis lacustris* (28.3%)	F
			Kirchneriella pseudoaperta (3.7%)	F
		Coccoid (4%)	*Monoraphidium pseudobraunii* (2.2%)	X1
			Tetraëdron minimum (1.8%)	J
Bacillariophyceae (3.2%)	Nanoplankton	Coccoid (3.2%)	*Cyclotella stelligera* (3.2%)	A
Chrysophyceae (1.8%)	Nanoplankton	Flagellate (1.8%)	*Mallomonas* sp. 1 (1.8%)	E
Dinophyceae (1.6%)	Microplankton	Flagellate (1.6%)	*Peridinium gatunense* (1.6%)	L_0

Table 3. Abundant species and respective percentage of contribution at the IAG reservoir during the period 22–28/January/1997 (rainy)

Class	Fraction	Habit	Species	Association
Chlorophyceae (71.1%)	Nanoplankton	Flagellate (46.5%)	*Chlamydomonas planctogloea* (41.3%)	X2
			Chlamydomonas sordida (1.3%)	X2
			Chlamydomonas sp. (3.9%)	
		Colonial (5.9%)	*Oocystis lacustris* (4.3%)	F
			Crucigenia tetrapedia (1.6%)	J
		Coccoid (6.8%)	*Monoraphidium tortile* (4.8%)	X1
			Tetraëdron caudatum (2%)	J
	Microplankton	Coccoid (10.4%)	*Monoraphidium arcuatum* (10.4%)	X1
		Colonial (1.5%)	*Elakatothrix gelatinosa* (1.5%)	F
Cyanophyceae (17.7%)	Picoplankton	Coccoid (15.5%)	*Synechococcus* aff. *nidulans* (15.5%)	Z
	Nanoplankton	Colonial (2.2%)	*Chroococcus minor* (2.2%)	Z

Table 4. Results of ANOVA performed to establish vertical and temporal variation significance of representative species at IAG reservoir during the period 20–26/August/1996 (dry)

Statistics	Depths		Hours		Days	
	F	*p*	*F*	*p*	*F*	*p*
Chlamydomonas planctogloea	0.356	0.839	0.534	0.588	16.596	0.000
Oocystis lacustris	0.291	0.884	0.040	0.961	19.586	0.000
Chlamydomonas sp.	0.053	0.995	1.390	0.254	15.432	0.000
Kirchneriella pseudoaperta	0.242	0.914	1.398	0.252	10.541	0.000
Cyclotella stelligera	0.377	0.825	6.990	0.001	2.419	0.032

species. The eigenvalues for axis 1 ($\lambda = 0.601$) and 2 ($\lambda = 0.044$) explain 49.1% of variance of the biological data. Correlation species-environment for axes 1 (0.988) and 2 (0.688) was high, indicating a strong correlation between phytoplankton species distribution and environment variables used for ordination (Table 6). Canonical coefficients for axis 1 showed that temperature was the most important environment variable for the ordination. Intraset correlation indicated, in turn,

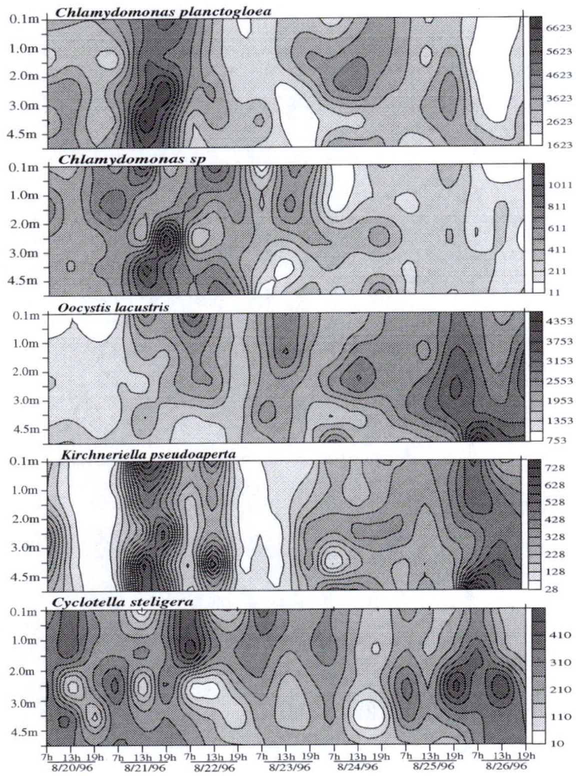

Figure 3. Depth–time diagrams of absolute density (ind. ml^{-1}) of phytoplankton representative species at the IAG reservoir during the period 20–26 August 1996 (dry).

that besides temperature, other variables such as DO, CO_2, pH, and TP effectively contributed to the species-environment relationship (Table 7).

All rainy period sampling units are shown on the negative side of axis 1 in Figure 5 and are associated to the highest temperature ($r = -0.923$), TP ($r = -0.923$), free CO_2 ($r = -0.898$), and N-NH_4 values ($r = -0.516$). However, all dry period sampling units are found on the positive side and are associated to the highest OD ($r = 0.975$) and pH values ($r = 0.914$).

Distribution of sampling units was based on the temporal variation of data and clearly showed influence of seasonal events on their ordination. CCA reinforced PCA results performed for the abiotic variables, which explained association of abiotic variables with the dry and rainy periods. Sampling units' stratification patterns during the rainy period were related to the reservoir's thermal stratification, which was long-lasting during that period and the opposite of its non-defined distribution during the dry period, when daily circulation occurred. Concluding, axis 1 represents seasonality.

N-NH_4 besides having high correlation with axis 2 ($r = -0.775$) also presented intra-set correlation (Table 7) indicating that this variable influenced ordination by separating surface and bottom, since the highest concentrations of the ion were detected reservoir's bottom.

Regarding the phytoplankton species, the central part of Figure 5 shows the species common to both climatic periods and which form a group not closely associated to the sampling units, but with a marked tendency towards the stratification periods (negative side) and mixture (positive side). Within this group, most species present high correlation with axis 1 representing seasonality. Located on the negative side are *Crucigenia tetrapedia* (Kirchner) West & West ($r = -0.569$), *Elakatothrix gelatinosa* Wille ($r = -0.539$), *Chlamydomonas planctogloea* ($r = -0.288$), *Kirchneriella pinguis* Hindák ($r = -0.416$), *Chlamydomonas sordida* Ettl ($r = -0.128$), and *Chlamydomonas* sp.

Table 5. Results of ANOVA performed to establish significance of vertical and temporal variation of representative species densities at the IAG reservoir during the period 22–28/January/1997 (rainy)

Statistics	Depths		Hours		Days	
	F	*p*	*F*	*p*	*F*	*p*
Chlamydomonas planctogloea	0.213	0.887	6.747	0.002	4.890	0.000
Synechococcus aff. *nidulans*	0.042	0.988	1.814	0.170	33.168	0.000
Monoraphidium arcuatum	2.661	0.054	2.679	0.075	3.923	0.002
Chlamydomonas sp.	11.124	0.000	0.271	0.763	1.258	0.288
Monoraphidium tortile	1.912	0.260	2.567	0.083	2.803	0.016
Oocystis lacustris	4.071	0.010	2.496	0.089	1.999	0.077

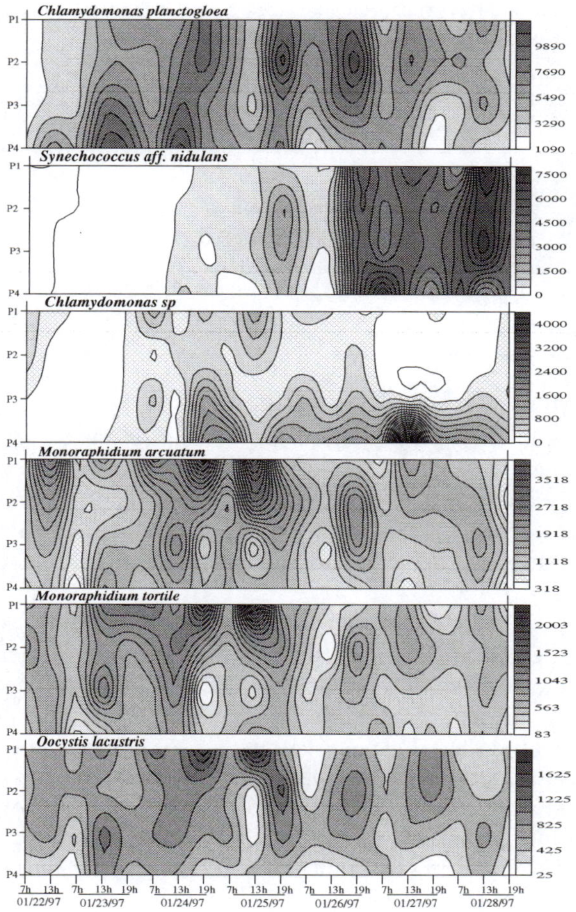

Figure 4. Depth–time diagram of absolute density (ind. ml^{-1}) of the representative phytoplankton species at the IAG reservoir during the period 22–28 January 1997 (rainy).

(r = −0.120). On the other hand, on the positive side of the same axis are *Cyclotella stelligera* Cleve & Grunow (r = 0.505), *Oocystis lacustris* (r = 0.546), and *Mallomonas* sp. 2 (r = 0.113).

Regarding the exclusive species distribution of each period, settled from the CCA, one observes, in Figure 5 (axis 1), that the rainy period was characterized by high temperature, TP, and CO_2 values and a tendency towards a close association of the species *Monoraphidium tortile* (r = −0.750), *Coccomonas orbicularis* Stein (r = −0.651), *Monoraphidium arcuatum* (r = −0.695), *Tetraëdron caudatum* (Corda) Hansgirg (r = −0.694), *Monoraphidium dibowski* (r = −0.607), *Chroococcus minor* (Kützing) Nägeli (r = −0.475), and *Synechococcus* aff. *nidulans* (r = −0.498) to the sampling units taken from between 1.5 and 3.5 m deep, i.e. exactly where the thermocline was located throughout the study period. Furthermore, presence of *Cryptomonas erosa* Ehrenberg (r = −0.432) was detected between 3 and 3.5 m deep (at the thermocline level). The formation of a group of species that appeared is evident due to that period's thermal stratification.

During the dry period, which was characterized by temporary stratification, daily circulation of the water column, and high DO and pH values, five species (*Tetraëdron minimum*, *Monoraphidium pseudobraunii* (Belcher & Swale) Heynig, *Peridinium gatunense* Nygaard, *Kirchneriella pseudoaperta* Komárek, and *Mallomonas* sp. 1) presented high correlation with axis 1 and close association with sampling units taken at 1–2 m deep. Consequently, it was possible to deduct that those species could easily represent the dry period which is characterized by mixing of the water column.

Discussion

Green microalgae are usually the greatest component of phytoplankton community of oligotrophic

Table 6. Result of CCA using 6 abiotic and 22 biotic variables (species) performed for the IAG reservoir and including both climatic periods studies: dry (20–26 August 96) and rainy (22–28 January 1997)

	Axis 1	Axis 2
Autovalue (λ)	0.601	0.044
Variance percentage explained	45.80	3.300
Variance percentage accumulated	45.80	49.10
Pearson Correlation (species-environment)	0.988	0.688
Monte Carlo test (autovalues) – *p*	0.010	0.010
Monte Carlo test (correlation species-environment)	0.010	0.010

Table 7. Canonical coefficient and intra-set correlation of environment variables with ordination axes 1 and 2 for the 22 biotic variables in the IAG reservoir during periods 20–26 August 1996 (dry) and 22–28 January 1997 (rainy)

Variable	Canonical coefficient		Correlation coefficient (intraset)	
	Axis 1	Axis 2	Axis 1	Axis 2
Temperature	−0.695	0.723	−0.926	0.251
Dissolved oxygen (DO)	0.450	0.836	0.975	0.177
Free CO_2	−0.068	−0.938	−0.894	−0.313
PH	0.410	−0.959	0.917	0.039
Ammonium (NH_4)	−0.074	−0.546	−0.483	−0.783
Total phosphorus (PT)	−0.034	0.436	−0.865	0.015

systems, and such success is partly due to their tiny size that represents high surface:volume rates, greater membrane area per cell, and photosynthesis rates that are favored by high CO_2 diffusion rates and nutrients. Consequently, tiny-sized Chlorophyceae may be at a selective advantage in oligotrophic waters, where essential nutrients like nitrogen and phosphorus may be scarce. Phytoplankton communities with high surface:volume rates (nano and picoplankton) are well adapted to low nutrient concentrations (Tundisi et al., 1981; Tundisi, 1983; Happey-Wood, 1988). Hence, abundant presence of nanoplankton species, along with low nutrient concentrations and chlorophyll *a* contents, reinforces the oligotrophic characteristics of IAG reservoir.

In a thermally stratified system, flagellate organisms possess selective advantage over the Chlorococcales. Being entirely deprived of self-motility, the latter are totally dependent on water turbulence to keep them suspended in the water column. Consequently, they exhibit their greatest population net growth rates in turbulent waters (Happey-Wood, 1988). High density of such species during the first three sampling days in the IAG reservoir was most probably related to the greater availability of nutrients and free CO_2 detected during that part of the study period.

Turbulent diffusion and sedimentation are the main mechanisms of vertical displacement of phytoplankton that lack self-motility capability. This situation is contrary to that of flagellates that, under stratified conditions, may actively migrate up and down in the water column (Tilzer & Goldman, 1978). Presence of flagella contribute to the organism's motility (Reynolds, 1984), as well as to the water renovation around the organism, thus easing its contact with nutrients by breaking gradients around the cell (Wetzel, 1993). Consequently, motility is advantageous even in homogeneous environments because it allows organisms to move from nutrient-depleted areas, thus renewing the surrounding environment (Margalef, 1978).

The more effective and faster the locomotion is, the lower the energetic cost will be. Such quick motility would be especially useful, with reference to nutrient availability, in systems with significant spatial heterogeneity (Boney, 1975). These two factors would explain the great success of *Chlamydomonas planctogloea* at the IAG reservoir during the two climatic periods studied.

According to Branco (1978), *Chlamydomonas* is a characteristic genus of eutrophic waters. However according to Lembi (1980), although the genus is referred as commonly occurring in highly eutrophic and organic polluted waters, there are species that are typical of oligotrophic acidic waters. Likely among those species, referred to by Lembi (1980) as characteristic of acidic oligotrophic environments, are included the extremely small *Chlamydomonas* presently identified for the IAG reservoir which, as it was already mentioned, is an acidic oligotrophic system since its pH average was never above 6.3 during the dry period and never above 5.83 during the rainy one.

During the dry period (winter), temperature ranged from 8.4 to 12.7 °C (minimum) and from 25.3 to 29.4 °C (maximum). During the day heating led to the development of a weak, but measurable, thermocline at 13:00 and 19:00 h. In fact, it was possible to detect a greater tendency towards thermal stratification and a strong resistance to water mixing proven by the very high

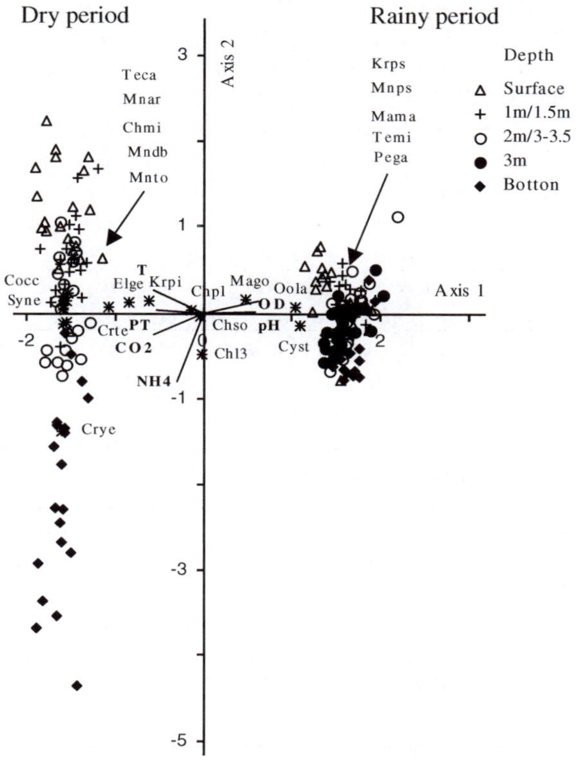

Figure 5. CCA ordination of sampling units at the IAG reservoir produced from 6 abiotic and 22 biotic (species) variables during periods 20–26 August 1996 (dry) and 22–28 January 1997 (rainy). Abiotic variables: T (temperature), TP (total phosphorus), CO_2 (free CO_2), NH_4 (ammonium), DO (dissolved oxygen), and pH. Species: Chpl (*Chlamydomonas planctogloea*), Chl3 (*Chlamydomonas* sp.), Chso (*Chlamydomonas sordida*), Cocc (*Coccomonas orbicularis*), Mnar (*Monoraphidium arcuatum*), Mnto (*Monoraphidium tortile*), Mndb (*Monoraphidium dibowski*), Mnps (*Monoraphidium pseudobraunii*), Teca (*Tetraëdron caudatum*), Temi (*Tetraëdron minimum*), Crte (*Crucigenia tetrapedia*), Elge (*Elakatothrix gelatinosa*), Krpi (*Kirchneriella pinguis*), Krps (*Kirchneriella pseudoaperta*), Oola (*Oocystis lacustris*), Crye (*Cryptomonas erosa*), Pega (*Peridinium gatunense*), Mama (*Mallomonas* sp. 1), Mago (*Mallomonas* sp. 2), Cyst (*Cyclotella stelligera*), Chmi (*Chroococcus minor*), and Syne (*Synechococcus* aff. *nidulans*).

relative thermal resistance values calculated for those same times (Lopes & Bicudo, 2001). Nocturnal cooling was considerable and consequently the surface water cooled down, thus increasing its density and breaking down the thermocline. Such distratification of the water column was repeated daily and the reservoir alternated stratifications (day time) with destratifications (night time) on a daily basis during the entire dry period studied. This phenomenon was called atelomixis by Lewis (1973) and was already recorded in Brazil, at Lake Carioca by Barbosa & Tundisi (1980) and again at Lake Carioca and also Lake Dom Helvécio by Barbosa & Padisák (2002).

The above mentioned periods of turbulence may have been long enough to allow an unprecedented growth of *Oocystis lacustris* at the IAG pond, to the extent that the latter species surmounted *Chlamydomonas planctogloea*'s density. Dynamics of *Chlamydomonas planctogloea* is very much according to Moss' (1972) observations regarding thermal stratification and water column's mixture at the IAG reservoir. Moss (1972) worked with material from Gull Lake, Michigan, U.S.A., and related growth of *Chlamydomonas* spp. to the stratification periods and their decline during the mixing periods. Calijuri (1999) reported dominance and persistence of *Chlamydomonas* sp. at the Barra Bonita reservoir, state of São Paulo, due to the intensity of the mixing periods, the long retention time of water in the reservoir, and the depth of the euphotic zone.

Greater abundance of *Cyclotella stelligera* during the dry period (3.2% of total density) compared to the rainy one (0.5% of total density) and its distribution along the entire water column during all sampling days was possibly related to the mixing conditions during that climatic period, i.e. to atelomixis. According to Lewis (1973), atelomixis is the only possible explanation for how a non-motile group like the diatoms and desmids can be so widespread and abundant in tropical lakes during the stratified period. Not considering atelomixis, however, Reynolds (1997) stated that growth of the species of *Cyclotella* is better in mixed waters and are typically dominant in oligotrophic environments. Presence of *Cyclotella stelligera* during the dry period was followed by a homogeneous distribution of orthosilicate in the reservoir, whose highest concentrations were present during the latter period.

Of the four most common species to both study periods, only *Oocystis lacustris* underwent a drastic density reduction during the rainy period. The latter species density was 4.5 times less during the rainy than in the dry period. This fact is related to the much better adaptation of *Oocystis lacustris* to turbulence since under thermal stratification conditions, representatives of the Chlorococcales tend to sediment since they rely on the water turbulence to keep themselves in suspension (Tilzer & Goldman, 1978; Happey-Wood, 1988). Higher turbulence was observed during the dry period, the opposite of the rainy one, when the species maintained their stability in the water column throughout the period.

During the dry period no significant blue-green species was detected, the highest abundance remaining with the greens. However, *Synechococcus* aff. *nidulans* (picocyanophyte) was abundant during the rainy period. This fact may be related to the nutritional condition of the system. Contrary to Shapiro's (1990) observations that blue-green algae are favored under low CO_2 availability and/or high pH condition, during the entire study period at the IAG reservoir greatest growth of *Synechococcus* aff. *nidulans* was observed under high free CO_2 (13.9–20.7 mg l^{-1}) concentration and low pH values (5.5–5.8).

The water column stability is an important condition for development of blue-green algae. These algae are able to store phosphorus and fix atmospheric nitrogen, enabling them to survive in systems with different trophic levels. Duration of thermal stability in the system is fundamental to the success and formation of 'blooms' (Reynolds, 1984; Paerl, 1988). Blue-greens are usually much more tolerant to high temperatures than eukariont algae that inhabit the same system (Wetzel, 1993). Temperatures above 20 °C (Tilman et al., 1986) and increase of turbidity (Smith, 1986) also contribute to the maximum competitiveness and subsequent increase of blue-greens' density in the system. During the rainy period at the IAG reservoir, water temperature varied between 19.7 and 24.9 °C, and water transparency dropped to about 60% compared to that of the dry period. Consequently, higher values of turbidity were measured and vertical light attenuation coefficient calculated.

Paerl (1988) reported that *Synechococcus* is a common genus in oligotrophic environments where it makes up a considerable part of the phytoplankton biomass, however, it is seldom dominant. Representatives of *Synechococcus* are inhabitants of the epi and hypolimnion, in which they form distinct strata in oligotrophic lakes. However, very seldom is it the dominant alga. Besides, the genus has a surface:volume ratio makes it very well adapted to low nutrient available environments. Paerl's (1988) observations are considered in Reynolds et al. (2002) that included *Synechococcus* in the Z assemblage proposed for planktonic organisms of oligotrophic systems.

Density decline of *Chlamydomonas planctogloea* could also be due to the mixing of the water column during the night proceeding the last sampling day. Such mixing was evident at 07:00 h on that day when the greatest depth of the mixing zone was also observed (Fig. 2). Species density (except for *Chlamydomonas* sp.) and abundance were much greater at the upper levels of the reservoir, between 0.2 and 1.5 m depth. Such depths corresponded to the mixing zone depth with thermocline at all sampling times. This distribution was very much expected considering that phototrophic organisms tend to accumulate in the eutrophic zone and above the thermocline during stratification conditions in the reservoir (Fig. 4). Tendency towards a greater abundance of all species in the upper layers of the reservoir, especially during the stratification period, was observed by Vinner & Kemp (1983) at the Rotongaio Lake, New Zealand.

Tendencies pointed out by the CCA indicated that water temperature was the main determinant factor of temporal and spatial variability of the phytoplankton community at the IAG reservoir during the two climatic periods studied. Presence of species that were common to both periods and species that were exclusive to each period was observed. The latter could be considered the stratification and mixture periods describing species.

Species presently studied denoted a marked seasonal numerical density variation that was associated to the highest nutrient concentrations and to the greatest water temperature values which were, in turn, influenced by the alternation of a dry and a rainy season. However, seasonality did not show an effect on the vertical phytoplankton zonation, which would indicate a tendency towards having the greatest densities in the upper reservoir layers as well as at its bottom during both climatic periods studied.

During the entire rainy period, under thermal stratification condition, although the species were found throughout the water column, the greatest density was observed between 1.5 and 3.5 m depth. Greatest concentration of *Chlamydomonas* sp. was observed at the bottom and that of *Cryptomonas erosa* at about 3.5 m depth, i.e. at the interface between the end of thermocline and the most superficial bottom layers (Figs. 4, 5). On the other hand, during the dry period under frequent mixture conditions, species were found in the entire water column and *Kirchneriella pseudoaperta*, *Oocystis lacustris*, and *Cyclotella stelligera* showed their maximum densities near the reservoir bottom.

It is important for the development of a comprehensive community assemblage and phytoplankton selection scheme to determine which environmental conditions enable consistent predictions (Huszar et al., 2000). IAG reservoir's oligotrophic conditions and thermal structure associated to morphological, functional, and ecological characteristics of the of the phytoplankton community descriptive species for the stratification and mixture periods led us to identify nine assemblages based on the phytoplankton functional classification proposed by Reynolds (1997), Reynolds et al. (2002), and Kruk et al. (2002).

Acccording to Kruk et al. (2002), genus *Monoraphidium* was included in assemblage X1, associated to environments that are rich in TP, PO_4, and NH_4. Reynolds et al. (2002) included into X1 species that occur in eutrophic–hipereutrophic systems. In this group we included *Monoraphidium arcuatum*, *M. dibowskii* (Woloszýnska) Hindák & Komárková-Legnerová, *M. pseudobraunii*, *M. tortile*, *Tetraëdron caudatum*, and *T. minimum* (A. Braun) Hansgirg are species associated with high PT values during the rainy period.

Chlamydomonas planctogloea, *Chlamydomonas sordida*, *Chlamydomonas* sp., and *Coccomonas orbicularis* are motile nanoplanktonic species. According to Reynolds et al. (2002), assemblage X2 is made up of species that occur in meso-eutrophic lakes, and is characterized by including mobile nanoplanktonic species from eutrophic systems (Reynolds, 1997). In the present study, *Chlamydomonas* and *Coccomonas* species were associated to high TP values during the rainy period.

Chroococcus minor and *Synechococcus* aff. *nidulans* are, in accordance to Reynolds (1997, et al. 2002) who classified them in assemblage Z, representative of clear oligotrophic lakes. However, it is important to emphasize that Huszar et al. (2000) included *Synechococcus aquatilis* in assemblage X1, saying that it is not an invader species, with rapid growth, and very common in Brazilian coastal hypereutrophic lagoons. Kruk et al. (2002) reported that X1 assemblage is associated with high *p* values and although in the present study *Synechococcus* aff. *nidulans* has shown high correlation with TP greatest values during the rainy period, we presently preferred to agree with Reynolds (1997) taking into consideration the oligotrophic character of the system.

Reynolds et al. (2002) and Reynolds (1997) included *Crucigenia tetrapedia*, *Tetraëdron caudatum*, and *T. minimum* in assemblage J based on what typifies this assemblage, i.e. presence of Chlorococcales (non-motile) species without mucilage in shallow enriched lakes. Considering that the system studied is a shallow reservoir and that the species was associated with high TP values during the rainy period, we identified assemblage J in the reservoir.

The inclusion of *Cryptomonas erosa* in the assemblage Y, according to Kruk et al. (2002), was based on its tolerance to high light attenuation coefficient values, which indicate its adaptation to light deficient environments. In the present study, CCA suggested that this assemblage is best associated with TP and PO_4 values. Melo & Huszar (2000) referred to Batata Lake dominance of invading nanoplanktonic species that are tolerant to X1 and Y flushing assemblages (*Cryptomonas pyrenoidifera* Geitler). In the present study, *C. erosa* was associated with the greatest TP values during the rainy period, when it occurred mainly at 3–3.5 m deep and at the bottom of the water column (at the thermocline level), i.e. at depths in which light was less available. *Cryptomonas erosa* was associated to the greatest TP values during the rainy period, a fact that may be related to the opportunistic character of all Cryptophyceae. Woring with the Garças Lake, an eutrophic system belonging to the same hydrographic basin, Huszar et al. (2000) detected the occurrence of a *Cryptomonas* species whose presence was justified by the genus ubiquity in systems displaying moderate nutrient enrichment.

Elakatothrix gelatinosa, *Kirchneriella pinguis*, *K. pseudoaperta*, and *Oocystis lacustris*, i.e. Chlorococcaleans of medium size covered with abundant colonial mucilage that are present in enriched systems during its clear phase (Reynolds et al., 2002). They were included in assemblage F species that are not tolerant to low nutrient availability. In the present study, *Elakatothrix gelatinosa* and *Kirchneriella pinguis* were associated with the greatest TP values during the rainy period. However, *K. pseudoaperta* and *Oocystis lacustris* were present during the dry period, in which they were associated with the least TP and N-NH_4 values and the greatest values of turbulence. It must be emphasized that the peculiarity of this association is the presence of colonial mucilage. In this sense, *K. pseudoaperta* is an exception since its representative individuals may also occur isolated and *Oocystis lacustris*, according to Happey-Wood (1988), exhibit greatest population net growth rates in turbulent water systems.

Mallomonas foi were included in the assemblage E by Reynolds et al. (2002), who typified this assemblage by being composed of species inhabiting mesotrophic systems and thus tolerate low nutrient availability. In the present study, *Mallomonas* sp. 1 and *Mallomonas* sp. 2 were associated to the least TP and NH_4 values during the dry period, thus fitting in assemblage E.

The inclusion of *Cyclotella stelligera* in the assemblage A is according to Reynolds (1997) due to its occurrence in clear oligotrophic well mixed lakes with low nutrient availability. These characteristics were observed at IAG reservoir during the dry period, i.e. at the same time as was detected the greatest number of *C. stelligera* cells.

Peridinium gatunense was included in the assemblage L_0 because it is a typical dinoflagellate of small oligotrophic lakes, and it is tolerant to low pH values (Reynolds, 1997) as well as low *p* values (Kruk et al., 2002). Such conditions agree with those at the IAG reservoir. The species is associated with the low nutrient availability during the dry period.

Finally, IAG reservoir (oligotrophic) during dry period in which there was constant mixing and the least nutrient availability (dry period) assemblages X2 (*Chlamydomonas planctogloea*, *Chlamydomonas* sp.) and F assemblages (*Oocystis lacustris* and *Kirchneriella pseudoaperta*) were relevant, contributing with 81.2% total density. During the rainy season, period of daily stratifications and greatest nutrient availability, the same species of X2, Z (*Synechoccocus nidulans*, *Chroococcus minor*), F (*Oocystis lacustris*, *Elakatothrix gelatinosa*), J (*Crucigenia tetrapedia* and *Tetraedron caudatum*) and X1 assemblages, with small Chlorococcales (*Monoraphidium arcuatum* and *Monoraphidium tortile*) totalised 83.3% of total community.

It is important to mention, however, that much more study is needed in order to provide a consistent list of representative species for each assemblage above, mainly of the Chlorococcales, whose diversity and strategies vary greatly. Present research allowed identification of distinct functional species groups for each climatic period. This calls for attention to the fact that identification of species assemblages that would typify stratification and mixing periods may be an important tool for environment changes prediction.

Acknowledgements

We are much indebted to CAPES, Coordenadoria de Aperfeiçoamento de Pessoal Superior and to CNPq, Conselho Nacional de Desenvolvimento Científico e Tecnológico, for partial financial support.

References

Barbosa, F. A. R. & J. Padisák, 2002. The forgotten lake stratification pattern: atelomixis, and its ecological importance. Verhandlungen der internationale Vereinigung für Theoretische und Angewandte Limnologie 28: 1385–1395.

Barbosa, F. A. R. & J. G. Tundisi, 1980. Primary production of phytoplankton and environmental characteristics of a shallow quaternary lake at eastern Brazil. Archiv für Hydrobiologie 90: 139–161.

Bicudo, C. E. M., C. F. Carmo, D. C. Bicudo, R. Henry, A. C. S. Pião, C. M. Santos & M. R. M. Lopes, 2002. Morfologia e morfometria de três reservatórios do PEFI. In Bicudo, D. C. & M. C. C. E. M. Forti Bicudo (eds.). Secretaria do Meio Ambiente do Estado de São Paulo. São Paulo: 350 p.

Boney, A. D., 1975. Phytoplankton. Edward Arnolds Publications, London 116. (Studies in Biology no. 52).

Branco, S. M., 1978. Hidrobiologia aplicada à Engenharia Sanitária (2nd.nd ed.). Companhia Estadual de Saneamento Básico, São Paulo 620.

Calijuri, M. C., 1999. A comunidade fitoplanctônica em um reservatório tropical (Barra Bonita, SP). Associate Professor Dissertation. Universidade de São Paulo, São Carlos 211.

Hall, R. I & J. P. Smoll, 1992. A weight-averaging regression and calibration model for inferring total phosphorus concentration from diatoms in British Columbia (Canada) lakes. Freshwater Biology 27: 417–434.

Happey-Wood, C. M., 1988. Ecology of freshwater planktonic green algae. In Sandgren, C. D. (ed.). Cambridge University Press, Cambridge: 103–133.

Huszar, V. L. M. & L. H. S. F. A. Silva Esteves, 1990. Estrutura das comunidades fitoplanctônicas de 18 lagoas da região do baixo rio Doce, Linhares, Espírito Santo, Brasil. Revista Brasileira de Biologia 50: 585–598.

Huszar, V. L. M., L. H. S. Silva, M. M. Marinho, P. Domingos & C. L. Sant'Anna, 2000. Cyanoprokayote assemblages in eight productive tropical Brazilian waters. Hydrobiologia 424: 67–77.

Hutchinson, G. E., 1967. A Treatise on Limnology: Introduction to Lake Biology and the Limnoplankton (2nd ed.). John Wiley & Sons Inc, New York 1115.

Lembi, C. A., 1980. Unicellular chlorophytes. In Cox, E. R. (ed.). Elsevier, Amsterdam: 5–59.

Kruk, C. & N. G. C. S. Mazzeo Lacerot Reynolds, 2002. Classification schemes for phytoplankton: a local validation of a functional approach to the analysis of species temporal replacement. Journal of Plankton Research 24(5): 901–912.

Lewis, W. M. Jr., 1973. The thermal regime of Lake Lanao (Philippines) and its theoretical implications for tropical lakes. Limnology and Oceanography 18: 200–217.

Lobo, E. & G. Leighton, 1986. Estruturas de las fitocenoses planctónicas de los sistemas de desmbocaduras de rios y esteros de la zona central de Chile. Revista de Biología Marina 22: 143–170.

Lopes, M. R. M. & C. E. M. Bicudo, 2001. Short term variation of physical characteristics of a shallow subtropical oligotrophic reservoir, southeast Brazil. Acta Limnologica Brasiliensia 13: 87–98.

Lund, J. W. G. & C. E. D. Kipling LeCren, 1958. The inverted microscope method of estimating algal number and the statistical bases of estimating by counting. Hydrobiologia 11: 143–170.

Margalef, R., 1978. Life-forms of phytoplankton as survival alternatives in an unstable environment. Oceanologica Acta 1(4): 493–509.

McCune, B. & M. J. Mefford, 1997. PC-ORD Multivariate Analysis of Ecological Data. Version 3.0. MJM. Software design, Oregon. 47 p.

Melo, S. & V. L. M Huszar, 2000. Phytoplankton in an Amazonian flood-plain lake (Lago Batata, Brazil): diel variation and species strategies. Journal of Plankton Research 22(1): 63–76).

Moss, B., 1972. Studies on Gull Lake, Michigan, 1: seasonal and depth distribution of phytoplankton. Freshwater Biology 2: 289–307.

Paerl, H. W., 1988. Growth and reproductive strategies of freshwater blue-green algae (Cyanobacteria). In Sandgreen, C. D. (ed.). Cambridge University Press, Cambridge: 261–305.

Pielou, E. C., 1984. The Interpretation of Ecological Data: A Primer on Classification and Ordination. John Wiley & Sons, New York 263.

Reynolds, C. S., 1984. The Ecology of Freshwater Phytoplankton. Cambridge University Press, Cambridge 384.

Reynolds, C. S., 1997. Vegetation process in the pelagic: a model for ecosystem theory. In Kinne, O. (ed.), Excellence in Ecology. (9). Ecology Institute. 371 p.

Reynolds, C. S., V. L. M. Huszar, C. Kruk, L. Naselli-Flores & S. Melo, 2002. Towards a functional classification of the freshwater phytoplankton. Journal of Plankton Research 24(5): 417–428.

Shapiro, J., 1990. Currents beliefs regarding dominance by cluegreens: the case for the importance of CO_2 and pH. Verhandlungen der internationale Vereinigung für Theoretische und Angewandte Limnologie 24: 38–54.

Smith, V. H., 1986. Predicting the proportion of blue-green algae in lake phytoplankton. Canadian Journal of Fisheries and Aquatic Sciences 43: 148–153.

Ter-Braak, C. J. F., 1986. Canonical Correspondence Analysis: a new eigenvector technique for multivariate direct gradient analysis. Ecology 67: 1167–1179.

Ter-Braak, C. J. F. & I. C. Prentice, 1988. A theory of gradient analysis. Advanced Ecology Research 18: 271–317.

Tilman, D., R. Kiesling, R. Sterner, S. Kilham & F. A. Johnson, 1986. Green, blue-green and diatom algae: taxonomic differences in competitive ability for phosphorus, silica and nitrogen. Archiv für Hydrobiologie 106: 473–485.

Tilzer, M. M., 1973. Diurnal periodicity in the phytoplankton assemblage of a high mountain lake. Limnology and Oceanography 18: 15–30.

Tilzer, M. M. & C. R. Goldman, 1978. Importance of mixing, thermal stratification and light adaptation for phytoplankton productivity in Lake Tahoe (California, Nevada). Ecology 59: 810–821.

Tundisi, J. G., 1983. A review of basic ecological processes interacting with production and standing-stock of phytoplankton in lakes and reservoirs in Brazil. Hydrobiologia 100: 223–243.

Tundisi, J. G., T. Matsumura-Tundisi, M. C. F. Pontes & J. G. Gentil, 1981. Limnological studies at quaternary lakes in Eastern Brasil, 1: primary production of phytoplankton and ecological factors at Lake D. Helvecio. Revista Brasileira de Botânica 4: 5–14.

Utermöhl, H., 1958. Zur Vervolkomnung der quantitativen Phytoplankton: Methodik. Mitteilungen der internationale Vereinigung für Theoretische und Angewandte Limnologie 9: 1–38.

Vinner, A. B. & L. Kemp, 1983. The effect of vertical mixing on the phytoplankton of the Lake Rotongaio (July 1979–January 1981). New Zealand Journal of Marine and Freshwater Research 17: 407–422.

Wetzel, R. G., 1993. Limnologia. Fundação Calouste Gulbenkian, Lisboa 919. (Translation to Portuguese by Maria José L. Boavida).

Hydrobiologia (2005) 542:249–259
H. Segers & K. Martens (eds), Aquatic Biodiversity II
DOI 10.1007/s10750-004-3950-z

The female reproductive organ in podocopid ostracods is homologous to five appendages: histological evidence from *Liocypris grandis* (Crustacea, Ostracoda)

Renate Matzke-Karasz[1,*] & Koen Martens[2]

[1] *Department of Environmental and Geosciences, Palaeontology, Ludwig-Maximilian-Universitaet Muenchen, and Geo-Bio-Center^{LMU}, Richard-Wagner-Str. 10, 80333 Munich, Germany*

[2] *Koninklijk Belgisch instituut voor Natuurwetenschappen, Afdeling Zoetwaterbiologie (Royal Belgian Institute of Natural Sciences, Freshwater Biology), Vautierstraat 29, 1000 Brussels, Belgium*

(**Author for correspondence: Tel.: +49-8921806649, Fax: +49-8921806601, E-mail: r.matzke@lrz.unimuenchen.de*)

Key words: homology, copulatory appendages, histology, segmentation

Abstract

The giant ostracod, *Liocypris grandis*, found in South African temporary pools, has five pairs of additional appendage-like structures, seemingly associated with the Female Reproductive Organ (FRO). Original investigations of these structures by light and scanning electron microscopy could not unequivocally determine whether or not these structures are homologous to genuine appendages and if so, to how many. The present paper investigates morphology and attachment of these structures by frontal and transversal micro-sectioning (thickness = 1 μm) of the caudal soft parts of mature females of *L. grandis*. It is found that all five appendages (R1–R5) have one individual attachment point where they enter into the FRO. Also R3(I) and R3(II) have a single attachment only, these two parts thus represent one single appendage. Whereas R1–R4 have clear lateral insertions, R5 is caudally positioned. The fusion of this latter appendage with the FRO occurs internally, at the posterior edge of the bulge of the FRO and is not externally visible. It is herewith thus accepted that R1–R5 are genuinely endogenous structures (and not, for example, exogenous parasites), that the five structures are homologous to five appendages, and not for example to parts of biramous appendages. It follows from this that the female reproductive organ is homologous to at least five, not to one segment. This would bring the number of (original) thoracic segments in podocopid ostracods to at least 8. An additional pair of triangular structures, inserted in between the two FRO is here described as structures 'X'. Their homology and function remain unknown.

Abbreviations: A1 – Antennula; A2 – Antenna; Cp – Carapace; CR – Caudal Ramus; FRO – Female Reproductive Organ; Md – Mandibula; Mx1 – Maxillula; R1–R5 – additional appendages in adult females; St – Sternum; T1 – first thoracopod; T2 – second thoracopod; T3 – third thoracopod; X – Structures X
Higher taxonomy of the Ostracoda follows the synopsis by Horne et al. (2002).

Introduction

Ostracods are most conservative with regard to the number and form of their appendages. Most podocopid ostracods have seven pairs of appendages (A1, A2, Md, Mx1 and three pairs of thoracopods), a pair of copulatory organs and a pair of caudal structures, previously named furca, now referred to as caudal rami (or uropods, Meisch, 2000). There is thus a reduction in total number of appendages as compared to typical (for example decapod) Crustacea. Swanson (1989, 1990) described the presence of an eighth limb in the punciid ostracod *Manawa* (now seen as the only living representatives of the Palaeocopida – Horne et al., 2002) and found that this 'uropod' was associated with the copulatory appendages in both males and females. Cohen & Morin (1993) homologised the paired male copulatory limb in Cypridinidae (Myodocopida) with an eighth limb, most likely thoracic, and later (Cohen & Morin, 1997) extended this homology to the female genital limbs. Tsukagoshi & Parker (2000) analysed traces of trunk segmentation in Podocopida and concluded that podocopid ostracods originally had 11 trunk segments (this body plan still occurs in the Platycopida (Schultz, 1976) and in the punciids – Swanson, 1989, 1990), that segments 1–6 are thoracic, while segments 7–11 are most likely abdominal. They associated male copulatory organs with the 10th trunk segment (abdominal); the female copulatory organs with the fifth segment (thoracic). These positions would indeed associate the female copulatory organ with the eighth limb, as postulated by Cohen & Morin (1997), but would homologise the male copulatory organs with (a) more caudally positioned appendage(s). In all of these cases, however, the copulatory organs were either described as being associated with, or were fully homologised with, only one pair of somatic appendages.

Recently, the giant (defined as having a length >3 mm) temporary pool ostracod *Liocypris grandis* (Sars, 1924) from South Africa was rediscovered for the first time since its original description by Sars (1924) (Martens, 2001, 2003). Detailed analyses with both light and scanning electron microscopy showed that this species had additional caudal appendages, seemingly closely associated with the female copulatory organ. This was most unexpected and led to the erection of a new subfamily with the Cyprididae, the Liocypridinae (Martens, *loc. cit.*). The following observations were made: (1) these structures (named R1–R5) occurred in all adult females; (2) they did not occur in males, nor in A-1 females, all structures thus appearing fully formed during the final moult; (3) the structures were symmetrical in all investigated adult female individuals; (4) they were situated either in between R1 and R4 or directly caudal to R5 the female copulatory organs. This lead Martens (*loc. cit.*) to conclude that (1) these structures were genuine appendages, and not, for example, exogenous parasites. (2) They were closely associated with some reproductive function (e.g. brood care or mate recognition). (3) These structures represented 3–5 true appendages, depending on whether or not some of these structures formed the exo- and endopodite of one paired appendage.

All three conclusions remained speculative and required further testing, either by direct behavioural observation, or by determining the insertion position of the five pairs of structures through sequential microtome sectioning of the posterior part of the adult female body. The present paper performs the latter test.

Material and methods

Material

Specimens used for the present study were collected from Blinkvlei, near Van Rhynsdorp (Cape Province, South Africa). Coordinates: 31°44′ 21″ S, 18° 55′ 23″ E. Blinkvlei is a large, circular (diameter ca. 250 m) temporary pan, rather shallow (mostly less than 1 m deep), on clay and mud. Water is very turbid, with near zero transparency and is devoid of macrophytes, either emergent or submerged. At the time of collection, conchostracans abounded. Five males and seven females were collected by K. Martens & L. Hoenson on 7 September 2001. The slides derived from the studied specimens are deposited in the Bavarian State Collection of Palaeontology, Munich (BSP 2003 XXIV 1-35).

Sectioning

Posterior halves of adult two female soft-bodies have been embedded in Spurr medium. Sectioning has been carried out with a Reichert Microtome OM U3. Serial sections of 1 μm were cut either transversally or frontally, stained with Methylene Blue and mounted with Entellan mounting medium.

Light-microscopical investigations were carried out with a Leica DMLB Microscope and photographs were taken with a Fuji HC 2000 digital camera.

Results

Brief redescription of R-appendages in L. grandis *(modified from Martens, 2003) (Fig. 1A)*

R1 a well-sclerotised, plate-like structure, with rounded dorsal and straight ventral margins, anteriorly bluntly pointed, posteriorly with a long, elongated point; concavely rounded towards the antero-lateral sides.

R2 a worm-like, tubular structure, not well-sclerotised, distally bluntly pointed and with apparent pseudo-segmentation, not continued internally.

R3 resembling a true appendage more closely than any of the other R-appendages, with both a palp-like (I) and a respiratory plate-like structure (II). Structure II resembling a respiratory plate of the Mx1, with rays not arranged linearly, but rather in unclear subgroups; these rays mono-or bifurcated and completely smooth. Palp (I) and plate (II) incompletely separated. R4 rod-like, with pronounced pseudo-segmentation, distally pointed, hollow up to the most distal point.

R5 strongly resembling a limb *Anlage* as illustrated for other Cyprididae (Smith & Martens, 2000), fully hollow, with about the same size as a full-grown T2 and pointing in a caudal direction.

None of these additional structures R1–R5 has internal musculature.

Transversal series of sections (ventro-dorsal)

The first part of the series (= the ventral most part) was not sectioned. The series here described thus starts halfway up appendage R1.

Figure 2A: right FRO with internal cell mass, R1 and R2 opening into FRO. Left of the FRO (= right side in the picture) transversal cuts of the rami of the CR; the latter bent towards the anterior side.

Figure 2B: right FRO, R1 only marginally visible, with larger part of R2 visible.

Figure 2C: right FRO, with dorsal-most part of R2 showing full connection to FRO; ventral most tips of ray-like structures of R3, still with CR.

Figure 2D: oviduct (OD) in FRO, with egg in the middle, CR present, parts of ray-like structures of R3 in between these.

Figure 2E: detail of Figure 2D, showing location where R3 opens into FRO; egg in OD with sculptured outer surface, inner surface of OD also sculptured.

Figure 2F: FRO with egg in OD, larger part of R3, CR present, anteriorly with one of two additional structures X (see discussion).

Figure 3A: detail of egg in OD, showing sculptured external surface of egg and internal lining of OD.

Figure 3B: left and right FRO; right FRO with egg in oviduct; R3 present and connected to both FRO; R5 present near and almost connected to right FRO.

Figure 3C: detail of beginning of connection between R5 and right FRO.

Figure 3D: both FRO with R3, right FRO with egg in OD and with R5 fully connected.

Figure 3E: basal part of R5 fully fused with right FRO; left FRO with R3.

Figure 3F: right FRO with distal part of R5 disconnected; left FRO with R4 connected, sperm present in the spermatheca.

Frontal series of sections (postero-anterior)

Figure 4A: posterior part of body, with attachment of CR; extremities of R5 on both sides.

Figure 4B: caudal part of body showing ventral 'sternum', with both R5, parts of CR and distal part of R4. Note dorsal triangular structure on body.

Figure 4C: detail of triangular structure on body.

Figure 4D: posterior part of body, with two R5, parts of R3 and R4 and ventral part of right FRO ventral of R5; CR with parts of claws.

Figure 4E: R5 fuses with body, i.e. main part of FRO

Figure 4F: Right FRO completely fused (main and ventral part and R5); left FRO with R5 fused to main part, ventral part separate; parts of R3 and R4 present.

Figure 5A: both FRO with basal parts of R3 merging, posterior parts of R2 also present.

Figure 5B: right R2 merging with right FRO, two structures X in between both FRO.

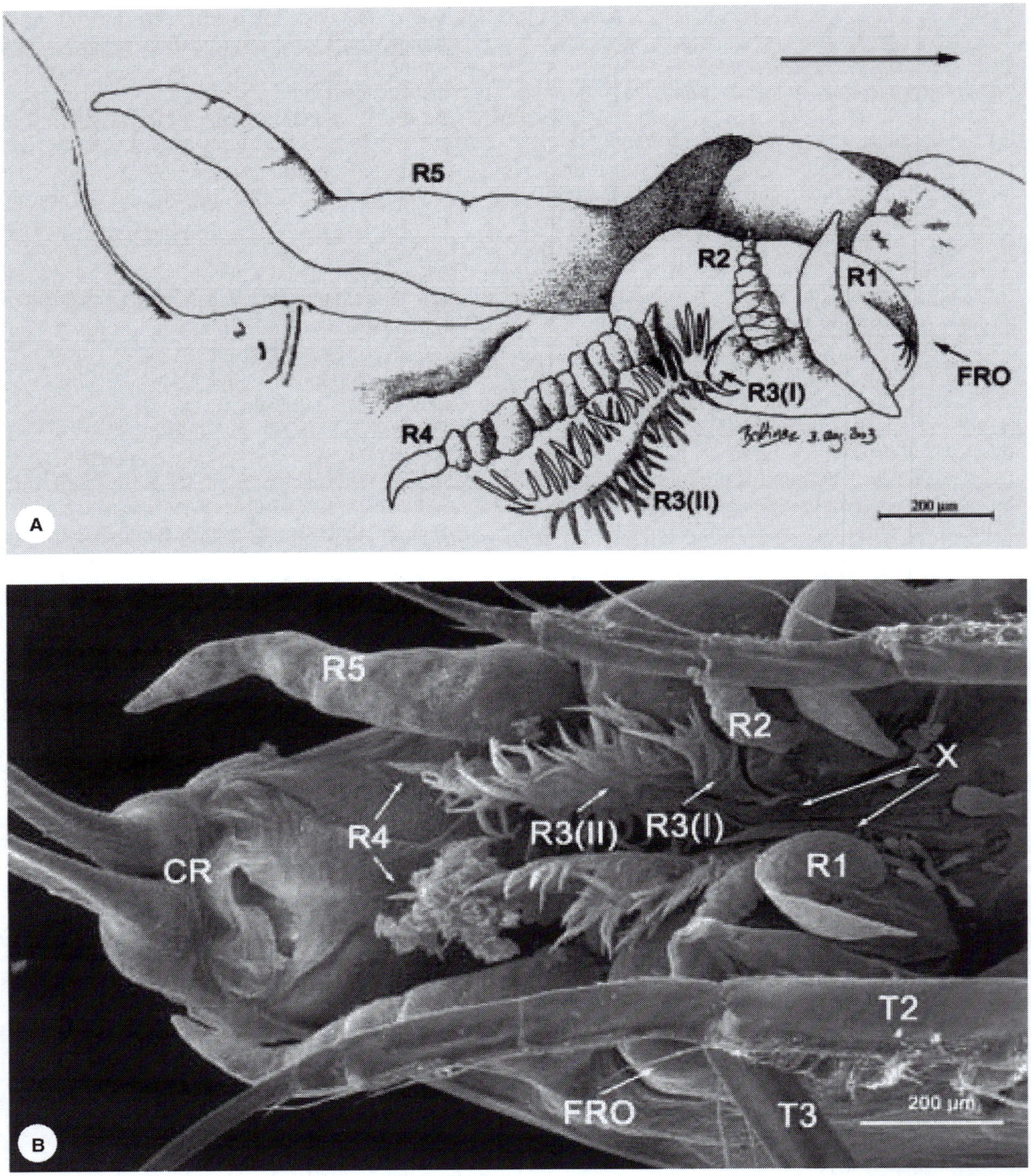

Figure 1. Caudal part of adult female *Liocypris grandis*: (A) Reconstruction of five R-appendages and their attachments to FRO. (B) Scanning electron microscopic micrograph of caudal part of adult female, ventral view.

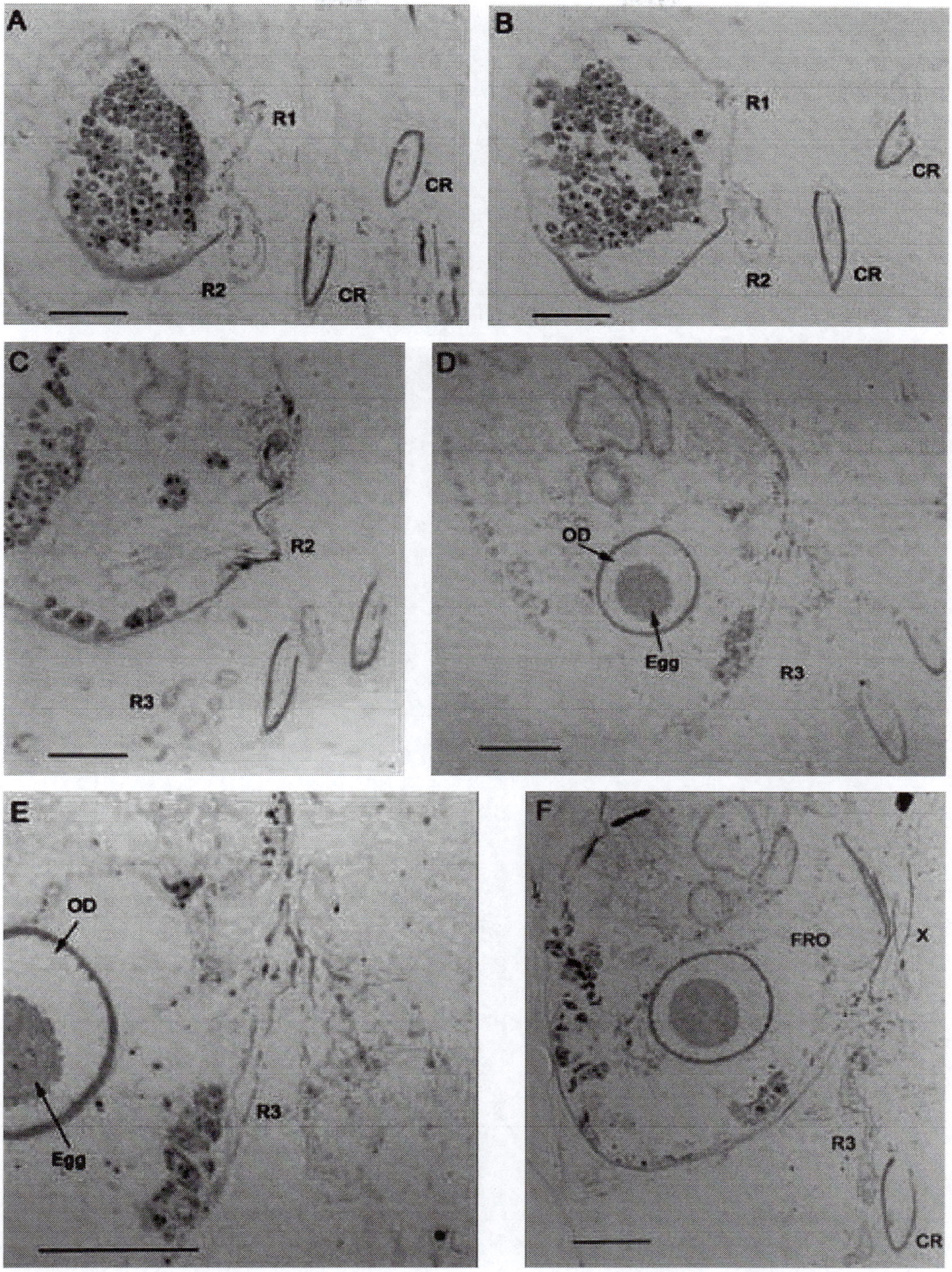

Figure 2. Selected transversal sections, starting from ventral side, of caudal part of adult female *Liocypris grandis* (scales = 100 μm): (A) R1 and R2 opening into medial side of right FRO (right side in the picture); transversal cuts of the rami of the CR. (B) Right FRO, R1 and R2. (C) Dorsal-most part of R2 fully connected to right FRO; ventral-most tips of ray-like structures of R3 visible. (D) Oviduct (OD) in FRO, with egg in the middle; R3 present. (E) Detail of D, R3 opens into FRO; egg in OD with sculptured outer surface, inner surface of OD also sculptured. (F) FRO with egg in OD, larger part of R3; structure X visible.

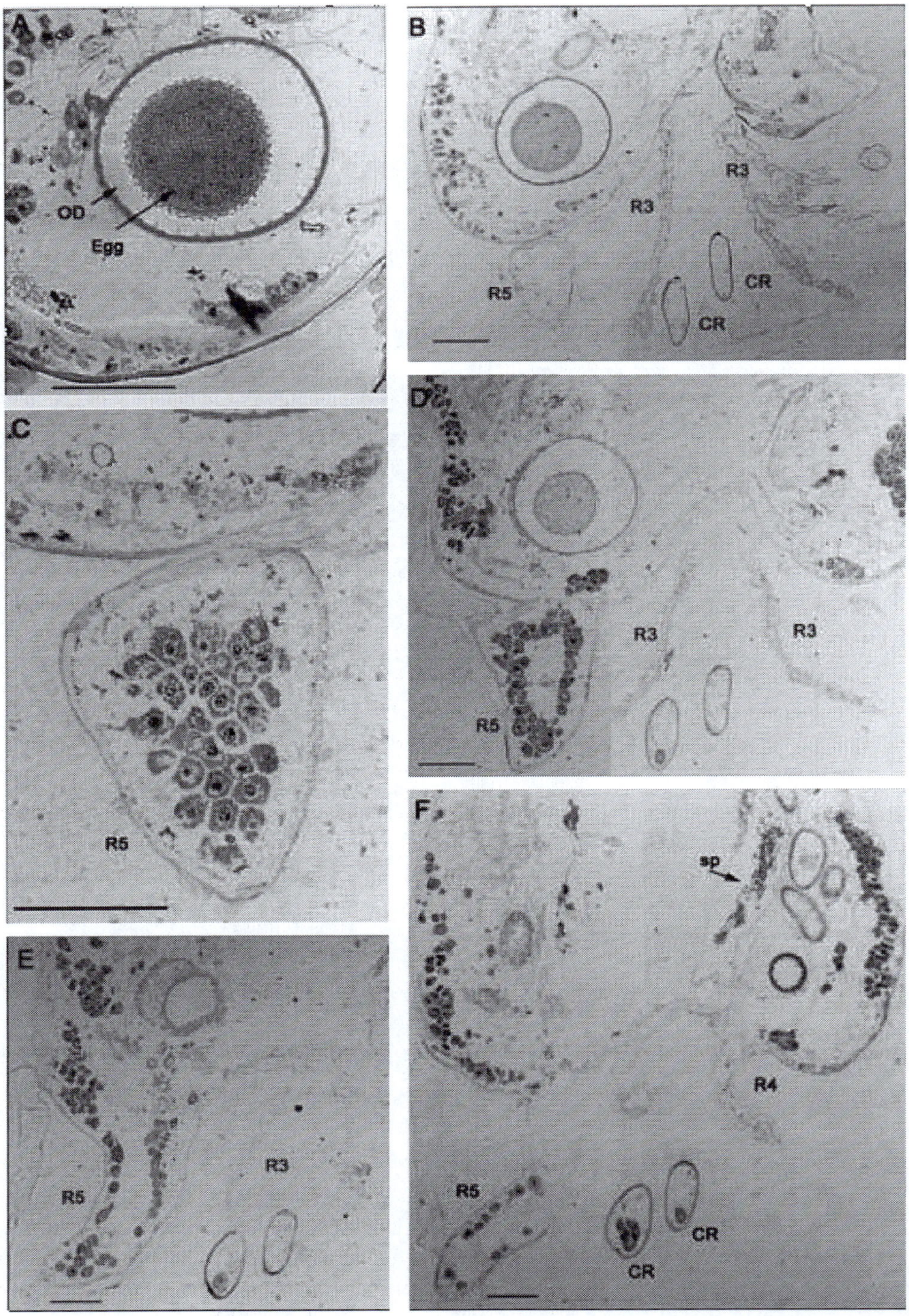

Figure 3. Selection of dorsal-most transversal sections of caudal part of adult female *Liocypris grandis* (scales = 100 μm): (A) Detail of egg in OD, showing sculptured external surface of egg and internal lining of OD. (B) Left and right FRO; both R3 connected to both FRO; R5 almost connected to right FRO. (C) Detail of beginning of connection between R5 and right FRO. (D) Both FRO with R3, with R5 fully connected. (E) Basal part of R5 fully fused with right FRO; left FRO with R3. (F) Right FRO with distal part of R5 disconnected; left FRO with R4 connected; cuts of sperm present in die spermatheca (sp).

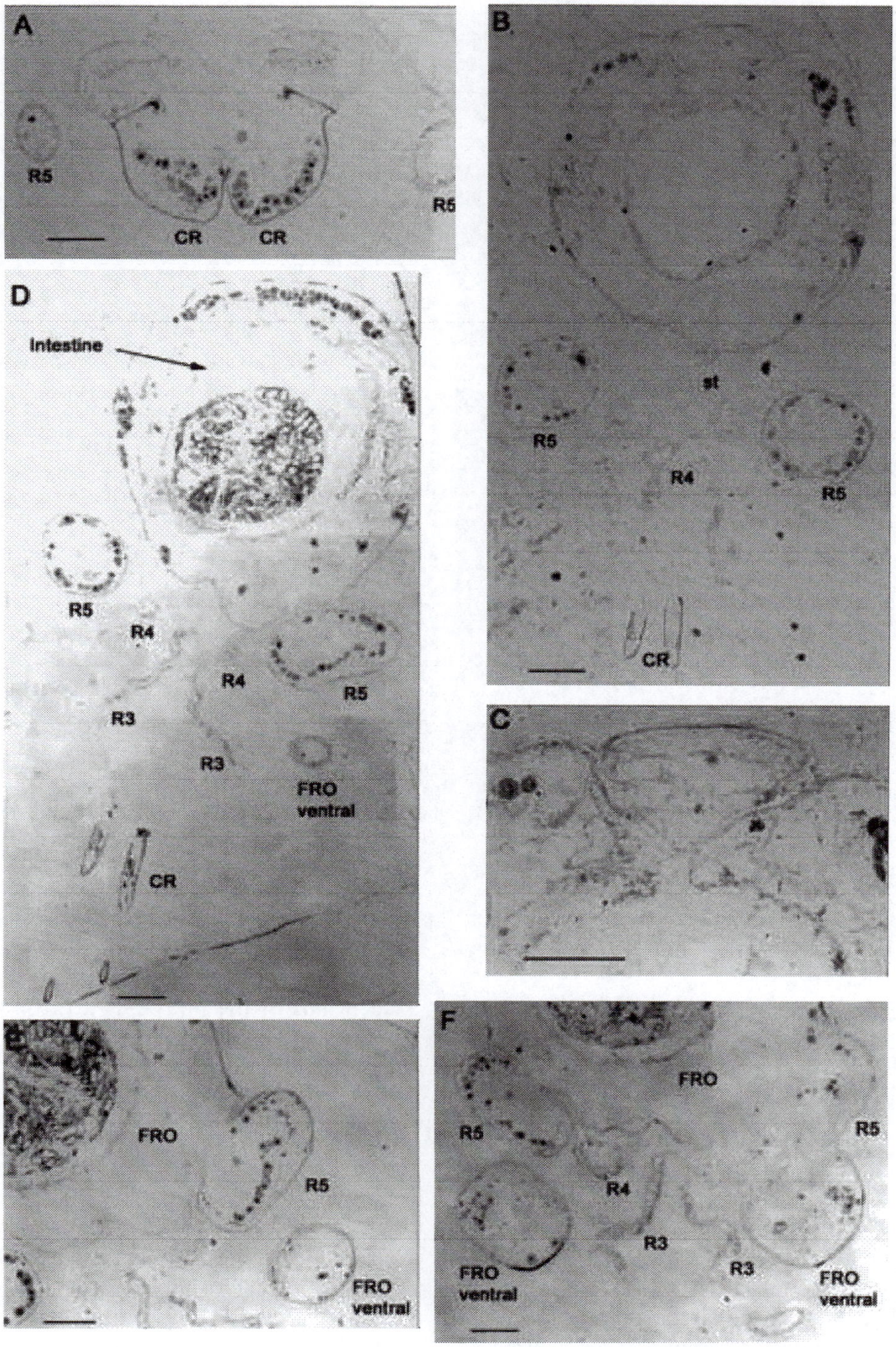

Figure 4. Selection of posterior-most sections of frontal series of caudal part of adult female *Liocypris grandis* (scales = 100 μm): (A) Posterior part of body, with attachment of CR; extremities of R5 on both sides. (B) Caudal part of body showing ventral 'sternum' (st), with both R5, parts of CR and distal part of R4. Note dorsal triangular structure on body. (C) Detail of B, showing triangular structure on body. (D) Posterior part of body, with two R5, parts of R3 and R4 and ventral part of right FRO ventral of R5; CRwith parts of claws. (E) R5 fused with body, i.e. main part of FRO (F) Right FRO completely fused (main and ventral part and R5); left FRO with R5 fused to main part, ventral part separate; parts of R3 and R4 present.

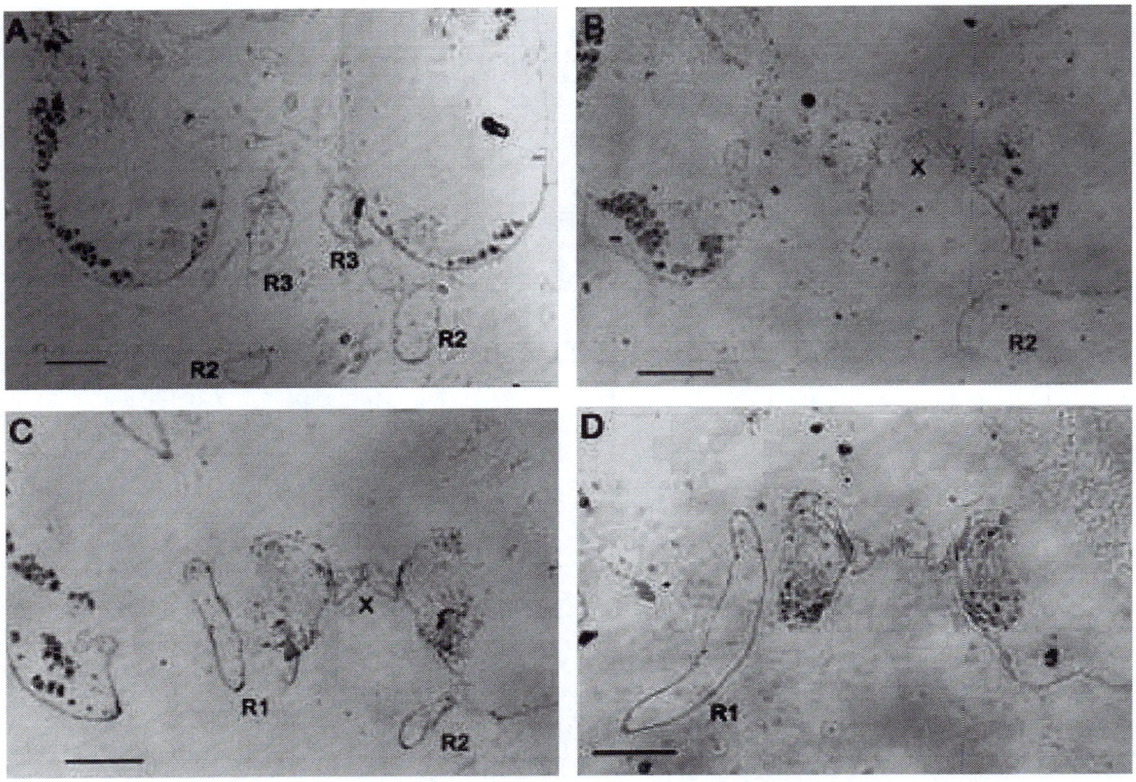

Figure 5. Selection of anterior-most sections of frontal series of caudal part of adult female *Liocypris grandis* (scales = 100 μm): (A) Both FRO with basal parts of R3 merging, posterior parts of R2 also present. (B) Right R2 merging with right FRO, two structures X in between both FROs. (C) Left FRO with posterior part of R1, right FRO with anterior part of connection of R2. (D) Left FRO with frontal section of the plate-like R1.

Figure 5C: left FRO with posterior part of R1, right FRO with anterior part of connection of R2. Figure 5D: left FRO with frontal section of the plate-like R1.

Discussion

Are structures R1–5 genuine appendages?

Structure R5 resembles a typical podocopid limb-*Anlage* (compare to Smith & Martens, 2000), and the ray-like extremities of R3 show similarities with the rays of respiratory plates of the Mx1. R1, R2 and R4, however, have a strikingly atypical morphology and do not resemble true ostracod appendages; they may even give the impression of being exogenous parasites. However, the present results confirm that structures R1–R5 are effectively endogenous to the adult female body, as they are all symmetrical and connect individually to the body and open into the female FRO. The present study indeed shows that each appendage has a single merging point with the FRO. To which part of the biramous arthropod appendage these structures are homologous, is still not unequivocal.

Are all of these appendages associated with the reproductive organ?

In exterior view, the FROs appear like two well-defined oval regions, lying posteriorly to the last 'normal' thoracic appendage (T3). It is clearly elevated, thus forming a bulge which is often referred to as a 'genital operculum' (Danielopol, 1978). However, the histological study shows that this organ extends into a more dorsal region interiorly, comprising a variety of internal structures such as ovi- and spermiducts.

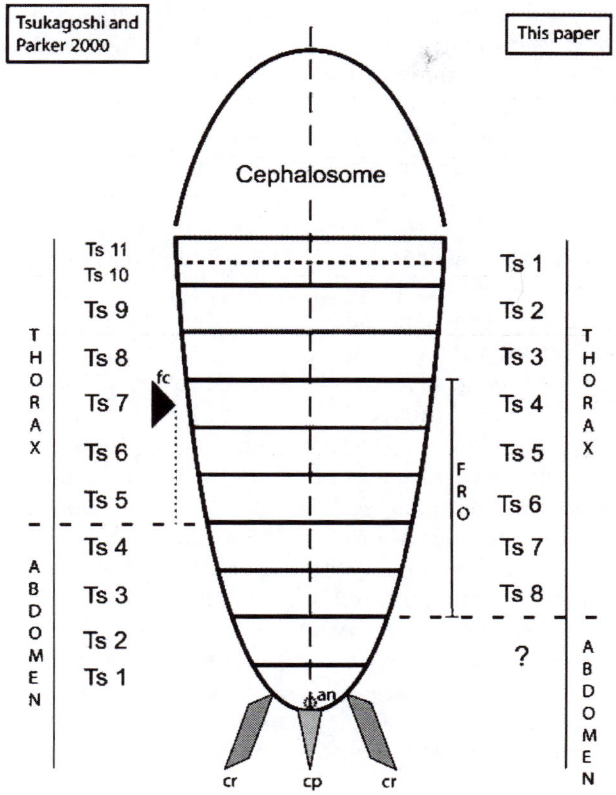

Figure 6. Schematic dorsal view of a podocopine ostracod combining the hypothetic body plan suggested by Tsukagoshi & Parker (2000) (left) and by this paper (right), modified after Tsukagoshi & Parker (2000). After the model of Tsukagoshi & Parker (2000: 25 ff. and Fig. 11), the female copulatory organ fc originates most probably from the segment, which is numbered T7 in their diagram; however, the dotted line below fc indicates a putative origin of fc also in T6 or T5. Furthermore, their model proposes the abdomen consisting of T1–T4 and the thorax thus comprising seven segments (T5–T11). In contrast, the here presented model describes a thorax formed by eight segments (T1–T8), of which the caudal most five segments contribute to the formation of the female reproductive organ FRO. Abbreviations: an – anus; cp – caudal process; cr – caudal ramus; fc – female copulatory organ; FRO – female reproductive organ; Ts – trunk segment.

The five paired structures that accompany the FRO in *Liocypris grandis* are all histologically associated with it. While R1–R4 are directly inserting from the bulge itself, R5 is somewhat displaced to the posterior edge of the bulge. Nevertheless, sections like in Figure 3B and D clearly show that R5 emerges from a body area containing a mature egg in its oviduct – an area representing the interior part of the FRO. All five appendages are therefore associated with the FRO.

The structures X

In addition to the structures inserting from the bulge of the FRO as described by Martens (2003), the present study revealed the presence of an additional pair of small, pointed, tongue-like structures close to R2 and R3 and inserted on the ridge between the two bulges of the FROs (Figs. 1B, 5B and C). These structures are not included into the complex of R-structures (=structures associated with the Female Reproductive Organ), because points of insertion (and thus their origin) cannot not clearly be assigned to the FROs. They are here tentatively referred to as ‘structures X’.

How many appendages and segments do these structures represent?

The five paired structures inserting from the FRO in adult *Liocypris grandis* feature five different morphologies, most of them not allowing a clear judgement about the kind of insertion just by scanning electron microscopy. In particular, it was not possible to define, if R3 has one single connection to the FRO or two independent connecting points. In the latter case, structures R3(I) and R3(II) would have represented two appendages rather than one. The sections now reveal that all five R-structures each have one well-defined opening into the FRO, so that we consider these five paired structures to represent five paired appendages. Furthermore, we tentatively conclude that the FRO can be homologised with five segments instead of with one single segment as proposed by Cohen & Morin (1997) and Tsukagoshi & Parker (2000). In addition, Martens (2003) named a triangular ventral plate-like structure, directly caudal to the FRO, the 'sternum' and considered it as the delimitation between thorax and abdomen. If this is true, the ancestral podocopid thorax must have had at least eight, not six, segments as was postulated by Tsukagoshi & Parker (2000). Indeed, the three segments associated with the presently recognised thoracopods Tl-3 and five segments homologous to the FRO together would form a thorax of at least eight segments (Fig. 6).

Putative function of the FRO-associated appendages

The plate-like morphology of R3, similar to respiratory plates of the Mx1, might indicate a possible role of the additional appendages in ameliorating the respiration of this giant ostracod, either of the adult female itself, or of the potential brood. Respiratory plates of podocopid ostracod appendages are known to increase the water current within the carapace in order to increase the food and oxygen inflow. However, as shown in this study, the FRO-associated appendages do not contain muscles and active movement therefore has to be excluded. The hypothesis that appendages R1–R5 have a possible function in respiration of the adult and/or increased oxygenation of eggs or young juveniles must thus be rejected. Nevertheless, *Liocypris grandis* is a giant ostracod in which the relationship between body mass and body surface (and hence oxygen uptake) might be unfavourable. The question thus arises whether or not giant freshwater ostracods have to increase their body surface to survive. The fact that neither in males of *Liocypris grandis*, nor in either males or females of *Megalocypris* spec ($L =$ ca. 8 mm, which is even larger than *L. grandis*) such additional appendages occur, again leads us to reject this idea.

We therefore conclude that the FRO-associated appendages are most likely playing a role in reproduction, possibly in mate-recognition or oviposition, but not in oxygenation of early instars, if ever brood care were to exist in this species.

Evolutionary significance

Liocypris grandis thus far remains the only ostracod in which such a series of FRO-associated appendages is known. The present data do not discriminate conclusively between the different possible evolutionary and phylogenetic scenarios as outlined by Martens (2003).

Nevertheless, the identification of five separate points of connection with the FRO increases the probability that the structures indeed represent remnants of ancestral appendages, rather than newly formed, functional structures. In the latter case, one could expect more randomly inserted structures (some connected to FRO, others not), although the possibility can as yet not be excluded. If R1–R5 are remnants of the original appendages making up the FRO, then it is possible that in lineages where they have disappeared (this is all other podocopid lineages as far as presently known) at least parts of them might have been incorporated in the internal anatomy of the FRO. This can be tested by analysing the internal structures of the FRO in various cypridinid lineages, which should then have a higher complexity than those in *L. grandis*.

Acknowledgements

Jenny Day and Genevieve Jones (University of Cape Town, RSA) are gratefully acknowledged for suggesting research on ostracods from the surroundings of Van Rhynsdorp and this led to the rediscovery of *Liocypris grandis*. Liz Hoenson (South African Museum, RSA) provided valuable assistance in the field. The Water Research Commission of South Africa provided financial support for the collection of the material. The Belgian National Science Foundation (FWO) contract no. G.0109.99 is acknowledged.

We furthermore acknowledge the opportunity to do the histological work in the Bavarian State Collection of Zoology, Munich (special thanks to Prof. Dr. Haszprunar and Mrs. Eva Lodde) and at the Zoological Institute of the University of Cologne (special thanks to Prof. Dr. Berking and Dr. Vedder).

RM-K acknowledges the support of the European Community – Access to Research Infrastructure Action of the Improving Human Potential Programme.

References

Broodbakker, N. W. & D. L. Danielopol, 1982. The chaetotaxy of Cypridacea (Crustacea, Ostracoda) limbs: proposals for a descriptive model. Bijdragen tot de Dierkunde 52: 103–120.

Cohen, A. C. & J. G. Morin, 1993. The cypridinid copulatory limb and a new genus *Kornickeria* (Ostracoda, Myodocopida) with four new species of bioluminiscent ostracods from the Caribbean. Zoological Journal of the Linnean Society 108: 23–84.

Cohen, A. C. & J. G. Morin, 1997. External anatomy of the female genital (eight) limbs and the setose openings in myodocopid ostracodes (Cypridinidae). Acta Zoologica 78: 85–96.

Danielopol, D. L., 1978. Über Herkunft und Morphologie der Süsswasser-hypogäischen Candoninae (Crustacea, Ostracoda). Sitzungsberichten der österreichische Akademie der Wissenschaften. Mathematisch-naturwissenschaftliche Klasse, Abteilung 1(187): 1–162.

Horne, D. J., A. Cohen & K. Martens, 2002. Biology, Taxonomy and Identification Techniques. In Holmes, J. A. & A. Chivas (eds), The Ostracoda: Applications in Quaternary Research. American Geophysical Union, Washington DC: 5–36.

Martens, K., 1987. Homology and functional morphology of the sexual dimorphism in the antenna of *Sclerocypris* Sars, 1924 (Crustacea, Ostracoda, Megalocypridinae). Bijdragen tot de Dierkunde 57: 183–190.

Martens, K., 2001. Ostracoda. Guides to the freshwater invertebrates of Southern Africa. In Day, J.A. et al. (eds), Water Research Commission Report TT 148/01: 9–77.

Martens, K., 2003. On a remarkable South African giant ostracod (Crustacea, Ostracoda, Cyprididae) from temporary pools, with additional appendages. Hydrobiologia 500: 115–130.

Meisch, C., 2000. Freshwater Ostracoda of western and central Europe. Süßwasserfauna von Mitteleuropa 8/3, Spektrum Akad. Verlag/Gustav Fischer, Heidelberg. 522 pp.

Sars, G. O., 1924. The fresh-water Entomostraca of the Cape Province (Union of South Africa). Part 2: Ostracoda. Annals of the South African Museum 20: 105–193.

Schultz, K. 1976. Das Chitinskelett der Podocopida (Ostracoda, Crustacea) und die Frage der Metamerie dieser Gruppe. Ph.D. Thesis, Hamburg. 167 pp.

Smith, R. J. & K. Martens, 2000. The ontogeny of the cypridid ostracod *Eucypris virens* (Jurine, 1820) (Crustacea, Ostracoda). Hydrobiologia 419: 31–63.

Swanson, K. M., 1989. *Manawa staceyi* n.sp. (Punciidae, Ostracoda): soft anatomy and ontogeny. Courier Forschungsinstitut Senckenberg 113: 235–249.

Swanson, K. M., 1990. The punciid ostracod – a new crustacean evolutionary window. Courier Forschungsinstitut Senckenberg 123: 11–18.

Tsukagoshi, A. & A. R. Parker, 2000. Trunk segmentation of some podocopine lineages in Ostracoda. Hydrobiologia 419: 15–30.

Hydrobiologia (2005) 542:261–274
H. Segers & K. Martens (eds), Aquatic Biodiversity II
DOI 10.1007/s10750-004-4952-6

Daphnia species diversity in Kenya, and a key to the identification of their ephippia

Joachim Mergeay[1,]*, Dirk Verschuren[2] & Luc De Meester[1]
[1] *Laboratory of Aquatic Ecology, Katholieke Universiteit Leuven, Ch. de Bériotstraat 32, B-3000 Leuven, Belgium*
[2] *Department of Biology, Ghent University, Ledeganckstraat 35, B-9000 Gent, Belgium*
*(*Author for correspondence : Tel.: +32-16-323856, Fax: +32-16-324575, E-mail: joachim.mergeay@bio.kuleuven.ac.be)*

Key words: Africa, *Daphnia magna*, *Daphnia pulex*, diapause, dormant eggs, tropical limnology

Abstract

The distribution of *Daphnia* species in tropical Africa is poorly known and understood. *Daphnia* are assumed rare in tropical regions, but systematic studies covering large areas are sparse. We sampled the active community (live zooplankton) and/or the dormant community (diapausing egg banks in the sediment) of 41 standing water bodies in Kenya in search for *Daphnia*. Overall the dormant communities yielded 11 species of *Daphnia*, a species richness more than twice the species richness found in the active communities. Dormant community species diversity better reflects the spatial, and particularly the temporal (multi-annual) variation in environmental conditions available to *Daphnia* in these tropical standing waters. Hence, we suggest that the dormant community be taken into account when assessing local zooplankton diversity, especially in fluctuating tropical lake ecosystems, where the presence of each local *Daphnia* species in the active community may be strongly seasonal or erratic. Geographic distribution data from this study are supplemented with previous records of *Daphnia* in East Africa to provide an overview of the known distribution of *Daphnia* in Kenya and neighbouring countries. We also present a detailed key for morphological identification of the ephippia of the 11 *Daphnia* species encountered, complemented with photographs and drawings of diagnostic characters.

Introduction

In many aquatic ecosystems, temporal habitat instability is a major factor determining zooplankton community composition at any given time. Different zooplankton taxa often show strong changes in abundances with time and space, depending on the prevailing conditions. As a result, the total zooplankton diversity can not be assessed by sampling at a single point in time or space. Rather, multiple sampling throughout the year is needed during the course of a growing season and, preferrably, covering several years. In order to cope with this temporal habitat instability many cladocerans produce resistant dormant stages that can survive unfavourable conditions and that can stay viable for years, even decades and longer (Cáceres, 1998). Sampling the dormant community from the sediment may therefore allow an assessment of the total species diversity of the local cladoceran community from a single sampling event (May, 1986; Havel et al., 2000; Vandekerkhove et al., submitted), provided that these dormant stages can be identified to species. Remains of cladocerans recovered from subrecent and older sediments include the ephippia of these dormant eggs but also mandibles and post-abdominal claws, and all have long been used in paleolimnological studies (e.g., Kitchell & Kitchell, 1980; Leavitt et al., 1989; Hann et al., 1994; Kerfoot et al., 1999). Unfortunately, since most local studies deal with ephippia of only a few species, which are related directly to the species present in the local active population, specific information regarding species identification of ephippia is mostly lacking or incomplete.

Most *Daphnia* species reproduce by cyclic parthenogenesis. Sexual reproduction occurs when living conditions are unfavourable or deteriorate (Carvalho & Hughes, 1983). During sexual reproduction the female carapace produces an ephippium around the brood chamber (Zaffagnini, 1987). Inside the ephippium a new cuticula is formed which encloses two embryos in arrested development. On both sides of the ephippium two convex swellings provide the necessary space for the dormant embryos. These swellings (hereforth called egg chambers) are more melanised than other parts of the ephippium and are often more sclerotised, and hence more resistant to physical damage. The ephippium is eventually released during the following moult, together with the old carapace (Zaffagnini, 1987). Because of their good preservation in lake sediments, ephippia are ideally suited for ecological and paleoecological studies.

Daphnia are considered rare in tropical regions, especially in the warm lowland tropics (Fernando et al., 1987; Dumont, 1994). Yet in equatorial East Africa *Daphnia* have been reported from a fairly large number of permanent lakes and ponds, with at least nine species found so far (Green, 1995). Previous regional studies (Green, 1971; Lehman, 1996) focused mostly on zooplankton diversity in the large East African Rift lakes. Small waterbodies, often of a semi-permanent or temporary nature, were often overlooked. Yet these habitats are often fishless, which is considered a key factor permitting the presence of *Daphnia* in tropical regions (Hebert, 1978; Dumont, 1994). We studied the dormant *Daphnia* community of 41 water bodies in Kenya, and compared its species diversity with that in the active community by simultaneous sampling of the live zooplankton. We also present a key for the identification of *Daphnia* species of Kenya and neighbouring regions, based on the morphology of ephippia.

Methods

Study area

Field surveys in August 2001 and January 2003 yielded collections of live zooplankton and recently deposited surface sediments from 41 standing water bodies in southern and southwestern Kenya. The sampled waters vary from small semi-permanent ponds and watering holes to larger lakes and dams (Table 1 and Fig. 1). The area is dissected from North to South by the Eastern Rift Valley. The surface elevation of the sampled waters varies from 700 to 1200 m a.s.l. in the southeastern lowlands (12 sites) to between 900 and 2800 m in the Rift Valley and adjacent escarpments (29 sites). In addition, we analysed ephippia extracted from previously studied sediment cores from the Lake Naivasha basin (Mergeay et al., submitted), and surface sediments of four lakes (Bandasa, Ellis, Nkunga, Simba) on Mt. Kenya (courtesy of Dr Philip Barker, Lancaster University, UK). This Mt. Kenya collection includes both high-elevation alpine lakes and lakes on the lower slopes of the mountain, with elevations ranging between 1800 and 4600 m. Overall, vegetation near the sampled waters ranges from dry *Acacia* bush land, grassland with scattered trees, highland forests and tropical rainforests to afromontane and alpine vegetation. Rainfall is strongly seasonal throughout the study region, predominantly bimodal to the East of the Kenyan Rift Valley and with a trimodal pattern in the West. Rainfall ranges from 350 mm in the arid southern savannah to 1100–1500 mm in the highlands (>1500 m). Mean temperature patterns are closely linked to altitude, and seasonal variation is very limited (Kiai et al., 2002).

Sampling

Water transparency was determined with a Secchi disc (20 cm diameter). Temperature and conductivity were determined in the middle of the water body using a Hydrolab multiprobe (Hydrolab Inc., USA). Geographical positioning and altitude measurement was done using a Garmin Global Positioning System (Garmin International Inc.).

Table 1. Sampling locations and presence of *Daphnia* species

Number	Name	Geographical coordinates	Alt (masl)	Secchi depth (m)	Temperature (°C)	Conductivity ($\mu S/cm^{-1}$)	Active community									Dormant community								
							barbata	laevis	curvirostris	pulex	sp. nov. type Limuru	magna	longispina s.l.	lumholtzi	dolichocephala	barbata	laevis	curvirostris	pulex	sp. nov. type Limuru	magna	longispina s.l.	lumholtzi	dolichocephala
1	Lake Challa	S 03° 19,0′ E37° 42,0′	842	13.00	27	324																		
2	Lake Jipe	S 03° 34.6′ E 37° 46,0′	735	0.30	28.5	1140																		
3	Temporary pond Taveta-Taita 1	S 03° 24,5′ E 38° 03,5′	1115	0.10	32.3	00										1								
4	Temporary pond Taveta-Taita 2	S 03° 24,6′ E 38° 002,2′	1103	0.10	28	150											1							
5	Temporary pond Taveta-Taita 3	S 03° 24,7′ E 37° 49,7′	1102	0.10	28.2	235																		
6	Permanent pond Voi-Nairobi 1	S 02° 45,4′ E 38° 15,3′	695	0.10	32.6	53																		
7	Permanent pond Voi-Nairobi 2	S 02° 21,0′ E 37° 53,3′	1031	0.10	31.5	236																		
8	Amboseli Olokenya swamp	S 02° 42,1′ E 37° 18,7′	1155	0.80	19.8	179																		
9	Amboseli Serena Lodge	S 02° 42,0′ E 37° 15,4′	1159	0.80	21.4	172																1		
10	Amboseli Enamishera	S 02° 40,2′ E 37° 14,6′	1150	0.80	24.7	255																		
11	Amboseli Lake Kioko	S 02° 38,5′ E 37° 13,5′	1153	0.50	25	539																		
12	Amboseli Lake Amboseli	S 02° 36,1′ E 37° 12,5′	1150	0.01	29.1	4690	1									1								
13	Temporary pond Kajiado	S 01° 50,2′ E 36° 47,9′	1757	0.25	24	150	1																	
14	Nairobi Nat. Park Narogomon dam	S 01° 21,0′ E 36°47,9′	1707	0.40	21.3	188		1									1							
15	Nairobi Nat. Park Hyena Dam	S 01° 20,3′ E 36° 48,6′	1709	0.30	21.7	260										1								
16	Nairobi Nat. Park 3	S 01° 21,9′ 36° 50,6′	1709	0.60	23	141										1								
17	Nairobi Nat. Park 4 (river)	S 01° 21,6′ E 36° 51,2′	1664	0.40	21	150																		
18	Nairobi Nat. Park Hyrax valley	S 01° 22.9′ E 36° 47,0′	1716	0.40	23.5	250																		
19	Temporary pond Limuru 1a	S 01° 7.0′ E 36° 37.9′	2306	0.15	21.5	201					1									1				
20	Temporary pond Limuru 1b	S 01° 7.0′ E 36° 37.9′	2306	0.15	21.5	201					1									1				
21	Permanent lake Limuru 2	S 01° 6,3′ E 36° 37,8′	2294	0.70	23.1	155		1			1						1		1	1				
22	Permanent Lake Limuru 3	S 01° 8,3′ E 36° 40.7′	2135	0.50	24.6	414																		
23	Lake Naivasha main	S 00° 46,3′ E 36° 21,7′	1897	0.20	22.9	292	1			1						1			1		1		1	
24	Lake Naivasha Crater	S 00° 45,8′ E 36° 24,6′	1897	1.20	22.5	375	1			1						1			1					

Table 1. (Continued)

Number	Name	Geographical coordinates	Alt (masl)	Secchi depth (m)	Temperature (°C)	Conductivity ($\mu S/cm^{-1}$)	Active community									Dormant community								
							barbata	laevis	curvirostris	pulex	sp. nov. type Limuru	magna	longispina s.l.	lumholtzi	dolichocephala	barbata	laevis	curvirostris	pulex	sp. nov. type Limuru	magna	longispina s.l.	lumholtzi	dolichocephala
25	Lake Oloidien	S 00° 48.9′ E 36° 15,9′	1897	0.30	23.3	2590															1			
26	Eldoret Sigawet Dam	N 00° 35,1′ E 35° 13,1′	2014	0.20	22.5	214										1	1		1					
27	Lake Katalin	N 00° 38,4′ E 35° 28,9′	2337	0.37	19.5	78																		
28	Permanent pond Eldoret Nakuru 1a	N 00° 26,4′ E 35° 18,3′	2185	0.50	14.1	196											1							
29	Permanent pond Eldoret Nakuru 1b	N 00° 26,4′ E 35° 18,3′	2185	0.30	16.2	381											1							
30	Permanent pond Eldoret Nakuru 2	N 00° 21,7′ E 35° 21,1′	2217	0.50	14.5	171											1							
31	Permanent pond Eldoret Nakuru 3	N 00°20,0′E 35° 21,9′	2214	0.40	16.2	221											1							
32	Lake Narasha	N 00° 2,6′ E 35° 32,4′	2764	2.20	16	34		1									1							
33	Permanent pond Eldoret Nakuru 5	S 00° 3.6′ E 35° 38,4′	2471	0.25	20.8	204											1							
34	Permanent pond Eldoret Nakuru 6	S 00° 4.9′ E 35° 39,5′	2536	0.10	24.5	117		1									1							
35	Permanent pond Nakuru Naivasha 1	S 00° 19,4′ E 36° 9.1′	1917	0.10	24	201																		
36	Permanent pond Nakuru Naivasha 2	S 00° 20,5′ E 36° 10,1′	1920	0.20	20.4	178																		
37	Lake Ol Bolossat	S 00° 9.9′ E 36° 25.9′	2358	0.20	20.6	1960	1				1					1	1		1	1	1			
38	Ol Bolossat pond 1	S 00° 10,1′ E 36° 26,0′	2350	0.20	19.5	250	1									1	1							
39	Ol Bolossat pond 2	S 00° 10,1′ E 36° 26,0′	2350	0.20	19.4	250	1									1	1							
40	Lake Banngo	N 00° 39,1′ E 36° 3,6′	967	0.05	26	1600	1									1	1		1					
41	Mogotio Kapchelukung Dam	N 00° 05′ E 36° 05′	1200	0.20	22.2	90	1									1	1							
42	Bandasa Mt Kenya	S 00° 09′ E 37° 28′	2900	> 1	18	20												1						1
43	Ellis Mt Kenya	?	3500	> 3	12.3	33																		1
44	Nkunga Mt Kenya	?	1840	> 1	16.2	316												1						
45	Simba Mt Kenya	?	4575	> 5	8.1	19																		

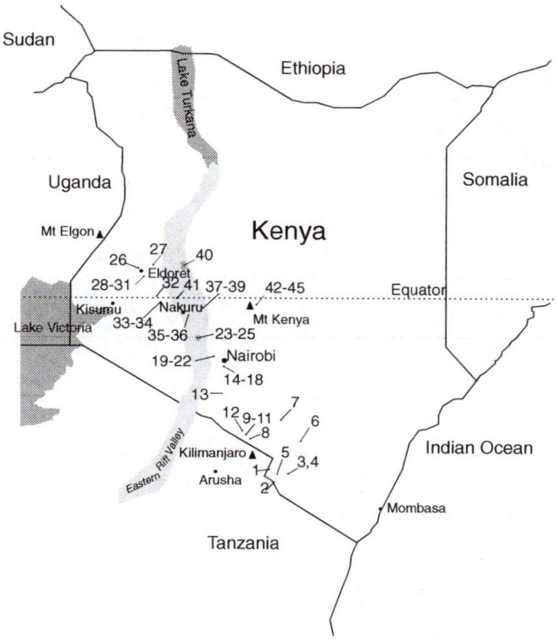

Figure 1. Map of Kenya with indication of sampling sites. Numbers refer to sample sites in Table 1.

Qualitative zooplankton samples were collected using a conical tow-net (25 cm diameter, 150 μm mesh), combining vertical hauls with horizontal hauls, and taking care to include zones with macrophytes and littoral vegetation. Samples were washed in the net and fixed in 100% ethanol. Offshore surface sediments were sampled from a boat using a weighted Wildco Fieldmaster® horizontal water sampler to ensure that only the topmost layer of flocculent mud was taken. In small ponds or very shallow water bodies, sediment was collected manually (~500 ml). These samples were stored without fixation.

In the laboratory, *Daphnia* in the zooplankton samples were identified using a compound microscope with reference to Kofinek (1999), and counted until 300 *Daphnia* had been processed; otherwise the whole sample was analysed. Subsamples of surface sediment were washed through a 250 μm mesh metal sieve, and the retained residue was scanned for *Daphnia* ephippia in a counting tray at 20–50 magnification. Ephippia were mainly identified using ephippial females in zooplankton reference collections.

Results

Of the 41 water bodies sampled, 27 contained *Daphnia*. Of these, 16 contained *Daphnia* in the live zooplankton, while in 25 of them *Daphnia* was found in the dormant egg bank (Table 1). In the active community only four species were found (*Daphnia barbata* Weltner, *Daphnia laevis* Birge, *Daphnia pulex* Leydig and a new undescribed species of the subgenus *Ctenodaphnia*, *Daphnia* sp. nov. type Limuru) while nine species were found in the sediment samples (*Daphnia barbata*, *Daphnia curvirostris* Eylman, *Daphnia longispina s.l.* O.F. Muller, *Daphnia laevis*, *Daphnia lumholtzi* Sars, *Daphnia magna* Straus, *Daphnia pulex*, *Daphnia cf. similis* Claus and *Daphnia* sp. nov. type Limuru). In addition *Daphnia dolichocephala* Sars was found in sediment samples from Mt Kenya. The average species richness was 1.6 species per lake that contained *Daphnia*. Four or more species were found in the recent sediment of three lakes (five in Lake Ol Bolossat and four in Lake Naivasha and Lake Limuru2) (Table 1). In many water bodies very few individuals were found however, in the active as well as in the dormant community. Those water bodies were mostly dominated by small cladocerans (*Moina*, *Ceriodaphnia*, *Diaphanosoma*) and small calanoid copepods.

Morphology of ephippia

In order to use an identification key based on ephippium morphology, a preliminary introduction to morphological characteristics of ephippia and features diagnostic for the different subgenerea is useful. Figure 2 shows the ultrastructural characteristics of ephippia and diagnostic features needed for identification. Depending on the species and subgenus some characteristics are absent or present on the ephippium.

Ephippia of species belonging to the subgenus *Ctenodaphnia* (except for species formerly belonging to the genus *Daphniopsis*, Hebert, 1995) are characterised by a pair of anterior appendages, each with a row of spines, which are derived from the dorsocephalic suture between the head shield and the carapace (Fig. 2). In most species (subgenera *Hyalodaphnia*, *Daphnia* and some *Ctenodaphnia*) the ephippial capsule reaches the ventral edge of the carapace, at the base of the terminal spine. In other

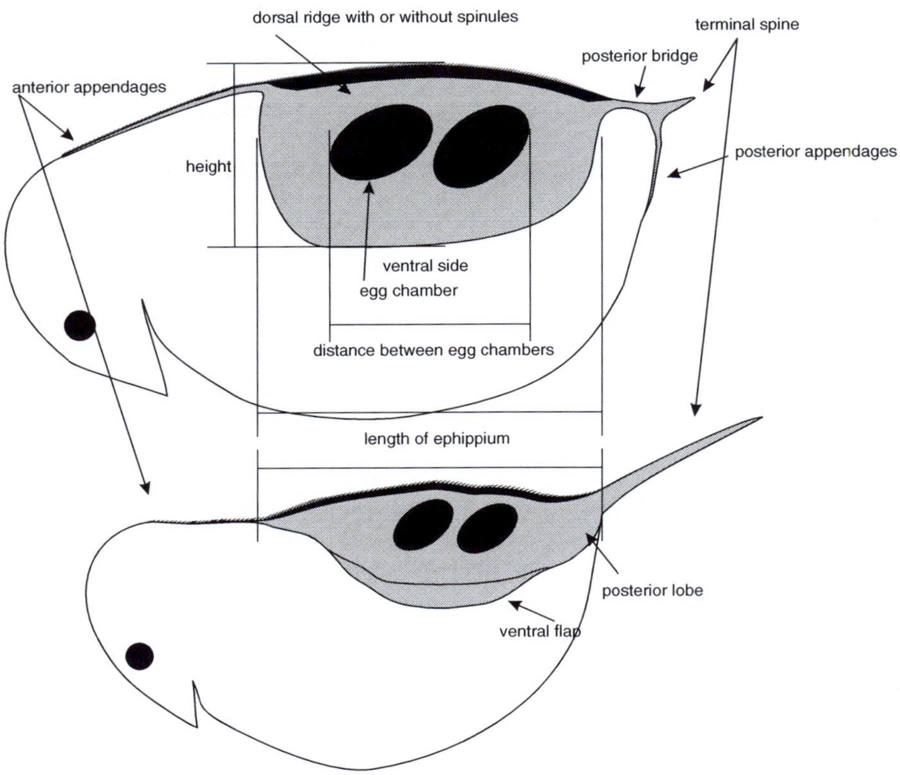

Figure 2. Morphological descriptors of ephippia.

Ctenodaphnia species (e.g., *D. lumholtzi*, *D. magna*, *D. similis*) the ephippial capsule does not reach the edge of the maternal carapace; in these cases the terminal spine of the carapace is connected to the ephippium by a narrow posterior bridge (Fig. 2a), or is not connected at all (e.g., *D. lumholtzi*). In a few species (e.g., *D. magna*) the sclerotised ventro-posterior margin of the maternal carapace is retained after shedding of the ephippium, forming the ventro-posterior appendages (Fig. 2a).In some species the two valves of the ephippium are always open at the ventral side, forming two ventral flaps standing outward, showing the internal cuticula surrounding the embryos (e.g., *D. barbata*). In these species the ephippium reaches the ventral margin of the maternal carapace (Fig. 2b). In *Ctenodaphnia* the two egg chambers are positioned obliquely relative to the dorsal margin, sometimes almost horizontally. Spinules are always present on the dorsal ridge (except for species formerly belonging to the genus *Daphniopsis*, Hebert, 1995) and are mostly large. At least six *Ctenodaphnia* species are present in equatorial East Africa: *D. barbata*, *D. dolichocephala*, *D. lumholtzi*, *D. magna*, *D.* cf. *similis* and *Daphnia* sp. nov. type Limuru).

Representatives of the subgenus *Hyalodaphnia* mostly have a smooth dorsal ridge. The egg chambers are positioned more or less perpendicularly to the dorsal ridge, but can be tilted a little towards the anterior side as well as the posterior side. The anterio-ventral corner is often bulging out a little beyond the dorsal ridge (Fig. 2d). No appendages are present. The terminal spine is often very long, but breaks off easily. At least three species are present: *D. curvirostris*, *D. longispina s.l.* and *D. laevis*.

Members of the subgenus *Daphnia s.s*, mostly have a dorsal ridge with relatively small but numerous spinules. A few species have, however, a smooth dorsal ridge (Hebert, 1995). The egg chambers are positioned more or less perpendicularly to the dorsal ridge. The anterior side is mostly straight, in some species bulging out like in many *Hyalodaphnia*. Only two species of this subgenus have been found in East Africa: *D. obtusa* Kurz and *D. pulex*.

Identification key to the ephippia of Daphnia *species found in Kenya*

1 (a) Ephippium with anterior appendages, however often broken off in damaged specimens. Egg chambers positioned obliquelyrelative to the dorsal ridge. Dorsal ridge with (large) spinules → 6 *(Ctenodaphnia)*

(b) Ephippium without anterior appendages. Egg chambers positioned more or less perpendicular to the dorsal ridge. Dorsal ridge with or without spinules. If present, relatively small. → 2

2 (a) Dorsal ridge with spinules → 3

(b) Dorsal ridge smooth, without spinules; L/H ratio >1.4 (1.3). Posterior end of ephippium body strongly tapered. The anterior end ften

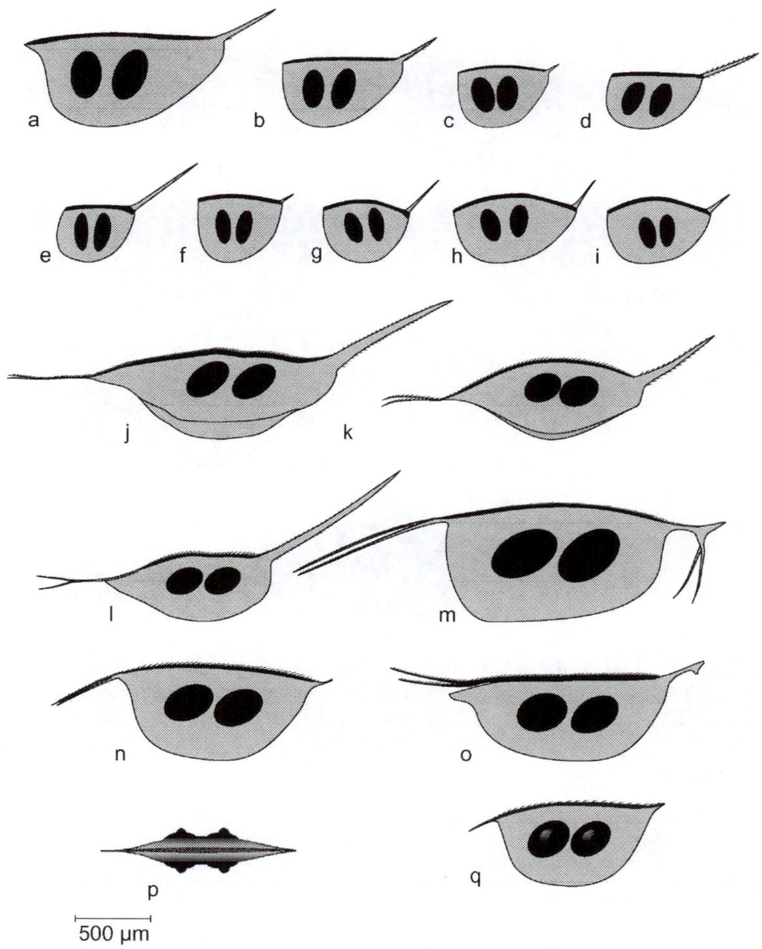

Figure 3. Schematic drawings of ephippia of *Daphnia* species found in Kenya. In brackets the locality from which the individuals originate that were used for the drawings. a, b: *D. pulex* (Lake Naivasha, Kenya); c: *D. obtusa* (Lake Limuru2, Kenya); d: *D. longispina* s.l. (Lake Naivasha, Kenya); e: *D. laevis* (Lake Naivasha, Kenya); f–i: *D. curvirostris*, different morphs. Only morph f found so far in Kenya (Lake Naivasha) (g–i: Czech republic, collection A. Petrusek); j: *Daphnia* sp. nov. (Limuru, Kenya); k: *D.dolichocephala* (Bandasa Lake, Mt Kenya); I: *D. barbata* (Lake Naivasha, Kenya); m: *D. magna* (Lake Naivasha, Kenya); n, o: D. *similis*, different morphs (n: Jordan, collection Prof H. Dumont; o: Germany, collection A. Petrusek); p: *D. lumholtzi*, dorsal view (Lake Naivasha, Kenya); q: *D. lumholtzi* (Lake Naivasha, Kenya).

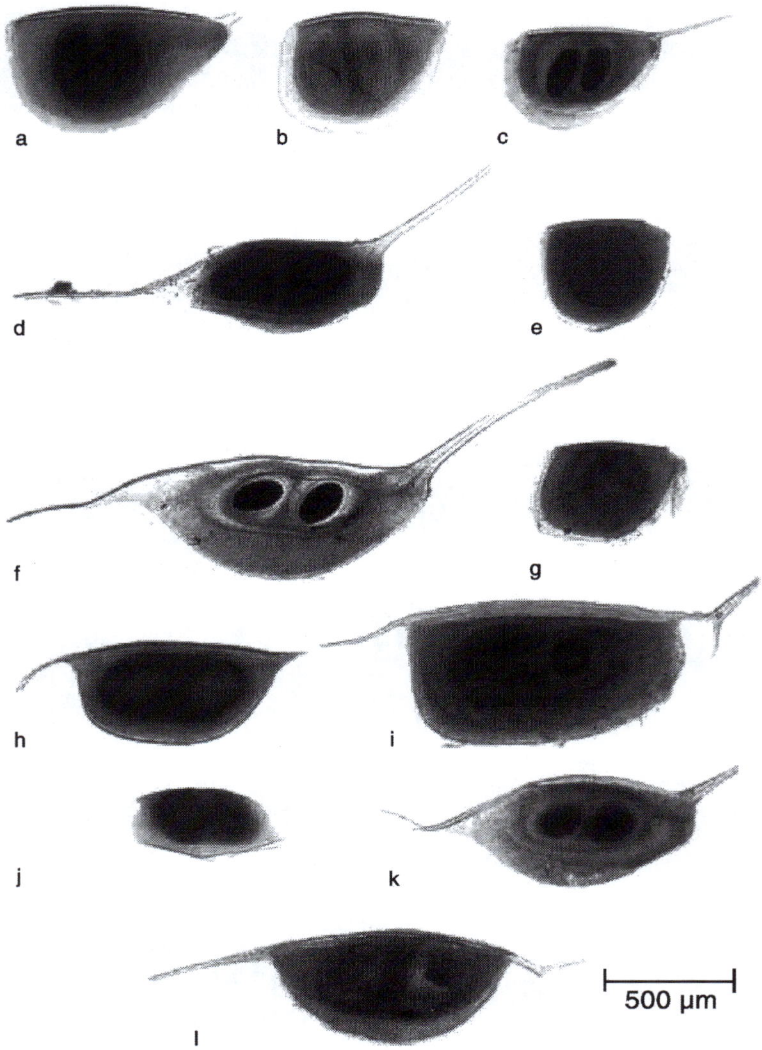

Figure 4. Photographs of ephippia *of Daphnia* species found in Kenya. a: *D.* pu/ex(Lake Naivasha, Kenya); b: *D. obtusa* (Lake Limuru2, Kenya); c: *D. longispina* s.l. (Amboseli swamps, Kenya); d: *D. barbata* (Lake Naivasha, Kenya); e: *D. curvirostris* (Lake Naivasha, Kenya); f: *D.* sp. *nov*. (Limuru, Kenya); g: *D. laevis* (Lake Naivasha, Kenya); h: *D. lumholtzi* (Lake Naivasha, Kenya); i: *D. magna* (Lake Naivasha, Kenya); j: *D. dolichocephala* (Lake Bandasa, Mt Kenya), internal cuticle; k: *D. dolichocephala* (Bandasa Lake, Mt Kenya); I: *D. cf. similis* (Kassala, Sudan, collection H. Dumont).

bulges out a little. → *D. longispina s.l.* (Figs. 3d and 4c).

3 (a) Largest spinules very small, often only visible microscopically (~1–5 µm). Ephippium hardly tapered towards posterior end (terminal spine). → 4

(b) Largest spinules clearly visible at low magnifications (>5 µm), very numerous. Anterior side never bulging out clearly. Ephippium always tapered → 5

4 (a) Anterior side clearly bulging out. Dorsal ridge highest towards posterior end. → *D. laevis* (Figs. 3e and 4g)

(b) Anterior margin never bulging out clearly, perpendicular to dorsal ridge or slightly oblique. Dorsal ridge not higher posteriorly. → *D. curvirostris* (Figs. 3f and 4e)

5 (a) L/H ratio 1.4–2.1, mostly >1.5. Dorsal margin only slightly convex or straight. Very variable in shape and size (0.6–1.7 mm). Dorsal ridge in large ephippia sometimes well extended anteriorly beyond rest of ephippium, making the anterior margin sigmoidal (Fig. 2b). → *D. pulex* (Figs. 3a, b and 4a)

(b) L/H ratio < 1.4. Rather small. Dorsal margin often a little convex, with a twist above the posterior egg chamber. Tapering of posterior end less pronounced than in *D. pulex* (compare Fig. 3b and c), and terminal spine always very short. → *D. obtusa* (Figs. 3c and 4b)

6 (a) Ephippium with small protuberances on both sides of each egg chamber (Fig. 3p). Ephippium without terminal spine; Anterior appendages short (~/4 of total length of ephippium) and curved downwards. Margin of posterior side elongated and sigmoidal; margin of anterior side sigmoidal; ventral margin rounded. Very few (25–50) but large (20–35 μm) spinules on dorsal ridge. L/H ratio 1.8–2.4. → *D. lumholtzi* (Figs. 3p, q and 4h)

(b) No bumps on lateral side of egg chambers. Anterior appendages longer than 1/4 of total length of ephippium. → 7

7 (a) Ephippium with posterior bridge and terminal spine not directly connected to the ephippium (Fig. 2a). Posterior appendages present or absent → 8

(b) No posterior bridge or posterior appendages present. Terminal spine directly connected to the ephippium (Fig. 2b). Ephippium elongated, anterior side tapered, posterior side with a lobe → 9

8 (a) Posterior appendages present. Very large, 1.0–1.7 mm. Rectangular shape, but greatest height nearly always at anterior end. L/H ratio 1.6–2.0. Dorsal ridge straight or slightly convex, very densely spined (>80 spinules per row). → *D. magna* (Figs. 3m and 4i)

(b) Ephippium without posterior appendages, always broken off at the base of the terminal spine near the bridge. Large to very large (0.9–1.5 mm). Greatest height mostly in the middle or near the posterior end of the ephippium. L/H ratio (1.9) 2–2.4. Dorsal margin with large spinules, < 80 per row. Anterior side sigmoidal, posterior side elongated sigmoidal. ventral side clearly rounded, ephippium not with rectangular shape, the top of the anterior side is often detached from the dorsal margin (Fig. 3o) → *D. similis* (Figs. 3n, o and 4l)

9 (a) Dorsal side of inner cuticle surrounding embryos convex and without depression between eggs, ventral side more or less straight (Fig. 4j). Dorsal ridge of ephippium clearly curved and convex. Large spines on dorsal margin, < 60 per row. Large ventral flaps present (20–35% of total heigth). ventro-anterior margin well defined (compare with *D. barbata*); L/H ratio 2.2–2.7 → *D. dolichocephala* (Figs. 3k and 4j, k) (in some cases *D. barbata* has a straight or slightly convex dorsal margin, without depression; check ventro-anterior margin and size of ventral flaps).

(b) Dorsal margin undulated due to depression between egg chambers, sometimes only very slightly, in rare cases straight or slightly convex. Inner cuticle surrounding embryos with strong depression between eggs (Fig. 4f) → 10

10 (a) Dorsal margin with very large (40–60 μm) but relatively few spines (20–60 per row). Ventro-anterior margin ill defined beyond the anteriormost egg chamber, very thin and translucent, often damaged (Figs. 3l and 4d). Ratio L/H 2–2.4. Ratio L/(total length of egg chambers) 1.5–2. Ventral flaps not so large, max. 25% of total height (mostly < 15–20%). total length of ephippium 0.6–1.1 mm → *D.* banbafa (Fig. 3l)

(b) Dorsal margin with 50–80 spines per row, mostly smaller than those of *D. barbata* (30–50 μm). ventro-anterior margin clearly defined, not thinner than other parts of ephippium (Fig. 4f). Anterior side elongated sigmoidal. Ratio L/H 2.4–2.7. Ratio L/(total length of egg chambers) 2–2.7. Ventral flaps very large, easily 30% of total heigth and often standing outwards (Fig. 4f). total size 1–1.6 (1.8) mm. → *D. nr. gibba* (Figs. 3j and 4f)

Discussion

Taxonomic considerations

Due to the recent accumulated knowledge of genetic relationships within the genus *Daphnia*, the taxonomy of this genus is not fully stabilised. One is often dealing with groups of sibling species, without consideration of the relationships with members of the type population. This is the case for many species considered in this paper, like *D. obtusa*, which is thought to consist of more than 12 sibling species worldwide (Hebert & Finston, 1996; Adamowicz et al., 2004; A. Petrusek, personal communication; V. Korinek, perssonal communication). Unless mentioned otherwise, identification in the key is solely based on Kenyan populations. Identification of *Daphnia obtusa* in this paper was based on Belgian populations because only one ephippium from Kenya was found, and no African populations were available. Daphnia *laevis* equally consists of more than four sibling species (Taylor et al., 1996; Adamowicz et al., 2004) while the African *D. laevis* is morphologically and genetically distinctive from American siblings (J. Mergeay, unpublished results). *Daphnia lumholtzi* also consists of a complex of at least two but probably four sibling species (Havel et al., 2000; Korinek, personal communication). Equally, *Daphnia pulex* consists of genetically very distinct sibling species (Colbourne et al., 1998), while *D. similis* is made up of at least three and probably five sibling species (A. Petrusek, personal communication). Identification of ephippia of *D. similis* was based on ephippia from Jordan (collection H. Dumont) and Germany (collection A. Petrusek). Regarding the species group of *D. hyalina*, *D. galeata*, *D. longispina* and *D.* rosea, which were all previously recorded in East-Africa (Lowndes, 1936; Green, 1995; Schwenk et al., 2000), a lot of confusion exists regarding the exact identity of these species (e.g. Schwenk et al., 2000). They are therefore recorded here as *D. longispina s.l.*

Given the uncertain taxonomic status of the sibling species mentioned above, we must stress that many of the species considered are in fact different from the sibling species from the type locality, or have an uncertain taxonomic relationship with specimens from the type locality.

The presence of Daphnia *in the active vs. dormant community*

More than twice as many species were found in the dormant community compared to the active community. This is consistent with other studies that assessed both the active and the dormant community (May, 1986; Havel et al., 2000; Vandekerkhove et al., submitted), confirming that species are easily overlooked when only the active community is sampled. Identification of dormant specimens is possible by hatching them and culturing them until the active specimens can be identified (Vandekerkhove et al., submitted), but this approach is time-consuming and elaborate. Identification of ephippia based on morphology, on the other hand, is fast and can also be practiced on damaged or unviable ephippia (Vandekerkhove et al., submitted). In case of doubt, hatching and identification of adults can still be done. Moreover, modern genetic techniques allow accurate species assignments based on minute amounts of DNA, and can be used to confirm morphological identifications (e.g., Duffy et al., 2000).

In all but one of the studied water bodies in which *Daphnia* was found in the active community, the same species were also found in the dormant community. This indicates that sexual reproduction readily occurs in equatorial East Africa and that the dormant community actually represents the species found in the active community. In a few situations, however, sexual reproduction is very rare; in Lake Victoria, e.g., ephippia of *Daphnia are* rarely found in the sediment (D. Verschuren, personal observation), yet at least three species are present (Jonna & Lehman, 2002). This is probably due to the state of 'endless summer' which prevails in such stable tropical lakes that are hardly influenced by seasonal changes (Kilham & Kilham, 1990; Lehman, 1996). It is therefore advised to sample both the dormant and the active community.

In many of the studied water bodies, *Daphnia* were very scarce, in the active as well as in the dormant community, with only a few individuals per sample. On the other hand, *Daphnia* was found in a high number of water bodies (~70%). A few lakes had four or more species of *Daphnia*. The sediment record of Lake Naivasha even shows that at least eight species have been present in Lake Naivasha since 1900 AD (Mergeay et al., submitted), indicating that dispersal is probably not the most important limitation. Most water bodies studied in which *Daphnia* were present, however, contained only 1–2 species.

Geographic distribution of Daphnia *species and ecological considerations*

Three species of *Daphnia* are considered typical for tropical African lowlands: *D. barbata*, *D. laevis* and *D. lumholtzi*, *D. laevis* and *D. barbata are* both widespread in sub-Saharan Africa but are absent from the central African tropical rainforests (Dumont, 1979; Green & Kling, 1988; Green, 1990; Egborge et al., 1994, Hart, 1999, 2001;). *D. barbata* is also found in the Nile system and a few North African water bodies (Dumont, 1979). *D. barbata* is ubiquitous in turbid and productive water bodies, with conductivities ranging from < 100 μS cm^{-1} to at least 5000 μS cm^{-1} (Mergeay et al., in preparation). *D. laevis* is typically found in relatively clear water bodies, but can also occur in more turbid systems (Secchi depth 0.4 m). Conductivity tolerance of *D. laevis* is probably less than 1000 μS cm^{-1}. *D. lumholtzi* is typically found in large lakes (Victoria, Albert, Edward, Naivasha, Tchad & Green, 1967; Dumont, 1979) in tropical Africa. It is common throughout the Nile basin (Sudan, Egypt), being found in ponds as well as in rivers (Dumont, 1979; Elhigzi et al., 1995). It is further distributed in Australia, South-Asia and Eastern Europe (Dumont, 1979), and was recently introduced in North America (Havel & Hebert, 1993). The Australian populations are however genetically significantly different from Asian and African populations (Havel et al., 2000). *D. lumholtzi* can withstand relatively high turbidity levels (Work & Gophen, 1999), but probably not as high as the levels *D. barbata* can tolerate. Just like *D. barbata* it can grow at temperatures up to 30 °C (Work & Gophen, 1999). *D. dolichocephala* is a rare and poorly known species. Up to now it has, in tropical Africa, only been found in water bodies on Mt. Kenya (Löffler, 1968). In the same water bodies, we found ephippia most probably belonging to this species. It is further found in southern Africa (Sars, 1895; Green, 1995), and is probably endemic to Africa. The relations between the different populations on the African continent (Kenyan vs. South-African) are however uncertain (V. Korinek, personal communication). *D. magna* is a typical temperate species that has a mainly Holarctic distribution (Hebert, 1995), but is also present in numerous North African temporary water bodies (Dumont, 1979). It is rather rare in East Africa but is yet found in a variety of habitats, ranging from alpine lakes on Mt. Kilimanjaro (Löffler, 1968) to large (eutrophic) lakes (Lake Naivasha, Lake Victoria, Lake Ol Bolossat) and eutrophic ponds with sewage waste water in Nairobi (B. Riddoch, personal communication). *D. magna* can be expected to be found more in southern Africa as well, as suggested by its presence in Botswana (B. Riddoch, personal communication). Competition with other species better adapted to high temperatures and high turbidity levels may however limit its distribution. *D. cf. similis* was found only once in Lake Oloidien in the 1990s as a single ephippium (D. Verschuren, personal observation). Its range in tropical Africa is poorly known. It is, however, common in the hydrological basin of the Sahara and Sahel, but is less resistant to high salinity levels than *D. magna* (Dumont, 1979). *D. pulex* is a nearly cosmopolitan species, but consists in fact of a species complex of which the internal boundaries are not yet completely known (Colbourne et al., 1998). It is uncommon in tropical Africa, occurring from 900 m a.s.l. (Lake Baringo) up to more than 3000 m (Green, 1995). Its range in Africa is not completely known, but its common occurrence in southern Africa (Green, 1995; Hart, 2001) and Zimbabwe (Green, 1990) suggests it may be widespread in permanent water bodies in subtropical regions and at higher latitudes and altitudes. *D. pulex* is found at conductivities up to 1000–1500 μS cm^{-1} and also withstands relatively high turbidity levels, given its common presence in Lake Naivasha today (Secchi depth ca. 25 cm in January 2003). It was also present at low densities in a sediment sample from Lake Baringo in 2001, but it was absent in sediment samples taken in 2003. The current turbidity and conductivity levels in Lake Baringo (Secchi depth < 5 cm; conductivity 1500–2000 μS cm^{-1}) may have caused the extinction of this species. D. *curvirostris* is a typical temperate species. In this study it was only found in two high-altitude clear lakes. It was present in

Lake Naivasha, however, at least between 1940 and 1955 (Mergeay et al., submitted). Apart from East Africa (Green, 1965, 1995) it has a mainly Holarctic distribution, occurring throughout Eurasia as well as northwestern Canada (Duffy et al., 2000). In tropical Africa it seems restricted to high altitudes (1800–4000 m; Green, 1995).

Direct and indirect effects of temperature are probably key factors limiting the distribution of *D. curvirostris* in the tropics. In our samples only one morphotype of *D. curvirostris* was found (Fig. 3f). There are, however, several very different morphs, more resembling *D. obtusa*, occurring in Europe. They might be found in Africa as well, and are represented in Figs 3g and h. *D. obtusa* was found only once during this study at an altitude of approx. 2300 m, but was previously recorded on Mt. Kenya and in the Ruwenzori mountain range (Uganda) at high altitudes (Löffler, 1968). Other members of this species complex are found in the Holarctic, Southern African and South-America. *D. longispina s.l.* was found only once in this study, in a relatively cool (20 °C) water body with abundant submerged and emergent vegetation, but only in the resting egg bank (Table 1). It was, however previously found in the sediment record of Lake Naivasha (Mergeay et al., submitted), as well as in a high altitude lake in the Ruwenzori range (J. Mergeay, personal observation.). It has further been found in a number of large East African lakes (Lake Victoria, Lake Edward and Lake Chad, and in Lake Tana, Ethiopia as *D. hyalina;* Green, 1965, 1995; Dumont, 1979; Schwenk et al., 2000; Jonna & Lehman, 2002). *Daphnia* sp. nov. type Limuru is a new species of *Ctenodaphnia* (V. Kořinek, personal communication). It was found in four water bodies in the active as well as the dormant community, at altitudes above 2200 m a.s.l. Additionally there are two isolated historical records of its presence in Lake Naivasha (~1900 m a.s.l.) (Mergeay et al., submitted). Except for one case, the water bodies in which this species was found were very turbid. In conclusion, the *Daphnia* fauna of Equatorial East Africa apparently is *a* mixture of typical African species (*D. laevis*, *D. barbata*, *D. lumholtzi*, *D. dolichocephala*, *D.* sp. nov. type Limuru), nearly cosmopolitan species (*D. pulex*) and Holarctic/Palearctic species that are restricted in the tropics to cold refugia at higher altitudes (*D. curvirostris, D. obtusa* species complex, *D. longispina s.l.*, *D. magna*).

Acknowledgements

This study was carried out with permission of the Permanent Secretary of the Ministry of Education, Science and Technology of Kenya under research permit 13/001/11C to D.V., and funded by the Katholieke Universiteit Leuven (project OT/00/14) and the Fund for Scientific Research – Flanders, Belgium (project G0086.00). The authors would like to thank Kenneth Mavuti (University of Nairobi, Kenya), William Shivoga (Egerton University, Kenya), Sarah Higgins (Kijabe Ltd., Naivasha), the Lake Naivasha Riparian Association, the people at Kampi Ya Samaki (Baringo) and the Lake Naivasha Yacht Club for logistic support, Hilde Eggermont and Katrien De Maeyer for assistance in the field, Philip Barker (Lancaster University, UK) for sediment samples from four lakes on Mt. Kenya, Adam Petrusek (University of Prague, Czech Republic) for help with ephippial morphology, Prof Vladimir Kořinek (University of Prague, Czech Republic) for help with identifications and for useful comments and corrections as a reviewer of the manuscript, and Prof Henri Dumont for access to his collection of African zooplankton samples. An anonymous reviewer provided useful comments on the manuscript. We are indebted to the Kenyan Wildlife Service for sampling permission in the National parks. J.M. is supported by a PhD fellowship from the Institute for the Promotion and Innovation of Science and Technology in Flanders, Belgium (IWT); D.V. is post-doctoral fellow with the Fund for Scientific Research (FWO) – Flanders, Belgium.

References

Adamowicz, S. J., P. D. N. Hebert & M. C. Marinone, 2004. Species diversity and endemism in the *Daphnia* of Argentina: a genetic investigation, Zoological Journal of the Linnean Society 140: 171–205.

Cáceres, C. E., 1998. Interspecific variation in the abundance, production, and emergence of *Daphnia* diapausing eggs, Ecology 79: 1699–1710.

Carvalho, G. R. & R. N. Hughes, 1983. The effect of food availability, female culture-density and photoperiod on ephippia production in *Daphnia magna* Straus (Crustacea, Cladocera), Freshwater Biology 13: 37–46.

Claska, M. E. & J. J. Gilbert, 1998. The effect of temperature on the response of *Daphnia* to toxic cyanobacteria, Freshwater Biology 39: 221–232.

Colbourne, J. K., T. J. Crease, L. J. Weider, P.D.N. Hebert, F. Dufresne & A. Hobaek, 1998. Phylogenetics and evolution of a circumarctic species complex (Cladocera: *Daphnia pulex*). Biological journal of the Linnean Society 65: 347–365.

Crispim, M. C. & T. Watanabe, 2001. What can dry reservoir sediments in a semi-arid region in Brazil tell us about Cladocera? Hydrobiologia 442: 101–105.

Duffy, M. A., L. J. Perry, C. M. Kearns, L J. Weider & N. G. Hairston, 2000. Paleogenetic evidence for a past invasion of Onondaga Lake, New York, by exotic *Daphnia cutvirostris* using mtDNA from dormant eggs. Limnology and Oceanography 45: 1409–1414.

Dufresne, F. & P. D. N. Hebert, 1994. Hybridization and origins of polyploidy. Proceedings of the Royal Society of London Series B – Biological Sciences 258: 141–146.

Dumont, H. J., 1979. Limnologie van Sahara en Sahel: bijdrage tot een beter begrip van de klimaatsveranderingen van het Laat-Pleistoceen en Holoceen. Thesis, University of Ghent.

Dumont, H. J., 1994. On the diversity of the Cladocera in the tropics. Hydrobiologia 272: 27–38.

Egborge, A. B. M., C. C. Onwudinjo & P. C. Chigbu, 1994. Cladocera of coastal rivers of western Nigeria. Hydrobiologia 272: 39–46.

Elhigzi, F. A. R., S. A. Haider & P. Larsson, 1995. Interactions between Nile tilapia (*Oreochromis niloticus*) and cladocerans in ponds (Khartoum, Sudan), Hydrobiologia 307: 263–272.

Fernando, C. H., J. C. Paggi & R. Rajpaksa, 1987. *Daphnia* in tropical lowlands. In Peters, R. H. & R. De Bernardi (eds), *Daphnia*, 45. Memorie dell'Instituto Italiano di Idrobiologia, Verbania: 107–141.

Flössner, D., 1972. Kiemenund Blatfusser, Branchiopoda; Fischlause, Branchiura. Die Tierwelt Deutschlands 60: 1–501.

Green, J., 1967. The distribution and variation of *Daphnia lumholtzi* (Crustacea: Cladocera) in relation to fish predation in Lake Albert, East Africa, Journal of Zoology 151: 181–197.

Green, J., 1971. Associations of Cladocera in the zooplankton of the lake sources of the White Nile, Journal of Zoology 165: 373–414.

Green, J., S. A. Corbet & E. Betney, 1974. Ecological studies on crater lakes in West-Cameroon – Debundsha Lake. Journal of Zoology 173: 199–223.

Green, J. & G. W. Kling, 1988. The Genus *Daphnia* in Cameroon, West-Africa, Hydrobiologia 160: 257–261.

Green, J., 1990. Zooplankton Associations in Zimbabwe. Journal of Zoology 222: 259–283.

Green, J., 1995. Altitudinal distribution of tropical planktonic cladocera. Hydrobiologia 307: 75–84.

Hann, B. J., P. R. Leavitt & P. S. S. Chang, 1994. Cladocera community response to experimental eutrophication in Lake-227 as recorded in laminated sediments, Canadian Journal of Fisheries and Aquatic Sciences 51: 2312–2321.

Hart, R. C., 1999. On the limnology of Spioenkop, a turbid reservoir on the upper Thukela river, with particular reference to the structure and dynamics of its plankton community, Water SA 25: 519–528.

Hart, R. C., 2001. A baseline limnological study of Wagendrift Dam (Thukela Basin, Kwazulu-Natal). Water Sa 27: 507–516.

Havel, J. E., E. M. Eisenbacher & A. A. Black, 2000. Diversity of crustacean zooplankton in riparian wetlands: colonization and egg banks. Aq. Ecol. 34: 63–76. 2000.

Havel, J. E. & P. D. N. Hebert, 1993 Daphnia lumholtzi in North America – another exotic zooplankter. Limnology and Oceanography 38: 823–1827.

Hebert, P. D. N., 1978. Population biology of *Daphnia* (Crustacea, Daphnidae) Biological Reviews of the Cambridge Philosophical Society 53 : 387–426.

Hebert, P. D. N., 1995. The Daphnia of North America: an illustrated fauna (on CD-rom).

Hebert, P. D. N. & T. L. Finston, 1996. Genetic differentiation in *Daphnia obtuse:* a continental perspective, Freshwater Biology 35: 311–321.

Jonna, R. & J. T. Lehman, 2002. Invasion of Lake Victoria by the large bodied herbivorous cladoceran *Daphnia magna*. In Odada, E. O. & D. O. Olago (eds), The East African Great Lakes: Limnology, Palaeolimnology and Biodiversity. Kluwer Academic Publishers, Dordrecht: 321–333.

Kerfoot, W. C., J. A. Robbins & L. J. Weider, 1999. A new experimental approach to historical reconstruction: combining descriptive and experimental paleolimnology. Limnology and Oceanography 44: 1232–1247.

Kiai, W., W. Mwangi & E. Bosire, 2002. The impact of HIV/AIDS on the land issue in Kenya. Nairobi, Kenya, Forest Action Network.

Kilham, P. & S. S. Kilham, 1990. Endless summer – internal loading processes dominate nutrient cycling in tropical lakes, Freshwater Biology 23: 379–389.

Kirk, K. L., 1991. Suspended clay reduces *Daphnia* feeding rate – behavioral mechanisms, Freshwater Biology 25: 357–365.

Kitchell, J. A. & J. F. Kitchell, 1980. Size-selective predation, light transmission, and oxygen stratification – evidence from the recent sediments of manipulated lakes. Limnology and Oceanography 25: 389–402.

Kořinek, V., 1999. A Guide to the Limnetic Cladocera in African Inland Lakes (Crustacea, Branchiopoda), Science Publishers, Stuttgart, Germany.

Leavitt, P. R., S. R. Carpenter & J. F. Kitchell, 1989. Whole-lake experiments – the annual record of fossil pigments and zooplankton. Limnology and Oceanography 34: 700–717.

Lehman, J. T., 1996. Pelagic food webs of the East African Great Lakes. In Johnson, T. C. & E. O. Odada (eds), The Limnology, Climatology and Paleoclimatology of the East African Lakes. Gordon and Breach, Toronto: 281–301.

Lennon, J. T., V. H. Smith & K. Williams, 2001. Influence of temperature on exotic *Daphnia lumholtzi* and implications for invasion success. Journal of Plankton Research 23: 425–434.

Löffler, H., 1968. Die Crustaceenfauna der Binnengewasser Ostafrikanischer Hochberge. Hochgebirgsforschung 1: 107–170.

Lowndes, M. A. F. L. S., 1936. Scientific results of the Cambridge Expedition to the East African Lakes 1930–1. N°. 16. The smaller Crustacea, Linnean Journal of Zoology XL: 1–33.

May, L, 1986. Rotifer sampling – a complete species list from one visit. Hydrobiologia 134: 117–120.

Moore, M. V., C. L. Folt & R. S. Stemberger, 1996. Consequences of elevated temperatures for zooplankton assemblages in temperate lakes. Archiv Fur Hydrobiologie 135: 289–319.

Sars, G. O., 1895. On some South African Entomostraca raised from dried mud., *Videnskabs-Selskabets Skrifter*. I., Mathematisk-naturv. Klasse No. 8.

Schwenk, K., D. Posada & P. D. N. Hebert, 2000. Molecular systematics of European *Hyalodaphnia*: the role of contemporary hybridization in ancient species. Proceedings of the Royal Society of London, Series B 267: 1833–1842.

Spaak, P., 1997. Hybridization in the *Daphnia galeata* complex: are hybrids locally produced? Hydrobiologia 360: 127–133.

Taylor, D. J. & P. D. N. Hebert, 1993. Habitat-dependent hybrid parentage and differential introgression between neighboringly sympatric *Daphnia* species. Proceedings of the National Academy of Sciences of the United States of America 90: 7079–7083.

Taylor, D. J., P. D. N. Hebert & J. K. Colbourne, 1996. Phylogenetics and evolution of the *Daphnia longispina* group (Crustacea) based on 12s rdna sequence and allozyme variation, Molecular Phylogenetics and Evolution 5: 495–510.

Verschuren, D. & L. F. Marnell, 1997. Fossil zooplankton and the historical status of westslope cutthroat trout in a headwater lake of Glacier National Park, Montana, Transactions of the American Fisheries Society 126: 21–34.

Work, K. A. & M. Gophen, 1999. Factors which affect the abundance of an invasive cladoceran, *Daphnia Lumholtzi*, in US reservoirs. Freshwater Biology 42: 1–10.

Zaffagnini, F., 1987. Reproduction in *Daphnia*. In Peters, R. H. & R. De Bernard (eds), *Daphnia*, 45. Memorie dell'Instituto Italiano di Idrobiologia, Verbania: 245–284.

Hydrobiologia (2005) 542:275–286
H. Segers & K. Martens (eds), Aquatic Biodiversity II
DOI 10.1007/s10750-004-1875-1

Some aspects of water filtering activity of filter-feeders

S.A. Ostroumov
Department of Hydrobiology, Faculty of Biology, Moscow State University, Moscow 119992, Russia
E-mail: saostro@online.ru

Key words: filter-feeders, water purification, bivalves, pollution, filtering activity

Abstract

On the basis of the previous publications, our new data and the existing scientific literature, we have formulated some fundamental principles that characterize the pivotal roles of the biodiversity of filter-feeders in ecosystems. Among those roles are: (1) the role of ecological repair of water quality, (2) the role of contributing to reliability and stability of the functioning of the ecosystem, (3) the role of contributing to creation of habitat heterogeneity, (4) the role of contributing to acceleration of migration of chemical elements. It is an important feature of the biomachinery of filter-feeders that it removes from water various particles of a very broad range of sizes. Another important principle is that the amount of the organic matter filtered out of water is larger than the amount assimilated so that a significant part of the removed material serves no useful function to the organism of the filter-feeder, but serves a beneficial function to some other species and to the ecosystem as a whole. The new experiments by the author additionally demonstrated a vulnerability of the filtration activity of filter feeders (e.g., bivalves and rotifers) to some xenobiotics (tetradecyltrymethylammonium bromide, heavy metals and some others). The inhibition of the filtration activity of filter-feeders may lead to the situation previously described as that of an ecological bomb of the second type.

Abbreviations: AFDW – ash-free dry weight; LD – liquid detergent; SD – synthetic detergent (laundry detergent in the form of powder); SDS – sodium dodecyl sulphate; TDTMA – tetradecyltrimethylammonium bromide

Introduction

Quantitative data on the filtering activity of various benthic filter-feeders have been obtained (e.g., Alimov, 1981; Gutelmaher, 1986; Newell 1988; Monakov 1998 and many other authors). A significant body of data on planktonic filter-feeders was produced (e.g., Sushchenya, 1975; Zankai, 1979; Starkweather & Bogdan, 1980). We have found new evidence of the vulnerability of that activity of filter-feeders to some chemicals (Ostroumov, 1998, 2002a, b, 2003a). In order to better understand the implications of those effects, it is worth analyzing the fundamental patterns in the data on the filtration activity of aquatic organisms.

The goal of this paper is to contribute to that analysis.

In the paper we address the following topics:

1. Some examples of the filtering activity and clearance time in some water bodies;
2. Diversity of filter-feeders: its role in reliability of the biomachinery involved in water filtration;
3. The concentration of particles: effects on filtration rates;

4. Production of pellets by filter-feeders. Role for the ecosystem;
5. Filter-feeders and regulation of the ecosystem;
6. Author's new experimental data on effects of chemicals on filter-feeders.

Examples of the filtering activity and clearance time in some water bodies

Among many examples of quantitative studies of various aquatic species, we mention here only a few examples. Thus, the filtration activity of *Eudiaptomus gracilis* (copepod) in Lake Balaton was in winter 0.11 ml animal^{-1} day^{-1} and in summer 1.44 ml animal^{-1} day^{-1}; their maximal activity was as high as 3.27 ml animal^{-1} day^{-1} (Zankai, 1979). The filtration activity of *Keratella cochlearis* was 5–6 μl animal^{-1} h^{-1} (Starkweather & Bogdan, 1980). A higher activity (5–30 μl animal^{-1} h^{-1}) was reported for *Brachionus calyciflorus* when they were feeding in the laboratory on ciliates (*Coleps* sp.) (see: Ostroumov et al., 2003b). Some other examples are in Tables 1 and 2 below.

The filtration rate for a number of benthic organisms was as high as 4.7–10.2 l g^{-1} h^{-1} (dry body weight; see Dame et al., 2001).

In detailed studies it was shown that the filtration rate FR (l h^{-1}) increased with increasing body weight W (dry weight of the soft parts) according to the equation (Sushchenya, 1975; Alimov, 1981; Kryger & Riisgård, 1988)

$$\mathrm{FR} = aW^b$$

The data for a and b for many species were determined (e.g., Alimov, 1981; Kryger & Riisgård, 1988). Thus, the a value for *Dreissena polymorpha* was 6.82 (feeding on *Chlorella vulgaris*), and for *Sphaerium corneum*. 2.14 (also feeding on *Chlorella vulgaris* (Kryger & Riisgård, 1988). As for b values, they are often within the range from 0.46 to 0.92 (Kryger & Riisgård, 1988).

Studies by many authors have shown that the total filtering activity of benthic populations of bivalves (e.g., Unionidae, Dreissenidae) is often within the range from 0.1 to 5 m^3 m^{-2} day^{-1} (see: Ostroumov, 2003a).

Table 1. Examples of the impact of fiter-feeders on the water column: clearance time

System	Organisms	Per what period of time the total volume of water is filtered	References
A *Sphagnum* bog-pond, the areas of high biomass of sponges 31.8 g m^{-2}	*Spongilla lacustris* (sponge)	Less than 24 h	Frost (1978)
Lake Tuakitoto (New Zealand)	*Hyridella menziesi* (bivalve)	32 h	Ogilvie & Mitchell (1995)
Königshafen 7.2 × 10^6 m^3	Among dominants *Mytilus edulis*, *Crassostrea virginica*	0.9–2.8 days	Asmus & Asmus (1991)
North Inlet (South Carolina, USA) 22 × 10^6 m^3	Among dominants *C. virginica*	0.8–6.1 days	Dame et al. (1980)
South San Francisco Bay 2500 × 10^6 m^3	Among dominants *Potamocorbula amurensis*	0.6 days	Cloern (1982)
Narragansett Bay 2724 × 10^6 m^3	Among dominants *Mercenaria mercenaria*	32.1 days	Pilson (1985)
Oosterschelde 2740 × 10^6 m^3	Among dominants *M.edulis*, *Cerastoderma edule*	3.7 days	Smaal et al. (1986)
Chesapeake Bay 27 300 × 10^6 m^3	Among dominants *Crassostrea virginica*	87.5 days	Newell (1988)
Marina da Gama 0.025 × 10^6 m^3	–	1.1 days	Davies et al. (1989)
Kertinge Nor, Denmark 11 000 × 10^6 m^3	Among dominants *Ciona intestinalis* (ascidian)	0.8–5 days	Petersen & Riisgård (1992)
Bay of Brest, France 1480 × 10^6 m^3	Multiple species	2.8–6 days	Hily (1991)

Table 2. Some examples of diversity of taxons of benthic organisms involved in removing seston from water

Organisms	Biomass or another quantitative parameter used	Filtration (removal) rate	References
Polichaeta, Sabellidae	Per 1 g of biomass (dry body weight)	0.5–1.8 l g^{-1} h^{-1} (sabellids remove particles 3–8 μm)	Dame et al. (2001)
Polichaeta, Sabellidae; *Sabella spallanzanii*	Per 1 g of biomass (ash-free dry weight; AFDW)	1 l g^{-1} h^{-1}	Dame et al. (2001)
Polichaeta, Serpulidae	Per 1 g of biomass (dry body weight)	4.7–10.2 l g^{-1} h^{-1} (they remove particles 2–12 μm)	Dame et al. (2001)
Spongia (Porifera) *Thenea abyssorum*	Per 1 g of biomass (ash-free dry weight; AFDW); in nature the biomass is upto 1524 mg AFDW m^{-2} (Norwegian-Greenland Sea, depth 2020–2630 m)	upto 12 l g^{-1} h^{-1}	Witte et al. (1997)
(sponge) *Spongilla lacustris*	Per 1 g of dry weight	2 ml g^{-1} sec^{-1}	Frost (1980)
Ascidia, *Ascidiella aspersa*	Per 1 g of ash free dry weight, AFDW	2.7 l g^{-1} h^{-1}	Dame et al., (2001)
Bryozoa, *Plumatella fungosa*	Per 1 g of dry weight	2.2 l g^{-1} h^{-1}	Monakov (1998)
Cirripedia, *Balanus crenatus*	Per 1 g of AFDW	2 l g^{-1} h^{-1}	Dame et al. (2001)
Mollusca, *Ostrea edulis*	Per 1 g of AFDW)	Upto 8.8 l g^{-1} h^{-1}	Dame et al. (2001)
Decapoda, *Porcellana longicornis*	Per 1 animal (crab)	0.1–0.27 l $crab^{-1}$ h^{-1}	Achituv & Pedrotti (1999)
Echinodermata, Ophiuroidea, *Ophiothrix fragilis*	Per 1 g of AFDW	10.4 l g^{-1} h^{-1}	Dame et al. (2001)
Corals, *Alcyonium digitatum*	Per 1 g of biomass	0.16 mg C g^{-1} h^{-1} (the suspension filtered: the culture of diatoms *Skeletonema costatum*)	Migne & Davoult (2002)
Brachiopoda, *Laqueus californianus*	Up to 3000 animals m^{-2}; depth: 2–1600 m	No data	Krylova (1997)
Brachiopoda, *Diestothyris frontalis*	Up to 900 animals m^{-2}; Depth 0–435 m,	No data	Krylova (1997)

It is important that the filtering activity is so high that the total volume of many water bodies is filtered within a time period of from 1–2 days to some dozens of days. Even if only benthic filter-feeders are taken into account, that time period is from one day to several dozens of days (see Table 1). If only planktonic filter-feeders are considered, similar results were reported (Gutelmaher, 1986).

Diversity of filter-feeders and its role for reliability

The biodiversity of organisms that are involved in filter-feeding covers uni- and multi-cellular, planktonic and benthic, freshwater and marine organisms (Alimov, 1981; Krylova, 1997; Monakov, 1998; Dame et al., 2001; Shuntov, 2001). Here we consider only invertebrates and mainly benthic organisms. However, in order to indicate the broad biodiversity involved, some examples of vertebrates could be mentioned: filter-feeding is a way of trophic activity of some fish (Clupeidae and many others) and even birds (flamingo, Phoenicopteridae).

Filtration rate was measured for a number of taxons of invertebrates (Alimov 1981; Kryger & Riisgård, 1988 and others). Some examples for benthic filter-feeders are given in Table 2. It is

Table 3. Effect of the increase in concentration of algae (*Chlorella vulgaris*) on the filtration rate and the amount consumed (C, %) by rotifers *Brachionus calyciflorus*

Level of the concentration of algae no.	Biomass of algae		Consumption of algae per day by one animal	
	(mg l^{-1}) A	A as % of that at level 1	% of the weight of the animal, B	B as % of that at level 1
1	38	100	31	100
2	75	197	37	119
3	150	397	72	232
4	300	789	59	190
5	450	1184	64	206

Calculated by the author on the basis of the data presented in Monakov (1998).

interesting that for a number of taxons the filtering rate is as high as up to 10 l g^{-1} h^{-1} (dry weight of the filter-feeder).

Such filter-feeders as brachiopods are important but the quantitative data on their filtration rates are not available. They are important in both shallow habitats (e.g., *Tythothyris rosimarginata*, 5–25 m) and deep bottom habitats – e.g., *Pelagodiscus atlanticus* (366–5530 m), *Multitentacula* (*Dentaria*) *amoena* (found at 5060 m), *Bathyneaera hadalis* (2970–8430 m) (Krylova, 1997).

The diversity of organisms involved in water filtration makes the entire ecosystem activity in water filtration and seston removal more reliable. It is interesting to note that each of the two main parts of the aquatic ecosystem (the pelagic and benthic parts) has several taxons of filter-feeders; it is noteworthy that many of the benthic filter feeders (e.g., bivalves) produce larvae that constitute another element of pelagic filter-feeders. The various taxons of filter feeders may have different food requirement and complement each other. However their ecological niches partially overlap, which increases the reliability of the entire water filtering activity of aquatic biota en bloc.

The concentration of particles: effects on filtration rates

Studies of filtering activities at various concentrations of particles were performed (e.g., Schulman & Finenko, 1990; Monakov, 1998). At relatively low concentration of particles, the dependence of the filtration rate on the concentration of particles is usually linear or almost linear. At the concentration of particles above certain threshold, a nonlinear dependence of the filtering rate on the concentration of particles was found in many cases (Schulman & Finenko, 1990). One of many examples of studies of that type is illustrated by the data of Table 3.

In several studies it was shown that the higher the concentration of suspended particles (the cells of algae, etc.), the lower is the filtration activity and (in relative terms) the grazing pressure on seston.

That pattern of the filtering activity and consumption activity was reported in many other cases. Among them are *Mytilus galloprovincialis* (feeding on *Gymnodinium kowalevskii*) (Shulman & Finenko, 1990), *Brachionus calyciflorus* (feeding on *Langerheimia ciliata, Scenedesmus acuminatus*), *Daphnia magna, D. rosea, Diaphanosoma brachyurum* and other filter-feeders (Sushchenya, 1975; Monakov, 1998).

It means that when for some reason the concentration of phytoplankton increases above certain threshold, the grazing pressure on it – at least in some cases – might not be increased proportionately, which means that some relative decrease of the control of the phytoplankton takes place. It might open the way for further growth of the phytoplankton. That pattern of cause and effect is an exemplification of the system of positive feedbacks. A detailed discussion of the role of positive feedbacks was given in (Dame et al., 2001).

As a final result, the above mentioned might contribute to the local increase in the relative phytoplankton abundances in some parts of aquatic ecosystems. The relative increase in phytoplankton changes some other parameters of the local aquatic environment (transparency of water, amount of detritus, etc.) so that the heterogeneity of the aquatic habitat increases.

The level of heterogeneity of the habitat is a very important parameter from the viewpoint of general ecology and it is necessary to take into consideration all possible mechanisms that contribute to its formation.

Production of pellets by filter-feeders. Role of the process for the ecosystem. Ecological taxation

The role of organic matter produced by filter-feeders as pellets and detritus is of so high an importance to the ecosystem that it can be viewed as a service done to ecosystem or a type of ecological tax paid by filter-feeders. To evaluate the amount of the tax or to make the picture more visible, it makes sense to compare the amount of pellets excreted and the productivity of populations of filter-feeders. That kind of comparison may be done by calculating the ratio *F*:*P* (*F* – excreted non-digested, non-assimilated organic material; *P* – productivity). It should be noted that the ratio gives only a rough (and perhaps indirect) estimate of that what might be considered the tax as the pellets are covered by a mucus layer. The mucus is a product of metabolism of invertebrates and some energy is spent on its synthesis. The formation of the mucus envelope for the pellets is another evidence of the special role of the function of pellet production. To pay the tax in an organized way, the organism sacrifices (or donates) valuable metabolic energy and effort to synthesize a special package for the material returned to the ecosystem. The mucus envelope contributes to better hydrodynamic attributes of the pellets so that they faster move downward to the deeper layers of water or to the bottom. The downward transfer of organic matter is important for the well-being and stability of the ecosystem.

What is said above generates interest in calculating the ratio *F*:*P*. An example of the calculation is given in Table 4, where the original data are part of the energy budget of the ecosystem of the central part of the Sea of Okhotsk in the summer time.

We may consider *P* (productivity) as an analogy of profit, and *F* might be viewed as an estimate of what the organism returns to the community as a tax towards some stability of environment.

Another way of calculating an estimate of ecological tax is the ratio *F*: (*P* + *R*), where *P* and *R* stands for productivity and respiration, respectively (Table 5). Again, we have to make some reservations and emphasize that this ratio is only a rough estimate as some metabolic energy was spent on the synthesis of the mucus envelope for the pellets.

Discussing the pellets, it is worth mentioning that the mucus that covers the pellets has several roles. It was shown that mucus binds various pollutants from water, including some metals (for discussion, see Ostroumov, 2001c).

As a result of pellet production, some amount of the material is deposited on the bottom. The amount of the biosediment formed was measured in a number of studies. Some examples are given in Table 6.

Table 4. The ratio *F*:*P* in some groups of organisms

Organisms	Productivity, *P*	Excreted non-digested, non-assimilated organic material, *F*	*F*:*P* (%)
Microzooplankton	1200	1000	83.33
Zooplankton (non-predatory)	1800	4000	222.22
Zooplankton (predatory)	500	600	120.00
Zoobenthos	600	1600	266.67
Fish	150	250	166.67
Mammals and birds	2	15	750.00

The ecosystem of the Sea of Okhotsk. Units: cal m^{-2} d^{-1}. Calculated by the author on the basis of the data on *P* and *F* from the work by Sorokin et al. (1997).

Table 5. The ratio $F{:}(P + R)$ in some filter feeders (calculated by the author per 1 unit of energy spent on the sum of productivity and respiration)

Organisms	$P + R$ (production + respiration)	F	$F{:}(P + R)$, %	References for the data on P, R, and F
Mytilus galloprovincialis 2 g (wet weight with shells); natural seston	4.8–51, average 25 % of calories of filtered matter	75%	300	Shulman & Finenko (1990)
M. galloprovincialis 10 g (wet weight with shells)	10–63, average 31	69%	222.58	
M. galloprovincialis 30 g (wet weight with shells)	15.4–90, average 41	59%	143.90	
Appendicularia	2.9 kcal m^{-2} d^{-1}	1.9 kcal m^{-2} d^{-1}	65.52	Vinogradov & Shushkina (1987)
Doliolidae	0.8 kcal m^{-2} d^{-1}	0.53 kcal m^{-2} d^{-1}	66.25	
Calanoida (small)	1.5 kcal m^{-2} d^{-1}	0.94 kcal m^{-2} d^{-1}	62.67	
Calanoida (large)	5.7 kcal m^{-2} d^{-1}	3.8 kcal m^{-2} d^{-1}	66.67	

Note: Vinogradov & Shushkina (1987) – the region of the upwelling (Peru, 2 miles from the seashore).

Filter-feeders and regulation of the ecosystem

We would like to emphasize that another important role of filter-feeders in ecosystems is the regulation of abundance of some key players in the ecosystem including not only algae but also bacteria. By doing so, filter-feeders control an extremely active process of oxidation of organic matter in the ecosystem as the bacteria are sometimes the most active actors in performing the oxidation of organic matter in aquatic ecosystems (see, e.g., Table 7). Also, the table shows that filter-feeders themselves (zooplankton plus part of zoobenthos) are also an important

Table 6. Results of the ecological tax

System	Organisms	Amount	Comments	References
Rocky shores	*Mytilus edulis*	11.9 kg m^{-2} y^{-1} (dry weight), of which faeces 9.2 kg, pseudofaeces 2.7	–	Tsuchiya (1980)
Norwegian-Greenland Sea, depth 2020-2630 m	Sponge *Thenea abyssorum*, biomass up to 1524 mg AFDW m^{-2}	0.6–2.2 mg C m^{-2} d^{-1}	The poriferan community possibly adds upto 10% to the vertical particle flux	Witte et al. (1997)
	Thenea abyssorum	Up to 0.7 g C m^{-2} y^{-1}	–	Author's estimate based on the data Witte et al. (1997)
Baltic coastal ecosystem	M. edulis	1092 g m^{-2} y^{-1} dry weight	Sedimentation was 3521 g m^{-2} y^{-1} dry weight	Kautsky & Evans (1987)
Marine/ estuarine	*M. edulis*	60 g m^{-2} h^{-1}	–	Widdows et al. (1998)
Netherlands	Bivalves	25 g m^{-2} h^{-1}	–	Smaal et al. (1986)
Marine/estuarine	M. chilensis	18 g m^{-2} h^{-1}	–	Jaramillo et al. (1992)

Biosediments formation.

Table 7. Contribution of various aquatic organisms to oxidation of organic matter in the ecosystem of central part of the Sea of Okhotsk (the period of time: the summer minimum of phytoplankton, the end of July to beginning of August) (Sorokin et al., 1997)

Organisms	Respiration (% of the total respiration of the community)
Macrophytes	0.3
Phytoplankton	8.9
Bacteria	55.6
Microzooplankton	7.7
Zooplankton (non-predatory)	12.2
Zooplankton (predatory)	4.45
Zoobenthos	8.3
Fish	2.5
Mammals and birds	0.05

part of the biomachinery for carbon oxidation in the ecosystem.

From the data of the Table 7, we clearly see the role of bacteria. Important are also the other organisms that help bacteria, e.g., by producing extracellular organic metabolites, recycling nutrients (excretion of various forms of nitrogen and phosphorus) that are available for bacteria. Moreover, the regulatory role of filter-feeders is also important towards some optimization of the rates of oxidation of organic molecules by bacteria and other small heterotrophic organisms.

Analyzing the role of filter-feeders in regulating the abundance of bacteria and therefore their oxidation activity, we should also underline another aspect of the influence of filter-feeders on heterotrophic bacteria. This aspect is the production by filter-feeders of the pellets rich in organic matter that provide the organic substrate for heterotrophic bacteria, and also provide the organic molecules that serve the function of electron acceptors for some anaerobic bacteria of sediments.

Author's new experimental data on the effects of chemicals on filter-feeders

It is important to better understand the implications of water pollution, including some potential effects of pollutants on filter-feeders and their functional activities. We have studied the effects of some chemicals (surfactants and others) on the filtering rate of invertebrates. The individual chemicals tested represented all three main classes of surfactants: anionic, cationic, non-ionic surfactants. The representatives of those that were tested were SDS, TDTMA, and Triton X-100, respectively. Some inhibition was found (Ostroumov et al., 1997; Kartasheva & Ostroumov, 1998, 2000a, b, c, 2002a, b; Ostroumov et al., 2003b). Surfactant SDS inhibited the filtration activity of *Mytilus edulis* (Ostroumov et al., 1997). Surfactants TDTMA and SDS (both at 0.5 mg l^{-1}) inhibited the filtration activity of *Crassostrea gigas* (Ostroumov, 2003a). TDTMA (0.5 mg l^{-1}) inhibited water filtration by rotifers (Kartasheva & Ostroumov, 1998; Ostroumov et al., 2003b). Quantitative data on the effects of some chemicals on filter-feeders are presented in Table 8. To measure the effect of the chemicals, we used monitoring the amount of suspended matter that stayed in water after a certain period of filtration by invertebrates. As a result of the filtration, some suspended matter was removed from the water. If the filtration activity was inhibited by the chemical, the amount of the removed suspended matter was less than the amount of that removed in the control. Therefore the concentration of the non-removed suspended matter in water was higher in the system with the chemical tested than in the system with uninhibited activity of filter feeders. We calculated the ratio of the concentration of suspended matter in the system with the chemical inhibitor to that in the control system where the filtration activity was uninhibited. The ratio was presented in Table 8. It is seen that both individual surfactants and the chemical products that are the mixtures of several chemicals including surfactants produced a significant effect on the filtration activity. The effect was concentration-dependent. Those data are in accord with previous studies by other authors who have found a similar inhibitory action of other chemicals or have used other organisms (e.g., Mitin, 1984; Widdows & Page, 1993; Stuijfzand et al., 1995; Donkin et al., 1997).

It should be noted that the response of some species to specific pollutants may have a complex (e.g., bi-phasic) character and we have to be careful

Table 8. Some chemicals that have an adverse effect on the filtering activity of the filter-feeders

Measurement no.	Chemical (described in the text)	Concentraton of the chemical, mg l^{-1}	Organism	Effect of the chemical (the ratio of the concentration of suspended matter in the system with the chemical to that in the control), %	Reference
1.	TDTMA	0.5	*Crassostrea gigas*	344.2	New data
2.	SD1 (L)	20	*C. gigas*	261.7	New data
3.	LD2 (F)	2	*M. galloprovincialis,*	218.8	Ostroumov (2001b)
4.	LD2 (F)	2	*C. gigas*	1790.0	New data
5.	SD2 (I)	10	*M. galloprovincialis*	157.8	Ostroumov (2002c)
6.	SD3 (D)	30	*C. gigas*	5800.0	Ostroumov (2002c)
7.	Triton X-100	1	*M. edulis*	236.2	Ostroumov et al. (1998)
8.	Triton X-100	4	*M. edulis*	1505.6	
9.	SDS	1	*M. edulis*	271.1	Ostroumov et al. (1997)
10.	SDS	4	*M. edulis*	1473.2	
11.	SD4 (OMO)	50	*Unio tumidus*	186.7	Ostroumov (2001a)

As a result, the amount of suspended matter removed from the water by the filter feeders decreased. Therefore the amount of suspended matter left in the water was more than that in control. (Ostroumov (2003a), with additions).
Abbreviations: LD1 (E) – liquid detergent E; LD2 (F) – liquid detergent Fairy; SD1 (L) – synthetic detergent Lanza; SD2 (I) – synthetic detergent IXI; SD3 (D) – synthetic detergent Deni; SD4 (OMO) – synthetic detergent OMO.

when interpreting our experiments. We have to avoid oversimplification of the situation when a pollutant affects aquatic organisms as it is known that some concentrations of toxicants might produce a stimulation of biological functions. We do not extend our conclusions on all types and all concentrations of pollutants and xenobiotics. However at least in case of surfactants we definitely observed the inhibition of the filtering activity both in marine and freshwater organisms.

It is interesting that the filering activity can be inhibited not only by the man-made chemicals, but also by some natural ecological chemoeffectors (kairomones). Thus, the filtration activity of *Ceriodaphnia* cf. *dubia* was inhibited by kairomones exuded by eastern rainbowfish *Melanotaenia duboulayi* at fish densities 0.125 fish l^{-1} or higher, when the fish were placed in culture water for one hour (Rose et al., 2003).

All in all, the data shows that the filtering activity is a labile and vulnerable process.

From facts to some principles and general conclusions

The material above leads to formulation of some conclusions on fundamental features of filter-feeders. The logic of thoughts that start with facts and lead toward fundamental conclusions is presented in Table 9. Each line of the table starts with a summary of factual data in the left cell of the line. The factual data are commented in the middle cell of each line, which pave the way for the short formulation of a more general principle in the right cell of each line.

The analysis given above provides some basis for a new vision of the risk of man-made disturbances in biota and ecosystems. This vision was formulated by us in the papers (Ostroumov, 2000a, 2003b) and is summarized in Table 10.

According to the approach presented in the table and discussed in more detail in (Ostroumov, 2003b), the man-made inhibition of the water fil-

Table 9. Some features of water-filtering biomachinery

Features (facts)	Comment/consequences	Fundamental principles (in short)
Significant amount of water filtered per unit of biomass of animals or per unit of area or per unit of time	Significant contribution to the ecological repair of water quality	Large-scale repair of water quality
Parallel functioning of several taxons of filter-feeders	Increase in reliability of the biomachinery of water filtration	Contribution to the reliability of the mechanism and stability of ecosystem
The higher the concentration of particles, the lower the filtration rate and relative grazing pressure	Positive feedbacks that in turn may lead to the increase in heterogeneity of parts of the water column	Potential contribution to creating habitat heterogeneity
The amount of suspension that is being filtered out of water is usually more than needed for metabolism	A significant amount of the formerly suspended matter is finally packed, ejected and/or excreted as pellets	Ecological taxation: filter-feeders pay ecological tax to the ecosystem
Filter-feeders produce pellets	The pellets gravitate towards the bottom or the lower layers of the water column	Acceleration (biocatalysis) of migration of elements
Filter feeders remove bacteria (inter alia)	Regulatory effect (control) on bacteria (the latter may perform over 50% of the total metabolism of the ecosystem)	Contribution to the regulation of the metabolism of ecosystem

tration rate is a significant disturbance to the ecosystem. The inhibition of the filtration activity of filter-feeders may lead to the situation previously described as that of ecological bomb of the second type (Ostroumov, 1999).

Conclusions

The studies of filter-feeders help to elucidate some important aspects of ecosystem stability and regulation, which are among the priorities in ecology (Ostroumov et al., 2003a).

1. The filter-feeders perform some functions that are important services to ecosystem and are characterized by the following:
   (a) Filter feeders participate in the large-scale repair of water quality;
   (b) Filter feeders contribute to the reliability of the mechanism and stability of ecosystem;
   (c) Filter feeders may potentially contribute to creating habitat heterogeneity;
   (d) Filter feeders by removing seston and excreting pellets are involved in 'ecological taxation': filter-feeders pay some ecological tax to the ecosystem;
   (e) Filter feeders contribute to acceleration (biocatalysis) of migration of elements;
   (f) Filter feeders contribute to the regulation of the metabolism of ecosystem.
2. Using several species of filter-feeders as a model system for studies, we have shown that the filtration rate may be inhibited by some xenobiotics and pollutants.
3. On the basis of our studies of various chemicals and organisms, we predict that new examples of pollutants and xenobiotics that

Table 10. The level–block approach to the analysis of ecological hazards of anthropogenic effects on the biota

Disturbance level	Examples of disturbances and their consequences (some of them may be assigned to different levels)
Individual responses	Toxic effects on individual species (increased mortality, decreased fertility, ontogenetic disturbances, diseases, etc.), changes in morphological and physiological variability, and behavioral changes
Aggregated (summarized) responses of a group of organisms	Changes in primary productivity, aggregated parameters of biomass, water chlorophyll, and dissolved O_2 concentrations
Ecosystem stability and integrity	Rearrangements and/or weakening of plankton–benthos connections (coupling); rearrangements and/or weakening of links in the food web; changes in the level of bacterial destruction; decrease in the filtration rates and elimination of suspended particles (seston) from water; decrease in water self-purification; decrease in some regulatory effects because of the loss, migration, or trophic inertness of organisms belonging to higher trophic levels
Ecosystem contribution to biospheric processes	Changes in C flows (e.g., sedimentation of pellets formed by filter-feeding organisms) and N flows (e.g., nitrogen fixation), as well as in flows and cycles of other elements, including S and P; changes in energy (heat etc.) flows

After Ostroumov (2003b) with some modifications.

inhibit the filtration activity of aquatic organisms will be found in future.

Acknowledgements

The author is grateful to many colleagues at Moscow University, the Institute of Biology of Southern Seas NANU (Sevastopol), and the Plymouth Marine Laboratory for assistance, to Prof. V.D. Fedorov, Prof. M.E. Vinogradov, Prof. A.F. Alimov, Prof. V.V. Malakhov, Prof. E.A. Kriksunov, Prof. V.N. Maximov, Prof. A.O. Kasumyan, Prof. G.E. Shulman, Prof. N. Walz and Prof. P. Wangersky for stimulating discussions, to Dr. N.N. Kolotilova, L.V. Podshchekoldina and L.I. Shpitonova for help, to Dr. R. Dame and D. Bushek for providing copies of useful publications, to the reviewers of the manuscript for valuable comments. The grants from the MacArthur Foundation (the Program on Global Security and Sustainability) and the Open Society Foundation (RSS Grant no. 1306/1999) supported the research with *M. galloprovincialis* and *Unio* sp. The work with *M. edulis* was done together with Dr. P. Donkin and Prof. J. Widdows.

References

Achituv, Y. & M. L. Pedrotti, 1999. Costs and gains of porcelain crab suspension feeding in different flow conditions. Marine Ecology Progress Series 184: 161–169.

Alimov, A. F., 1981. Functional Ecology of Freshwater Bivalves (Funktzionalnaja Ekologija Presnovodnykh Dvustvorchatykh Molluskov), Nauka press, Leningrad, 248 pp.

Asmus, R. M. & H. Asmus, 1991. Mussel beds: limiting or promoting phytoplankton? Journal Experimental Marine Biology Ecology 148: 215–232.

Bacher, C., 1989. Capacité trophique du bassin de Marennes-Oléron: couplage d'un modele de transport particulaire et d'un modele de croissance de l'huitre *Crassosrea gigas*. Aquatic Living Resources 48: 199–214.

Ball, B., R. Raine & D. Douglas, 1997. Phytoplankton and particulate matter in Carlingford Lough, Ireland: an assessment of food availability and the impact of bivalve culture. Estuaries 20: 430–440.

Bern, L., 1987. Zooplankton grazing on [methyl-3H]thymidine-labelled natural particle assemblages: determination of filtering rates and food selectivity. Freshwater Biology 17: 151–159.

Cloern, J., 1982. Does the benthos control phytoplankton biomass in South San francisco Bay? Marine Ecology Progress Series 9: 191–202.

Dame, R., D. Bushek & T. Prins, 2001. Benthic suspension feeders as determinants of ecosystem structure and function in shallow coastal waters. Ecological Studies 151: 11–37.

Dame, R., N. Dankers, T. Prins, H. Jongsma & A. Smaal, 1991. The influence of mussel beds on nutrients in the Western Wadden Sea and Eastern Scheldt estuaries. Estuaries 14: 130–138.

Dame, R., R. Zingmark, H. Stevenson & D. Nelson, 1980. Filter feeding coupling between the water column and benthic systems. In Kennedy, V. (ed) Estuarine Perspectives. Academic Press, New York: 521–526.

Davies, B., V. Stuart & M. de Villiers, 1989. The filtration activity of a serpulid polychaete population (*Ficopomatus enigmaticus* (Fauvel) and its effects on water quality in a coastal marina. Estuary Coast Shelf Science 29: 613–620.

Donkin, P., J. Widdows, S. V. Evans, F. Staff & T. Yan, 1997. Effects of neurotoxic pesticides on the feeding rate of marine mussels *Mytilus edulis*. Pesticide Science 49: 196–209.

Frost, T. 1978. Impact of the freshwater sponge *Spongilla lacustris* on a *Sphagnum* bog-pond. Verhandlungen Internationale Vereinigung für Theoretische und Angewandte Limnologie 20: 2368–2371.

Frost, T., 1980. Clearance rate determination for the freshwater sponge *Spongilla lacustris*: effects of temperature, particle type and concentration, and sponge size. Archiv für Hydrobiologie 90: 330–356.

Gutelmaher, B. L., 1986. Metabolism of Plankton as the Whole (Metabolizm Planktona Kak Edinogo Tzelogo), Nauka, Leningrad, 156 pp.

Hily, C., 1991. Is the activity of benthic suspension feeders a factor controlling water quality in the Bay of Brest? Marine Ecology Progress Series 69: 179–188.

Jaramillo E., C. Bertran, A. Bravo, 1992. Mussel biodeposition in an estuary in southern Chile. Marine Ecology Progress Series, 82: 85–94.

Kartasheva, N. V. & S. A. Ostroumov, 1998. Tetradecyltrimethylammonium bromide (tetradetziltrimetilammonij bromid). Toxicological Bulletin (Toksikologicheskii Vestnik) 5: 30–32.

Kautsky, N. & S. Evans, 1987. Role of biodeposition by Mytilus edulis in the circulation of matter and nutrients in a Baltic coastal ecosystems. Marine Ecology Progress Series 38: 201–212.

Krylova, E.M., 1997. Abyssal bivalves of the superfamily Cuspidarioidea (Septibranchia) from the Far-Easterrn seas and the North-Western Pacific. In Kuznetzov, A. P. & O. N. Zezina (eds), Composition and Distribution of Bottom Invertebrate Animals in the Seas of Russia and Adjacent Waters. Institute of Oceanology, Moscow: 80–84.

Kryger, J. & H. U. Riisgård, 1988. Filtration rate capacities in 6 species of European freshwater bivalves. Oecologia 77: 34–38.

Kryuchkova, N. M. & V. H. Rybak, 1989. Patterns of feeding of *Moina rectirostris* Leydig on phytoplankton of a hypertrophic water body (osobennosti pitaniya *Moina rectirostris* Leydig fitoplanktonom gipertrofnogo vodoema. Hydrobiological Journal (Gidrobiologicheskiy Zhurnal) 25(3): 28–31.

Migne, A. & D. Davoult, 2002. Experimental nutrition in the soft coral *Alcyonium digitatum* (Cnidaria: Octocorallia): removal rate of phytoplankton and zooplankton. Cahiers de Biologie Marine 43(1): 9–16.

Mitin, A., 1984. Effect of some environmental factors on the water-clearing activity of bivalves, Summary of Ph.D. Thesis, Moscow, 22 pp.

Monakov, A. V., 1998. Feeding of Freshwater Invertebrates (Pitanie Presnovodnykh Bespozonochnykh). Institute of Ecological and Evolutionary Problems, Moscow, 320 pp.

Newell, R., 1988. Ecological changes in Chesapeake Bay: are they the result of overharvesting the American oyster, *Crassostrea virginica*? Understanding the estuary: Advances in Chesapeake Bay Research. Proceedings of a Conference. 29–31 March 1988. Baltimore, Maryland. Chesapeake Research Consortium Publication 129: 536–546.

Ogilvie, S. & S. Mitchell, 1995. A model of mussel filtration in a shallow New Zealand lake, with reference to eutrophication control. Archiv für Hydrobiologie 133(4): 471–482.

Ostroumov, S. A., 1998. Biological filtering and ecological machinery for self-purification and bioremediation in aquatic ecosystems: towards a holistic view. Rivista di Biologia/Biology Forum 91: 247–258.

Ostroumov, S. A., 1999. Integrity-oriented approach to ecological biomachinery for self-purification and bioremediation in aquatic ecosystems: stopping an ecological time bomb. Limnology and Oceanography: Navigating into the Next Century. February 1–5, 1999, Santa Fe, New Mexico. ASLO, Waco, TX. 134 pp.

Ostroumov, S. A., 2000a. Criteria of ecological hazards due to anthropogenic effects on the biota: searching for a system (Kriterii ekologhicheskoj opastnosti antropoghennykh vozdejstvij na biotu: poiski sistemy). Doklady Biological Sciences 371: 204–206 (the Russian edition: Doklady Akademii Nauk 371: 844–846).

Ostroumov, S. A., 2000b. The concept of aquatic biota as a labile and vulnerable component of the water self-purification system (Kontzeptzija vodnoi bioty kak labil'nogo i ujazvimogo zvena sistemy samoochishchenija vody). Doklady Biological Sciences 372: 286–289 (the Russian edition: Doklady Akademii Nauk 372: 279–282).

Ostroumov, S. A., 2000c. Aquatic ecosystem: a large-scale diversified bioreactor with a water self-purification function (Vodnaja ekosistema: krupnorazmernyj diversifitzirovannyj bioreaktor s funktzijej samoochishchenija vody). Doklady Biological Sciences 374: 514–516 (the Russian edition: Doklady Akademii Nauk 374: 427–429).

Ostroumov, S. A., 2001a. Responses of *Unio tumidus* to a mixture of chemicals and the hazard of synecological summation of anthropogenic effects (Reagirovanie *Unio tumidus* pri vozdeistvii smesevogo himicheskogo preparata i opasnost sinekologicheskogo summirovaniya antropogennyh vozdeistviy). Doklady Akademii Nauk 380: 714–717.

Ostroumov, S. A., 2001b. Synecological foundation for the solution of the problem of eutrophication (Sinekologicheskie osnovy resheniya problemy evtrofirovaniya). Doklady Akademii Nauk 381:709–712.

Ostroumov S. A., 2001c. Amphiphilic chemical inhibits the ability of molluscs to filter water and to remove the cells of phytoplankton. Izvestia RAN. Seriya Biology (Bulletin of Russian Academy of Sciences, Biology Series). Issue 1: 108–116.

Ostroumov, S. A., 2002a Inhibitory analysis of top-down control: new keys to studying eutrophication, algal blooms, and water self-purification. Hydrobiologia 469: 117–129.

Ostroumov, S. A., 2002b. Polyfunctional role of biodiversity in processes leading to water purification: current conceptualizations and concluding remarks. Hydrobiologia 469 (1-3): 203–204.

Ostroumov, S. A., 2002c. Biodiversity and water quality: the role of feed-backs (Sokhranenie bioraznoobraziya i kachestvo vody: rol obratnykh svyazej v ekosistemakh) Doklady Akademii Nauk 382: 138–141.

Ostroumov, S. A., 2003a. Studying effects of some surfactants and detergents on filter-feeding bivalves. Hydrobiology 500: 341–344.

Ostroumov, S. A., 2003b. Anthropogenic effects on the biota: towards a new system of principles and criteria for analysis of ecological hazards. Rivista di Bilogia/Biology Forum 96: 159–169.

Ostroumov, S. A., P. Donkin & F. Staff, 1997. Inhibition by the anionic surfactant, sodium dodecyl sulphate, of the ability of mussels *Mytilus edulis* to filter and purify the sea water (Anionnoje poverkhnostno-aktivnoje veshchestvo inghibirujet sposobnost' midij filtrovat' i ochishchat' morskuju vodu). Vestnik Moskovskogo Universiteta. Seria 16. Biologija. (Bulletin of Moscow University. Series 16. Biology) 3: 30–36.

Ostroumov, S. A., P. Donkin & F. Staff, 1998. Filtration inhibition induced by two classes of synthetic surfactants in the bivalve mollusc (Narushenije filtracii dvustvorchatymi molluskami pod vozdejstvijem poverkhnostno-aktivnykh veshchestv dvukh klassov). Doklady Akademii Nauk 362: 574–576.

Ostroumov, S. A., N. Walz & J. Widdows, 2001. Measuring activity of filter-feeders: linking organismal and superorganismal levels of hydrobiology and biological oceanography. Ecological Studies, Hazards, Solutions 5: 26–27.

Ostroumov, S. A., S. Dodson, D. Hamilton, S. Peterson & R. G. Wetzel, 2003a. Medium-term and long-term priorities in ecological studies. Rivista di Biologia/Biology Forum 96: 327–332.

Ostroumov, S. A., N. Walz & R. Rusche, 2003b. Effects of a cationic amphiphilic chemical on rotifers. Doklady Akademii Nauk 390(3): 423–426.

Petersen, J. & H. U. Riisgård, 1992. Filtration capacity of the ascidian *Ciona intestinalis* and its grazing impact in a shallow fjord. Marine Ecology Progress Series 88: 9–17.

Pilson, M., 1985. On the residence time of water in Narragansett Bay. Estuaries 8: 2–14.

Rose, R., M. Warne & L. Lim, 2003. Exposure to chemicals exuded by fish reduces the filtration and ingestion rates of *Ceriodaphnia* cf. *dubia*. Hydrobiologia 501: 215–217.

Smaal, A. C., J. Verhagen, J. Coosen & H. Haas, 1986. Interactions between seston quantity and quality and benthic feeders in the Oosterschelde, The Netherlands. Ophelia 26: 385–399.

Sorokin, Yu. A., P. Yu. Sorokin, T. I. Mamaeva & O. V. Sorokina, 1997. The Sea of Okhotsk bacterioplankton and planktonic ciliates. In Sapozhnikov, V. V. (ed.), Complex studies of ecosystem of the Sea of Okhotsk. VNIRO Press, Moscow: 210–216.

Shulman, G. E. & G. A. Finenko, 1990. Bioenergetics of Aquatic Organisms, Naukova Dumka Press, Kiev, 246 pp.

Shuntov, V. P., 2001. Biology of Far-Eastern seas of Russia, Vol. 1. TINRO-Centre, Vladivostok, 580 pp.

Starkweather, P. & K. Bogdan, 1980. Detrital feeding in natural zooplankton communities: discrimination between live and dead algal foods. Hydrobiologia 73: 83–85.

Stuijfzand, S. C., M. H. S. Kraak, Y. A. Wink & C. Davids, 1995. Short-term effects of nickel on the filtration rate of the zebra mussel *Dreissena polymorpha*. Bulletin of Environmental Contamination and Toxicology 54: 376–381.

Sushchenya, L. M., 1975. Quantitative Trends in the Feeding of Crustaceans, Nauka i Tekhnika Press, Minsk, 208 pp.

Tsuchiya, M., 1980. Biodeposit production by the mussel *Mytilus edulis* L. on rocky shores. Journal Experimental Marine Biology and Ecology 47: 203–222.

Vinogradov, M. E. & E. A. Shushkina, 1987. Functioning of plankton communities of the epipelagial of the ocean, Nauka Press, Moscow, 240 pp.

Widdows, J. & D. Page, 1993. Effects of tributyltin and dibutyltin on the physiological energetics of the mussel, *Mytilus edulis*. Marine Environmental Research 35: 233–249.

Widdows, J., M. Brinsley, P. Salkeld & M. Elliott, 1998. Use of annular flumes to determine the influence of current vedlocity and bivalves on material flux at the sediment-water interface. Estuaries 21(4A): 552–559.

Witte, U., T. Brattegard, G. Graf & B. Springer, 1997. Particle capture and depostion by deep-sea sponges from the Norwegian-Greenland Sea. Marine Ecology Progress Series 154: 241–252.

Yablokov, A. V. & S. A. Ostroumov, 1991. Conservation of Living Nature and Resources: Problems, Trends, Prospects. Springer Verlag, Berlin, Heidelberg, New York, 271 pp.

Zankai, N., 1979. Jahreszeitliche Änderung der Filtrierrate der Copepoden *Eudiaptomus gracilis* (G.O.Sars) im Balaton-See. Verhandlungen Internationale Vereinigung für Theoretische und Angewandte Limnologie 20: 2551–2555.

Zezina, O. N., 1997. Brachiopods in a natural bottom biofilter at the shelves and slopes in the Far-Eastern Seas of Russia. In Kuznetzov, A. P. & O. N. Zezina (eds), Composition and Distribution of Bottom Invertebrate Animals in the Seas of Russia and Adjacent Waters. Institute of Oceanology, Moscow: 52–60.

Hydrobiologia (2005) 542:287–296
H. Segers & K. Martens (eds), Aquatic Biodiversity II
DOI 10.1007/s10750-004-2566-7

Recent ostracods (Crustacea, Ostracoda) found in lowland springs of the provinces of Piacenza and Parma (Northern Italy)

Giampaolo Rossetti*, Valentina Pieri & Koen Martens
Department of Environmental Sciences, University of Parma, Parco Area delle Scienze 33A, Parma 43100, Italy
*(*Author for correspondence: Tel.:/Fax: +39-0521-905977, E-mail: giampaolo.rossetti@unipr.it)*

Key words: lowland springs, water quality, ostracods, scanning electron microscopy, distribution, Po River plain

Abstract

The "fontanili" are artificial aquatic ecosystems, typical of the lowland plains of Northern Italy, exploiting natural resurgences of deep groundwater. These habitats are characterized by low variation in hydrologic, hydrochemical and thermal conditions throughout the year. Proper management is required to prevent the spring clogging by biomass accumulation. In spite of their importance as refugia for endangered species, many springs were completely abandoned in the last years and several of them will disappear. We report the results of a study carried out in 2001 on 31 springs of the provinces of Piacenza and Parma, distributed in seven areas defined on hydrological and geological considerations. Physical and chemical variables and parameters of waters were measured and ostracod samples were collected. Most of the springs showed high nitrate concentration, due to a diffuse pollution of agricultural origin. Twelve ostracod species in four families were identified. Ostracod valves were analysed by scanning electron microscopy. *Cypria ophtalmica* was found in all the springs; other relatively common species were *Cyclocypris laevis*, *Notodromas persica*, and *Prionocypris zenkeri*. The maximum number of species per site was four. *Cypridopsis vidua*, *P. zenkeri*, and *N. persica* showed a very localized distribution in the study area. The ostracod fauna of the "fontanili" was compared to other species assemblages found in spring habitats and to the available information on recent freshwater ostracods reported for Italy.

Introduction

Northern Italy hosts peculiar aquatic ecosystems, locally known as 'fontanili'. These habitats are clustered along an almost continous line on the Alpine side of the Po River Valley and more scattered on the Apennine side. They originate from natural resurgences occurring along the alluvial fans of the main watercourses, namely in the transition zone from the higher to lower plain which is characterized by changes in slope profile and sediment granulometry. According to Fetter (1988), most of them can be considered as 'fault springs' created where water in a permeable layer of coarse material (pebbles and gravel) encounters impermeable layers (sand and clay) and must flow upwards. Springs formed in this way are therefore fed by deep aquifers which guarantee a relatively constant flow and a little seasonal variation of water temperature. Only a limited number of these springs seems to be associated with the water table, showing variable flows influenced by the precipitation regime and marked changes in temperature and chemical features of waters over a yearly cycle (Petrucci et al., 1982).

The emergence of ground waters naturally tends to form rheocrene (discharging directly into a channel) or helocrene (discharging into marsh or swamp areas) springs, depending on the magnitude of the flow rate. Actually, the 'fontanili' of the Apennine side mostly appear as limnocrene habitats, in which one or more resurgences discharge in a pond (the so-called 'head', usually a few meters deep), from which water flows through a small channel (the 'stem'). This morphology is due to the excavation of a basin to facilitate the water uprising and it is maintained by periodic vegetation cutting and sediment removal.

The uninterrupted presence of human settlements in the lowland spring areas since prehistoric times (Bonardi et al., 1985) testifies on their importance as a source of drinking water and for agricultural purposes. In the last few decades, however, many "fontanili" have disappeared, due to the lack of management practices or have become deeply impacted by intense withdrawal of water for irrigation, disposal of waste water, infrastructure construction or spreading of alien species (e.g., the coypu *Myocastor coypus* and the red swamp crayfish *Procambarus clarkii*) (Barbaresi & Gherardi, 2000; Bracco et al., 2001). In spite of their artificial origin, these aquatic ecosystems are important refugia for biodiversity and may host stenotopic species and endemics (e.g., the gobiid fish *Knipowitschia punctatissima*), the persistence of which is strictly linked to the conservation of these peculiar habitats.

The present study forms part of a larger project aiming at assessing the conservation status and the chemical quality of lowland springs in the provinces of Parma and Piacenza (Western Emilia) (Bertolini, 2002). A representative subsample of springs was selected; the main hydrochemical and physical variables and parameters were analysed and the ostracod communities of these aquatic ecosystems were described for the first time.

Material and methods

Sampling took place between May and August 2001 and each spring was sampled once. On the whole, 58 sites were visited; hydrologically inactive systems or highly degraded habitats were not sampled and only 31 springs, 11 in the province of Piacenza and 20 in the province of Parma (Table 1, Fig. 1), were selected for this study. Springs were grouped within geographic districts reflecting distinctive hydrological and geological characteristics, due to the influence of different aquifers associated with the main rivers flowing from the Apennine chain and their paleocourses; the identification codes of springs are the same as in previous surveys carried out by Bernini & Torselli (1989, 1991) and ARPA (2001) (Table 1). All the springs considered here are mainly fed by deep waters, except PR30 and PR40, whose waters are likely to be lateral infiltrations from surface aquifers associated with adjacent rivers.

Water temperature was measured at a depth of 0.5 m using a digital thermometer (ILC). Water samples were collected from the same depth and kept refrigerated until analysed in the laboratory. Hydrochemical and physical variables and parameters were determined as follows: pH by potentiometry (TIM 90, Radiometer); total alkalinity by potentiometric end-point titration at pH 4.5 and 4.2 (TIM 90, Radiometer) and linearization according to Rodier (1978); electric conductivity at 20 °C by conductometry (CDM 83, Radiometer); ammonia (Koroleff, 1970), nitrite (APHA, AWWA, WPCF, 1975), nitrate (Rodier, 1978), chloride (Florence & Farrar, 1971), dissolved reactive silica (APHA, AWWA, WPCF, 1975), soluble reactive phosphorus (Valderrama, 1981), and chlorophyll-*a* (Golterman et al., 1978) by spectrophotometry (Beckman DU 65); calcium and magnesium by EDTA (0.01 M) end-point titration (Patton & Reeder, 1956). From the same water layer, samples for the determination of dissolved oxygen concentration via Winkler titration (Rodier, 1978) were also collected.

Ostracods were gathered by a 250 μm handnet pulled close to the sediment and through the vegetation in the "head" and in the proximal part of the "stem" of each spring. Living samples were transferred to the laboratory, where specimens were sorted under a binocular microscopy and then fixed in 90% ethanol. Both soft parts (dissected in glycerine and stored in sealed slides) and valves (stored dry in micropal slides and used for scanning microphotographs) were checked for species identification, using Danielopol (1980), Meisch (1984, 2000) and González-Mozo et al. (1996) as references. All the figured material is deposited in the Ostracod Collection (O.C.) of the Royal Belgian Institute of Natural Sciences, Brussels.

Table 1. Geographical characteristics of the springs considered in this study

Spring	Code	UTM Coordinates	m a.s.l.	Geographic district
Zappellazzo di sopra	PC03	32T 0565760 4976715	81	Chero – Pontenure
Santa Giustina	PC10	32T 0565372 4977810	74	Chero – Pontenure
Fontanone Baselicaduce	PC18	32T 0573965 4977260	61	Fiorenzuola d'Arda – Alseno
Cascina Fontana – Alseno	PC24	32T 0576790 4974143	58	Fiorenzuola d'Arda – Alseno
Forcelle nord A	PC32	32T 0577530 4977420	47	Fiorenzuola d'Arda – Alseno
Il Bosco 44	PC44	32T 0566695 4977190	76	Chero – Pontenure
Fontanone	PC46	32T 0536375 4990165	62	Castel San Giovanni
Sorgente Vitali	PC47	32T 0536500 4990115	65	Castel San Giovanni
Genovese 1	PC51	32T 0537030 4990150	65	Castel San Giovanni
Cavi San Lazzaro 75	PC75	32T 0563015 4972935	120	Chero – Pontenure
Cavi San Lazzaro 76	PC76	32T 0562935 4973020	120	Chero – Pontenure
Rimale	PR01	32T 0580786 4971315	65	Fidenza
Bastelli Fontanella 2	PR02	32T 0584371 4973557	50	Fidenza
Bastelli Granella	PR03	32T 0584547 4973354	51	Fidenza
Bastelli Fontanella 4 testa A	PR04	32T 0583410 4972502	57	Fidenza
La Commenda 1 testa B	PR06	32T 0601364 4971453	33	Trecasali
Cornazzano	PR09	32T 0600957 4970740	35	Trecasali
San Rocco	PR10	32T 0601697 4971298	32	Trecasali
Quartiere Piccolo	PR13	32T 0601420 4971020	33	Trecasali
Al Fontanon	PR14	32T 0601380 4971025	32	Trecasali
Alberato Cervara	PR17	32T 0603452 4969477	36	Trecasali
Due Grande	PR20	32T 0601869 4968768	36	Trecasali
Ricci	PR26	32T 0697980 4962606	44	Parma
Fontanone 2	PR27	32T 0610246 4961014	45	Parma
Marzola 2 testa A	PR29	32T 0609585 4962085	40	Parma
San Donato	PR31	32T 0610388 4961570	43	Parma
Gabbi testa A	PR32	32T 0610103 4961362	44	Parma
Casa Nuova	PR33	32T 0612160 4961986	40	Parma
Negri-Bertoni	PR34	32T 0608263 4960733	50	Parma
Sinzano	PR39	32T 0598806 4955940	115	Collecchio – Sala Baganza
Casale	PR40	32T 0598826 4953254	145	Collecchio – Sala Baganza

Statistical analyses were performed by means of the package PAST version 1.06 (Hammer et al., 2001). Principal Component Analysis (PCA) was applied to chemical data after a $\log(x + 1)$ transformation (except for pH). Temperature, dissolved oxygen, soluble reactive phosphorus, and chlorophyll-*a* concentrations were not included in this analysis, because these parameters are more influenced by specific conditions at each site (sampling time, shading by tree canopy, development of macrophyte cover, etc.) rather than reflecting the chemical characteristics of groundwaters. Similarity between species assemblages was assessed by Cluster Analysis, using the Jaccard coefficients derived from the matrix of presence/absence data.

Results

In most springs water temperature ranged between 13 and 17 °C; higher values were measured in PC76 (19.2 °C), PR26 (18.6 °C), PR34 (20.1 °C) and in the two sites of the Collecchio – Sala Baganza district (23.6 °C in PR39 and 17.6 °C in PR40). In all the sampled biotopes, pH values were slightly alkaline, from 7.16 (PC 10) to 8.02 (PR14) units. Total alkalinity was typical of well

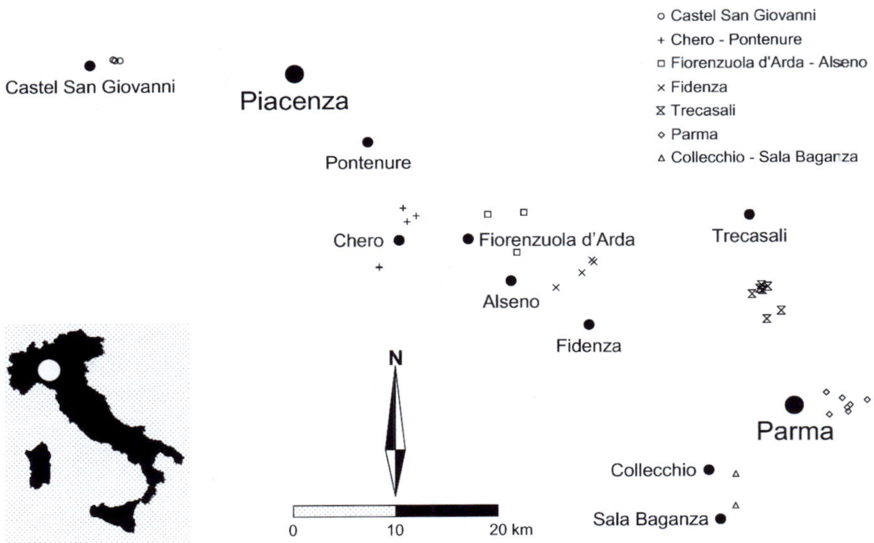

Figure 1. Geographical location of the studied springs.

buffered waters, with a minimum of 1.85 meq l^{-1} in PR39 and a maximum of 5.18 meq l^{-1} in PC46. Deficits in dissolved oxygen content were frequently found, with critical conditions (saturation below 20%) in PC31 (Castel San Giovanni), PR01 and PR03 (Fidenza) and PR09 (Trecasali). On the other hand, distinct oxygen oversaturation (>160%) was observed in PC75 (Chero – Pontenure) and PR39 (Collecchio – Sala Baganza). Conductivity showed marked differences among springs, also within the same geographic area. Most of the springs had values from 300 to 600 μS cm^{-1}, with maxima (up to 1880 μS cm^{-1} in PR03) measured in the Fidenza district. Chloride content was characterized by high variability: concentrations below 2 mg l^{-1} were observed in three springs of the Chero – Pontenure district (PC03, PC44 and PC76); in the other biotopes values ranged from 2 to 32 mg l^{-1}, except PR02 and PR03 (Fidenza district) in which concentrations of 119 and 203 mg l^{-1} were measured, respectively. Most of the analyzed samples contained calcium and magnesium concentrations ranging from 5 to 10 meq l^{-1}; the highest value was observed in PR01 (13.12 meq l^{-1}), the lowest in PR39 (3.46 meq l^{-1}). In all the springs, nitrate was the dominant form of dissolved inorganic nitrogen, with concentrations over 7 mg l^{-1} in several springs situated in the districts of Chero – Pontenure, Fiorenzuola d'Arda – Alseno, Trecasali. Nitrate peaks remained below 3 and 5 mg l^{-1} in the districts of Castel San Giovanni and Parma, respectively, whilst the lowest values were measured in the two springs of the Collecchio – Sinzano district (0.12 mg l^{-1} in PR39 and 0.59 mg l^{-1} in PR40). Ammonia reached a maximum of 319 μg l^{-1} in PR33; the highest nitrite concentration (577 μg l^{-1}) was observed in PC51. The highest values of dissolved reactive silica were found in the Fidenza district (>11 mg l^{-1} in PR02, PR03 and PR04); relatively low concentrations (<3 mg l^{-1}) were observed in the Chero – Pontenure, Fiorenzuola d'Arda – Alseno, and Parma districts. Soluble reactive phosphorus concentrations were generally low, and always ≤5 μg l^{-1} in the springs of the Parma and Sala Baganza – Collecchio districts; isolated peaks were recorded in the other districts, with a maximum of 72 μg l^{-1} in PC75. Chlorophyll-*a* concentrations were mostly <7 μg l^{-1}, with the exception of PC76 (17.8 μg l^{-1}), PR26 (10.7 μg l^{-1}), PR31 (43.8 μg l^{-1}), and above all PR39 (273.6 μg l^{-1}).

The results of the PCA for the first two components (which jointly account for 73.05 % of the total variance) are presented in the biplot of Figure 2. Springs located in different districts generally tend to form definite clusters, indicating marked variations in the mineral content of groundwaters also on relatively small geographic scales. On the first axis, which orders samples

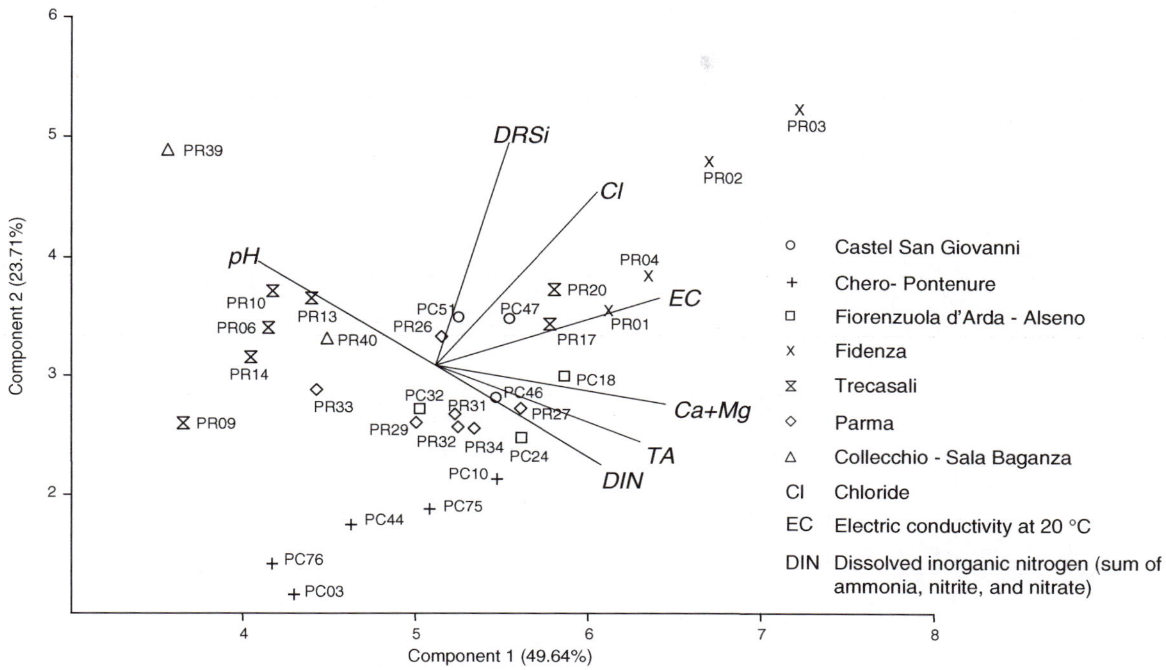

Figure 2. Principal Component Analysis diagram representing the ordination of springs in relation to chemical characteristics of waters. Ca + Mg: sum of calcium and magnesium; DRSi: dissolved reactive silica; TA: total alkalinity.

mainly in relation to total ionic concentrations and pH values, the springs of the Collecchio – Sala Baganza district and four springs of the Trecasali district are clearly opposed to those of the Fidenza district. PR17 and PR20 did not aggregate with the other springs of the same district, due to their significantly higher conductivity and content of nitrate and chloride. Along the second axis, which is primarily correlated with the concentrations of reactive silica and chloride, the most divergent clusters are those of Chero – Pontenure and, once again, Fidenza. The springs in the districts of Castel San Giovanni, Fiorenzuola d'Arda – Alseno, and Parma overlap each other, showing no clear separation among these sites with regards to the selected chemical variables.

Altogether, 12 ostracod species were identified, belonging to 4 Cypridoidea families (Table 2; Fig. 4). Ostracod species occurrences with respect to the different springs is shown in Figure 3. Similarity between communities usually did not reflect any clear geographical gradient, although some species showed definite distribution patterns (Fig. 3). *Cypria ophtalmica* was reported for all the springs, and in nine of them it was the only species found. Six species (*Pseudocandona pratensis*, *P. albicans*, *Ilyocypris decipiens*, *I. inermis*, *Herpetocypris brevicaudata* and *Potamocypris fulva*) only occurred each in one sampling site. The highest diversity was seen in the Trecasali district, where seven ostracod species were present. In this area, three springs (PR10, PR14, and PR20) had four species each, i.e. the maximum number of taxa per site observed in this study. Conversely, in the springs of the Fidenza district, characterized by high ionic content and particularly rich in chloride, only *C. ophtalmica* was reported. *Cyclocypris laevis*, the second most common species, was found in nine springs disseminated in different districts; the distribution of *Cyclocypris ovum* was restricted to only two springs, PC10 and PC32. These two congeneric species were never observed together. *Notodromas persica* showed a distinctly uneven distribution, inhabiting only springs located in the western (PC46 and PC47) and eastern (PR13, PR17, and PR26) extremes of the study area.

Table 2. Taxonomic status of the ostracods analysed in this study

Family Candonidae (Kaufmann, 1900)
Subfamily Cyclocypridinae (Kaufmann, 1900)
Genus *Cypria* (Zenker, 1854)
Cypria ophtalmica (Jurine, 1820)
Genus *Cyclocypris* (Brady & Norman, 1889)
Cyclocypris laevis (O.F. Müller, 1776)
Cyclocypris ovum (Jurine, 1820)
Subfamily Candoninae (Kaufmann, 1900)
Genus *Pseudocandona* (Kaufmann, 1900)
Pseudocandona pratensis (Hartwig, 1901)
Pseudocandona albicans (Brady, 1864)
Family Ilyocyprididae (Kaufmann, 1900)
Subfamily Ilyocypridinae (Kaufmann, 1900)
Genus *Ilyocypris* (Brady & Norman, 1889)
Ilyocypris decipiens (Masi, 1905)
Ilyocypris inermis (Kaufmann, 1900)
Family Notodromadidae (Kaufmann, 1900)
Subfamily Notodromadinae (Kaufmann, 1900)
Genus *Notodromas* (Lilljeborg, 1853)
Notodromas persica (Gurney, 1921)
Family Cyprididae (Baird, 1845)
Subfamily Eucypridinae (Bronshtein, 1947)
Genus *Prionocypris* (Brady & Norman, 1896)
Prionocypris zenkeri (Chyzer & Toth, 1858)
Subfamily Herpetocypridinae (Kaufmann, 1900)
Genus *Herpetocypris* (Brady & Norman, 1889)
Herpetocypris brevicaudata (Kaufmann, 1900)
Subfamily Cypridopsinae (Kaufmann, 1900)
Genus *Cypridopsis* (Brady, 1867)
Cypridopsis vidua (O.F. Müller, 1776)
Genus *Potamocypris* (Brady, 1870)
Potamocypris fulva (Brady, 1868)

Prionocypris zenkeri was found exclusively in five out of seven springs of the Trecasali district; the occurrence *of Cypridopsis vidua* was also limited to three springs of the same district (PR10, PR14 and PR17), all of them with pH $\geq$ 7.79.

Discussion

The lack of conservation measures, together with inappropriate management practices, has caused a progressive decline in water quality for a large number of lowland springs. In particular, as evidenced in this study, the presence of high concentration of nitrate due to leaching from cultivated areas has become a serious problem which affects most of the aquifers of the Po River plain (Russo & Zavatti, 2001).

Within each district the ionic composition of the spring water was relatively constant; remarkable differences were occasionally noted in those variables more strictly related to the hydrological flow (e.g., water temperature) or the ecosystem productivity (e.g., chlorophyll-*a* and dissolved oxygen). Management practices (rate of sediment removal, presence of buffer strips around the spring, type of crops in the neighbouring area, etc.) may also locally influence the water quality.

The ostracod check list was compiled from samples collected in one survey for each spring. This may lead to an underestimation of the actual ostracod diversity, although seasonal changes in species composition are likely to be minimal in the springs, where environmental conditions are supposed to vary to a lesser extent on a yearly basis than in other aquatic ecosystems.

Some of the identified species prefer springs or waters connected to springs (*Ilyocypris inermis*, *Herpetocypris brevicaudata*, *Prionocypris zenkeri*) or are stygophilic (*Potamocypris fulva*), but the largest part of the ostracod fauna was constituted by taxa highly tolerant towards a number of ecological conditions and colonising a wide variety of habitats (Meisch, 2000). This could be a possible cause of the lack of clear relationships between ostracod occurrence and hydrochemical features of the investigated springs. Also the peculiar distribution in the study area of three species, *Cypridopsis vidua*, *P. zenkeri*, and *persica*, does not seem to be related to any particular environmental gradient. The presence of only *Cypria ophtalmica* in the springs of the Fidenza district may probably depend on the tolerance of this species to salinity (up to 25‰ and chloride concentrations up to 13 g l^{-1}) or a higher degree of eutrophication (Meisch, 2000).

The biocoenotic structure was generally simple in the study area. The observed total number of taxa and mean number of taxa per site are generally comparable to those reported for ostracod communities found in other spring ecosystems (Table 3), notwithstanding the differences in sampling effort, geographic area, and habitat type. As expected, changes in taxon composition are evident

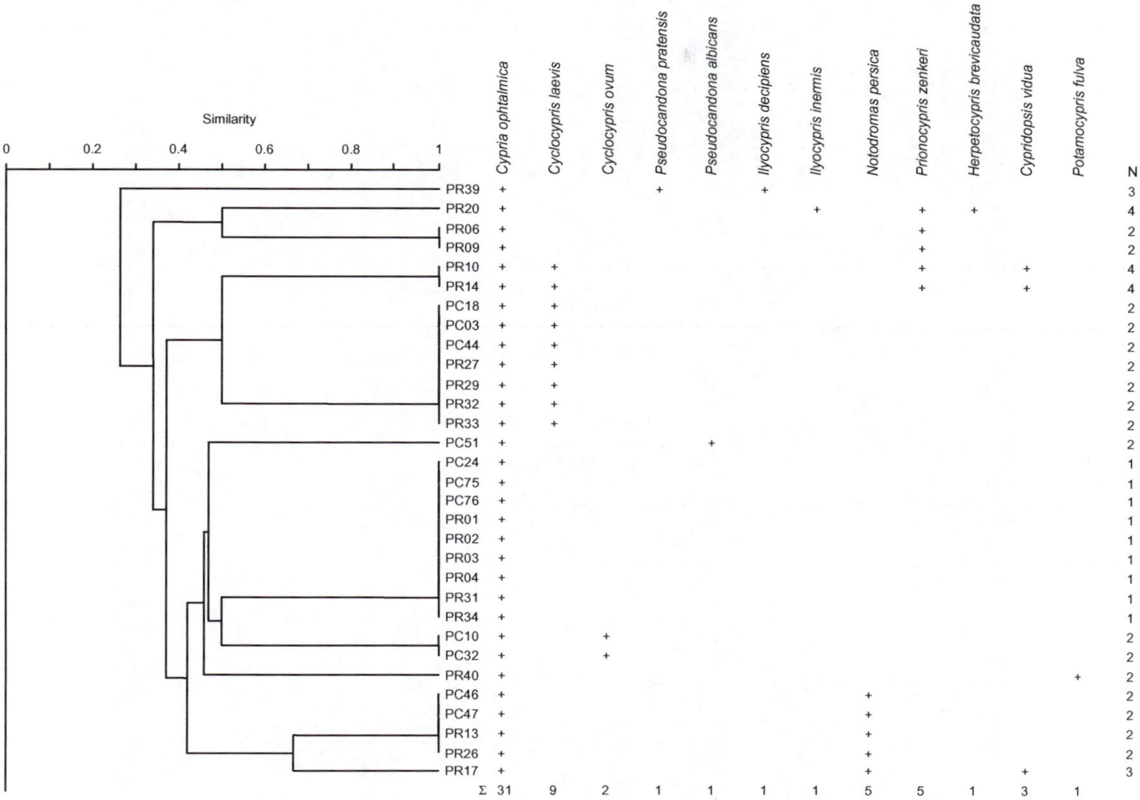

Figure 3. Dendrogram obtained from the Cluster analysis showing the similarity level between ostracod communities. The occurrence of ostracod species in each spring is also indicated. Σ: number of total records per species. N: number of species per site.

amongst different geographic regions, especially at the specific level. All genera found in the "fontanili", with the exception of *Prionocypris*, are present in the Iberian springs, also depending on the large number of sites sampled by Roca & Baltanás (1993) and Mezquita et al. (1999). Ostracod assemblages from Alpine springs in Italy (Stoch, 1998) and Germany (Gerecke et al., 1997) are very similar to each other, but they differ markedly from those of the springs considered in this study: only three genera (*Cypria*, *Pseudocandona*, and *Potamocypris*) occur in both Alpine and lowland habitats. *Pseudocandona* and *Potamocypris*, together with *Ilyocypris*, are also reported from Finnish springs (Särkkä et al., 1997). Surprising enough, four genera (*Ilyocypris*, *Cypria*, *Herpetocypris*, and *Prionocypris*) are shared between the Italian 'fontanili' and the springs investigated by Külköylüoğlu & Vinyard (2000) in the United States, although there are no species in common. On the other hand, *Herpetocypris* is the only genus in common among springs of the Po River plain and of the Western United States (Forester, 1991).

Most of the available information on the distribution of recent freshwater ostracods in Northern Italy is resumed in the synopsis by Ghetti & McKenzie (1981) and in the papers on ostracods from ricefields by McKenzie & Moroni (1986) and Rossi et al. (2003). All the species found in the "fontanili" were already known for the Italian fauna. Moreover, the absence of strictly cold-stenotherm species of ostracods in these springs does not support their role as refugia for this faunal group, contrarily to what is observed for a number of relict (palaeo-) endemics represented by plant and animal taxa which underwent altitudinal displacements from mountain areas during the Würm glacial expansion (Bernini & Mezzadri, 1990; Bracco & Sburlino, 2001).

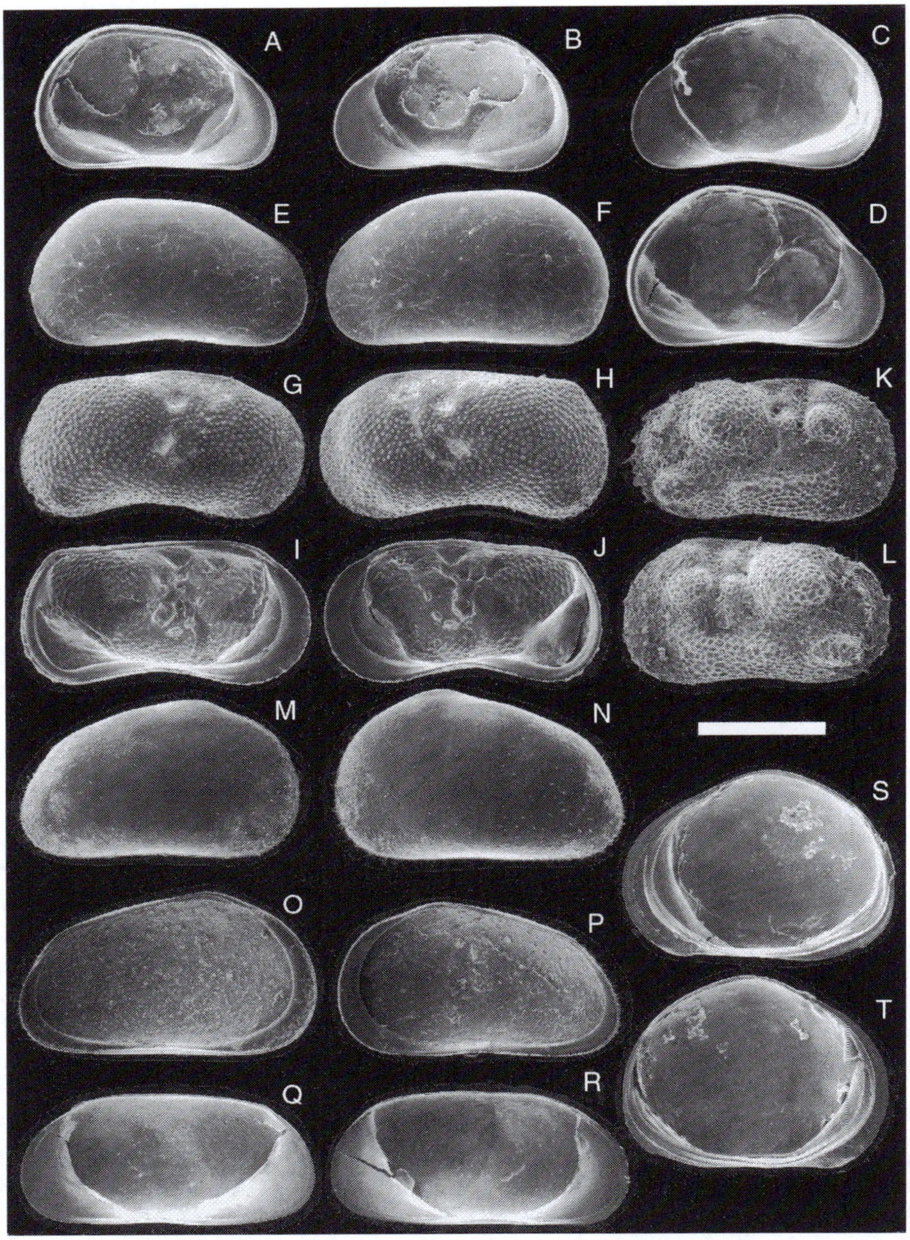

Figure 4. Scanning electron images of ostracods found in the study springs: *Pseudocandona pratensis* (A–D), *Pseudocandona albicans* (E, F), *Ilyocypris inermis* (G–J), *Ilyocypris decipiens* (K, L), *Prionocypris zenkeri* (M–P), *Herpetocypris brevicaudata* (Q, R), *Notodromas persica* (S, T). All adult specimens. Scale bar: 500 μm for A–D; 400 μm for E, F, S, T; 444 μm for G–J; 667 μm for M–P; 727 μm for Q, R. (A) O.C.2833, female, left valve, internal view; (B) idem, right valve, internal view; (C) O.C.2832, male, right valve, internal view; (D) idem, left valve, internal view; (E) O.C.2831, female, right valve, external view; (F) idem, left valve, external view; (G) O.C.2828, female, right valve, external view; (H) idem, left valve, external view; (I) O.C.2825, female, left valve, internal view; (J) idem, right valve, internal view; (K) O.C.2824, female, right valve, external view; (L) idem, left valve, external view; (M) O.C.2829, female, right valve, external view; (N) idem, left valve, external view; (O) O.C.2830, female, left valve, internal view; (P) idem, right valve, internal view; (Q) O.C.2826, female, right valve, internal view; (R) idem, left valve, internal view; (S) O.C.2827, female, right valve, internal view; (T) idem, left valve, internal view.

Table 3. Diversity of ostracod taxa in habitat springs from literature data

Source	Geographic area	S	N
Stoch (1998)	Adamello-Brenta National Park, Northern Italy	30	9
Gerecke et al. (1997)	Alps of Berchtesgaden, Upper Bavaria, Germany	19	11
Roca & Baltanás (1993)	Central Pyrenees	149	21
Mezquita et al. (1999)	Eastern Iberian Peninsula	117	28
Särkkä et al. (1997)	Southern Finland	31	16
Forester (1991)	Western United States	45	6
Külköylüoğlu & Vinyard (2000)	Nevada, Idaho, Oregon	24	14
This study	Piacenza and Parma provinces, Northern Italy	31	12

S: number of investigated springs. N: number of ostracod taxa.

Acknowledgements

Simona Bertolini and Prof. Pierluigi Viaroli (D.E.S., Parma) are gratefully acknowledged for their participation in the fieldwork and for the chemical analysis of the water samples. Juien Cilis (R.B.I.N.Sc., Brussels) offered technical assistance with the scanning electron micrographs. The material was largely analysed during several visits of GR to KM, financed by the EU Project ABC granted to the R.B.I.N.Sc.

References

APHA, AWWA & WPCF, 1975. Standard Methods for the examination of water and wastewater. 14th edn. APHA, Washington.

ARPA, 2001. Censimento dei fontanili della provincia di Parma. Azienda Regionale Prevenzione ed Ambiente dell'Emilia Romagna, Sezione Provinciale di Parma. Internal Report.

Barbaresi, S. & Gherardi, 2000. The invasion of the alien crayfish *Procambarus clarkii* in Europe, with particular reference to Italy. Biological Invasions 2: 259–264.

Bernini, F. & S. Mezzadri, 1990. Su una popolazione di *Triturus alpestris apuanus* (Bonaparte, 1839) in pianura (Amphibia, Urodela, Salamandridae). Atti della Società Italiana di Scienze Naturali e del Museo Civico di Storia Natuale di Milano 130: 253–259.

Bernini, F. & A. Torselli, 1989. Le risorgive della pianura piacentina. Caratterizzazione di unità igrofile di particolare pregio (prima parte). Rivista Museo di Storia Naturale Castell'Arquato 4: 27–110.

Bernini, F. & A. Torselli, 1991. Le risorgive della pianura piacentina. Caratterizzazione di unità igrofile di particolare pregio (seconda parte). Rivista Museo di Storia Naturale Castell'Arquato 5–6: 43–160.

Bertolini, S. 2002. Individuazione di indicatori di qualità delle acque finalizzati alla caratterizzazione e alla gestione dei fontanili di Parma e Piacenza. Graduation thesis, University of Parma, 55 pp.

Bonardi, S., P. L. Dall'Aglio & G. Marchetti, 1985. Geomorfologia e vicende storiche: la pianura Piacentina tra T. Nure e T. Ongina. Padusa 21: 133–153.

Bracco, F., S. Paradisi, G. Sburlino & F. Stoch, 2001. Degrado e conservazione. In Minelli, A. (ed.), Risorgive e fontanili. Acque sorgenti di pianura dell'Italia Settentrionale. Ministero dell'Ambiente, Museo Friulano di Storia Naturale: 123–1 39.

Bracco, F. & G. Sburlino, 2001. Aspetti floristici e vegetazionali. In Minelli, (ed.), Risorgive e fontanili. Acque sorgenti di pianura dell'Italia Settentrionale. Ministero dell' Ambiente, Museo Friulano di Storia Naturale: 29–61.

Danielopol, D. L., 1980. Sur la biologie de quelques Ostracodes Candoninae épigés et hypogés d'Europe. Bulletin du Musée national d'Histoire naturelle, Paris 2: 471–506.

Fetter, C. W., 1988. Applied Hydrogeology. 2nd edn, Merrill Publishing Co., Columbus, Ohio.

Florence, T. M. & Y. I. Farrar, 1971. Spectrophotometric determination of chloride at parts-per-billion level by mercury(II)thiocyanate method. Analytica Chimica Acta 54: 373–377.

Forester, R. M., 1991. Ostracode assemblages from springs in the western United States: implications from paleohydrology. Memoirs of the Entomological Society of Canada 155: 181–201.

Gerecke, R., C. Meisch, F. Stoch, F. Acri & H. Franz, 1997. Eucrenon-hypocrenon ecotone and spring typology in the Alps of Berchtesgaden (Upper Bavaria, Germany). A study of microcrustacea (Crustacea: Copepoda, Ostracoda) and water mites (Acari: Halacaridae, Hydrachnellae). In. Botosaneanu, L. (ed.), Studies in Crenobiology. Backhuys Publishers, Leiden: 167–182.

Ghetti, P. F. & K. McKenzie, 1981. Ostracodi (Crustacea: Ostracoda). Guide per il riconoscimento delle specie animali delle acque interne italiane. Consiglio Nazionale delle Ricerche.

González Mozo, M. E., K. Martens & A. Baltanás, 1996. A taxonomic revision of European *Herpetocypris* Brady and Norman, 1889 (Crustacea, Ostracoda). Bulletin de l'institut Royal de Sciences Naturelles de Belgique 66: 93–132.

Golterman, H. L., R. S. Clymo & M. A. M. Ohmstand, 1978. Methods for physical and chemical analysis of fresh water. I.B.P. Handbook No. 8. Blackwell, Oxford.

Hammer, Ø., D. A. T. Harper & P. D. Ryan, 2001. PAST: Paleontological Statistics Software Package for education and data analysis. Palaeontologia Electronica 4: 9 pp, http.//palaeo-electronica.org/2001_1/past/issue 1_01.htm.

Koroleff, F., 1970. Direct determination of ammonia in natural waters as indophenol blue. Information on Techniques and Methods for Seawater Analysis, Interlaboratory Report No. 3, I.C.E.S., Charlottenlund: 19–22.

Külköylüoğlu, O. & G. L. Vinyard, 2000. Distribution and ecology of freshwater Ostracoda (Crustacea) collected from springs of Nevada, Idaho, and Oregon: a preliminary study. Western North American Naturalist 60: 291–303.

McKenzie, K. G. & A. Moroni, 1986. Man as an agent of crustacean passive dispersal via useful plants – exemplified by Ostracoda *ospiti esteri* of the Italian ricefiels ecosystem – and implications arising therefrom. Journal of Crustacean Biology 6: 181–198.

Meisch, C., 1984. Revision of the recent Western Europe species of genus *Potamocypris*. Part I: Species with short swimming setae on the second antennae Travaux scientifiques du Musée national d'histoire naturelle de Luxembourg 3: 1–55.

Meisch, C., 2000. Freshwater Ostracoda of Western and Central Europe. Spektrum Akademischer Verlag GmbH, Heidelberg, Berlin.

Mezquita, F., G. Tapia & J. R. Roca, 1999. Ostracoda from springs of the Eastern Iberian Peninsula: ecology, biogeography and palaeolimnological implications. Palaeogeography, Palaeoclimatology, Palaeoecology 148: 65–85.

Patton, J. & W. Reeder, 1956. New indicator for titration of calcium with (Ethylenedinitrilo)-tetraacetate. Analytical Chemistry 28: 1026.

Petrucci, F., B. Bigi, M. Pecorari & M. E. Vidonti Tani, 1982. Le risorgive della pianura parmense e piacentina. Geografia Fisica e Dinamica Quaternaria 5: 277–284.

Roca, J. R. & A. Baltanás, 1993. Ecology and distribution of Ostracoda in Pyrenean springs. Journal of Crustacean Biology 13: 165–174.

Rodier, J., 1978. L'analyse de l'eau. Dunod, Orleans.

Rossi, V., G. Benassi, M. Veneri, C. Bellavere, P. Menozzi, A. Moroni & K. G. McKenzie, 2003. Ostracoda of the Italian ricefields thirty years on: new synthesis and hypothesis. Journal of Limnology 62: 1–18.

Russo, E. & A. Zavatti (eds), 2001. Nitrati. Acqua e suolo da salvaguardare. Prevenzione dell'inquinamento dei sistemi idrogeologici. I quaderni di Arpa, Agenzia Regionale Prevenzione e Ambiente dell'Emilia Romagna.

Särkkä, J., L. Levonem & J. Mäkelä, 1997. Meiofauna of springs in Finland in relation to environmental factors. Hydrobiologia 347: 139–150.

Stoch, F., 1998. Fauna a microcrostacei (copepodi e ostracodi). In Cantonati, M. (ed.), Le sorgenti del Parco Adamello-Brenta. Ricerche idrobiologiche su fonti non captate. Parco Documenti: 151–159.

Valderrama, J. C., 1981. The simultaneous analysis of total nitrogen and total phosphorus in natural waters. Marine Chemistry 10: 109–122.

Hydrobiologia (2005) 542:297–314
H. Segers & K. Martens (eds), Aquatic Biodiversity II
DOI 10.1007/s10750-004-1873-3

Representation of aquatic invertebrate communities in subfossil death assemblages sampled along a salinity gradient of western Uganda crater lakes

Bob Rumes*, Hilde Eggermont & Dirk Verschuren
Department of Biology, Section Limnology and Animal Ecology, Ghent University, K.L. ledeganckstraat 35, Ghent 9000, Belgium
*(*Author for correspondence: E-mail: Bob.rumes@Ugent.be)*

Key words: Africa, aquatic insects, Bryozoa, *Chaoborus*, Diptera, Ephemeroptera, paleo-environmental reconstruction, paleolimnology, Trichoptera, Turbellaria

Abstract

We analysed subfossil death assemblages of aquatic invertebrate communities in a salinity series of 35 western Uganda maar-crater lakes to evaluate their potential as biological indicators of past habitat conditions in paleo-environmental research. The study region encompasses the climatological and hydrological gradient between the dry floor and moist shoulders of the Edward-George branch of the East African Rift Valley, and includes mesotrophic to hyper-eutrophic, and shallow unstratified to deep meromictic lakes with a surface-water salinity range between 101 and 135 400 μS/cm. Focusing on non-chironomid aquatic invertebrates with good fossil preservation, we found that fossil larval remains of the Dipteran families Culicidae, Ephydridae, and Stratiomyidae are good indicators of saline environments. Our data further suggest that the abundances of Bryozoan statoblasts and Chaoboridae are indicative of, respectively, the fraction of the littoral zone covered by aquatic macrophytes and of lake trophic state, but a lake reference data set more specifically designed to cover variation in these environmental factors will be needed to determine the strength of these relationships. In these small, simple lake basins, recent death assemblages recovered from a single mid-lake surface-sediment sample provides a more complete inventory of local aquatic invertebrate communities and the distribution of species among lakes than exploratory live sampling of those taxa in a selection of littoral, benthic and pelagic habitats.

Introduction

Featuring great ecological diversity, rapid generation turnover, and efficient dispersal, aquatic invertebrates are sensitive indicators of natural and anthropogenic changes in lake ecosystems. The use of biological indicators in applied limnological research is well established in north-temperate regions (Rosenberg, 1993), but in most tropical regions our knowledge of the ecology of individual species of aquatic biota is too fragmentary or anecdotal to use them as reliable indicators of specific environmental conditions. This is unfortunate, because sound management of scarce freshwater resources is critical to the socio-economic development of many tropical countries, and biological assessment of surface waters can be a powerful yet low-cost tool to resolve water quality issues caused by water extraction, pollution, or other human activities.

The species composition and abundance of aquatic invertebrate communities vary in space and time (Timms & Moss, 1984; Lampert, 1993; Burks et al., 2002), making intensive sampling schemes a basic requirement of any applied ecological research based on them. One reason for the current scarcity of ecological data on lake biota in tropical lakes is that such sampling and monitoring programmes are difficult to organise and maintain in study sites remote from well-equipped laboratories. Thus, not only is it difficult to assess current biological diversity in tropical lakes, but also little historical data exist to determine how aquatic ecosystems have changed under the influence of human activities in their drainage basins. However, most aquatic invertebrate taxa leave at least some morphologically identifiable remains in lake sediments (Frey, 1964, 1988). Paleolimnological techniques can therefore help to overcome the above problems. First, death assemblages extracted from recently deposited surface sediments can be used to obtain a spatially and temporally integrated view of present-day invertebrate species diversity and community structure. Second, data on the distribution of particular taxa in surface sediments of a series of lakes along environmental gradients can be used to assess their habitat preferences and ecological tolerances in a quantitative manner, and thus to calibrate their potential as biological indicators of environmental change. Third, analyses of sub-recent fossil assemblages in dated sediment cores permit reconstruction of environmental change through time. This compensates for the lack of historical data on how individual lakes have responded to human impact on the aquatic ecosystem itself and the surrounding terrestrial landscape, and how they functioned in their pristine state when human impact was non-existent or negligible (Smol et al., 1991).

Provided that the environmental optima and tolerances of the studied organisms are known, analysis of aquatic invertebrate fossils in sediment cores is also a powerful method to reconstruct past climate history, and to elucidate climate-change processes at time scales from decades to millennia. However, reconstructing climate by documenting the history of climate-sensitive lakes is a highly complex task, due to indirect and non-linear relationships between the biological (or sedimentological, or geochemical) indicators and lake-level or water-chemistry fluctuations, and between lake-evel fluctuation and climate change (Fritz, 1996). This is why lake-based climate reconstruction is intrinsically multi-disciplinary, relying on inferences from a range of independent climate-proxy indicators preserved in lake sediments. Aquatic biota currently used as water-chemistry indicators for climate reconstruction in Africa are the diatoms (Gasse et al., 1983, 1995), Chironomidae (Verschuren, 1997; Verschuren et al., in press) and to a lesser extent the Ostracoda (Cohen et al., 1983). The potential of other aquatic invertebrates as indicators of past climate-driven environmental change in African surface waters remains to be explored.

This study is an exploratory assessment of the preservation potential of various taxa of aquatic invertebrates in the sediment record of African lakes, and of their indicator value for specific environmental conditions. These include taxa which in north-temperate regions are already established tools in Quaternary paleoecology [e.g. the testate Amoebae (Warner, 1990a), Bryozoa (Francis, 1997, 2001), Trichoptera (Solem & Birks, 2000), and Chaoboridae (Uutala, 1990; Uutala & Smol, 1996)], while other taxa have not previously received much attention. As this study's main objective is to discover new potential among these poorly studied taxa in an African context, our analysis excluded the Chironomidae, Cladocera and Ostracoda. We also excluded several soft-bodied taxa, such as copepods, leeches and oligochaete worms, whose remains are poorly preserved in lake sediments.

Methods

Study area

The Lake Edward-George branch of the African Rift Valley and adjacent uplands in western Uganda contain about 80 volcanic crater lakes, distributed over four clusters identified as the Fort Portal, Kasenda, Katwe-Kikorongo, and Bunyaruguru lake districts (Melack, 1978; Fig. 1). Natural vegetation surrounding the lakes mainly reflects the local rainfall regime and thus water balance, and varies from dry grassland on the floor of the Rift Valley to moist upland forest on its shoulders. In most wetter, more densely populated areas, natural vegetation around the crater basins has now been replaced by an agricultural landscape, but the steep inner crater walls usually remain relatively undisturbed, except by firewood

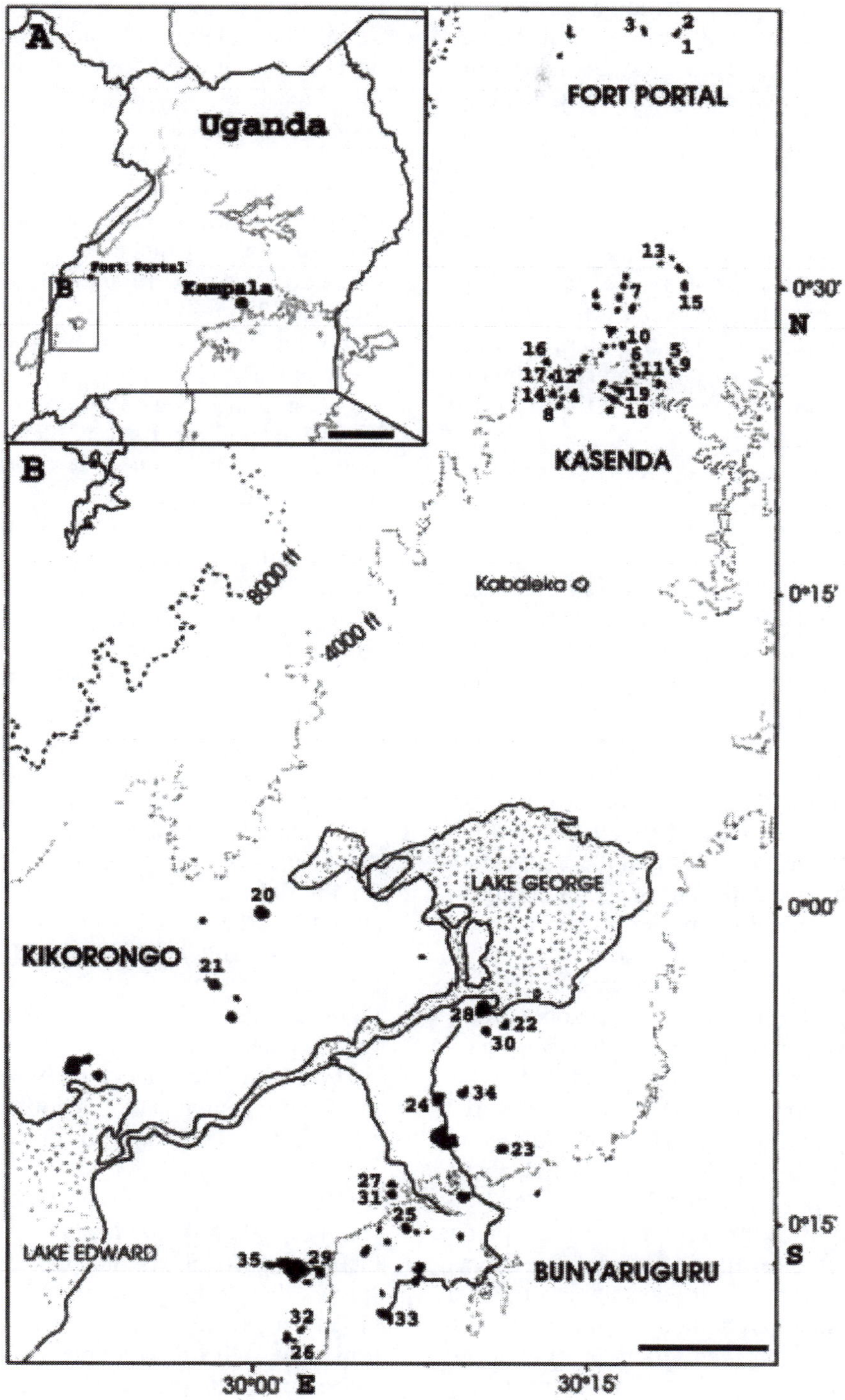

Figure 1. (A) Map of Uganda. Scale bar equals 100 km. (B) Part of the western arm of the African Rift Valley and neighbouring uplands in western Uganda, with the four clusters of volcanic crater lakes (adapted from Melack, 1978). Scale bar equals 10 km. N: North; S: South; E: East.

collection and cattle trails. Routine burning of secondary vegetation and intense subsistence agriculture inside some of the more gently sloping crater basins augments their phosphorus loading, and is then a cause of eutrophication (Kizito et al., 1993). This study is based on biological and paleolimnological exploration of about half of all crater lakes in western Uganda, spread along the regional Rift Valley ecotone between 975 and 1300 m elevation. These small lakes (surface area 0.01–3.84 km^2; Claessens, 2002) cover a salinity gradient from 101 to 135 400 $\mu S/cm$, and range from mesotrophic to highly eutrophic, and from shallow polymictic to deep and permanently stratified (Table 1).

Sampling and analytical procedures

Four field campaigns between March 2000 and July 2002 yielded live collections and surface-sediment death assemblages of the aquatic invertebrate communities in 35 crater lakes. We collected field data on the physical and chemical limnology of the lakes, and mapped their bathymetry using echo-sounding linked to satellite-based positioning (GPS). In 16 of the freshwater lakes, profiles of temperature, specific conductivity, oxygen and pH through the offshore water column were used to determine the relative depths of the epilimnion and hypolimnion (or monimolimnion in permanently stratified lakes), as indication of the fractions of water mass and lake bottom sufficiently aerated to permit presence of zooplankton and zoobenthos. Live sampling was done by repeatedly sweeping a 50-μm mesh net across planktonic and epibenthic habitats in near-shore (littoral) environments, and the off-shore planktonic (pelagic) environment. Samples were preserved in a 5% formaldehyde solution neutralized with powdered calcite. Bottom sediments were collected intact using either a piston corer operated with drive rods or a gravity corer suspended from a wire, depending on water depth. In deep meromictic lakes, sediment recovery by gravity corer was often poor due to expansion of methane gas in the sediment during ascent through the water column, which forced the sediment out of the tube. In these lakes, unstratified samples of flocculent surface sediments were collected using a horizontal water sampler equipped with extra weight. Where lake depth did not exceed the capability of our gravity-coring equipment (~100 m), sediment samples were taken in the deepest part of the lake, which in these simple crater basins is almost always near their geographical center. In Lake Kyaninga, where maximum depth exceeded our measuring capacity, surface sediments were sampled at 57 m depth from the floor of the shallower south basin. Surface cores were extruded upright at the lake in 1-cm or 2-cm increments with a fixed-interval sectioning device, and the mud was transferred to polyethylene bags for transport. In the laboratory, wet weight, dry weight and organic-matter content of surface sediments were determined by volumetric Loss-On-Ignition methods (Bengtsson & Enell, 1986).

Preserved samples of the living aquatic invertebrate fauna were rinsed on a 50-μm mesh sieve to replace the formaldehyde solution by distilled water. They were then transferred to a transparent counting tray and scanned for identification at 30× under a combination of incident and transmitted light. Our reference literature on African aquatic invertebrates included Verbeke (1957) and McGowan (1972) for the Chaoboridae, Rickenbach (1981) for the Culicidae, Philippon (1981) for the Simuliidae, Teskey (1984) for other Diptera, Dethier (1981) for the Hemiptera, Westfall (1984) for the Odonata, Demoulin (1981) and Edmunds (1984) for the Ephemeroptera, Morse & Holzenthal (1984) and Williams (1988) for the Trichoptera, Balogh (1972) and Smith & Cook (1991) for the Acari, Borg (1936) and Wiebach (1970) for the Phylactolaemata, Warner (1990b) for the Neorhabdocoela and Chardez (1964) for the Rhizopoda. Some remains of Trichoptera were identified by J. Dean, Environment Protection Authority, New South Wales, Australia.

Voucher specimens were picked and preserved in 70% ethanol to facilitate identification of incomplete fossil specimens. Evaluation of live species distribution among the 35 study lakes is based on presence–absence data from 88 planktonic and epibenthic samples.

Sediment samples were prepared for analysis of invertebrate remains by deflocculation in a 10% solution of potassium hydroxide (KOH) at 70 °C (Walker & Paterson, 1985). Between 2 and 33 cm^3 of mud from the upper 5 cm of the sediment cores

Table 1. Study lakes in western Uganda ordered by cluster, with data on surface area (SA), maximum depth (Zmax), surface-water conductivity (SCond.), and mixing regime (Me: Meromictic, Mo: Monomictic, P: Polymictic)

No.	Lake/cluster	SA (in ha)	Zmax (in m)	SCond. (in μS/cm)	Mixing regime
	Fort Portal				
1	Ekikoto	1.6	72.0	464	Me
2	Kayihara	0.8	55.0	498	Me
3	Kyaninga South	24.0	57.0	423	Me
	Kasenda				
4	Kanyamukale	2.1	10.8	909	Mo
5	Kanyanchu	0.7	4.7	598	P
6	Kasenda	9.3	13.5	275	Mo
7	Kifuruka	13.0	5.3	347	P
8	Kyanga	11.8	55.0	1055	Me
9	Kyerbwato	1.4	12.7	463	Mo
10	Lugembe	8.0	18.6	395	Mo
11	Murusi	21.4	65.4	366	Me
12	Njarayabana	12.2	38.0	879	Mo
13	Nkuruba	1.3	34.8	361	Mo
14	Ntambi	31.7	110.0	5821	Me
15	Nyabikere	39.5	47.0	263	Me
16	Nyamiteza	33.6	100.0	1024	Me
17	Nyamugosani	11.4	37.0	971	Mo
18	Wandakara	3.2	12.1	1125	Mo
19	Wankenzi	16.5	58.0	496	Me
	Kikorongo				
20	Kikorongo	92.0	10.7	21 700	Mo
21	Kitagata	62.1	8.9	135 400	Mo
	Bunyaruguru				
22	Bagusa	32.8	0.9	61 100	P
23	Bugwagi	60.2	71.0	440	Me
24	Chibwera	76.4	11.7	453	Mo
25	Kamweru	20.3	45.0	170	Me
26	Karolero	5.1	15.5	143	Mo
27	Katinda	43.5	17.0	743	Me
28	Kibengo	88.1	5.4	197	P
29	Kyasunduka	50.3	2.0	156	P
30	Maseche	31.9	0.1	68 200	P
31	Mirambi	53.0	22.0	652	Me
32	Murabyo	26.6	14.0	157	Mo
33	Nkugute	89.0	58.0	101	Me
34	Nshenyi	44.3	0.1	31 600	P
35	Nyamusingere	384.0	4.4	620	P

was processed to yield at least 50 identifiable fossils (range 7–278, mean 90). The density of invertebrate remains present was strongly related to organic-matter content, suggesting that it is mainly controlled by the degree of dilution with clastic mineral inputs. Deflocculated samples were rinsed on a 150-μm mesh sieve. The retained organic residue was transferred to a transparent counting tray and scanned at 30× magnification under transmitted light, using a binocular dissecting microscope. All distinctive invertebrate remains were picked out with a fine needle, mounted in glycerine (50% glycerol and water) on microscope slides, and analysed at up to 400× magnification with an Olympus BH-2 compound microscope. The fossil remains were identified using the same literature as for the living fauna, and comparison with our reference collection of intact specimens. Taxonomic resolution of the fossil identifications is taxon-specific, as it depends on the quality of preservation and the diagnostic value of preserved components. To ensure consistency of identification throughout the collection, the location of all fossils on the microscope slides was noted for future reference, and representative well-preserved specimens were photographed with a Nikon Coolpix 990 digital camera. Fossil densities are expressed as absolute numbers of fossils per gram dry organic matter (Deevey et al., 1977), which compensates for clastic dilution and has been found to correlate well with live population biomass (Hann et al., 1994).

In this exploratory study, we analysed the distribution of aquatic invertebrate taxa among Uganda crater lakes in relation to salinity measured as surface-water conductivity, and Secchi-disk transparency as a proxy for lake trophic status. Specific conductivity of lake water, measured in the field with a Hydrolab Quanta CTD profiler and expressed in μS/cm at 25 °C, reflects the osmotic stress that biota are exposed to. Available data on epilimnetic nutrient concentrations and algal biomass in the study lakes are insufficiently homogeneous to analyse invertebrate species distribution in relation to these factors. Our use of Secchi-disk transparency as a first-order proxy for lake trophic status is based on the general decrease in water transparency with increasing total phosphorus (TP) and algal biomass (Carlson, 1977), allowing the status of oligotrophic–mesotrophic (TP < 30 mg/m^3) and hyper-eutrophic lakes (TP > 120 mg/m^3) to be predicted from Secchi-disk depth. The relationship between TP and Secchi-disk transparency has been found to be weak in the range of eutrophic lakes (30 mg/m^3 $<$ TP $<$ 120 mg/m^3) (Seip et al., 1992), but given the exceptionally wide range of trophic status encountered among our study lakes this should not impede first-order analysis of species distribution patterns. In most Uganda crater lakes, suspended mineral matter has a negligible influence on water transparency, and thus Secchi-disk depth.

Results

Excluding Chironomidae, Ostracoda and Cladocera, our analysis of the 35 surface-sediment samples yielded a total of 3163 fossils which could be sorted into 32 distinct types of invertebrate remains. Five of these are small exoskeletal fragments of various aquatic insect larvae, but could not be identified due to a lack of diagnostic features. The other 27 fossil types can be assigned to 21 different taxa of Protozoa, Turbellaria, Bryozoa, Acari and Insecta. The large majority (86%) of all identified fossils are also chitinous exoskeletal components of aquatic insect larvae and pupae, representing 6 orders and 14 families.

Class Insecta

Order Diptera

Chaoboridae (sub-order Nematocera) or phantom midges have a life cycle consisting of four aquatic larval instars, a pupa and short-lived aerial adults. The larvae are predators on benthic and planktonic invertebrates (Mumm, 1997), and may constitute a major part of the benthic and planktonic fauna in lakes (Håkanson & Jansson, 1983). The mandibles of larval Chaoboridae preserve well in lake sediments (Frey, 1964) and can be identified to the species level (Uutala, 1990). The mandibular fans with their strongly curved setae are also found (Fig. 2), but not identifiable at the species level. In North America, *Chaoborus* mandibles preserved in lake sediments are established indicators of past changes in fish community structure (e.g., Uutala, 1990; Lamontagne & Schindler, 1994), based on the different ability of species to coexist with zooplanktivorous fish. In East Africa, three *Chaoborus* species are known: *Chaoborus* (*Neochaoborus*) anomalus Edwards, *C.* (*Sayomyia*) *ceratopogenes* Theobald, and *C.* (*Sayomyia*) *edulis* Edwards (Verbeke, 1957; McGowan, 1972). In our study lakes, *C. anomalus* (Fig. 3) was found only in Lake Kibengo, while *C. ceratopogenes* (Figs 2

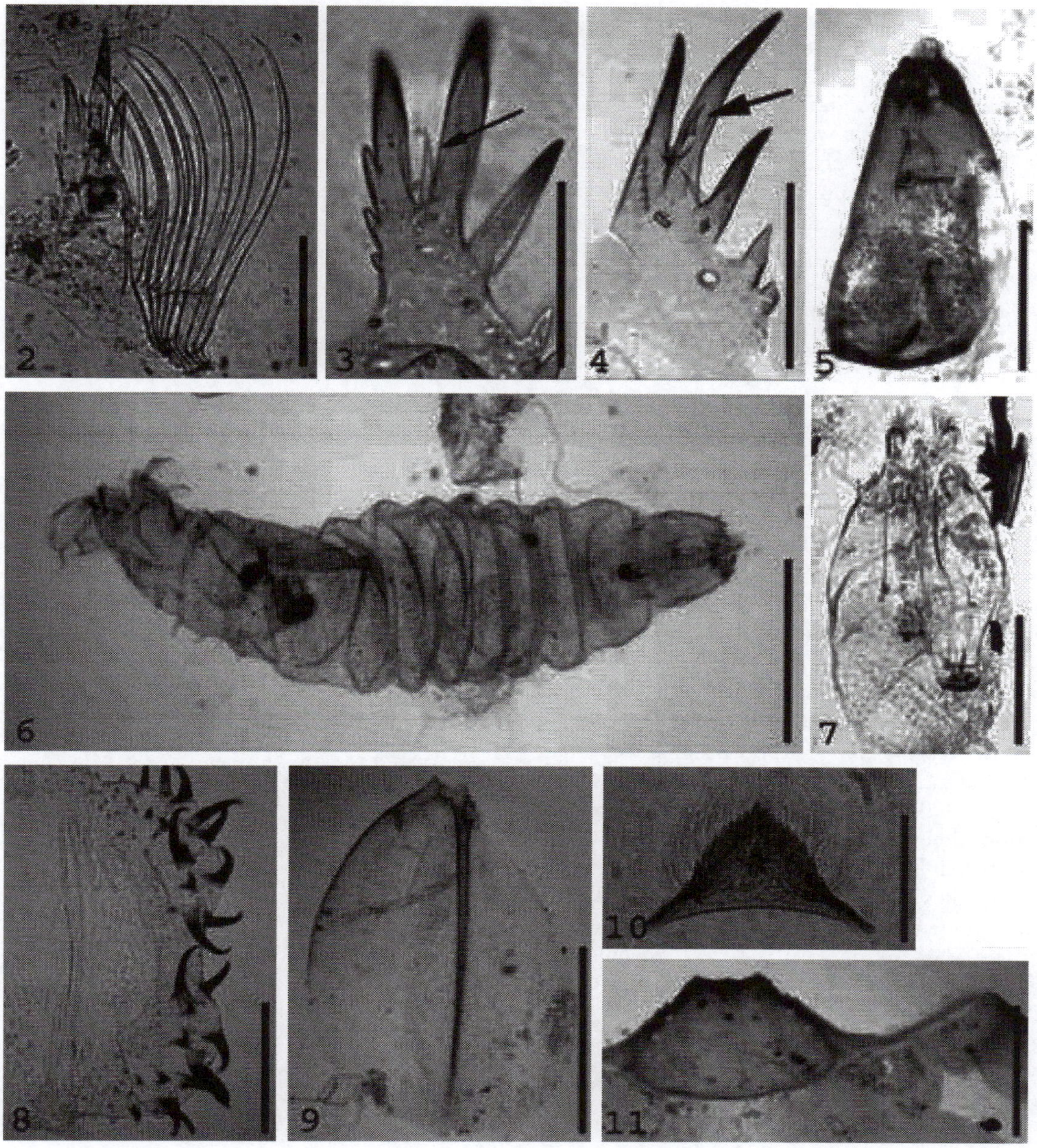

Figures 2–11. Subfossil remains of aquatic invertebrates from western Uganda crater lakes. 2. *Chaoborus ceratopogenes* mandibular fan. 3. *Chaoborus anomalus* mandible. 4. *Chaoboms ceratopogenes* mandible. 5. Ceratopogonidae head capsule. 6. Stratiomyidae larva. 7. Stratiomyidae, head capsule with characteristic mandibular-maxillary complex. 8. Ephydridae parapods. 9. Culicidae flattened joints of the posterior torsi. 10. Culicidae mouthparts. 11. Simuliidae hypostoma. Scale bars equal 50 μm for Figures 2–4, 10, and 11; 200 μm for 5 and 8; 500 μm for 6; 100 μm for 7; and 250 μm for 9.

and 4) occurred in all freshwater lakes, including Kibengo, and in saline Lake Ntambi (5821 μS/cm).

Ceratopogonidae (sub-order Nematocera, biting midges; Fig. 5), Stratiomyidae (sub-order Brachycera, soldier flies; Figs 6 and 7) and Ephydridae (sub-order Brachycera, shore or brine flies; Fig. 8) occur in a wide variety of inland aquatic environments (Timms, 1983; Teskey, 1984), but are typical components of the benthic community in saline lakes (Hammer, 1986), because they osmoregulate better than most other aquatic insects (Bayly, 1972). Ephydrid larvae and adults are often the most abundant inhabitants of saline lakes throughout the world (Herbst, 1986; Stephens, 1990). This is also reflected in this study. Ephydridae were represented by their larval parapods (Fig. 8) in Lake Ntambi (5821 μS/cm) and all three hypersaline lakes (61 100–135 400 μS/cm), but in none of the freshwater lakes. Stratiomyidae were represented by complete larval skins (Fig. 6), or the head capsule with its characteristic mandibular-maxillary complex (Fig. 7), found in lakes Maseche (68 200 μS/cm) and Kitagata (135 400 μS/cm). We only found a single specimen of Ceratopogonidae, a head capsule of *Bezzia* sp. (Fig. 5) in Lake Kikorongo (21 700 μS/cm).

Culicidae (sub-order Nematocera, true mosquitoes; Figs 9 and 10) have four aquatic larval instars and a pupa, which usually floats upside down from the water surface (Newson, 1984). The Culicidae include the well-known genera *Culex*, *Aedes*, and *Anopheles*, distributed worldwide in all types of standing water and in Africa important vectors of waterborne diseases such as malaria (*Anopheles*), yellow fever (*Aedes*) and West Nile virus (mainly *Culex*). In our collection their remains (larval mouthparts and the flattened joints of the posterior torsi or pupal 'paddles'; Figs 9 and 10) were found only in the hypersaline lakes Maseche, Bagusa and Nshenyi. The presence of *Aedes natronius* in Lake Maseche was previously noted by Beadle (1932).

Larvae of the Simuliidae (sub-order Nematocera, black flies; Fig. 11) are restricted to running-water environments (Peterson, 1984). They have a sturdy head capsule with a hypostoma, typically consisting of nine teeth, at the ventral side (Philippon, 1981). In certain genera the hypostomal teeth can be strongly reduced (Fig. 11; Currie & Walker, 1992). A single simuliid hypostoma was recovered from the surface sediments of Lake Wankenzi (496 μS/cm), which distinguishes itself from all other surveyed freshwater lakes in being fed by a stream. Hence, the Simuliid fossil extracted from mid-lake sediments likely derives from the inflowing stream rather than from the lake fauna.

Order Hemiptera

Adults of the family Corixidae have a very short, triangular, unsegmented rostrum, and a front tarsus consisting of a single scoop-like segment fringed with stiff setae (Polhemus, 1984). This characterizes Hemipteran remains (Fig. 12) found in the sediments of saline lakes Ntambi (5821 μS/cm) and Kikorongo (21 700 μS/cm). A total of 275 Corixidae fossils were found in the single sediment sample from Lake Kikorongo, where they comprise 98% of all diagnostic faunal remains. In both lakes the species involved is likely *Micronecta scutellaris* Stål, which is abundant in live samples from Lake Ntambi. Analysis of live samples from freshwater lakes yielded large numbers of Notonectidae and in lesser numbers Gerridae, Belostomatidae, Pleidae and Macroveliidae, but no fossil remains from these lakes could be assigned with confidence to any of these taxa.

Order Odonata

Exoskeletal elements of dragonfly (sub-order Anisoptera) and damselfly (sub-order Zygoptera) nymphs no doubt constitute part of the unidentified aquatic-insect remains found in the freshwater lakes, but clearly identifiable specimens were rare. A single zygopteran nymph palpal lobe was found in Lake Ntambi (Fig. 13), despite a salinity of 5820 μS/cm. Palpal lobes are the fangs on the lower lip with which dragonfly nymphs catch their prey, and are armed with hooks, spines, teeth and raptorial setae that vary with family and genus (Westfall, 1984).

Order Ephemeroptera

Ephemeroptera (mayflies) are generally stenotypical of freshwater, although some Baetidae and Ephemeridae do occur in waters with elevated salinity (Brener & Sloan, 1954; Chadwick & Feminella, 2001). In this collection, fossil Ephemeroptera mandibles (Figs 14 and 15) and maxillae (Fig. 16) were found in both fresh and saline lakes with conductivity ranging from 101 to 31 600 μS/cm. One type of mandible belonging to the Ephemeridae (Fig. 14) was found in 19 lakes, all freshwater environments with conductivity ranging from 101 to 1125 μS/cm. The second type of mandible (Fig. 15) was found in 22 lakes, both fresh and saline, with conductivity ranging from

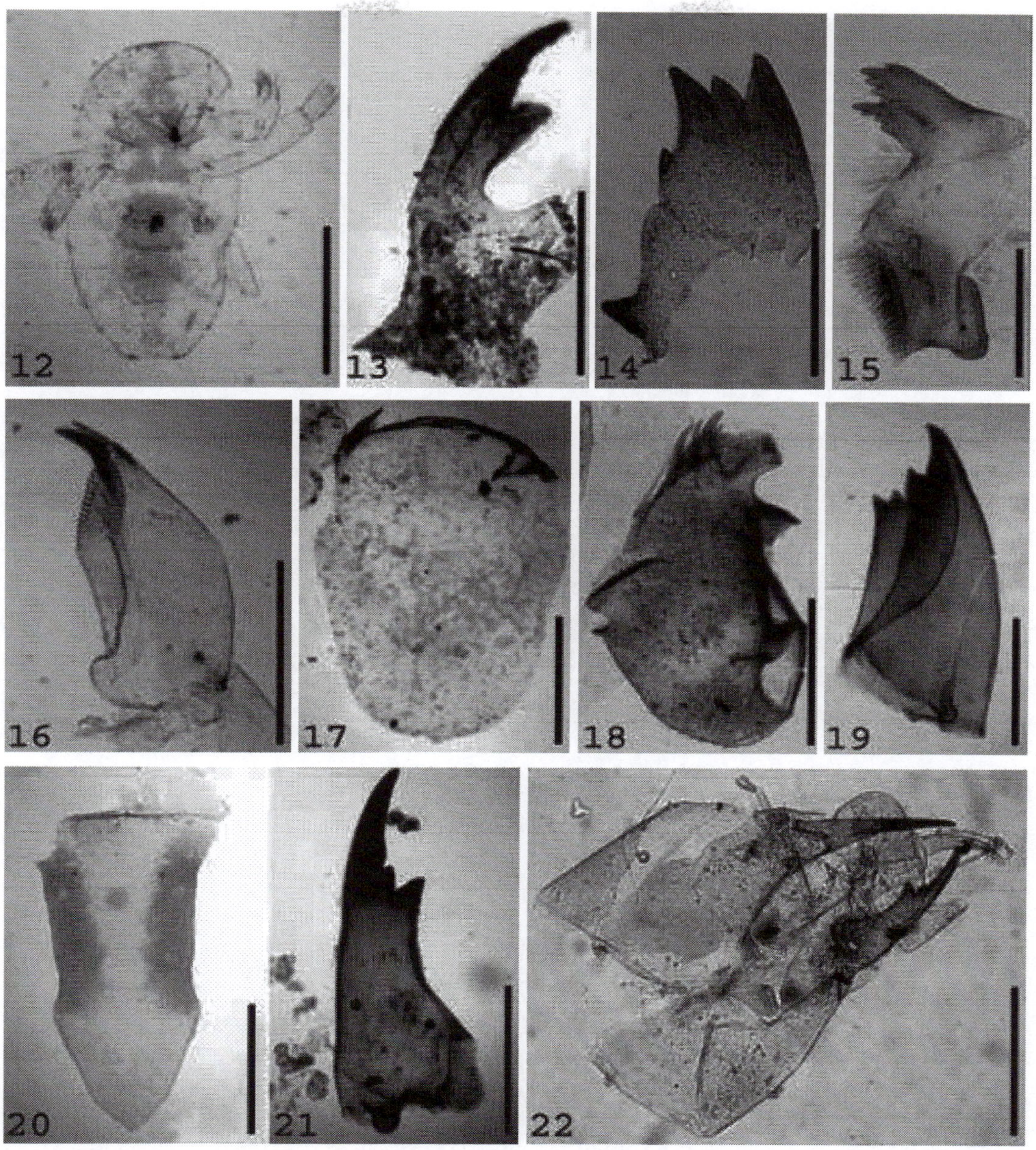

Figures 12–22. Subfossil remains of aquatic invertebrates from western Uganda crater lakes. 12. Corixidae front tarsus. 13. Odonata mandible. 14. Ephemeridae mandible. 15. Ephemeroptera mandible. 16. Ephemeroptera maxilla. 17. Leptoceridae frontoclypeus. 18. Leptoceridae anal claw. 19. Hydropsychidae mandible. 20. *Ecnomus* sp. frontoclypeus. 21. *Ecnomus* sp. mandible. 22. *Oecetis* sp. head capsule. Scale bars equal 1000 μm for Figure 12; 100 μm for 13, 14 and 16; 150 μm for 15; 200 μm for 17–22; and 500 μm for 22.

170 to 31 600 μS/cm, often together with an Ephemeropteran maxilla (Fig. 16). Unfortunately, the generalised morphology of these skeletal fragments precluded assignment to a particular mayfly family. Live samples from these lakes included Larvae of the Caenidae, Baetidae, and Leptophlebiidae, all of which possess this type of mandible.

Order Trichoptera

Most Trichoptera (caddisflies) are characteristic of fresh waters and usually associated with streams or

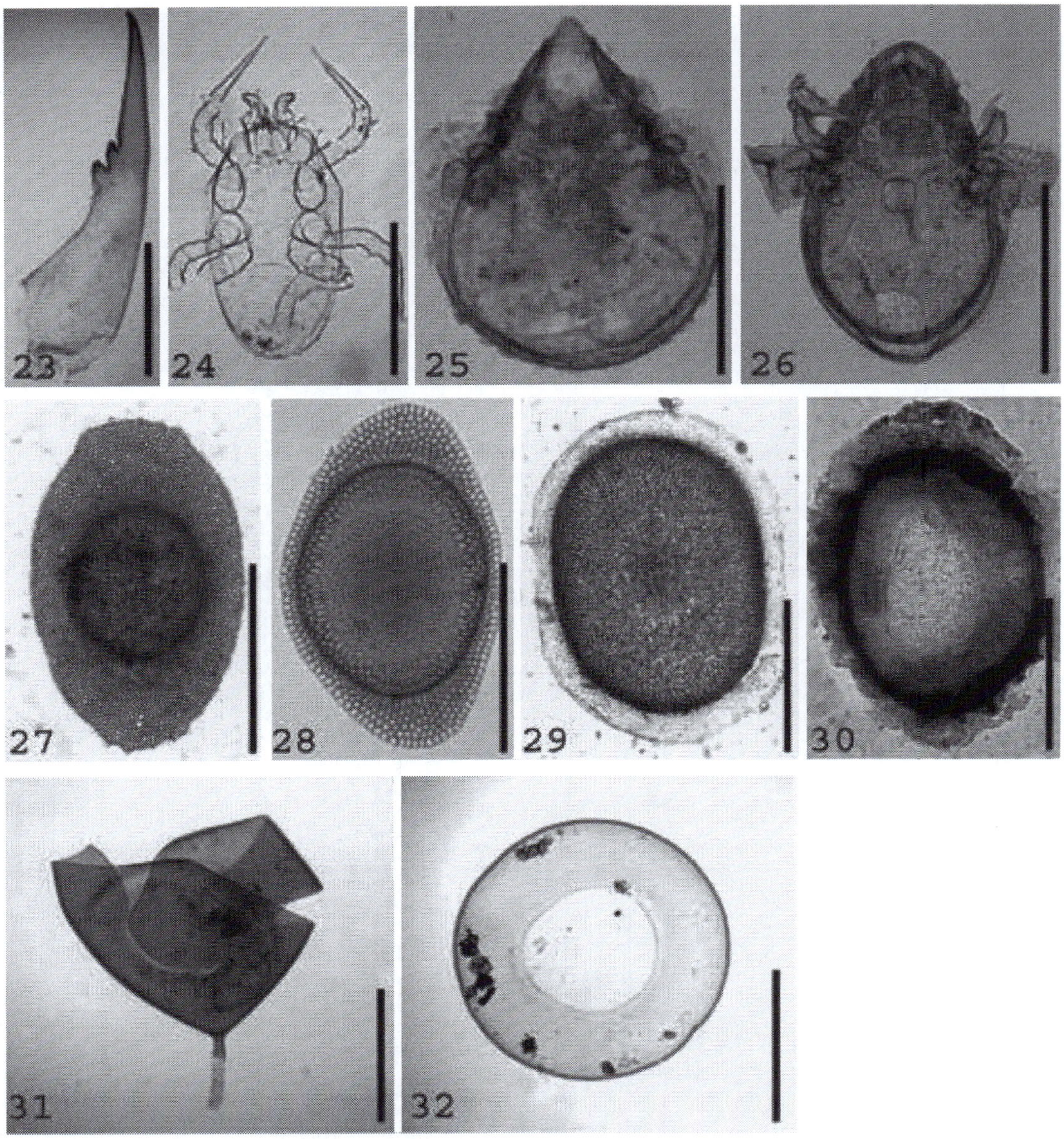

Figures 23–32. Subfossil remains of aquatic invertebrates from western Uganda crater lakes. 23. *Oecetis* mandible. 24. Hydracarina. 25. Pterogasterina. 26. Apterogasterina. 27. *Lophopodella carteri* flottoblast. 28. *Plumatella* sp. flottoblast. 29. *Plumatella* sp. frontal valve of sessoblast. 30. *Plumatella* sp. basal valve of sessoblast. 31. Neorhabdocoela egg capsule. 32. *Arcella* sp. test. Scale bars equal 200 μm for 23–26 and 28–30; 400 μm for 27; 50 μm for 31; and 100 μm for 32.

the littoral zone of lakes (Hammer, 1986). Trichoptera are represented by three exoskeletal components: the frontoclypeus (Figs 17, 20 and 22), the mandibles (Figs 19, 21 and 23), and the anal claw (Fig. 18), which is a part of the anal prolegs at the rear of the abdomen. All remains from the saline lakes (Figs 17 and 18) belonged to the Leptoceridae, a family with several representatives in saline habitats (Hammer, 1986). Remains from the freshwater lakes were more diverse, and include Hydropsychidae (Fig. 19), *Ecnomus* sp. of the Ecnomidae (Figs 20 and 21) and *Oecetis* sp. (Figs 22 and 23), a genus of Leptoceridae restricted to freshwater habitats (J. Dean, Environment Pro-

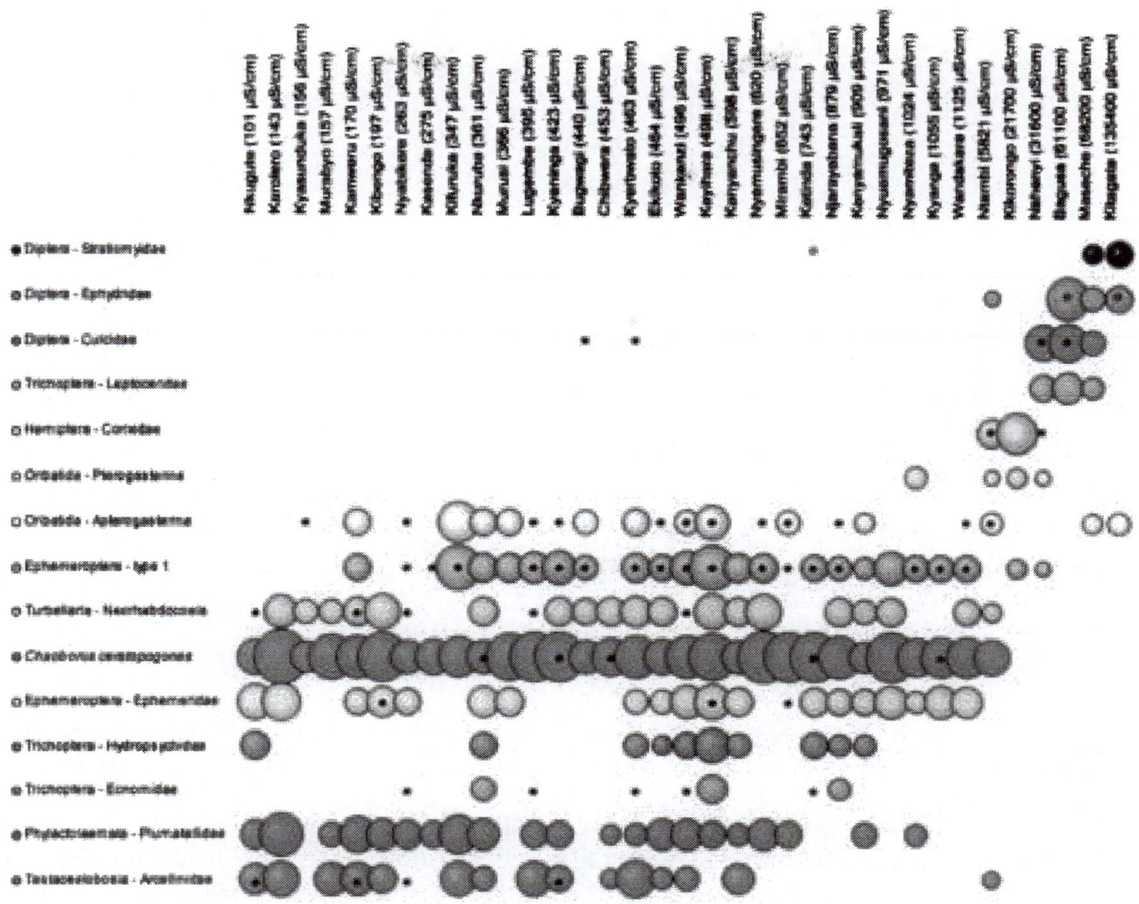

Figure 33. Absolute abundance of selected aquatic invertebrate taxa along the salinity gradient in western Uganda lakes. Coloured circle surface area is a fourth root transformation of fossil abundance per gram dry sediment. Small black dots indicate presence of the same taxon in live samples.

tection Authority, New South Wales, Australia, personal communication).

Subclass Acari

Remains of two groups of (semi-)aquatic mites were found, the order Hydracarina (true water mites), and the suborder Oribatida (oribatid mites) of the order Acariformes. Hydracarina (Fig. 24) are mostly soft-bodied and nearly always completely decompose shortly after death, which limits their potential in paleo-environmental studies (Frey, 1964). They were present in low numbers in the surface sediments of lakes spanning the entire salinity gradient, including lakes Ntambi (5820 μS/cm) and Kitagata (135 400 μS/cm), but since they are unlikely to be found in down-core sediments we do not further consider them here.

Except for a few primitive superfamilies, most oribatid mites have a characteristic tear drop-shaped body, with fused cephalothorax and abdomen, and a hard, sclerotized cuticula, which ensures their remains to be preserved in terrestrial and lacustrine sediments (Sølhoy & Sølhoy, 2000). Aquatic species abundantly occur among submerged and emergent vegetation (Warner, 1990b). Remains of two infraorders of the Oribatida were found: Pterogasterina, which can be recognised by their pteromorphae, characteristic wing-shaped lateral appendages of the cephalothorax; and Apterogasterina, which lack these structures.

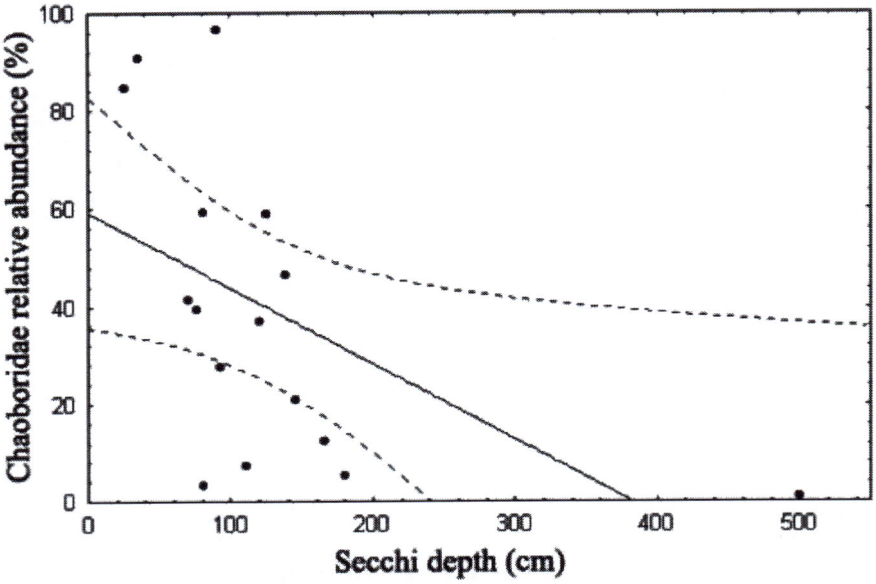

Figure 34. Correlation between the relative abundance of Chaoboridae mandibles and Secchi-disk depth ($r^2 = 0.39$; $p < 0.05$; $n = 16$). Dotted lines mark the 95% confidence interval.

Fossil Pterogasterina (Fig. 25) were found in saline lakes Kikorongo (21 700 μS/cm) and Nshenyi (31 600 μS/cm) and in freshwater Lake Nyamiteza (1024 μS/cm). Fossil Apterogasterina (Fig. 26) were found in over half of all freshwater lakes, and in saline lakes Ntambi (5820 μS/cm), Maseche (68 200 μS/cm) and Kitagata (135 400 μS/cm). No clear distinction could be made between the Apterogasterina occurring in saline lakes and those occurring in freshwater lakes.

Class Phylactolaemata (Bryozoa)

Freshwater bryozoans are colonial and sessile organisms, living on the underside of hard substrates such as stems and leaves of submerged macrophytes, floating wood, rocks, molluscs, and aquatic insects (Wiebach, 1980). Adult bryozoans hardly ever preserve in lake sediments (Warner, 1990b), but they produce encapsulated resting stages called statoblasts, that can be identified to the species level (Wiebach, 1970). They come in two types, free-floating flottoblasts (Figs 27 and 28), and sessoblasts (Figs 29 and 30) which are fixed to hard a substrate. Frontal and basal valves of sessoblasts possess a distinct morphology, due to the adhesive function of the basal valve. In this Uganda collection we found three types of Bryozoan fossils belonging to the Plumatellidae, but only in the freshwater lakes. The large, conspicuous flottoblasts of *Lophopodella carteri* Hyatt (Fig. 27) were found only in Lake Kanyamukali (909 μS/cm). Both flottoblasts (Fig. 28) and frontal (Fig. 29) and basal (Fig. 30) sessoblast valves of one or more *Plumatella* sp. were found in 21 freshwater lakes.

Order Neorhabdocoela

Flatworms of the order Neorhabdocoela (Platyhelminthes: Turbellaria) are soft-bodied animals that would not be expected to leave any well-preserved remains in lake sediments, but the capsules of their dormant eggs are remarkably resistant (Pennak, 1978), and have been noted even from interglacial sediments of ~125 000 years old (Frey, 1964). Some Neorhabdocoelan egg capsules are simply globular, while others possess a stalk-like extension (Fig. 31) for attachment to hard substrates (Warner, 1990b). In this collection, the large majority of Neorhabdocoela were found in freshwater lakes, but some were recovered from saline Lake Ntambi (5820 μS/cm).

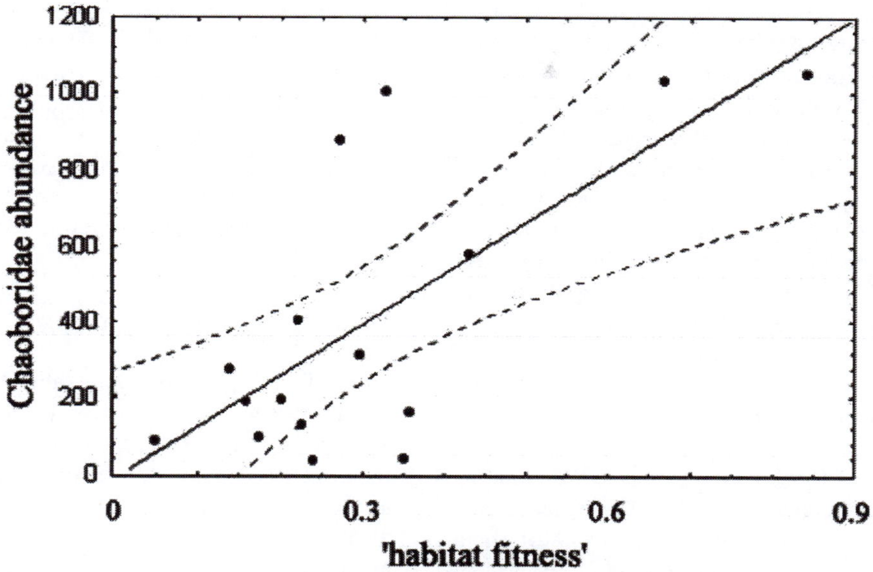

Figure 35. Correlation between the absolute abundance of Chaoboridae mandibles and a parameter for habitat fitness defined as the ratio between relative hypolimnion depth and Secchi-disk transparency ($r^2 = 0.51$; $p < 0.01$; $n = 16$). Dotted lines mark the 95% confidence interval.

Class Rhizopoda

Testate amoebae (order Testacida) are unicellular animals with a discrete shell or test enclosing the cytoplasm. These tests are readily preserved and abundant in peat and lake sediments (Warner, 1990a). Testate amoebae of the genus *Arcella* (Arcellinidae; Figs 32 and 33) were found mostly in fairly dilute freshwater lakes (101–598 μS/cm) again with the exception of saline Lake Ntambi (5821 μS/cm).

Discussion

Better knowledge of the ecology of individual aquatic invertebrate taxa in African lakes would allow them to be developed as biological indicators in applied limnological research, and provide a powerful yet low-cost method to monitor the quality of often scarce, valuable surface waters. Aquatic invertebrates with a good fossil record in lake sediments deserve particular attention, because they can also be used as biological indicators of past environmental change. Our exploratory assessment of the preservation potential and ecological indicator value of aquatic invertebrates in African lakes shows that the sediments of these lakes contain a great diversity of identifiable remains, besides the three major taxa traditionally used in paleoecology (Chironomidae, Ostracoda, and Cladocera). Further, as in comprehensive regional surveys of live aquatic invertebrate communities in African lakes (e.g., Tudorancea et al., 1989), the subfossil death assemblages show clear differentiation between freshwater and saline lake communities (Fig. 33), demonstrating that the taxonomic resolution achieved suffices to reveal the ecological specificity of individual species and group taxa in relation to selected environmental variables.

Notwithstanding adequate preservation of the investigated taxa, comparison of their distribution along the salinity gradient in live and fossil samples also reveals systematic differences: some taxa appear under-represented in the living fauna, while other taxa appear under-represented in the accompanying death assemblages. Among the first category are several subfossil taxa that entirely lack a counterpart in the living fauna, most notably the Thecamoeba, Bryozoa, *Chaoborus*, Trichoptera, and the mayfly family Ephemeridae. In most of these cases the discrepancy can be attributed to the selective nature of our live sampling, and has no bearing on the relative preservation potential of their remains. For example, our net sampling did not include searching for sessile colonies of *Plumatella* and *Lophopodella* on hard submerged substrates. Similarly, poor representation of *Chaoborus* in the live samples is likely due to our pelagic sampling having been restricted to the daytime epilimnion. Among the second category are the Mollusca, Odonata, and all Hemiptera excluding the Corixidae, which were often abundant in littoral epibenthos of freshwater lakes, but not (Mollusca and Hemiptera) or only rarely (Odonata) found as subfossils in surface sediments from the center of these lakes.

In the case of the Mollusca, lack of remains in the offshore sediments of Uganda crater lakes can be due to limited post-mortem transport of their relatively heavy shells, or dissolution of aragonite in the acidic, anoxic monimolimnia of the many meromictic lakes in our data set. Limited post-mortem transport of littoral taxa may also influence the fossil representation of aquatic insect larvae. Analysis of surface-sediment death assemblages from both near-shore (8 and 42 m depth) and offshore (108 m depth) locations in the relatively large (643 m diameter). Lake Ntambi showed higher densities of subfossil Hydracarina and Oribatida near-shore than in the center of the lake. Near-shore sediments were also found to contain numerous fossils of the Corixid *Micronecta scutellata*, but the mid-lake sample contained no such fossils. However, the mid-lake fossil sample from Lake Ntambi contained more benthic invertebrate taxa, including ostracodes, than the two near-shore samples combined (14 and 10, respectively). This indicates that although offshore post-mortem transport is limited in some taxa, it is compensated by better spatial integration from different littoral habitats around the lake shore.

The spatial and temporal integration of aquatic invertebrate species inhabiting a lake in surface-sediment death assemblages (Iovino, 1975; Walker et al., 1984; Dodson & Frey, 1991) is clearly evident from the higher number of taxa recovered than in live samples [6.2 ± 2.2 of preserved groups in fossil samples ($n = 35$) vs. 2.1 ± 3.0 in live samples ($n = 88$)], even when the latter are pooled per lake (3.0 ± 1.8; $n = 35$). In the small and simple depositional basins of these maar-crater lakes, one offshore surface-sediment sample includes fossils from all local aquatic microhabitats, and integrates seasonal and interannual faunal variation. This has clear advantages over live sampling, which can be time-consuming to collect if all micro-habitats must be sampled proportionally and throughout the year.

Indicator value of the fossil fauna

Apart from rare Culicidae remains, Chaoboridae larvae were the only truly pelagic zooplankton analysed in this study. *Chaoborus ceratopogenes* in fact occurred in all 29 sampled freshwater lakes, and in the oligosaline Lake Ntambi (5821 μS/cm). Their absolute abundance in mid-lake fossil assemblages showed a significant negative correlation with Secchi-disk depth ($r^2 = 0.39$; $p < 0.05$; $n = 16$) (Fig. 34). This may indicate that the decrease in visibility associated with high algal biomass improves their capacity to co-exist with zooplanktivorous fish (Wissel et al., 2003), and suggests a positive correlation between Chaoboridae abundance and lake trophic status. The absolute and relative abundances of Chaoboridae remains also increased in the presence of an anoxic hypolimnion (or monimolimnion in meromictic lakes). This may be related to their tolerance for oxygen depletion (Rine & Kesler, 2001) providing them with a deepwater refugium against fish predation. These two relationships together translate in a highly significant positive correlation in Uganda crater lakes between Chaoborid fossil abundance and a *Chaoborus* 'habitat fitness' parameter defined as the ratio of relative hypolimnion depth to Secchi-depth transparency ($r^2 = 0.51$; $p < 0.01$; $n = 16$) (Fig. 35). Assuming lake level (hydrological balance) and fish community structure remain stable through time, fossil Chaoborid abundance in the sediment record of Uganda crater lakes could potentially be developed as an indicator of chanes in lake trophic status.

In this study, all fossil remains and most living specimens of the Culicidae were limited to shallow (maximum depth 0.05–0.9 m) saline lakes (31 600–68 200 μS/cm), except for two Culicidae pupae found in live samples from the deeper freshwater lakes Bugwagi (440 μS/cm) and Kyerbwato (463 μS/cm). While mosquitoes can occur in a wide range of waters from springs and small temporary pools to both fresh and saline permanent waters (Newson, 1984), Culicidae abundance in permanent freshwater lakes is probably strongly limited by fish zooplanktivory. Mosquito eggs and pupae float on the water surface, and larvae must come to the water surface at frequent intervals for oxygen, which makes both life stages highly vulnerable to visual predation. Fossil fish scales and live juvenile fishes sampled in most surveyed freshwater lakes document the ubiquitous presence of fish in the western Uganda crater lakes. Stratiomyidae showed a similar pattern, with high fossil densities in the hypersaline lakes Maseche (68 200 μS/cm) and Kitagata (135 400 μS/cm) and only a single live larva found in all freshwater lakes combined (Lake Katinda; 743 μS/cm). Stratiomyidae and Culicidae as well as Ephydridae, Ceratopogonidae (all Diptera), Leptoceridae (Trichoptera) and Corixidae (Hemiptera) all share the ability to concentrate water in their rectal glands, allowing hypo-osmotic regulation in saline waters (Bayly, 1972; Timms 1983) and much better tolerance for elevated salinities than most other aquatic invertebrates. Consequently, they can survive and develop significant population densities in saline waters lacking zooplanktivorous fish.

Recovery of fossil mandibles belonging to the mayfly family Ephemeridae totaled 75 specimens from 19 lakes, whereas analysis of live samples yielded only four specimens from three lakes. In contrast, all other Ephemeroptera families combined yielded 346 live specimens vs. 94 fossil remains. This marked difference in fossil representation probably reflects a combination of behavioural and morphological differences. Firstly, Ephemeridae are burrowers whereas representatives of the Baetidae, Caenidae, and Leptophlebiidae are swimmers, sprawlers or clingers (Edmunds, 1984). This evidently results in underrepresentation of Ephemeridae in live hand-net samples. Secondly, Ephemeridae have large and sturdy mandibular tusks that are probably more resistant to decomposition, and physical damage after ingestion by vertebrate predators, than mandibles of other Ephemeroptera.

Among fossil Oribatid mites, the noted difference in the distribution of Apterogasterina (present in most freshwater lakes and several saline lakes) and Pterogasterina (found only in saline lakes) is difficult to explain, because no representatives of the Pterogasterina were found in the live samples. It does not appear to be directly related to different tolerance for high salinity, as Apterogasterina also occur in the most saline of lakes. It could reflect unknown habitat differences in the littoral vegetation of fresh and saline lakes, implying that many recovered Oribatid fossils actually belong to terrestrial species.

No bryozoan statoblasts were found in surface-water conductivities higher than 1100 μS/cm. Among the freshwater lakes, the major factor limiting the presence of Bryozoa is likely the presence of suitable substrates. The present set of survey lakes was not selected for maximum contrast in substrate variation, hence quantitative assessment of the Bryozoa-substrate relationship is not currently possible. Still, highest concentrations of bryozoan statoblasts were found in the sediments of lakes Karolero and Kifuruka, two shallow lakes with high water transparencies, and abundant submerged macrophytes. This is consistent with data from shallow Florida lakes, where fossil statoblast abundance was found to be proportional to the fraction of lake floor covered by macrophytes (Crisman et al., 1986).

It has been suggested that Neorhabdocoel egg capsules could be used as indicators of eutrophication (Sreenivasa, 1973; Warner, 1990b). Our Uganda data show no straightforward relationship between egg-capsule abundance and Secchi-disk depth or any other environmental variable, except perhaps a negative relationship with high salinity. No egg capsules were found in lakes with surface-water conductivity higher than 5821 μS/cm.

The Thecamoeba are similar to Bryozoa in that variation in the relative abundance of fossil *Arcella* tests among the freshwater study lakes seems to be determined by the extent of submerged vegetation in littoral areas. Highest densities of *Arcella* tests were found in lakes Kifuruka and Karolero, which both have abundant submerged vegetation. The relative scarcity of such habitat in more concentrated tropical waters of high pH likely explains the absence of *Arcella* fossils in all freshwater lakes with a conductivity above 598 μS/cm. Still, the presence of *Arcella* in saline Lake Ntambi (5821 μS/cm) would suggest that at least some species possess broad salinity tolerance.

To conclude, this study demonstrates that quick screening of large-sized (>250 μm) aquatic-invertebrate fossils in offshore surface-sediment samples from small African lakes can produce a solid assessment of local taxonomic richness and community composition. Given good spatial integration of invertebrate remains from diverse littoral and offshore habitats in these lakes prior to burial, analysis of subfossil death assemblages yields higher values of local species richness than live samples taken on a single day at selected locations within the basin.

Applied to lake sediment records, rapid screening of aquatic invertebrate fossils can easily distinguish between fresh and saline episodes in former lake history, and give indications of the relative extent of the littoral zone at the time of deposition. Preliminary data on the distribution of *Chaoborus* among fresh Uganda lakes of varying transparency further suggest that analysis of fossil Chaoborus mandibles can also yield information on past changes in lake trophic status, provided that lake level and fish community structure can be assumed constant through time.

Specifically for the purpose of reconstructing past salinity variations due to climate change, the results of this study suggest that addition of non-chironomid Dipteran taxa such as Stratiomyidae, Culicidae, Ceratopogonidae, and Ephydridae to chironomid-based paleoenvironmental calibration data sets may improve the performance of quantitative salinity inference models for African lakes. Current chironomid-based models (Verschuren et al., in press; H. Eggermont et al., unpublished data) suffer from the low species diversity of Chironomidae in saline East African lakes (four). This low diversity and limited species turnover along the salinity gradient beyond the freshwater range limits the potential for accurate salinity inference, as is also the case in Canada where only two chironomid species occur in salinities exceeding 10 000 μS/cm (Walker et al., 1995). Expanding the East African calibration data set to include non-chironomid Diptera would improve the statistical performance of derived salinity-inference models by doubling the number of contributing halobiont taxa, and by extending the modelled salinity range to 135 000 μS/cm.

Acknowledgements

Fieldwork for this research was conducted under Uganda National Council for Science and Technology research clearance EC540, and sponsored by the Fund for Scientific Research of Flanders (FWO-Vlaanderen, Belgium) and Ghent University. We thank the Uganda Wildlife Authority for permission to study lakes in Kibale and Queen Elisabeth National Parks, the Uganda Petroleum Exploration and Production Department for logistic support, and all participants in the Uganda field campaigns. We also thank J. Dean (Environment Protection Authority, New South Wales, Australia) for assistance in the identification of fragmentary Trichoptera larvae. DV is postdoctoral fellow with the Fund for Scientific Research of Flanders (FWO-Vlaanderen, Belgium), and HE is research assistant with the Flemish Institute for the Advancement of Scientific-Technological Research in Industry (IWT, Belgium).

References

Balogh, J., 1972. Oribatid Genera of the World. Akdémiai kiadó, Budapest, 188 pp.

Bayly, I. A. E., 1972. Salinity tolerance and osmotic behavior of animals in athalassic saline and marine hypersaline waters. Annual Review of Ecology and Systematics 3: 233–263.

Beadle, L. E., 1932. Scientific results of the Cambridge Expedition to the east African lakes, 1930-1-4. The waters of some east African lakes in relation to their fauna and flora. The Journal of the Linnean Society of London (Zoology) 38: 157–211.

Bengtsson, L. & M. Enell, 1986. Chemical analysis. In Berglund, B. E. (ed.), Handbook of Holocene Paleoecology and Paleohydrology. John Wiley & Sons, Chichester: 423–451.

Borg, F., 1936. Ueber die Süsswasser-Bryozoen Afrikas. Senckenbergiana 18: 20–36.

Brener, L. & W. C. Sloan, 1954. The occurrence of a mayfly nymph in brackish water. Ecology 35: 98.

Burks, R., E. Jeppesen, D. M. Lodge & T. L. Lauridsen, 2002. Diel horizontal migration of zooplankton: costs and benefits of inhabiting littoral zones. Freshwater Biology 47: 343–365.

Carlson, R. E., 1977. A trophic state index for lakes. Limnology and Oceanography 22: 361–368.

Chadwick, M. A. & J. W. Feminella, 2001. Influence of salinity and temperature on the growth and production of a freshwater mayfly in the Lower Mobile River, Alabama. Limnology and Oceanography 46: 532–542.

Chardez, D., 1964. Thécamoebiens (Rhizopodes Testacés). In Symoens, J. J. (ed.), Exploration Hydrobiologique du Bassin du Lac Bangweolo et du Luapula 10(2): 1–77.

Claessens, I., 2002. Morphological, physical and chemical features of crater lakes in western Uganda. Masters Thesis, Ghent University, Ghent, 33 pp.

Cohen, A. S., R. Dussinger & J. Richardson, 1983. Lacustrine paleochemical interpretations based on eastern and southern African ostracodes. Palaeogeography, Palaeoclimatology, Palaeoecology 43: 129–151.

Crisman, T. L., U. A. M. Crisman & M. W. Binford, 1986. Interpretation of bryozoan microfossils in lacustrine sediment cores. Hydrobiologia 143: 113–118.

Currie, D. C. & I. R. Walker, 1992. Recognotion and palaeohydrologic significance of fossil black fly larvae, with a key to the Neartic genera (Diptera: Simuliidae). Journal of Paleolimnology 7: 37–54.

Deevey, E. S., H. Vaughan & G. B. Deevey, 1977. Lakes Yaxha and Sacnab, Peten, Guatemala: planktonic fossils and sediment focusing. In Golterman, H. R. (ed.), Interactions Between Sediments and Fresh Water. Junk, The Hague: 189–196.

Demoulin, G., 1981. Ephéméroptères. In Durand, J. R. & C. Leveque (eds), Flore et Faune aquatiques de l'Afrique sahelo-soudanienne. Orstom Press, Paris: 407–443.

Dethier, M., 1981. Hétéroptères. In Durand, J. R. & C. Leveque (eds), Flore et Faune aquatiques de l'Afrique sahelo-soudanienne. Orstom Press, Paris: 661–685.

Dodson, S. I. & D. G. Frey, 1991. Cladocera and other Branchiopoda. In Thorp, J. H. & A. P. Covich (eds), Ecology and Classification of North American Freshwater Invertebrates. Academic Press, Toronto: 723–786.

Edmunds, G. F. Jr., 1984. Ephemeroptera. In Merrit, R. W. & K. W. Cummins (eds), An Introduction to the Aquatic Insects of North America. Kendall/Hunt, Dubuque: 94–125.

Eggermont, H., 2004. Fossil Chironomidae (Insecta, Diptera) as biological indicators for past salinity variation in African lakes. Ph.D. Dissertation, Ghent University, Ghent, 497 pp.

Francis, D. R., 1997. Bryozoan statoblasts in the recent sediments of Douglas Lake, Michigan. Journal of Paleolimnology 17: 255–261.

Francis, D. R., 2001. Bryozoan statoblasts. In Smol, J. P., H. J. B. Birks & W. M. Last (eds), Tracking Environmental Change Using Lake Sediments. Zoological Indicators, Vol. IV. Kluwer Academic Publishers, Dordrecht: 105–123.

Frey, D. G., 1964. Remains of animals in Quaternary lake and bog sediments and their interpretation. Archiv für Hydrobiologie Ergebnisse der Limnologie 2: 1–114.

Frey, D. G., 1988. Littoral and offshore communities of diatoms, cladocerans and dipterous larvae, and their interpretation in paleolimnology. Journal of Paleolimnology 1: 179–191.

Fritz, S. C., 1996. Paleolimnological records of climatic change in North America. Journal of Freshwater Ecology 11: 882–889.

Gasse, F., J. F. Talling & P. Kilham, 1983. Diatom assemblages in East Africa: classification, distribution and ecology. Revue d'Hydrobiologie Tropicale 116: 3–34.

Gasse F., S. Juggins & L. Ben Khelifa, 1995. Diatom-based transfer functions for inferring past hydrochemical characteristics of African lakes. Palaeogeography, Palaeoclimatology, Palaeoecology 117: 31–54.

Håkanson, M. J. & M. Jansson, 1983. Lake Sedimentology. Springer-Verlag, Berlin, 316 pp.

Hammer, U. T., 1986. Saline Lake Ecosystems of the World. Junk Publishers, Dordrecht, 616 pp.

Hann, B. J., P. R. Leavitt & P. S. S. Chang, 1994. Cladocera community response to experimental eutrophication in Lake 227 as recorded in laminated sediments. Canadian Journal of Fisheries and Aquatic Sciences 51: 2312–2321.

Herbst, D. B., 1986. Comparative studies of the population ecology and life history patterns of an alkaline salt lake insect: *Ephydra* (*Hydropyrus*) *hians* Say (Diptera: Ephydridae). Ph.D Thesis, Oregon State University, Corvallis, 206 pp.

Iovino, A. J., 1975. Extant chironomid larval populations and the representativeness and nature of their remains in lake sediments. Ph.D. Dissertation, Indiana University, Bloomington, 54 pp.

Kizito, Y. S., A. Nauwerck, L. J. Chapman & W. Koste, 1993. A limnological survey of some Western Uganda Crater Lakes. Limnologica 23: 335–347.

Lamontagne, S. & D. W. Schindler, 1994. Historical status of fish populations in Canadian Rocky Mountain lakes inferred from subfossil *Chaoborus* mandibles. Canadian Journal of Fisheries and Aquatic Sciences 51: 1376–1383.

Lampert, W., 1993. Ultimate causes of diel vertical migration of zooplankton: new evidence for the predator avoidance hypothesis. Archiv für Hydrobiologie Ergebnisse der Limnologie 39: 79–88.

McGowan, L. M., 1972. Description of the larvae of *Chaoborus* (*Neochaoborus*) *anomalus* Edwards and *Chaoborus* (*Sayomyia*) *ceratopogenes* Theobald (Diptera Chaoboridae) from Lake George, Uganda, and their morphological variation in other African Lakes. Revue de Zoologie et de Botanique Africaine 85: 357–368.

Melack, J. M., 1978. Morphometric, physical and chemical features of the volcanic crater lakes of western Uganda. Archiv für Hydrobiologie 84: 430–453.

Morse, J. C. & R. W. Holzenthal, 1984. Trichoptera Genera. In Merritt, R. W. & K. W. Cummins (eds), An Introduction to the Aquatic Insects of North America. Kendall/Hunt, Dubuque: 312–347.

Mumm, H., 1997. Effects of competitors and *Chaoborus* predation on the cladocerans of a eutrophic lake: an enclosure study. Hydrobiologia 360: 253–264.

Newson, H. D., 1984. Culicidae. In Merritt, R. W. & K. W. Cummins (eds), An Introduction to the Aquatic Insects of North America. Kendall/Hunt, Dubuque: 515–533.

Pennak, R. W., 1978. Fresh-water Invertebrates of the United States, 2nd edn. John Wiley & Sons, New York, 803 pp.

Peterson, B. V., 1984. Simuliidae. In Merritt, R. W. & K. W. Cummins (eds), An Introduction to the Aquatic Insects of North America. Kendall/Hunt, Dubuque: 534–550.

Philippon, B., 1981. Simuliidae. In Durand, J. R. & C. Leveque (eds), Flore et Faune aquatiques de l'Afrique sahelo-soudanienne. Orstom Press, Paris: 643–659.

Polhemus, J. T., 1984. Aquatic and Semiaquatic Hemiptera. In Merritt, R. W. & K. W. Cummins (eds), An Introduction to the Aquatic Insects of North America. Kendall/Hunt, Dubuque: 231–260.

Rickenbach, A., 1981. Culicidae. In Durand, J. R. & C. Leveque (eds), Flore et Faune aquatiques de l'Afrique sahelo-soudanienne. Orstom Press, Paris: 570–581.

Rine, H. E. & D. H. Kesler, 2001. A long-term study of larval *Chaoborus punctipennis* (Diptera: Insecta) vertical migration and the role of critical oxygen concentration. Journal of Freshwater Ecology 16: 145–150.

Rosenberg, D. M., 1993. Freshwater Biomonitoring and Benthic Macroinvertebrates. Chapman & Hall, New York, 488 pp.

Seip, K. L., H. Sas & S. Vermij, 1992. Changes in Secchi disk depth with eutrophication. Archiv für Hydrobiologie 124: 149–165.

Smith, I. M. & D. R. Cook, 1991. Water mites. In Thorp, J. H. & A. P. Covich (eds), Ecology and Classification of North American Freshwater Invertebrates. Academic Press, Toronto: 523–592.

Smol, J. P., I. R. Walker & P. R. Leavitt, 1991. Paleolimnology and hindcasting climate trends. Verhandlungen Internationale Vereinigung für Limnologie 24: 1240–1246.

Solem, O. & H. H. Birks, 2000. Late-glacial and early-Holocene Trichoptera from Kråkenes Lake, western Norway. Journal of Paleolimnology 23: 49–56.

Sølhoy, I. W. & T. Sølhoy, 2000. The fossil oribatid mite fauna (Acari: Oribatida) in late-glacial and early-Holocene sediments in Kråkenes Lake, western Norway. Journal of Paleolimnology 23: 35–47.

Sreenivasa, B. A., 1973. Paleoecological studies of Sunfish lake and its environs. Ph.D Dissertation, University of Waterloo, Ontario, 185 pp.

Stephens, D. W., 1990. Changes in lake levels, salinity and the biological community of Great Salt Lake (Utah, USA), 1847–1987. Hydrobiologia 197: 139–146.

Teskey, H. J., 1984. Larvae of aquatic Diptera. In Merritt, R. W. & K. W. Cummins (eds), An Introduction to the Aquatic Insects of North America. Kendall/Hunt, Dubuque: 448–466.

Timms, B. V., 1983. A study of benthic communities in some shallow saline lakes of western Victoria, Australia. Hydrobiologia 105: 165–178.

Timms, R. M. & B. Moss, 1984. Prevention of growth in *Daphnia* populations: co-action in perspective. Memorie del'Istituto Italiano di Idrobiologia 45: 413–423.

Tudorancea, C., A. D. Harrison & C. H. Fernando, 1989. A comparative limnological study of zoobenthic associations in lakes of the Ethiopian Rift Valley. Archiv für Hydrobiologie Supplement 83: 121–174.

Uutala, A. J., 1990. *Chaoborus* (Diptera: Chaoboridae) mandibles – paleolimnological indicators of the historical status of fish populations in acid-sensitive lakes. Journal of Paleolimnology 4: 139–151.

Uutala, A. J. & J. P. Smol, 1996. Paleolimnological reconstructions of long-term changes in fisheries status in Sudbury area lakes. Canandian Journal of Fisheries and Aquatic Sciences 53: 174–180.

Verbeke, J., 1957. Chaoboridae (Diptera Nematocera) stades immatures et adultes. Résultats scientifiques de l'exploration hydrobiologique (1952–1954) des lacs Kivu, Edouard et Albert 3: 183–201.

Verschuren, D., 1997. Taxonomy and ecology of subfossil Chironomidae (Insecta, Diptera) from Rift Valley lakes in central Kenya. Archiv für Hydrobiologie Supplement 107: 467–512.

Verschuren, D., B. F. Cumming & K. R. Laird, in press. Quantitative reconstruction of past salinity variations in African lakes: assessment of chironomid-based inference models (Insecta: Diptera) in space and time. Canadian Journal of Fisheries and Aquatic Sciences.

Walker, I. R. & C. G. Paterson, 1985. Efficient separation of subfossil Chironomidae from lake sediments. Hydrobiologia 122: 189–192.

Walker, I. R., C. H. Fernando & C. G. Paterson, 1984. The chironomid fauna of four shallow, humic lakes and their representation by subfossil assemblages in the surficial sediments. Hydrobiologia 112: 61–67.

Walker, I. R., S. E. Wilson & J. P. Smol, 1995. Chironomidae (Diptera): quantitative palaeosalinity indicators for lakes of western Canada. Canadian Journal of Fisheries and Aquatic Sciences 52: 950–960.

Warner, B. G., 1990a. Testate Amoebae (Protozoa). In Warner, B. G. (ed.), Methods in Quaternary Ecology. Love Printing Service Ltd., Stittsville: 65–74.

Warner, B. G., 1990b. Other Fossils. In Warner, B. G. (ed.), Methods in Quaternary Ecology. Love Printing Service Ltd., Stittsville: 149–162.

Westfall, M. J., 1984. Odonata. In Merritt, R. W. & K. W. Cummins (eds), An Introduction to the Aquatic Insects of North America. Kendall/Hunt, Dubuque: 126–176.

Wiebach, F., 1970. Süswasser-Bryozoen aus Brasilien und Zentralafrika. Revue de Zoologie et de Botanique Africaines 81(1–2): 62–69.

Wiebach, F., 1980. Lophophoriens (Bryozoaires et Kamptozoaires). In Durand, J. R. & C. Leveque (eds), Flore et Faune aquatiques de l'Afrique sahelo-soudanienne. Orstom Press, Paris: 273–282.

Williams, N. E., 1988. The use of caddisflies (Trichoptera) in palaeoecology. Palaeogeography, Palaeoclimatology, Palaeoecology 62: 493–500.

Wissel, B., N. D. Yan & C. W. Ramcharan, 2003. Predation and refugia: implications for *Chaoborus* abundance and species composition. Freshwater Biology 48: 1421–1431.

Hydrobiologia (2005) 542:315–333
H. Segers & K. Martens (eds), Aquatic Biodiversity II
DOI 10.1007/s10750-004-3247-2

Life history strategies of cladocerans: comparisons of tropical and temperate taxa

S. S. S. Sarma[1,*], S. Nandini[2] & R. D. Gulati[3]
[1]*Laboratory of Aquatic Zoology, Unit U.M.F., Division of Research and Postgraduate Studies, National Autonomous University of Mexico, Campus Iztacala, AP 314, CP 54090, Los Reyes, Tlalnepantla, State of Mexico, Mexico*
[2]*UIICSE, Division of Research and Postgraduate Studies, National Autonomous University of Mexico, Campus Iztacala, AP 314, CP 54090, Los Reyes, Tlalnepantla, State of Mexico, Mexico*
[3]*The Netherlands Institute for Ecology, Centre of Limnology, Rijksstraatweg 6 3631 AC Nieuwersluis, The Netherlands*
*(*Author for correspondence: Fax: 552-55-5623-1256; E-mail: sarma@servidor.unam.mx)*

Key words: demography, competition, predation, food quality, threshold concentration, *Daphnia*, *Moina*, *Diaphanosoma*, lifespan

Abstract

We review recent works on different life history variables of cladoceran taxa in tropical and temperate freshwater bodies, comparing the strategies that cladocerans have evolved to adapt to contrasting environmental conditions in the two geographical regions. These life-history parameters relate to age and size at maturity, survival, fecundity, life-expectancy at birth, lifespan, gross, and net reproductive rates, generation time, the rate of population increase, peak population density and day of peak abundance. We also discuss the role of photoperiod and temperature on some of these life history parameters. We found a general paucity of experimental work and field data in tropics on cladocerans. There is very limited information on the few *Daphnia* species found in the tropics. The misconception of low species diversity of cladocerans in the tropics arose due to several reasons including lack of extensive and intensive field collections. Higher water temperatures apparently promote permanent infestation of tropical waters with toxic cyanobacteria, which reduce the zooplankton diversity. In addition to higher temperatures in the tropics, the year-round high predation pressure of planktivorous fish probably causes the tropical species, particularly in pelagic habitats, to reach maturity earlier (<3 days) than in temperate regions. Species of *Daphnia* in temperate regions are particularly adapted to living at food concentrations that are much lower and seasonably more variable than those for tropical genera such as *Diaphanosoma*. This is further corroborated by the more than an order of magnitude higher threshold food concentration (TFC) for tropical Cladocera than for their temperate counterparts. Fecundity patterns differ between tropical and temperate cladoceran taxa: cultured under optimal temperature regimes, tropical taxa have fewer eggs than temperate species of a comparable body size. Predation pressure may act differently depending on the size of the cladoceran neonates and thus on their population size structure. Global warming and climate changes seem to affect the behaviour (migration), distribution, and abundance of cladocerans. Apparently, in direct response to these changes, the possibility of encountering the tropical cladocerans in the northern, temperate hemisphere (bioinvasions) is on the rise.

Introduction

Cladocera (Crustacea: Anomopoda) are an important group of organisms in aquatic ecosystems, particularly in freshwaters. In lakes and ponds cladocerans act as a link in the food chain: most of them are herbivorous, feed on phytoplankton and, in turn, are preyed upon by certain invertebrate and fish predators. Thus, they play an important role in the transfer of energy from primary producers to secondary and tertiary consumers within the aquatic food-web (Dodson & Frey, 2001; Dumont & Negrea, 2002). Here we review and compare recent works on different life history variables of cladoceran taxa in tropical and temperate freshwater bodies with particular emphasis on the effects of abiotic and biotic factors such as light, temperature, food quality, competition and predation on the diversity and life history strategies of cladocerans found in the two climatically different regions.

Cladocerans inhabit diverse habitats and are at times exposed to a great variety of harsh and extreme environmental conditions (Dodson & Frey, 2001). Phenotypic plasticity (Cousyn et al., 2001) and flexible age-specific reproductive output and survival, collectively, form a greater part of life history strategies of Cladocera (Innes & Singleton, 2000), which facilitate their successful colonization in diverse habitats. A study of cladoceran life-history strategies requires frequent seasonal, plankton sampling under field conditions, and experimental verification under laboratory conditions of the important phenomena observed in the field. Data based on field collections yield information on several important aspects on the main life history variables such as body size, morphological adaptations, reproductive output (e.g., number of neonates female^{-1}) and population density, among others (Saunders et al., 1999). However, the complex interactions in the field among the life-history variables mentioned above make interpretations of field data difficult. Laboratory experiments under well-defined conditions help us resolve some of these complex interactions by providing insight into the effect of selected variables at a time, e.g., studying the effect of a desired parameter under well-defined food or temperature conditions, in order to model and draw generalizations.

The life-history strategies of tropical and temperate cladoceran taxa differ in response to several abiotic (temperature, light, and oxygen saturation(%)) and biotic factors (predation and inter and intra-specific competition). Most work on life-history characteristics of cladocerans is derived from detailed studies on species belonging to the genus *Daphnia* (Koivisto, 1995; Lampert & Sommer, 1997). Therefore, daphnids serve as model organisms for not only understanding the life history parameters but also comparing the strategies under different food, predation, and temperature conditions (see e.g., Gliwicz, 2003).

While dealing with tropical and temperate taxa of cladocerans, one often encounters a disagreement in defining the regional constraints of these two zones. The tropics lie in the geographical region between the tropic of Capricon and Cancer and taxa found in this region are considered tropical (e.g., Kořinek, 2002). Another consideration is to use the temperature range of water bodies in a limnological sense (Hutchinson, 1967), or taxonomical species listings considering altitudes or latitudes, or both (Dumont, 1994; Fernando & Paggi, 1997). This, however, is also not free from criticism for defining tropical or temperate taxa (see for details on this aspect and for Central America and the Carribbean Archipelago Kořinek, 2002). While considering the typical tropical or temperate taxa, we use the criterion of dominant distribution or occurrence, or both, in most parts of the year in a particular region. Some taxa undoubtedly have wider geographical distribution (e.g., *Simocephalus*) and thereby transcend the distinction into tropical and temperate regions (Kořinek, 2002). The distributional aspects of tropical and temperate taxa are beyond the scope of this review. However, such information is available in taxonomical works or other such specialized literature (e.g., Hebert & Finston, 1993). According to Kořinek (2002), *Diaphanosoma* is a tropical limnetic genus that has a considerably high species diversity (20 species). *Daphnia* has the highest diversity in temperate waters. For example, Taylor et al. (1996) have noted as many as 12 valid species within *D. longispina* group alone from temperate regions. Some species of *Daphnia* also occur in the tropical region. For example, Kořinek (2002) mentions 10 species of *Daphnia* including *D. carinata*, *D. lumholtzi*, *D. cephala*, and *D. similis* in the tropical region.

Obviously, temperature disparity appears to be the most important and contrasting environmental variable between the tropical and temperate regions. Within the same region, e.g., tropical region, altitudinal differences may mimic the regional differences and are not considered here. It may appear that many of the life-history variables of tropical or temperate taxa may be explained on the basis of temperature influence. But this view has a restricted connotation as it often fails to explain species adaptations to the local conditions. For example, Michell & Lampert (2000) failed to culture *D. magna* clones collected from seven geographically widespread locations in Europe (from Finland to Spain) above 30 °C. There were no significant differences in the growth rates of those reared at 26 and 29 °C. On the other hand, *D. carinata* collected from tropical India successfully reproduced at both 20 and 30 °C (Nandini, 2000). In addition to life history parameters, the threshold food concentration (TFC) which is the food level at which all the energy assimilated is utilized to meet the maintenance requirements or catabolic demands so that there is zero growth, has often been used to draw comparisons. Due to discernibly lower prevailing temperatures in temperate regions than in the tropics, most species of *Daphnia* have a longer lifespan than that of the common tropical planktonic genera, e.g., *Moina* (Petrusek, 2002). Some recent experimental studies in the temperate region (e.g., Moore et al., 1996; Achenbach & Lampert, 1997) have attempted to predict the effects of elevated temperatures on zooplankton. The main conclusion of the first study is that a temperature increase can produce a reduction in body size implying that in tropical waters smaller-bodied forms will dominate zooplankton assemblages, an effect that is seemingly mediated by differences in growth rate *vis-à-vis* increased respiration at higher temperature. A size-related discrepancy in growth may help explain the general dominance of tropical zooplankton by smaller-bodied forms. In the study of Achenbach & Lampert (1997), on the other hand, the main conclusion was that shifts from large-sized to small-sized species in lakes due to elevated temperature may be due to selective factors other than competition, e.g., predation and food quality. Thus, larger zooplankton species are competitively superior. For the tropical situation where smaller species are relatively more abundant, it suggests that factors other than competition such as food quantity, food quality, and predation may confer an advantage to these smaller forms.

Chapelle & Peck (1999) have demonstrated that maximum potential size (MPS) associated with low temperature but mediated through oxygen saturation, in some aquatic animals is much higher than in warm tropical areas where oxygen saturation (%) is much lower. Their study relates to length data of some 1850 species of amphipod crustaceans. Thus, tropical species are smaller due to both higher temperatures and relatively lower saturation (%) of dissolved oxygen in tropical waters. Although oxygen saturation (%) has been reported to affect the MPS it is also mentioned that oxygen saturation may have no effect in modifying the minimum size.

Cladoceran diversity: tropical vs. temperate regions

Dumont & Negrea (2002) have questioned the view about scarcity of zooplankton species diversity in tropics (Fernando & Paggi, 1997). They state that tropical water bodies, in contrast, may contain more cladoceran taxa than temperate water bodies (see also Dumont, 1994). Regardless of the total number of taxa found per water-body in a temperate or tropical situation, predatory genera are usually absent in lowland tropical aquatic systems (Fernando, 2002). However, the effects of absence of predators on the life-history strategy of potential prey species of zooplankton are not known. Thus, the misconception of low species diversity of cladocerans in the tropics appears to have arisen due to several factors: (a) lack of extensive and intensive field collections (Kořinek, 2002); (b) ephemeral or temporary nature of many water bodies (e.g., seasonal disappearance of ponds due to high evaporation rates) (Timotius & Goltenboth, 1995) makes it difficult to determine the zooplankton diversity, (c) permanent infestation of many water bodies with toxic cyanobacteria (*Microcystis* spp.) (Zafar, 1986), or bloom conditions which generaly inhibit feeding and reduce the zooplankton diversity (Gulati et al., 2000), (d) the lack of expertise and trained personnel in taxonomic identifications (Kořinek, 2002), (e) restriction of field collections to planktonic samples rather than including littoral or benthic areas in the work (Dumont, 1994), (f) year-round high predation pressure of planktivorous fish on zooplankton causing reduction in the diversity and abundance of relatively large cladocerans (Fernando, 2002), (g) relatively high industrial pollution in water bodies (and the general lack of rigorous pollution control and management strategies) (Kulshrestha et al., 1991), and (h) loss of potential diversity of all organisms including zooplankton due to increased urbanization and human settlements (Kulshrestha et al., 1991). Table 1 shows some of the salient features of tropical vs. temperate cladoceran taxa.

Effects of abiotic and biotic factors on tropical and temperate taxa

Role of light and temperature

The effect of temperature on cladocerans has implications from the viewpoint of species distribution, body size, and abundance (Stockwell & Johannsson, 1997). Most temperate water bodies contain cladoceran genera like *Daphnia* and *Bosmina* among herbivore taxa, and *Leptodora, Bythotrephes*, and *Cercopagis* among predatory taxa (Dumont & Negrea, 2002). In contrast, tropical water bodies typically contain *Moina, Ceriodaphnia, Macrothrix*, and *Diaphanosoma*, all of which are predominantly herbivorous taxa (Dodson & Frey, 2000). Although information from the tropics is far from complete, apparently there are fewer *Daphnia* species inhabiting water bodies in tropical regions (Kořinek, 2002). In other words, any random comparison of the water bodies will show that *Daphnia* are more common in temperate regions. Tropical and temperate species may also differ in their response to light intensity, daily light duration and the seasonal changes in irradiance; however, quantitative data are lacking. In general, tropical cladocerans experience both higher light intensities and longer periods of exposure to irradiance and lesser seasonal changes in these factors than temperate species for most of the year (Hutchinson, 1967).

Both well-defined seasons and marked seasonal changes in the light: dark cycles, together with sharp decline in food levels in late spring/early summer, trigger the formation of ephippia in many cladoceran species in temperate regions (Hutchinson, 1967; Alekseev & Lampert, 2001; Caceres & Schwalbach, 2001). Though ephippia production is a survival strategy (i.e., an adaptation to survive the periods of unfavourable conditions) for both temperate and tropical taxa (Dumont & Negrea, 2002), it is not clearly known what triggers their

Table 1. Comparison of general characteristics of tropical and temperate cladocerans

Variable	Temperate taxa	Tropical taxa	References
Known species diversity from the region	Lower	Higher	Dumont (1994)
Common plankton taxa	*Daphnia, Bosmina*	*Diaphanosoma, Ceriodaphnia*	Kořinek (2002)
Planctonic genus with high diversity	*Daphnia*	*Diaphanosoma*	Taylor et al. (1996) and Kořinek (2002)
Relative average body size	Large	Small	Fernando (2002)
Body length of largest known cladoceran	5 mm	<3 mm	Dumont & Negrea (2002)
Presence of predatory taxa of cladocerans	High	Nearly absent	Fernando (2002)
Average lifespan	Long	Short	De Meester (1994) and Nandini & Sarma (2002)
Age at maturity	>week	<week	De Meester 1994 and Nandini & Sarma (2002)
Mean number of offspring per brood	High	Low	De Meester (1994) and Nandini & Sarma (2002)
Relative threshold food requirement	Low	High	Duncan (1989) and Gliwicz (1990)
Mean generation time	Long	Short	De Meester (1994) and Nandini & Sarma (2002)
Rate of population increase per day	<1.0	>1.0	De Meester, 1994; Nandini & Sarma (2002)
Possible nutritional deficiency from algal food	High	Low	Wacker et al. (2001)

production in different cladoceran species. Factors like photoperiod, temperature, food availability, population density, and kairomone are all known to be involved in ephippial production (Zadereev, 2003).

There are also anthropogenic factors such as increased temperature due to global warming (Yurista, 1999), ozone layer depletion (Rhode et al., 2001; Rojas et al., 2001; De Lange & Van Reeuwijk, 2003), and acidification of freshwaters (Zimmer & Storr, 1984) that affect zooplankton behaviour (vertical migration), distribution, abundance, body size, and age at maturity of cladocerans in temperate regions (McKee et al., 2002). As observed for many other organisms cladoceran taxa not previously reported from the northern hemisphere are now increasingly being observed in routine zooplankton samples; one of the main causative factors may be global warming (Kolar et al., 1997). Typically tropical species *Daphnia lumholtzi*, native to Australia, Asia, and Africa, is the latest planktonic invader of Great Lakes in North America (Christopher, 2000). Increasingly late summer conditions observed in temperate lakes and ponds, i.e. elevated temperatures, appear to be suitable for certain tropical daphnids such as *D. lumholtzi* (Lennon et al., 2001). This is among the very plausible explanations for the more recent incidences of tropical species in temperate waters. Invasion of tropical cladocerans to the Americas has been of considerable interest in considering the role of temperature in colonization (Hebert & Cristescu, 2002).

Algal food limitation through nutrient deficiencies

Feeding on nutritionally deficient algae causes reduced somatic growth and reproductive rates of cladocerans and eventually lowers the secondary productivity in natural water bodies (Urabe et al., 1997; Ferrao-Filho & Azevedo, 2003). This, however, has been largely documented for temperate species (Wacker et al., 2001). Eutrophication is mainly caused by rise in nutrient concentrations, especially phosphates, in inland fresh waters due to increased inputs of human, industrial, and agricultural wastes. However, phosphate inputs to inland waters in many European countries and in North America have been reduced by stringent nutrient reduction measures. Such control measures have, however, so far not been initiated in many tropical countries. Therefore, eutrophication is not only rampant but still on the rise in tropical waters. Aquatic systems in the tropics also suffer from a greater degree of nitrogen than phosphorus limitation (Harper, 1992) than do temperate waters. These factors combine to make blooms of cyanobacteria, which cannot be utilized by or are harmful to zooplankton, dominate the phytoplankton. Water hardness also affects the growth of cladocerans. There is some indication that large cladocerans such as *Daphnia* need higher specific quantity of calcium than smaller taxa and are, therefore, probably poor competitors in soft water lakes (Wærvågen et al., 2002) but its implications for the zooplankton in general are virtually unknown.

The nutritional quality of phytoplankton will depend on the relative composition of species and dominance of, e.g., toxic cyanobacteria or limitation of essential macronutrients as phosphorus. On the other hand, edible phytoplankton in nature if limited by phosphorus will limit growth and abundance of Cladocera. This is especially true for species of *Daphnia* in lakes, which among cladocerans, are the most sensitive to P limitation in food (Gulati & DeMott, 1997; Whitton & Potts, 2000; Nelson et al., 2001). It is not known if such elemental limitations contribute to food quality differences for Cladocera in tropics and temperate waters and if they influence the zooplankton dynamics in general. Hessen et al. (2002) studied the growth of juvenile *Daphnia magna* fed using different concentrations of *Selenastrum capricornutum* grown using different light conditions and phosphorus concentrations. Their study showed that the physiological responses of algae in terms of chlorophyll content and elemental composition depended on ambient light and nutrient conditions. Further, up to 40% reduction in the growth rate of *D. magna* could be attributed to the increased C:P ratios of algal food. Such detailed studies for tropical cladoceran taxa have not been carried out. Also, small changes occurring naturally in the quality of algal diet in the tropical water bodies may be difficult to quantify using demographic techniques (Sterner, 1998).

Control of cyanobacteria by Cladocera: a regional comparison

Compared with temperate lakes, tropical water bodies experience higher light intensities, longer periods of sunlight hours, and higher and seasonally less variable water temperature (Timotius & Goltenboth, 1995). These regional disparities, together with differences in nutrient availabilities, contribute to the relatively greater proliferation of toxic cyanobacteria such as *Microcystis* spp. in tropical inland waters than in the temperate ones (Zafar, 1986). The availability of large data sets on the life-history variables of temperate species sometimes may also influence lake management strategies (Gulati et al., 1990) of tropical water bodies. For example, experiences in the use of *Daphnia* spp. for controlling toxic cyanobacteria such as *Microcystis* in the temperate regions are being utilized occasionally in the tropics (Mohamed & Smith, 2001). Eventhough certain species of *Daphnia* (e.g., *D. laevis*) are present in tropical or subtropical waters, their use in controlling the toxic *Microcystis* spp. may not be as efficient as those reported from temperate daphnids. For example, De Clerk et al. (1997) have calculated that *D. pulex* at a density of 1 ind. l^{-1} was sufficient to eliminate cyanobacterial blooms in a temperate lake (Lake Blankaart, Belgium). However, Nandini et al. (2000) have noted that for the subtropical waterbody (Valle de Bravo, Mexico), densities of *D. laevis* as high as 490 ind. l^{-1} failed to eliminate toxic cyanobacteria. It should be interesting to examine if efficiency of cladocerans to eradicate cyanobacteria from water bodies differs latitudinally.

Food consumption and threshold requirements

Information on food consumption rates of tropical taxa is virtually lacking, which makes comparisons with temperate species almost impossible. An important measure for comparing the minimum food concentration required for growth among filter-feeding cladocerans is the TFC. TFC may be quantified at individual or population level (Lampert, 1977). For an individual, it is the ambient food concentration that is sufficient to meet the metabolic requirements but not growth. For a population, TFC is the food concentration at which animals can maintain a zero growth rate. In demographic studies, the TFC is equal to a zero net reproductive rate ($R_0 = 0$). The TFC in cladocerans generally decreases with increasing body size of the animals: the larger-bodied taxa have an apparent advantage over the smaller-bodied taxa (Gliwicz, 1990; Gliwicz & Lampert, 1990). Nandini & Sarma (2003) have shown that this relationship is valid for individuals about 1300 μm long and above, e.g., *Daphnia laevis* (about 8.7 μg C (from unpublished data of R. D. Gulati), daphnids of comparable size from temperate waters), but below this size, the trend was reversed, i.e., TFC increased with increasing body size. Duncan (1989: Fig. 6) compared TFC of tropical and temperate cladocerans, mainly *Daphnia* spp. Several interesting observations emerge from her comparisons. First, the values for TFC are more than order of magnitude higher for the tropical forms (>0.1 mg dw^{-1}) than for the temperate forms. Secondly, regardless of size and food conditions, taxa in different geographical regions are at variance in this respect. An overall comparison among the different taxonomic groups, including ciliates, shows that the TFCs among the different taxonomic groups increase in the order Cladocera, Rotifera and Ciliata (Duncan, 1989). Data on the TFC of some cladoceran species are presented in Table 2. Duncan's (1989) comparisons imply that to maintain zero growth tropical cladocerans need much higher food concentrations to be able to ingest and assimilate enough food to compensate for their much higher metabolic rates. This can also be inferred from studies on feeding of daphnids by Gulati (1977) who found a Q_{10} of >2.0 for food uptake and assimilation rates measured for *Daphnia* spp. using a range of temperatures that covered both temperate and tropical waters.

Prevailing temperatures influence the TFC. For example, Hardy & Duncan (1994) have evaluated the combined effects of food (*Scenedesmus acutus*) densities ranging from 0.03 to 1.00 mg C l^{-1} at different temperatures (22, 27, and 32 °C) on growth, body size and reproduction of tropical species (*Daphnia gessneri*, *Diaphanosoma sarsi*, and *Moina reticulata*). They found that the TFC of postembryonic growth of *D. gessneri* was 0.1 mg C l^{-1} at 22 ° and 27 °C but much higher (0.25 to 0.50 mg C l^{-1}) at 32 °C. At 27 °C, the food threshold level varied between species: it was

Table 2. Threshold food concentrations (TFC) of selected cladoceran species. The different methods of expression of TFC (at individual or population level) using diverse approaches (chemostats, static renewal with fixed ration etc) and food quality (e.g., P-limited etc.) are not detailed here

Taxa	Food type	TFC (mg l^{-1} dw) (dry weight) or mg C (carbon) l^{-1})	Temperature (°C)	Reference
Daphnia magna, D. pulex, D. hyalina, D. thorata	Different types	0.01–0.02 dw	14–20	Duncan (1989)
D. gessneri, D. lumholtzi, Diaphanosoma brachyurum, D. excisum, Moina micrura, Moina reticulata	Different types	0.06–0.2 dw	22–32	Duncan (1989)
D. magna	*Scendesmus obliquus*	0.064–0.311 C	20	Boersma & Kreutzer (2002)
D. galeata	Natural diet	0.26	–	Hülsmann (2001)
Moina macrocopa	*Chlorella sorokiniana*	0.13 dw		Benider et al. (2002)
A. rectangula	*Chlorella vulgaris*	0.112 dw	24	Nandini & Sarma (2003)
C. dubia	*C. vulgaris*	0.162 dw	24	Nandini & Sarma (2003)
D. laevis	*C. vulgaris*	0.118 dw	24	Nandini & Sarma (2003)
Diaphanosoma brachyurum	*C. vulgaris*	0.289 dw	24	Nandini & Sarma (2003)
Moina macrocopa	*C. vulgaris*	0.204 dw	24	Nandini & Sarma (2003)
Scapholeberis kingi	*C. vulgaris*	0.016 dw	24	Nandini & Sarma (2003)

higher (0.25 mg C l^{-1}) for *D. sarsi* and lower (0.05 mg C l^{-1}) for *M. reticulata*.

Competition and predation

Competition influences the diversity, abundance and some life history characteristics of cladocerans in both tropical and temperate regions (Manca & de Bernardi, 1985). Competition occurs mainly when different taxa utilizing the same food source co-occur and food is limiting. Since many large tropical waterbodies are eutrophic (Infante & Riehl, 1984), the competition for food among zooplankton may be less intense compared with that in temperate regions during the *clear-water phase* (Lampert & Sommer, 1997). However, reduced cladoceran diversity and density in some tropical waterbodies may be due to the year-round persistance of toxic cyanobacterial blooms (Zafar, 1986) and to some extent to predation from fishes (Fernando, 2002). Only in ancient tropical water bodies (e.g., Lake Tanganyika) has the competition-related reduction in zooplankton species richness been reported (Dumont, 1994). Competition has several implications on the life history of cladocerans. Orcutt (1985) conducted life table demographic experiments on the competitive abilities of two cladoceran species (*Diaphanosoma brachyurum* and *Daphnia ambigua*) at different food levels. Among these species, *D. ambigua* was found to be competitively superior when food was not limiting while *D. brachyurum* was dominant when food level was limiting.

Matveev (1987) evaluated the effect of competition on the demography of planktonic cladocerans. Thus, when *Daphnia hyalina* and *D. brachyurum* were cultured together, *Daphnia* shortened the time lags of fecundity, birth rate, and population growth rate. Furthermore, the competitive ability in exploiting the food conditions was not related to the percentage of adults in the population.

Predation affects not only the life history of cladocerans but also the morphology and behaviour of cladocerans. Diel vertical migration (DVM) of cladocerans is a kind of anti-predator defence strategy (De Meester et al., 1998). It is a trade-off between predation risk from vertebrates such as fish in surface waters and reduced food availability in deeper waters. Fish, through the release of *kairomones* in the surrounding water strongly influence the DVM of zooplankton. In some characteristics, e.g., predator-induced morphology, both tropical and temperate cladocerans respond similarly. In other characteristics, however, the tropical and temperate cladocerans differ under predation (Arnott & Vanni, 1993).

Regardless of taxa from tropical or temperate regions, predation acts differently on the size of neonates of cladocerans (Lynch, 1980; Œlusarczyk, 1997; Manca et al., 2000) and the size structure of prey populations (Bungartz & Branstrator, 2003). However, generally, size of cladoceran neonates is larger in temperate waters than in tropical waters (Luening, 1992). Invertebrate predators such as flatworms, copepods, and insect larvae find it difficult to handle large rather than small prey (Dumont et al., 1994). On the other hand, reduced body size of neonate of tropical cladocerans not only permits higher population densities of prey population but is energetically an unprofitable choice for vertebrate predators such as fishes. Consequently, the survival chance of neonates is improved (Tessier, 1986; Stibor, 1992). The mean size as well as the minimal size of egg-bearing individuals of *Daphnia pulex* become smaller in the presence of fish kairomones compared with controls (Engelmayer, 1995). Reduced neonate size due to fish kairomones has also been observed for the hybrid *Daphnia galeata* × *hyalina* (Reede & Ringelberg, 1995). Gliwicz & Boavida (1996) observed high inter- and intra-population variability in maturation size (body length at first reproduction) and number of eggs per clutch in *Daphnia pulicaria* in lakes stocked with rainbow trout. The *Daphnia* length at first maturation was reduced by half compared with fishless lakes (body length 0.86–0.95 and 1.55–1.81 mm, respectively). The mean number of eggs per clutch was also reduced by similar degree, and was comparable with the reductions caused by copepod predation on eggs in brood cavities of mature daphnids.

Life history strategies of Cladocera: a comparison

The life-history strategies of cladocerans are generally compared using life table demography and population growth studies (de Bernardi & Manca, 1982; Orcutt, 1985; Lynch, 1992; Nandini & Sarma, 2002). Using the life-table method, most of the life history parameters mentioned above can be calculated. On the other hand, using population growth studies, it is possible to measure peak population density, day of peak abundance and the rate of population increase. Thus, these two approaches are considered complementary for understanding life-history strategies of zooplankton (Sarma & Nandini, 2002). Lynch (1980) has reviewed the general life history characteristics of cladocerans. He reports that: (a) small-sized cladoceran species are vulnerable to invertebrate predators but minimize this by producing relatively large offspring, and (b) large-sized cladoceran species cannot be easily handled by invertebrate predators and, therefore, they maximize early growth to larger size. Recently, some of these life history traits have been evaluated using different food types and concentrations, predator types and different temperature regimes, largely using temperate taxa (Lass & Spaak, 2003; Mooij et al., 2003; Weetman & Atkinson, 2004).

Size and reproductive maturity

Size and age at first reproduction have great ecological significance for a population (Roff, 2001). Delayed maturity may permit cladoceran populations to conserve their limited resources, and thus on return of favourable conditions the population may recuperate. Although cladocerans are considered as opportunistic, the differences in age at first reproduction within this group can be as large as 10 days (Dumont & Negrea, 2002). In most temperate waters, pelagic cladoceran species may reach maturity in about days after birth (Vanni & Lampert, 1992). On the other hand, species dominant in tropical habitats, particularly those in the pelagia of lakes, may reach maturity in < 3 days. This early maturity not only contributes to a higher reproductive output, but also confers to the species in question a competitive edge over other species under food-limited conditions. Size at maturity is to some extent directly related to the age at first reproduction. In both, the temperate and tropical regions, in addition to the temperature effects, the size differences may affect the time needed to attain maturity. As a general rule, at a given temperature, large species, such as *Simocephalus,* need more time to reach maturity than small ones, e.g., *Moina* (Babu & Nayar, 1997; Nandini & Sarma, 2000). Differences among littoral and benthic cladocerans based on latitude are less pronounced than among pelagic taxa. Most chydorids, being essentially benthic and littoral, despite their relatively small size, require a longer time to reach maturity. This may be partly due to their thick carapace, which takes more time for formation compared with soft-bodied taxa such as *Moina* (Dodson & Frey, 2001). The role of predator types (fish and invertebrates) in determining the duration or age at maturity of their prey differs; species more vulnerable to predation invest more in reproduction compared with relatively less vulnerable species (Kerfoot & Sih, 1987). We now know that whereas in temperate waters both types of predators are important, in tropics fishes are generally more important as predators, especially in permanent water bodies (Fernando, 2002). This could lead to differences in age at maturity in addition to the differences in effect of temperature in the two regions.

Size and age at maturity influence the life-history variables of organisms (Roff, 2001). In many a large branchiopod such as anostracans, age is related to maturity only to some extent (Dumont & Negrea, 2002). If a branchiopod is under-nourished or fed a nutritionally inadequate diet, the individuals often fail to mature even on attaining reproductive age. In *Daphnia* too, a 'threshold body size' must be attained to reach maturity (McKee & Ebert, 1996). Regional comparisons under different temperature regimes would certainly improve our understanding on the role of age vs. size on the maturity of tropical and temperate zooplankton.

Patterns of survivorship, offspring production, and growth rates

In temperate ponds, fecundity of cladocerans is normally high in spring but declines during the clear-water phase, which starts during the second half of June. Increase in water temperature and predation pressure and decrease in food quality and quantity are considered to be the causal factors for reduced fecundity of cladocerans in the clear-water period (Lampert & Sommer, 1997). In an elaborate model, Mooij et al. (2003) have evaluated the influence of temperature and food on the population dynamics and demographic characteristics of temperate *Daphnia* species. Their results showed: (a) in April water temperature is limiting, (b) in May–June and September–October, food was limiting and (c) in August, neither food nor temperature limits the cladoceran fecundities. In tropical water bodies such seasonally phased effects on offspring production are not documented (Lampert & Sommer, 1997), nor considered likely due to less clearly defined food and temperature changes.

Survivorship curves of most cladocerans under non-stressful conditions follow a rectangular pattern when plotted against age (time), where there is little mortality in the early age but the mortality increases as the individuals get older. This pattern depends a great deal on the rearing conditions (Nandini & Sarma, 2000). Both tropical and temperate taxa behave similarly when cultured under optimal conditions, although the duration of lifespan varies in that temperate species generally live longer than tropical taxa (Dumont & Negrea, 2002). Even within the same genus, the patterns of senescence and survivorship vary markedly (Dudycha, 2003). Changes in food concentration affect this general pattern (Urabe, 1991). It appears that moderately high food levels ($\sim$1.35 μg dw ml^{-1} or above) of even edible algae are less favourable for many cladocerans (Nandini & Sarma, 2000). Certain temperate species of *Daphnia* are particularly adapted to living at food concentrations much lower (Gliwicz, 1990) than those for tropical cladocerans such as *Diaphanosoma* (Nandini & Sarma, 2003). The mean lifespan of many temperate taxa mainly *Daphnia* is usually longer than 3 weeks but that of typical tropical species is relatively shorter (Dumont & Negrea, 2002). For example, for the temperate *D. magna* the average lifespan varied from 3 to 11 weeks (DeMeester, 1994) while for the tropical *Daphnia lumholtzi*, it is less than 3 weeks (Lennon et al., 2001) (Table 3).

In addition to the differences in mean lifespan , generation times of tropical and temperate cladoceran taxa also differ. In most cladocerans generation time ranges from 1 to 4 weeks (Table 4), with the shortest duration for tropical taxa such as *Moina* and the longest for temperate taxa, e.g., *Daphnia* and *Eurycercus* (Gillooly, 2000). The pattern of offspring production or fecundity is yet another important variable for comparing tropical and temperate taxa. In crustaceans including most cladocerans, when food is not limiting, fecundity increases with increasing body size: large-bodied

Table 3. Mean lifespan (days) of selected cladoceran taxa. Data on the reduced lifespan due to the stress from toxicants (heavy metals, pesticides) are excluded

Taxa/Variable	Range	References
Alona rectangular	17–18	Muro-Cruz et al. (2002)
Pleuroxus aduncus	8–68	Flossner (1972) in Dumont & Negrea (2002) and Nandini & Sarma (2000)
Ceriodaphnia cornuta	13–25	Nandini (2000) and Nandini & Sarma (2000)
C. dubia	20–25	Nandini & Sarma (2002)
C. cf. dubia	17–43	Rose et al. (2000)
Daphnia lumholtzi	4–20	Lennon et al. (2001)
D. magna	23–80	Botnaruic et al. (1960) in Dumont & Negrea (2002) and DeMeester (1994)
D. mendotae	>150	Flossner (1972 in Dumont & Negrea (2002)
Diaphanosoma birgei	13–16	Sipaúba-Tavares & Bachion (2002)
Diaphanosoma brachyurum	15–29	Lemke & Benke (2003)
Diaphanosoma celebensis	6–13	Shrivastava et al. (1999)
Macrothrix triserialis	27–31	Muro-Cruz et al. (2002)
Moina micrura	5–6	Sipaúba-Tavares & Bachion (2002)
M. macrocopa	5–13	Nandini & Sarma (2000) and Benider et al. (2002)
Scapholeberis mucronata	53–79	Lemke & Benke (2003)
Simocephalus serrulatus	24–55	Lemke & Benke (2003)
Simocephalus vetulus	30–40	Nandini & Sarma (2000)

Table 4. Data on the generation times (days) of some cladoceran species based on laboratory cultures

Taxa	Gen. Time	Reference
A. rectangular	10–11	Muro-Cruz et al. (2002)
Ceriodaphnia cornuta	4–18	Nandini (2000), Nandini & Sarma (2000) and Ferrão-Filho et al. (2000)
C. dubia	18–24	Nandini & Sarma (2002)
C. cf. *dubia*	10–18	Rose et al. (2000)
D. pulex	8–10	Ferrão-Filho et al. (2000)
D. ambigua	8–44	Lei & Armitige (1980) in Gillooly (2000)
D. magna	16–28	Korpelainen (1986) in Gillooly (2000)
D. obtuse	18–21	Nix & Jenkins (2000)
Diaphanosoma brachyurum	9–40	Lemke & Benke (2003)
Eurycercus lamellatus	19–100	Bottrell (1975) in Gillooly (2000)
M. triserialis	17–19	Muro-Cruz et al. (2002)
Moina macrocopa	4–8	Nandini & Sarma (2000, 2002)
M. micrura	10–11	Ferrão-Filho et al. (2000)
Moinodaphnia macleayi	10–12	Ferrão-Filho et al. (2000)
Pleuroxus aduncus	7–14	Nandini & Sarma (2000)
P. dentuculatus	25	Hann (1984) in Gillooly (2000)
P. uncinatus	14–78	Bottrell (1975) in Gillooly (2000)
Scapholeberis mucronata	13–43	Lemke & Benke (2003)
Sida crystallina	16–48	Bottrell (1975) in Gillooly (2000)
Simocephalus serrulatus	10–28	Lemke & Benke (2003)
Simocephalus vetulus	10–34	Nandini & Sarma (2000, 2003)

daphnid, *D. magna*, may have fecundity close to 1000 offspring per female during her lifetime (Dumont & Negrea, 2002). On the other hand, smaller species, *Chydorus* and *Alona*, may have <20 eggs per female during the lifespan (Muro-Cruz et al., 2002); cladocerans of intermediate sizes may have intermediate net reproductive rates (Venkataraman, 1990). When cultured under optimal (25–30 °C) temperature, tropical taxa appear to have lower number of eggs than their temperate counterpart of comparable body size (Nandini, 2000; Dumont & Negrea, 2002). Gross and net reproductive rates (Table 5) are also higher for large cladocerans than for smaller taxa. Increase in food level results in increased offspring production, regardless of the taxa (Rose et al., 2000). However, temperate taxa are less sensitive to low food levels than tropical taxa. The rate of population increase is yet another important variable affected by temperature and food level as well as their interaction (Maier, 1993). In tropical species population

Table 5. Data on the net reproductive rates (offspring female^{-1}) of some cladoceran species

Taxa	Net repro. rate	References
A. rectangula	13–15	Muro-Cruz et al. (2002)
Ceriodaphnia cornuta	1–45	Nandini, (2000) and Ferrão-Filho et al. (2000)
C. dubia	8–43	Nandini & Sarma (2002)
C. cf. dubia	38–46	Rose et al. (2000)
Daphnia lumholtzi	4–62	Lennon et al. (2001) and Pattison et al. (2003)
D. magna	59–83	Porter & Orcutt (1977)
D. obtuse	152–198	Nix & Jenkins (2000)
D. parvula	1–11	Pattison et al. (2003)
D. pulex	3–41	Ferrão-Filho et al. (2000)
Diaphanosoma brachyurum	1–20	Lemke & Benke (2003)
Moina macrocopa	11–26	Nandini & Sarma (2000)
M. micrura	2–43	Ferrão-Filho et al. (2000)
Moinodaphnia macleayi	9–28	Ferrão-Filho et al. (2000)
Macrothrix triserialis	19–27	Muro-Cruz et al. (2002)
Pleuroxus aduncus	4–6	Nandini & Sarma (2000)
Scapholeberis mucronata	20–51	Lemke & Benke (2003)
Simocephalus serrulatus	12–20	Lemke & Benke (2003)
S. vetulus	29–34	Nandini & Sarma (2000)

growth rate increases with increasing temperature upto a certain level above which it begins to decline. For example, the rate of population increase of *Moina salina* increased with increasing temperature from 20° to 30 °C but started to decline as the temperature rose to 33 °C (Wang & He, 2001). This is also true for temperate taxa, but at relatively lower temperatures. Temperate species of *Daphnia* generally have lower growth rates (0.1–0.4: Lennon et al., 2001) than *Moina* spp. cultured under tropical conditions (Table 6).

Some workers have also considered the inter-relationships within the life history variables. King (1982) has hypothesized that for iteroparous taxa including zooplankton, median lifespan is nearly twice as long as generation time. We used the literature data on average life span and generation time for 13 species of cladocerans and found a linear regression relationship of these two

Table 6. Rate of population increase (r) (day^{-1}) of selected cladoceran species. Values are derived from either life table or population growth studies

Taxa	Range of r	Reference
L 2134 ife table study		
A. rectangula	0.31–0.34	Muro-Cruz et al. (2002)
Ceriodaphnia cornuta	0.03–0.25	Nandini & Sarma (2000) and Ferrão-Filho et al. (2000)
C. dubia	0.12–0.27	Nandini & Sarma (2002)
C. cf. *dubia*	0.39–0.54	Rose et al. (2000)
D. cucullata	0.15–0.31	Lürling et al. (1997)
D. galeata	0.17–0.35	Weers & Gulati (1997) and Weber et al. (2003)
D. lumholtzi	0.21–0.42	Lennon et al. (2001) and Pattinson et al. (2003)
D. magna	0.40–0.58	Arbačiauskas & Lampert (2003) and De Lange & Van Reeuwijk (2003)
Daphnia parvula	0.04–0.35	Pattison et al. (2003)
D. pulex	0.08–0.51	Lürling & Van Donk (1997) and Walls et al. (1997)
D. obtusa	0.25–0.28	Nix & Jenkins (2000)
Diaphanosoma birgei	0.37–0.57	Sipaúba-Tavares & Bachion (2002)
D. brachyurum	0.04–0.32	Lemke & Benke (2003)
M. triserialis	0.25–0.26	Muro-Cruz et al. (2002)
M. macrocopa	0.06–0.52	Nandini & Sarma (2002) and Hurtado-Bocanegra et al. (2002)
M. micrura	0.11–0.69	Ferrão-Filho et al. (2000) and Sipaúba-Tavares & Bachion (2002)
Moinodaphnia macleayi	0.22–0.34	Ferrão-Filho et al. (2000)
Pleuroxus aduncus	0.13–0.15	Nandini & Sarma (2000)
Scapholeberis mucronata	0.07–0.30	Lemke & Benke (2003)
Simocephalus serrulatus	0.09–0.29	Lemke & Benke (2003)
S. vetulus	0.12–0.28	Nandini & Sarma (2000)
Population growth study		
A. pulchella	0.28	Nandini et al. (1998)
A. rectangula	0.02–0.28	Nandini et al. (2002) and Nandini & Sarma (2003)

Continued on p. 326

Table 6. (Continued)

Taxa	Range of *r*	Reference
C. cornuta	0.24–0.32	Nandini & Rao (1998)
C. dubia	0.04–0.27	Nandini & Sarma (2003)
D. carinata	0.10–0.30	Nandini & Rao (1998)
D. laevis	0.01–0.14	Nandini & Sarma (2003)
D. pulex	0.12–0.25	Alva-Martinez et al. (2001) and Mangas-Ramírez et al. (2002)
D. pulicaria	0.11–0.39	Kilham et al. (1997)
Diaphanosoma brachyurum	0.02–0.20	Nandini & Nandini (2003)
M. triserialis	0.02–0.26	Nandini et al. (1998, 2002)
M. macrocopa	0.02–0.60	Nandini & Rao (1998) Alva-Martínez et al. (2001) and Nandini & Sarma (2003)
Scapholeberis kingi	0.08–0.28	Nandini & Rao (1998) and Nandini & Sarma (2003)
Simocephalus vetulus	0.02–0.31	Nandini & Rao (1998) and Nandini & Sarma (2003)

Negative growth rates as a result of toxicant exposure have not been included.

parameters to be significantly related (Fig. 1: $p < 0.05$), although the ratios between average life span and generation time in the regression analysis varied widely, from 0.5 to 4.3. Moreover, there are large intra-specific variations of the ratios within both planktonic taxa, e.g., *Daphnia laveis*, and littoral taxa, *Scapholeberis mucronata* (Fig. 1). What causes the lifespan to be shorter than the generation time is not clear, i.e., a ratio of less < 1 implies that individuals die before completing a generation. This is only likely if all the individuals die, due to predation before they reach maturity. Further analysis of literature data and experimental investigation will reveal if there are subtle regional differences within or between the taxa of tropical and temperate regions.

Also, there seems to exist a relationship between the cladoceran body size and the maximum clutch size. In general, smaller taxa (e.g., chydorids) have lower number of eggs per brood than larger taxa such as *Daphnia*. Based on field data, Hülsmann (2001) reported for *D. galeata* a brood size varying from 2 to 20 per female with body size (length ca. 700–2100 μm). Clutch size of individual females, in the preserved field samples, may not represent true fecundity because not all eggs in the brood pouch are viable. Therefore, from the life-table method, one can obtain data on viable offspring for each clutch. The maximum body length for different cladoceran genera and the maximum number of offspring produced per clutch, show a direct regression relationship (Fig. 2). Thus, larger cladoceran taxa have a greater number of neonates per clutch than the smaller taxa. Since herbivorous species of planktonic cladocerans are generally larger in temperate

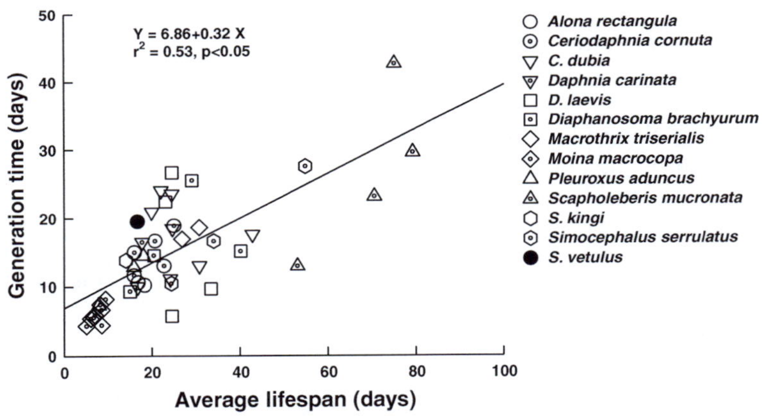

Figure 1. Relation between average lifespan and generation time of selected cladoceran species (Source: *A. rectangula*: Muro-Cruz et al., 2002; *C. cornuta*: Nandini & Sarma, 2000; *C. dubia*: Rose et al., 2000 and Nandini & Sarma, 2002; *D. carinata*: Nandini, 2000; *D. laevis*: Nandini et al., 2000; *D. brachyurum*: Lemke & Benke, 2003; *M. triserialis*: Muro-Cruz et al., 2002; *M. macrocopa*: Nandini & Sarma, 2000 and 2002; *P. aduncus* Nandini & Sarma, 2000; *Scapholeberis mucronata*: Lemke & Benke, 2003; *S. kingi*: Nandini: 2000; *S. serrulatus*: Lemke & Benke, 2003; *S. vetulus*: Nandini, 2000).

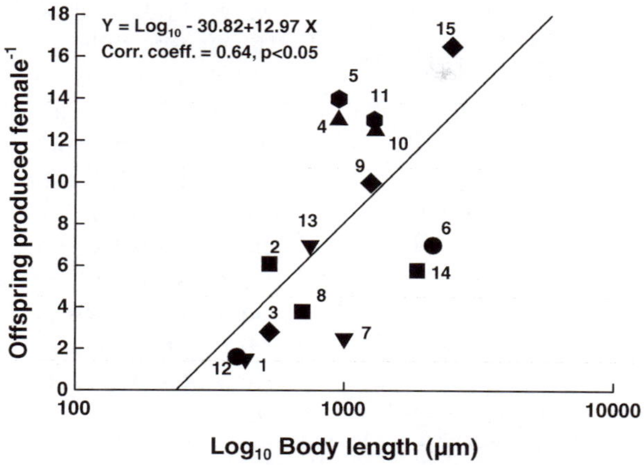

Figure 2. Relation between carapace length and the number of offspring produced per clutch in some cladoceran species. For the purpose of brevity, only maximum body size and maximum clutch size for each cladoceran species was presented. 1: *Alona rectangula*, 2–3: *Ceriodaphnia regaudi*, 4–5: *C. dubia*; 6. *Daphnia carinata*; 7: *D. laevis*; 8. *Macrothrix triserialis*; 9–11: *Moina macrocopa*; 12: *Pleuroxus aduncus*; 13: *Scapholeberis kingi*; 14–15: *Simocephalus vetulus*. Data source: 1,8: Muro-Cruz et al. (2002); 2,3,6,9,13,14: Nandini (2000); 4,10,12,15: Nandini & Sarma (2000); 5-11: Nandini & Sarma (2002); 7: Nandini et al. (2000).

waters (Dodson & Frey, 2001), one can expect higher number of neonates per individual (i.e., enhanced egg ratio) than for tropical taxa.

Population growth, body size, and numerical abundance

Under optimal food and temperature conditions, for a given volume, smaller-sized taxa are more abundant than larger-sized ones (Nandini & Sarma, 2003). Since tropical water bodies may generally contain smaller cladocerans, it is reasonable to speculate that numerically they reach greater numbers per unit volume than do temperate taxa. In relation to this, some information on the peak abundances of both tropical and temperate species is presented in Table 7.

Other differences

There are other factors, which if better investigated and understood will be helpful to contrast the two geographical areas. For example, there is a greater abundance of rotifers and smaller-sized cladocerans in tropics (Fernando, 2002) than in temperate areas, which implies a greater exploitation of smaller food particles in tropical waters than in temperate waters. Presently we know little about the extent of interference competition between rotifers and cladocerans and differences herein tropical and temperate regions. Because the larger-bodied pelagic cladocerans are less common in tropics (Fernando, 2002), smaller-bodied taxa may have advantage for resource exploitation. Perhaps there are more ephemeral water bodies in tropics than in temperate regions, which necessitates the completion of development and growth in a relatively shorter duration or the formation of ephippia or other resting stages to prevent desiccation of the fauna (Williams, 1987). Metabolic differences in tropical and temperate cladocerans could be of considerable interest since this may depend on the type of food available in the ambient water. For example, when *D. magna* was fed on experimentally manipulated diets (high carbon : phosphorus ratio), the animals disposed of the excess dietary carbon *via* respiration and excretion of dissolved organic carbon (Darchambeau et al., 2003). Also resistance to starvation may affect tropical and temperate cladocerans differently. Usually, at or near optimal temperature, larger species resist starvation for longer periods than the smaller taxa (see Kirk, 1997 for details). This is further influenced by the nutritional status of the individuals before undergoing starvation. Cladocerans fed nutritionally balanced diets are expected to resist starvation longer than those raised on unbalanced or poor quality food (Dodson & Frey, 2001).

Table 7. Peak/maximum densities of cladocerans (ind ml^{-1}) observed under field or laboratory conditions

Taxa	Density (no. ml^{-1})	Reference
Alona pulchella	55	Nandini et al. (1998)
A. rectangula	1–71	Nandini et al. (2002) and Nandini & Sarma (2003)
Bosmina longirostris	0.4	Hrbáček & Novotná-Dvořáková (1965)
Ceriodaphnia cornuta	14	Nandini & Rao (1998)
C. dubia	5–37	Alva-Martinez et al. (2001) and Mangas-Ramírez et al. (2002)
Macrothrix triserialis	1–50	Nandini et al. (1998, 2002)
Moina macrocopa	5–16	Mangas-Ramírez et al. (2002)
Moina micrura	6	Sipaúba-Tavares & Bachion (2002)
Scapholeberis kingi	1–18	Nandini & Rao (1998) and Nandini & Sarma (2003)
Diaphanosoma birgei	5–6	Sipaúba-Tavares & Bachion (2002)
Diaphanosoma brachyurum	1–13	Nandini & Sarma (2003)
D. carinata	2–4	Nandini & Rao (1998)
D. laevis	1–4	Nandini & Sarma (2003)
D. pulex	6–10	Alva-Martinez et al. (2001)
D. rosea	0.15	Shei et al. (1988) cited in Hülsmann (2003)
D. hyaline	0.32	Vijverberg & Richter (1982) cited in Hülsmann (2003)
D. galeata	0.14	Hülsmann & Voigt (2002)
D. longispina	0.18	Krylov & Arbačiauskas (1994) cited in Hülsmann (2003)

The taxa are presented roughly in body size sequence; there is a traceable trend of an inverse relationship between size and peak density.

Differences in trophic status of lakes, together with regional temperature disparity, may help us further understand differences in density and diversity of cladoceran species. Except water bodies at high altitudes, most ponds and lakes in tropical regions are mesotrophic, eutrophic or even hypereutrophic (Harper, 1992). On the other hand, relatively a much larger number temperate water bodies, especially at more northern latitudes, especially those at and higher altitudes, are oligotrophic or even ultra-oligotrophic. In such water bodies, small, slow-growing cladocerans are known to be superior competitors, in contrast with eutrophic waters where large rapid-growing species dominate and can outcompete the small cladocerans (Romanovsky, 1984). Eutrophic lakes may also contain larger number of smaller-sized cladoceran species (Hillbricht-Illkowska, 1977). Many of these aspects have not been investigated for tropical cladocerans.

Conclusions

It is striking to note that most work on life-history characteristics of cladocerans, especially *Daphnia* species, has been carried out in temperate waters. Similar studies from tropical regions are scarce. Therefore, regional comparisons are difficult. Temperature appears to be the most important contrasting environmental variable between the tropical and temperate regions. This is caused by marked seasonal differences in solar radiation between the two regions: greater seasonal changes in both light intensity and light duration distinguish the temperate regions from the tropical ones. Other environmental factors that differ are the food type and density, and a more ephemeral nature of water bodies in tropics. Whereas both fish and invertebrate predators are considered as important in determining the abundance, composition, and diversity of cladocerans in temperate waters, little is known about the importance of invertebrate predators in tropical waters.

Temperate and tropical cladocerans differ in many aspects of their life-history parameters including the size of same or similar taxa of comparable age. Major life-history traits that distinguish tropical cladocerans from their temperate counterpart are: relatively higher rates of population increase, higher threshold food levels, higher filtering and ingestion rates and greater tolerance to higher food levels and to temperature maxima. Introduction or the so-called bio-invasion of tropical taxa into temperate regions due to mainly anthropogenic activities could alter the cladoceran regional distributional patterns and reduce the existing differences regional diversity. Although more cladoceran taxa have been recorded and described from the temperate regions, more recent evidence shows that cladoceran diversity in tropical water bodies may be high as well. In some characteristics, e.g., predator-induced morphology, both tropical and temperate cladocerans respond similarly. The impacts of such morphological changes on the life-history variables of tropical and temperate cladocerans need to be evaluated in the light of environmental differences and compared more critically in future studies. Also, morphological changes in response to abiotic factors that affect survival and reproduction of cladocerans need greater attention. Comparative regional studies on tropical and temperate taxa will certainly help to understand the role of abiotic factors in inducing morphological changes and the impact of such adaptations on the life history characteristics of cladocerans. In the present review we have both shown that tropical and temperate cladocerans differ in several life-history characteristics and attempted to explain the environmental factors that contribute to these differences.

Acknowledgments

This work was supported by a project from CONACyT (C01-41786). S.S.S.S. and S.N thank authorities of UNAM Campus Iztacala for travel grants.

References

Achenbach, L. & W. Lampert, 1997. Effects of elevated temperature on threshold food concentrations and possible competitive abilities of differently sized cladoceran species. Oikos 79: 469–476.

Alekseev, V. & W. Lampert, 2001. Maternal control of resting-egg production in *Daphnia*. Nature 41: 899–901.

Alva-Martínez, A. F., S. S. S. Sarma & S. Nandini, 2001. Comparative population dynamics of three cladoceran species (Cladocera) in relation to different levels of *Chlorella vulgaris* and *Microcystis aeruginosa*. Crustaceana 78: 749–764.

Arbaèiauskas, K. & W. Lampert, 2003. Seasonal adaptation of ex-ephippio and parthenogenetic offspring of *Daphnia magna*: differences in life history and physiology. Functional Ecology 17: 431–437.

Arnott, S. E. & M. J. Vanni, 1993. Zooplankton assemblages in fishless bog lakes: Influence of biotic and abiotic factors. Ecology 74: 2361–2380.

Babu, S. & C. K. G. Nayar, 1997. Laboratory studies on the life cycle of *Simocephalus serrulatus* Koch 1881 (Cladocera: Crustacea). Journal of the Bombay Natural History Society 94: 317–321.

Benider, A., A. Tifnouti & R. Pourriot, 2002. Growth of *Moina macrocopa* (Straus, 1820) (Crustacea, Cladocera): influence of trophic conditions, population density and temperature. Hydrobiologia 468: 1–11.

Boersma, M. & C. Kreutzer, 2002. Life at the edge: is food quality really of minor importance at low quantities? Ecology 83: 2552–2561.

Bungartz, B. & D. K. Branstrator, 2003. Morphological changes in *Daphnia mendotae* in the chemical presence of *Bythotrephes longimanus*. Archiv für Hydrobiologie 158: 97–108.

Caceres, C. E. & M. S. Schwalbach, 2001. How well do laboratory experiments explain field patterns of zooplankton emergence? Freshwater Biology 46: 1179–1189.

Chapelle G. & L. S. Peck, 1999. Polar gigantism dictated by oxygen availability. Nature 399: 114–115.

Christopher, J. M., 2000. First record of *Daphnia lumholtzi* Sars in the Great Lakes. Journal of Great Lakes Research 26: 352–354.

Cousyn, C., L. De Meester, J. K. Colbourne, L. Brendonck, D. Verschuren & F. Volckaert, 2001. Rapid, local adaptation of zooplankton behavior to changes in predation pressure in the absence of neutral genetic changes. Proceedings of the National Academy of Sciences, USA 98: 6256–6260.

Darchambeau, F., P. J. Faerovig & D. O. Hessen, 2003. How *Daphnia* copes with excess carbon in its food. Oecologia 136: 336–346.

De Bernardi, R. & M. Manca, 1982. The consequences of life history strategies on competition between two cladocerans. Memorie dell Istituto italiano di idrobiologia 40: 145–161.

De Clerk, S., L. De Meester, N. Podoor & J. M. Conde-Porcuna, 1997. The relevance of size efficiency to biomanipulation theory. Hydrobiologia 360: 265–275.

De Lange, H. & P. L. Van Reeuwijk, 2003. Negative effects of UVB-irradiated phytoplankton on life history traits and fitness of *Daphnia magna*. Freshwater Biology 48: 678–686.

De Meester, L., 1994. Life histories and habitat selection in *Daphnia*: divergent life histories of *D. magna* clones differing in phototactic behaviour. Oecologia 97: 333–341.

De Meester, L., P. Dawidowicz, E. Van Good & C. J. Loose, 1998. Ecology and evolution of predator-induced behaviour of zooplankton: depth selection behaviour and diel vertical migration. In R. Tollrian & Harvell, C. D. (eds), Ecology and Evolution of Inducible Defences. Princeton University Press, New Jersey: 160–176.

Dodson, S. I. & D. G. Frey, 2001. Cladocera and other Branchiopoda. In Thorp, J. H. & Covich, A. P. (eds), Ecology and Classification of North American Freshwater Invertebrates, 2nd edn. Academic Press, San Diego, San Francisco: 849–913.

Downing, J. A. & F. H. Rigler (eds), 1984. A Manual for the Methods of Assessment of Secondary Productivity in Fresh Waters, 2nd edn. IBP Handbook 17. Blackwell Scientific Publications, London.

Dudycha, J. L., 2003. A multi-environment comparison of senescence between sister species of *Daphnia*. Oecologia 135: 555–563.

Dumont, H. J., 1994. On the diversity of the Cladocera in the tropics. Hydrobiologia 272: 27–38.

Dumont, H. J., J. G. Tundisi & K. Roche (eds), 1990. Intrazooplankton Predation. Developments in Hydrobiology No. 60. Kluwer Academic Publishers, Dordrecht, The Netherlands.

Dumont, H. & S. Negrea 2002. Introduction to the Class Branchiopoda. Guides to the Identification of the Microinvertebrates of the Continental Waters of the World, Backhuys Publishers, The Netherlands.

Duncan, N., 1989. Food limitation and body size in the life cycle of planktonic rotifers and cladocerans. Hydrobiologia 186/187: 11–28.

Engelmayer, A., 1995. Effects of predator-released chemicals on some life history parameters of *Daphnia pulex*. Hydrobiologia 307: 203–206.

Enríquez-García, C., S. Nandini & S. S. S. Sarma, 2003. Food type effects on the population growth patterns of littoral rotifers and cladocerans. Acta Hydrochimica et Hydrobiologica 31: 120–133.

Fernando, C. H., 2002. In Fernando, C. H. (ed.), Guide to Tropical Freshwater Zooplankton. Identification, Ecology and Impacts on Fisheries. Backhuys Publishers, The Netherlands: 255–280.

Fernando, C. H. & J. C. Paggi, 1997. Cosmopolitanism and latitudinal distribution of freshwater planktonic Rotifera and Crustacea. Verhandlungen der internationale Vereinigung für Limnologie 26: 1916–1917.

Ferrão-Filho, A. S. & S. M. F. O. Azevedo, 2003. Effects of unicellular and colonial forms of toxic *Microcystis aeruginosa* from laboratory cultures and natural populations on tropical cladocerans. Aquatic Ecology 37: 23–35.

Gillooly, J. F., 2000. Effect of body size and temperature on generation time in zooplankton. Journal of Plankton Research 22: 241–251.

Gliwicz, Z. M., 1990. Food thresholds and body size in cladocerans. Nature 343: 638–640.

Gliwicz, Z. M., 2003. Between hazards of starvation and risks of predation: the ecology of offshore animals. In Kinne, O. (ed.), Excellence in Ecology No. 12. International Ecology Institute, Oldedorf/Luhe, Germany: 379 pp.

Gliwicz, Z. M. & W. Lampert, 1990. Food thresholds in *Daphnia* species in the absence and presence of blue-green filaments. Ecology 71: 691–702.

Gliwicz, Z. M. & M. J. Boavida, 1996. Clutch size and body size at first reproduction in *Daphnia pulicaria* at different levels of food and predation. Journal of Plankton Research 18: 863–880.

Gulati, R. D., 1977. Influence of temperature on animal life with special reference to food uptake and metabolism of zooplankton. In Marois, M. (ed), Towards a Plan of Action for Mankind: Proceedings of a World Conference, Pergamon, Oxford, Vol. 3: Biological balance and thermal modifications: 205–218.

Gulati, R. D. & W. R. DeMott, 1997. The role of food quality for zooplankton: remarks on the state-of-the-art, perspectives and priorities. Freshwater Biology 38: 753–768.

Gulati, R. D., E. H. R. R. Lammens, M. L. Meyer & E. van Donk, 1990. Biomanipulation–tool for water management. Hydrobiologia 200/201: 1–628.

Hardy, E. R. & A. Duncan, 1994. Food concentration and temperature effects on life cycle characteristics of tropical cladocera (*Daphnia gessneri* Herbst, *Daiphanosoma sarsi* Richard, *Moina reticulata* (Daday)): 1. Development time. Acta Amazonica 24: 119–134.

Harper, D., 1992. Eutrophication of freshwaters. Principles, Problems and Restoration. Chapman & Hall, London.

Hebert, P. D. N. & T. L. Finston, 1993. A taxonomical reevaluation of North American *Daphnia* (Crustacea: Cladocera). *D. similis* complex. Canadian Journal of Zoology 71: 908–925.

Hebert, P. D. N. & M. E. A. Cristescu, 2002. Genetic perspectives on invasions: the case of the Cladocera. Canadian Journal of Fisheries and Aquatic Sciences 59: 1229–1234.

Hessen, D. O., P. J. Faeroevig & T. Andersen, 2002. Light, nutrients, and P:C ratios in algae: grazer performance related to food quality and quantity. Ecology 83: 1886–1898.

Hillbricht-Illkowska, A., 1977. Trophic relations and energy flow in pelagic plankton. Polish Ecological Studies 3: 3–98.

Hrbáček, J. & M. Novotná-Dvořáková, 1965. Plankton of four backwaters related to their size and fish stock. Řada matematických a Přírodních věd 75: 3–64.

Hülsmann, S., 2001. Reproductive potential of *Daphnia galeata* in relation to food conditions: implications for a changing size-structure of the population. Hydrobiologia 491: 35–46.

Hülsmann, S., 2003. Recruitment patterns of *Daphnia*: a key for understanding midsummer declines? Hydrobiologia 491: 35–46.

Hülsmann, S. & H. Voigt, 2003. Life history of *Daphnia galeata* in a hypertrophic reservoir and consequences of non-consumptive mortality for the initiation of a midsummer decline. Freshwater Biology 47: 2313–2324.

Hurtado-Bocanegra, M. D., S. Nandini & S. S. S. Sarma, 2002. Combined effects of food level and inoculation density on competition between *Brachionus patulus* (Rotifera) and the cladocerans *Ceriodaphnia dubia* and *Moina macrocopa*. Hydrobiologia 468: 13–22.

Hutchinson, G. E., 1967. A treatise on limnology, 2. John Wiley, New York. 1115 pp.

Infante, A. & W. Riehl, 1984. The effects of Cyanophyta upon zooplankton in a eutrophic tropical lake (Lake Valencia, Venezuela). Tropical zooplankton. Developments in Hydrobiology No. 23: 293–298.

Innes, D. J. & D. R. Singleton, 2000. Variation in allocation to sexual and asexual reproduction among clones of cyclically parthenogenetic *Daphnia pulex* (Crustacea: Cladocera). Biological Journal of the Linnean Society 71: 771–787.

Kerfoot, W. C. & A. Sih (eds), 1987. Predation: direct and indirect impacts on aquatic communities. University Press of New England, Hanover, N. H.

Kilham, S. S., D. A. Kreeger, C. E. Goulden & S. G. Lynn, 1997. Effects of algal food quality on fecundity and population growth rates of *Daphnia*. Freshwater Biology 38: 639–647.

King, C. E., 1982. The evolution of lifespan. In Dingle, H. & Hegmann, J. P. (eds). Evolution and Genetics of Life Histories. Springer Verlag, New York: 121–128.

Kirk K. L. 1997. Life-history responses to variable environments: Starvation and reproduction in planktonic rotifers. Ecology 78: 434–441.

Koivisto, S., 1995. Is *Daphnia magna* an ecologically representative zooplankton species in toxicity tests? Environmental Pollution 90: 263–267.

Kolar, C. S., J. C. Boase, D. F. Clapp & D. H. Wahl, 1997. Potential effect of invasion by an exotic zooplankter, *Daphnia lumholtzi*. Journal of Freshwater Ecology 12: 521–530.

Kořinek, V. 2002. In Fernando, C. H. (ed.), 2002. Guide to Tropical Freshwater Zooplankton. Identification, Ecology and Impacts on Fisheries. Backhuys Publishers, The Netherlands: 69–97.

Kulshrestha, S. K., U. N. Adholia, A. Bhatnagar, A. A. Khan & M. Baghail, 1991. Community structure of zooplankton at River Chambal near Nagda with reference to industrial pollution. Acta Hydrochimica et Hydrobiologica 19: 181–191.

Lampert, W., 1977. Studies on the carbon balance of *Daphnia pulex* de Geer as related to environmental conditions. IV. Determination of "threshold" concentration as a factor controlling the abundance of zooplankton species. Archiv für Hydrobiologie Supplement 48: 361–368.

Lampert, W. & U. Sommer, 1997. Limnoecology. The Ecology of Lakes and Streams. Oxford University Press, New York, 1–382.

Lass, S. & P. Spaak, 2003. Chemically induced anti-predator defences in plankton: a review. Hydrobiologia 491: 221–239.

Lemke, A. M. & A. C. Benke, 2003. Growth and reproduction of three cladoceran species from a small wetland in the south-eastern USA. Freshwater Biology 48: 589–603.

Lennon, J. T., V. H. Smith & K. Williams, 2001. Influence of temperature on exotic *Daphnia lumholtzi* and implications for invasion success. Journal of Plankton Research 23: 425–434.

Luening, J., 1992. Phenotypic plasticity of *Daphnia pulex* in the presence of invertebrate predators: morphological and life history responses. Oecologia 92: 383–390.

Lürling, M. & E. Van Donk, 1997. Life history consequences for *Daphnia pulex* feeding on nutrient-limited phytoplankton. Freshwater Biology 38: 619–628

Lürling, M., H. J. De Lange & E. Van Donk, 1997. Changes in food quality of the green alga *Scenedesmus* induced by *Daphnia* infochemicals: biochemical composition and morphology. Freshwater Biology 38: 693–709.

Lynch, M., 1980. The evolution of cladoceran life histories. Quarterly Reviews of Biology 55: 23–42.

Lynch, M., 1992. The life history consequences of resource depression in *Ceriodaphnia quadrangula* and *Daphnia ambigua*. Ecology 73: 1620–1629.

Maier, G., 1993. The life histories of two temporarily coexisting, pond dwelling cladocerans. Internationale Revue der gesamten Hydrobiologie 78: 83–93.

Manca, M. & R. de Bernardi, 1985. Energy budget and evolutive strategies in two cladocerans: *Daphnia obtusa* kurtz and Simocephalus vetulus (O. F. Müller). Memorie dell Istituto italiano di idrobiologia 43: 119–145

Manca, M., C. Ramoni & P. Comoli, 2000. The decline of *Daphnia hyalina galeata* in Lago Maggiore: a comparison of the population dynamics before and after oligotrophication Aquatic Sciences 62: 142–153.

Mangas-Ramírez, E., S. S. S. Sarma & S. Nandini, 2002. Combined effects of algal (*Chlorella vulgaris*) density and ammonia concentration on the population dynamics of *Ceriodaphnia dubia* and *Moina macrocopa* (Cladocera). Ecotoxicology and Environmental Safety 51: 216–222

Matveev, V. F., 1987. Effect of competition on the demography of planktonic cladocerans – *Daphnia* and *Diaphanosoma*. Oecologia 74: 468–477.

McKee, D. & D. Ebert, 1996. The effect of temperature on maturation threshold body-length *Daphnia magna*. Oecologia 108: 627–630.

McKee, D., D. Atkinson, S. Collings, J. Eaton, I. Harvey, T. Heyes, K. Hatton, D. Wilson & B. Moss, 2002. Macrozooplankter responses to simulated climate warming in experimental freshwater microcosms. Freshwater Biology 47: 1557–1570

Mohamed, Z. A. & D. H. Smith, 2001. Accumulation of cyanobacterial hepatotoxins by *Daphnia* in some Egyptian irrigation canals. Ecotoxicology and Environmental safety 50: 4–8.

Mooij, W. M., S. Hülsmann, J. Vijverberg, A. Veen & E. H. R. R. Lammens, 2003. Modeling *Daphnia* population dynamics and demography under natural conditions. Hydrobiologia 491: 19–34.

Moore, M. M., C. L. Folt & R. S. Stemberger, 1996. Consequences of elevated temperature for zooplankton assemblages in temperate lakes. Archiv für Hydobiologie 135: 289–319.

Muro-Cruz, G., S. Nandini & S. S. S. Sarma, 2002. Comparative life table demography and population growth of *Alona rectangula* and *Macrothrix triserialis* (Cladocera: Crustacea) in relation to algal (*Chlorella vulgaris*) food density. Journal of Freshwater Ecology 17: 1–11.

Nandini, S., 2000. Responses of rotifers and cladocerans to *Microcystis aeruginosa* (Cyanophyceae): a demographic study. Aquatic Ecology 34: 227–242

Nandini, S. & T. R. Rao, 1998. Somatic and population growth in selected cladoceran and rotifer species offered the cyanobacterium *Microcystis aeruginosa* as food. Aquatic Ecology 31: 283–298.

Nandini, S. & S. S. S. Sarma, 2000. Lifetable demography of four cladoceran species in relation to algal food (*Chlorella vulgaris*) density. Hydrobiologia 435: 117–126.

Nandini, S. & S. S. S. Sarma, 2002. Competition between *Moina macrocopa* and *Ceriodaphnia dubia*: a life table demography study. International Review of Hydrobiology 87: 85–95.

Nandini, S. & S. S. S. Sarma, 2003. Population growth of some genera of cladocerans (Cladocera) in relation to algal food (Chlorella vulgaris) levels. Hydrobiologia 491: 211–219

Nandini, S., G. Muro-Cruz & S. S. S. Sarma, 2002. Competition between littoral cladocerans *Macrothrix triserialis* and *Alona rectangula* (Cladocera) in relation to algal food level and inoculation density. Acta Hydrochimica et Hydrobiologica 30: 16–23

Nandini, S., S. S. S. Sarma & T. R. Rao, 1998. Effect of coexistence of the population growth of rotifers and cladocerans. Russian Journal of Aquatic Ecology 8: 1–10.

Nelson, W. A., E. McCauley & F. J. Wrona, 2001. Multiple dynamics in a single predator-prey system: experimental effects of food quality. Proceedings of the Royal Society, London B 268: 1223–1230.

Nix, M. H. & D. G. Jenkins, 2000. Life history comparisons of *Daphnia obtusa* from temporary ponds, cultured with low quality food. Aquatic Ecology 34: 19–27.

Orcutt, J. D. Jr., 1985. Food level effects on the competitive interactions of two co-occurring cladoceran zooplankton: *Diaphanosoma brachyurum* and *Daphnia ambigua*. Ergebnisse der Limnologie 21: 465–473.

Pattinson, K. R., J. E. Havel, R. G. Rhodes, 2003. Invasibility of a reservoir to exotic *Daphnia lumholtzi*: experimental assessment of diet selection and life history responses to cyanobacteria. Freshwater Biology 48: 233–246

Petrusek, A., 2002. *Moina* (Crustacea: Anomopoda, Moinidae) in the Czech Republic: a review. Acta Societas Zoologicae Bohemicae 66: 213–220.

Porter, K. G. & J. D. Orcutt, Jr. 1977. Nutritional adequacy, manageability, and toxicity as factors that determine the food quality of green and blue–green algae for *Daphnia*. In Kerfoot, W.C. (ed.), Evolution and Ecology of Zooplankton communities. University press of New England, Hanover, NH: 268–281.

Reede, T. & J. Ringelberg, 1995. The influence of a fish exudate on two clones of the hybrid *Daphnia galeata* x *hyalina*. Hydrobiologia 307: 207–212.

Rhode, S. C., M. Pawlowski & R. Tollrian, 2001. The impact of ultraviolet radiation on the vertical distribution of zooplankton of the genus *Daphnia*. Nature 412: 69–72.

Roff, D. A., 2001. Age and size at maturity. In Fox, C. W., Roff, D. A. & Fairbairn, D.J. (eds), Evolutionary Ecology. Concepts and Case Studies. Oxford University Press, New York: 99–127.

Rojas, N. E. T., M. A. Marins & O. Rocha, 2001. The effect of abiotic factors on the hatching of *Moina micrura* Kurz, 1874 (Crustacea: Cladocera) ephippial eggs. Brazilian Journal of Biology 61: 371–376.

Romanovsky, Y. E., 1984. Individual growth rate as a measure of competitive advantages in cladoceran crustaceans. Internationale Revue der gesamten Hydrobiologie 69: 613–632.

Rose, R. M., M. S. J. Warne & R. P. Lim, 2000. Life history responses of the cladoceran *Ceriodaphnia* cf. *dubia* to variation in food concentration. Hydrobiologia 427: 59–64.

Saunders, P. A., K. G. Porter & B. E. Taylor, 1999. Population dynamics of *Daphnia* spp. and implications for trophic interactions in small, monomictic lake. Journal of Plankton Research 21: 1823–1845.

Shrivastava, Y., G. G. Mahambre, C. T. Achuthankutty, B. Fernandes, S. C. Goswami & M. Madhupratap, 1999. Parthenogenetic reproduction of *Diaphanosoma celebensis* (Crustacea: Cladocera). Effect of algae and algal density on survival, growth, life span and neonate production. Marine Biology 135: 663–670.

Sipaúba-Tavares, L. H. & M. A. Bachion, 2002. Population growth and development of two species of Cladocera, *Moina micrura* and *Diaphanosoma birgei*, in laboratory. Brazilian Journal of Biology 62: 701–711.

Ślusarczyk, M., 1997. Impact of fish predation on a small-bodied cladoceran: Limitation or stimulation? Hydrobiologia 342/343: 215–221.

Sterner, R. W., 1998. Demography of a natural population of *Daphnia retrocurva* in a lake with low food quality. Journal of Plankton Research 20: 471–489.

Stibor, H., 1992. Predator induced life-history shifts in a freshwater cladoceran. Oecologia 92: 162–165.

Stockwell, J. D. & O. E. Johannsson, 1997. Temperature-dependent allometric models to estimate zooplankton production in temperate freshwater lakes. Canadian Journal of Fisheries and Aquatic Sciences 54: 2350–2360.

Taylor, D. J., P. D. N. Hebert & J. K. Colbourne, 1996. Phylogenetics and evolution of the *Daphnia longispina* group (Crustacea) based on 12S rDNA sequence and allozyme variation. Molecular Phylogenetics and Evolution 5: 495–510.

Timotius, K. H. & F. Goltenboth, 1995. Tropical Limnology. Salatiga, Indonesia.

Tessier, A. J., 1986. Life history and body size evolution in *Holopedium gibberum* Zaddach (Crustacea, Cladocera). Freshwater Biology 16: 279–286.

Urabe, J., 1991. Effect of food concentration on growth, reproduction and survivorship of *Bosmina longirostris* (Cladocera): an experimental study. Freshwater Biology 25: 1–8.

Urabe, J., J. Clasen & R. W. Sterner, 1997. Phosphorus limitation of *Daphnia* growth: Is it real? Limnology and Oceanography 42: 1436–1443.

Vanni, M. J. & W. Lampert, 1992. Food quality effects on life history traits and fitness in the generalist herbivore *Daphnia*. Oecologia 92: 48–57.

Venkataraman, K., 1990. Life-history studies on some Cladocera under laboratory conditions. Journal of the Andaman Science Association, Port Blair 6: 127–132.

Wacker, A., E. Von Elert & M. D. Bertness, 2001. Polyunsaturated fatty acids: evidence for non-substitutable biochemical resources in *Daphnia galfeata*. Ecology 82: 2507–2520.

Wærvågen, S. B., N. A. Rukke & D. O. Hessen, 2002. Calcium content of crustacean zooplankton and its potential role in species distribution. Freshwater Biology 47: 1866–1878.

Walls, M., C. Lauren-Määttä, M. Ketola, P. Ottra-Aho, M. Reinikainen & S. Repka, 1997. Phenotypic plasticity of *Daphnia* life history traits: the roles of predation, food level and toxic cyanobacteria. Freshwater Biology 38: 353–364.

Wang, Y. & Z. He, 2001. Effect of temperature and salinity on intrinsic increasing rate of *Moina mongolica* Daddy (Cladocera: Moinidae) population. Yingyong Shengtai Xuebao 12: 91–94.

Weber, A., S. Vesela & S. Repka, 2003. The supposed lack of trade-off among *Daphnia galeata* life history traits is explained by increased adult mortality in *Chaoborus* conditioned treatments. Hydrobiologia 491: 273–287.

Weers, P. M. M. & R. D. Gulati, 1997. Effect of the addition of polyunsaturated fatty acids to the diet on the growth and fecundity of *Daphnia galeata*. Freshwater Biology 38: 721–729.

Weetman D. & D. Atkinson, 2004. Evaluation of alternative hypotheses to explain temperature-induced life history shifts in *Daphnia*. Journal of Plankton Research 26: 107–116.

Whitton, B. A. & M. Potts (eds), 2000. The Ecology of Cyanobacteria. The diversity in time and space. Kluwer Academic Publishers, Dordrecht.

Williams, D. D., 1987. The Ecology of Temporary Waters. Croom Helm. London/Sidney, Timber Press, Oregon.

Yurista, P. M., 1999. Temperature-dependent energy budget of an Arctic cladoceran, *Daphnia middendorffiana*. Freshwater Biology 42: 21–34.

Zadereev, Y. S., 2003. Maternal effects, conspecific chemical cues and switching from parthenogenesis to gametogenesis in the cladoceran *Moina macrocopa*. Aquatic Ecology 37: 251–255.

Zafar, A. R., 1986. Seasonality of phytoplankton in some South Indian lakes. Hydrobiologia, 138: 177–187.

Zimmer, D. J. & J. F. Storr, 1984. The effects of environmental pH on life history parameters of *Daphnia parvula* (Crustacea; Cladocera). Proceedings of the second New York State symposium on atmospheric deposition, New York: 209–216.

Hydrobiologia (2005) 542:335–366
H. Segers & K. Martens (eds), Aquatic Biodiversity II
DOI 10.1007/s10750-004-4664-y

Phylogeographic history of the genus *Tropheus*, a lineage of rock-dwelling cichlid fishes endemic to Lake Tanganyika

Christian Sturmbauer*, Stephan Koblmüller, Kristina M. Sefc & Nina Duftner
Department of Zoology, University of Graz, Universitätsplatz 2, 8010 Graz, Austria
*(*Author for correspondence: Tel.: +43 316 380-5595, Fax: +43 316 380 9875,*
E-mail: christian.sturmbauer@uni-graz.at)

Key words: mtDNA sequences, control region, molecular phylogeny, allopatric speciation, adaptive radiation

Abstract

Lake Tanganyika contains the oldest and most complex flock of cichlid fishes counting about 200 endemic species. It is comprised of 16 ecologically, morphologically and genetically highly distinct tribes. Many species are further subdivided into arrays of geographic morphs, each colonizing particular sections of the shore line. The genus *Tropheus* represents the most spectacular and best studied example for this phenomenon, counting more than 100 distinctly colored populations and sister species, some living in sympatry. Their present distribution and genetic structure was shaped by a series of lake level fluctuations which caused cycles of isolation and secondary admixis. The present study extends previous work on the phylogeography of *Tropheus* and aims at the fine-scale reconstruction of the origin and spread of lineages in the central and southern basin of the lake. The previously defined mtDNA lineages were evaluated on the basis of statistical parsimony networks. Haplotype networks were created for each lineage and related to their centers of diversity in terms of present distribution. A linearized tree analysis and a mismatch distribution analysis corroborate two of the three radiation waves suggested in earlier works, but the new data suggest a different primary colonization scenario for the southern basin.

Introduction

The cichlid species flocks of the three East African Great Lakes are the most spectacular examples of adaptive radiation and explosive speciation within a single group of organisms (Kosswig, 1947; Fryer & Iles, 1972; Greenwood, 1984; Sturmbauer, 1998). With an age of 9–12 myr and about 200 described species the cichlid fishes of Lake Tanganyika represent the oldest and morphologically, behaviorally, and genetically most diverse cichlid species flock of the three East African Great Lakes (Fryer & Iles, 1972; Greenwood, 1984; Mayr, 1984; Poll, 1986; Meyer, 1993; Snoeks et al., 1994; Konings, 1998). Following the spatial segregation into rock- and sand-habitats, many of the littoral species are subdivided into genetically and phenotypically distinct populations that vary mainly in their coloration. The genus *Tropheus*, of which six nominal species (Poll, 1986) and about 100 distinctly colored local variants are currently described (Konings, 1998; Schupke, 2003), represents perhaps the best example for this phenomenon. *Tropheus* is abundant in the upper littoral zone in all types of rocky habitats, where it feeds on epilithic algae and takes shelter from predators, whereas sandy or muddy shores and river estuaries are strictly avoided. There is strong evidence from ecological and genetic studies that *Tropheus* is not capable of crossing greater distances over unsuitable habitats or open water, as a consequence of its pronounced habitat specificity, site fidelity, and territorial organization (Brichard, 1978; Kawanabe, 1981; Yanagisawa & Nishida, 1991; Sturmbauer & Meyer, 1992; Sturmbauer & Dallinger, 1995; Sturmbauer et al., 1997; Rüber et al., 1998; Baric et al., 2003).

A series of previous studies on the evolution and colonization history of this genus set out to reconstruct the origin and spread of this highly specialized rock-dwelling species-complex (Sturmbauer & Meyer, 1992; Sturmbauer et al., 1997; Baric et al., 2003). These data suggest an almost contemporaneous origin of eight major mitochondrial lineages about 697 000–945 000 (control region) or 868 000 (cytochrome *b*) years ago. The geographic distribution observed today was explained as the consequence of a 'primary radiation' by which ancestral *Tropheus* rapidly colonized all rocky shores and established a 'primary pattern of distribution' which was subsequently modified by several fluctuations of the lake level in the more recent history. Baric et al. (2003) delineated three intensive phases of cladogenesis, named the 'primary radiation' (about 700 Ka, datings of the three radiation events according to the 'fast rate', Sturmbauer et al., 2001), the 'secondary radiation' (about 400 Ka) and the 'tertiary radiation' (about 200 Ka). The most recent major lake level fluctuation 18 000 to 12 000 Ya shaped the present-day distribution (Sturmbauer et al., 2001). While the present-day distribution of seven lineages is restricted to particular sections of the lake shore (named lineages B to H in Baric et al., 2003), one lineage (sublineage A2) was found to have a much more widespread distribution. For the lineages restricted to particular shore sections it was hypothesized that these did slightly expand their distribution in relation to the original range, while lineage A2 was able to colonize several additional sections along the southeastern shoreline to replace or admix with the original colonizers. Moreover, at least three drops of the lake level were large enough to allow *Tropheus* to cross the lake along the ridges between the three lake basins in the central region of the lake. The dispersal events had different consequences for several adjacent populations. In some cases in which individuals of different major lineages came into secondary contact, they hybridized to form one genetically heterogeneous population, in other cases they remained reproductively isolated, so that more than one species of *Tropheus* now lives in sympatry. The genetic structure of other populations (e.g. *T.* 'Kirschfleck' at the Kungwe Mountain shore) suggests introgression at the onset of admixis, but fast establishment of reproductive isolation (Sturmbauer et al., 1997). Due to the now disjunct distribution of members of lineage F (at the Kungwe Mountain Range in *T.* 'Kirschfleck' and in the fish from Cameron Bay area at the southwestern coast of the southern basin), it was hypothesized that this lineage was originally spread over the entire southeastern shoreline (except for the very south where lineage G is most likely indigenous), to be replaced by invading members of lineage A. The study of Baric et al. (2003) could not satisfactorily explain the sometimes puzzling findings of island distribution of genotypes assigned to geographically distant major mtDNA lineages in the southern basin, which suggest a massive invasion event of one particular lineage (A) leading to replacement or hybridization with indigenous individuals.

Here we present data for 14 new populations which allow further insights into the colonization scenario shaping the complex pattern of lineage distribution in the central and southern basin of Lake Tanganyika. Furthermore, we present a new definition of major lineages based upon statistical parsimony. Patterns of geographic distribution of major lineages and levels of genetic divergence are then related to geological data and historic lake level fluctuations (Lezzar et al., 1996; Cohen et al., 1997; Scholz et al., 2003), with the aim to further develop the hypothetical scenario of the origin and spread of the genus *Tropheus*.

Materials and methods

Samples and molecular techniques

The present study builds on a previous work by Baric et al. (2003), which was based on 243 haplotypes (446 bp of the mitochondrial control region) found in 365 specimens of the genus *Tropheus* from 55 localities, covering almost the entire shoreline of Lake Tanganyika. For the new study, the sample size was increased to 462 individuals by including available sequences of the most variable section of the control region (367 bp), and by sequencing additional specimens. The first 79 bp representing the tRNA-Thr and tRNA-Pro were truncated from the 446 bp sequences for compatibility with the shorter control region sequences. This loss of information had no effect on lineage definitions, but several of the haplotypes that differed in the tRNA sequences in Baric et al. (2003) were not distinguished by the truncated control region sequences. The denotion of the 251 different haplotypes presented in this paper based on 367 bp does therefore not fully correspond to that in Baric et al. (2003).

The increase of sample size entailed the inclusion of 14 localities, so that the number of populations was raised to 69 (Table A1; see map in Fig. 1a for locality names and new locality coding). The sample locality identification presented in this paper includes geographic coordinates where available, and is therefore more precise than that in Baric et al. (2003). The majority of the *Tropheus* samples was collected during expeditions in 1991, 1992, 1995 and 1999, and several more recently imported specimens from remote areas of the lake were provided by importers and aquarists. All specimens used from the aquarium trade were either direct imports or pond-bred fish from a company at Bujumbura. Voucher specimens are available upon request.

The methods of DNA extraction, protocols for polymerase chain reaction, DNA sequencing and sequence alignment followed that described in Baric et al. (2003). We sequenced a 368 segment of the most variable section of the mtDNA control region using published primers (Kocher et al., 1989; Meyer et al., 1994). All nucleotide sequences are available from GenBank under the accession numbers listed in Table A1.

Phylogenetic analyses

Phylogenetic analyses were conducted using the computer program PAUP* (β-version 4.0 for Macintosh; Swofford 2000). Due to the large number of taxa imposing computational limitations, the phylogenetic relationships among taxa were reconstructed by neighbor-joining analyses (Saitou & Nei, 1987). Bootstrapping was applied as a measure of branch support. The initial analysis included all available DNA sequences of *Tropheus* (462 individuals; 251 haplotypes) and was conducted to assign the newly sequenced individuals to the major lineages defined by Baric et al. (2003). *T. duboisi* was used as outgroup (Sturmbauer & Meyer, 1992). Distance matrices were based on Kimura-distances (K2P) for compatibility with Baric et al. (2003; tree not shown) on the one hand, and on the HKY + I + Γ model of molecular evolution suggested by the hierarchical likelihood ratio test implemented in the computer program Modeltest (Posada & Crandall, 1998), on the other hand (base frequencies, A, 0.34, C, 0.23, G, 0.14, T, 0.30; invariable sites, 0.40; Γ, 0.62; Hasegawa et al., 1985).

Linearized tree and mismatch distribution analysis

A branch length test was carried out using the computer program LINTRE (Takezaki et al., 1995) and a linearized tree was constructed for all 251 haplotypes (none of the taxa violated the assumption of equal clock rate) using MEGA (Kumar et al., 2001) on the basis of the TrN + Γ model of molecular evolution (base frequencies, A, 0.35, C, 0.23, G, 0.12, T, 0.30; Γ, 0.27; Tamura & Nei, 1993) since the HKY + I + Γ model is not available in this program. This choice seemed justified since Modeltest Log likelihood scores of the two models were highly similar (HKY + I + Γ, 3.857.49, TrN + Γ, 3865.08). As alternative approach to detect major cladogenesis events, a mismatch distribution analysis was carried out on the basis of absolute mutation distances using the Arlequin software (Schneider et al., 2000). To relate the modes of the mismatch distribution to the pairwise K2P distances calculated in Baric et al. (2003), the cumulative numbers of sequence pairs (y-axis) were plotted against their K2P distances (x-axis), and the resulting curve was laid over the mismatch distribution. The height of the vertical steps in the cumulative curve reflects the number of sequence pairs with similar K2P distances, and large steps coincide with the modes of the mismatch distribution.

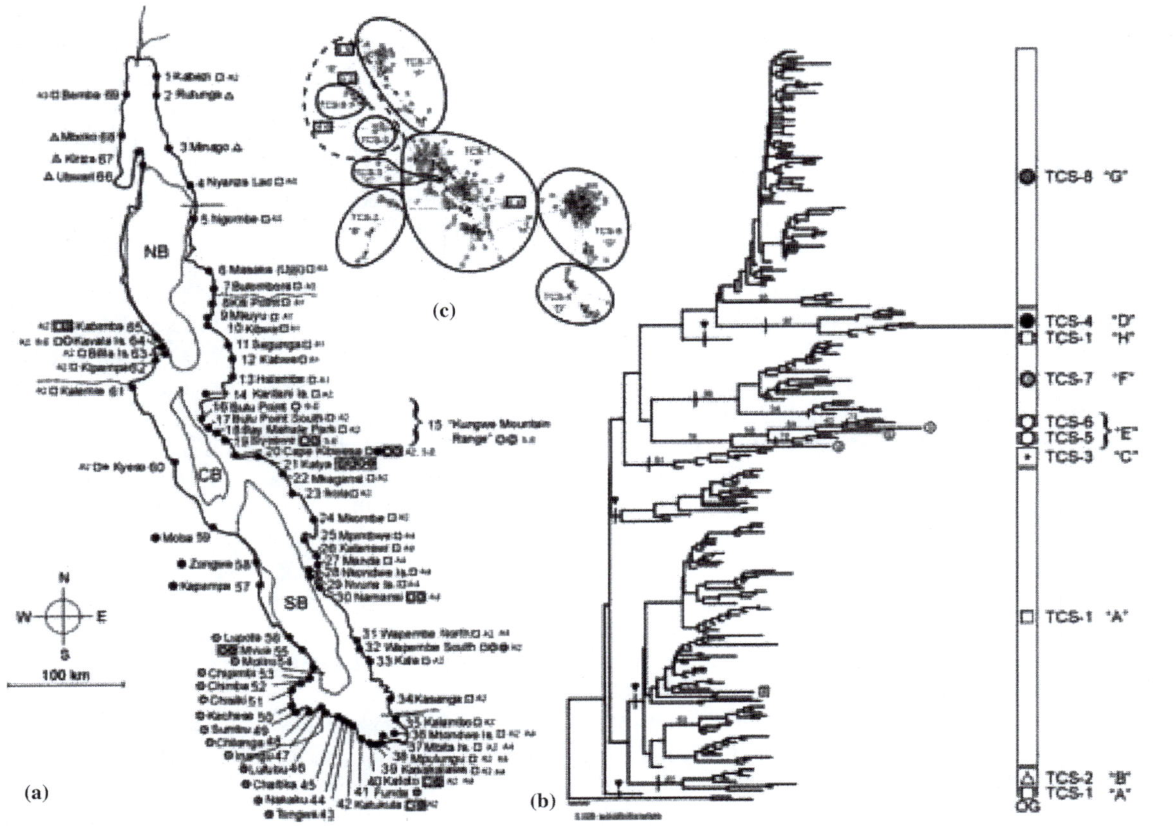

Figure 1. (a) Map of Lake Tanganyika showing the sampling sites and the distribution of haplotype lineages. Clockwise locality numbers are followed by the locality name (Is., island) and the symbols for mtDNA TCS-lineages (TCS-1-A to TCS-8-G, where capital letters refer to the lineages defined in Baric et al. (2003), see (b)). The four subdivisions of lineage TCS-1-A are indicated as A1 to A4 next to the square (symbol of the members of clade 1-A), and new TCS-lineages within the previously defined lineage E are indicated by 5, 6, b, c, d next to the open circle symbolizing lineage E. Single haplotypes that were resolved as independent lineages by TCS are represented by lower case letters a–d. Where morphologically uniform populations contain haplotypes of several clades, this is indicated by boxed haplotype symbols. Multiple unboxed symbols at a locality indicate that lineages are divided among sympatric species. The dotted lines delineate the three deepwater basins of the lake (NB, northern basin; CB, central basin; SB, southern basin). (b) Neighbor-joining tree based on 367 bp of the mtDNA control region comprising 251 haplotypes of *Tropheus*. Two individuals of *T. duboisi* were used as outgroup taxa. The model of sequence evolution applied was HKY + I + Γ (base frequencies, A, 0.34, C, 0.23, G, 0.13, T, 0.30; invariable sites, 0.40; Γ, 0.63). Bootstrap values are shown above the branches. The right-hand bars indicate separate clades identified by TCS (see text), and introduce the clade symbols used throughout the manuscript. Lower case letters a–d refer to single haplotypes resolved as independent clades by TCS. Capital letters in quotation marks refer to haplotype lineages defined by Baric et al. (2003). OG, outgroup. (c) Median-joining network comprising all 251 mtDNA haplotypes of *Tropheus* (open circles). Filled circles indicate branching points (median vectors). The length of the lines connecting haplotypes and branching points correlates approximately with the number of base substitutions. Clusters of haplotypes defined by TCS analysis are encircled and labeled as TCS-1-A to TCS-8-G. The four haplotypes that could not be assigned to any of the eight TCS-clades are marked with lower case letters (a–d) and highlighted by a square. Note that haplotype H241 from the locality Wapembe North, indicated by an asterisk, was previously assigned to an independent lineage ('H') by Baric et al. (2003) and clustered with clade TCS-1-A in the present analysis.

Assignment of major lineages and phylogeographic analysis

Due to computational limitations, the definition of major mtDNA lineages within the genus *Tropheus* was derived from subsets of the available haplotypes in Baric et al. (2003). In the present study, we applied two network construction alorithms, the median-joining network approach (Bandelt et al., 1999) implemented in the computer program Network (version 2.0 available at http://www.fluxus-engineering.com/

sharenet.htm), and the method of statistical parsimony (Templeton et al., 1992) implemented in the computer program TCS (Clement et al., 2000). The median-joining approach first combines all haplotypes in a single minimum-spanning network and then, using the parsimony criterion, creates intermediate haplotypes (on the basis of median vectors) and adds them to the network, in a way to minimize the overall tree length. The TCS-analysis is based upon a matrix of absolute pairwise distances, for which the probability of parsimony (as defined in Templeton et al., 1992) up to a threshold probability of 0.95 is calculated. Then those haplotypes are joined to a network that do not violate the parsimony criterion (i.e. up to the maximum number of substitutions below which there is no multiple hit effect). In this way haplotypes are grouped into separate clusters.

The partitioning of haplotypes in eight unconnected clades by TCS analysis was used as a basis for haplotype lineage definition. The topology of each lineage is presented as a neighbor joining tree using the model of sequence evolution suggested by Modeltest (Posada and Crandall, 1998, for details see figure legends) and as a haplotype network constructed by TCS.

Results

Assignment of major lineages

The first analyses included all 462 individuals (251 haplotypes, 69 localities) and used neighbor joining phylogram reconstruction and haplotype networks to identify major haplotype lineages and to relate the new data to the lineage definitions established by Baric et al. (2003). The TCS-analysis yielded 12 lineages, of which eight comprised several haplotypes (TCS-1 to TCS-8), while four lineages were formed by single haplotypes (a–d). The TCS-lineages are denoted in the neighbor-joining tree of all 251 haplotypes shown in Figure 1b and on the median-joining network depicted in Figure 1c. The TCS-based lineage assignment largely agrees with that of Baric et al. (2003), which defined eight major mtDNA lineages (termed lineages A–H) based on K2P distances among 29 selected haplotypes (848 bp of the control region and cytochrome *b*). As depicted in Figure 1b and c, all individuals except one (Kabimba, locality 65, H59) previously assigned to lineage A were placed in TCS-clade 1, which now also included the single individual from Wapembe North (locality 31, H241) previously representing lineage H of Baric et al. (2003). The following TCS clades and previously defined lineages are completely congruent: TCS-2 and lineage B, TCS-3 and lineage C, TCS-4 and lineage D, TCS-7 and lineage F, TCS-8 and lineage G. Lineage E of Baric et al. (2003) was subdivided into clades TCS-5, TCS-6 and three single haplotypes that were not connected to any of the TCS clades (TCS-b, haplotype H73; TCS-c, H142; and TCS-d, H70). Thus, five of the eight lineages (B, C, D, F, G) defined by Baric et al. (2003) were fully corroborated by the TCS analysis, while lineage A was almost entirely recovered in clade TCS-1, and lineage E was split into several, albeit related, haplotype clades. To be compatible with the lineage assignment of Baric et al. (2003) we will use the following lineage designations combining the TCS clade numbers with the lineage identifier of Baric et al. (2003): lineage 1-A, lineage 2-B, lineage 3-C, lineage 4-D, lineage 5-E, lineage 6-E, lineage 7-F, lineage 8-G, and the newly defined TCS-lineages a, b, c and d.

The median-joining network in Figure 1b shows that lineage 1-A represents the central haplotype cluster from which all other lineages derive. This finding in fact suggests that haplotypes of lineage 1-A mark the most ancestral cluster.

Linearized tree and mismatch distribution analysis

Both, the linearized tree (Fig. 2a) and the mismatch distribution analysis (Fig. 2b) point to two major cladogenesis events. The first radiation corresponds to about 5.30% TrN + Γ distance in the linearized tree and to 6.50% K2P distances according to the mismatch distribution analysis. Baric et al. (2003) calculated an average of 6.14% (SD, 1.37%) K2P distance for a slightly longer segment of the control region. The second radiation corresponds to about 1.75% TrN + Γ distance in the linearized tree and to 1.70% K2P distance according to the mismatch distribution analysis, which compares to an average K2P distance of 1.76% (SD, 0.51%) for a slightly longer

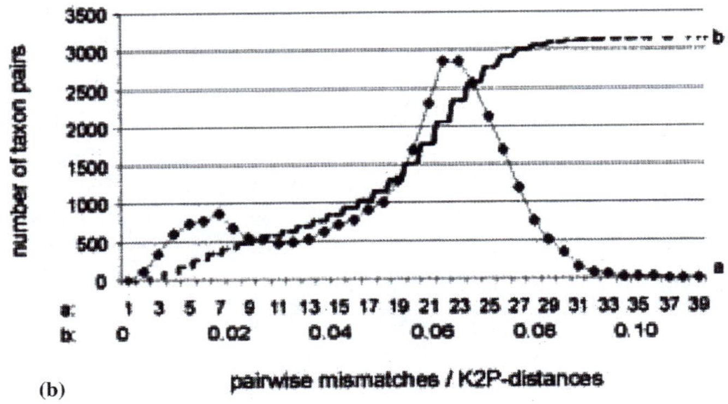
number of taxon pairs
3500
3000
2500
2000
1500
1000
500
0
a: 1 3 5 7 9 11 13 15 17 19 21 23 25 27 29 31 33 35 37 39
b: 0 0.02 0.04 0.06 0.08 0.10
a
b
pairwise mismatches / K2P-distances

(b)

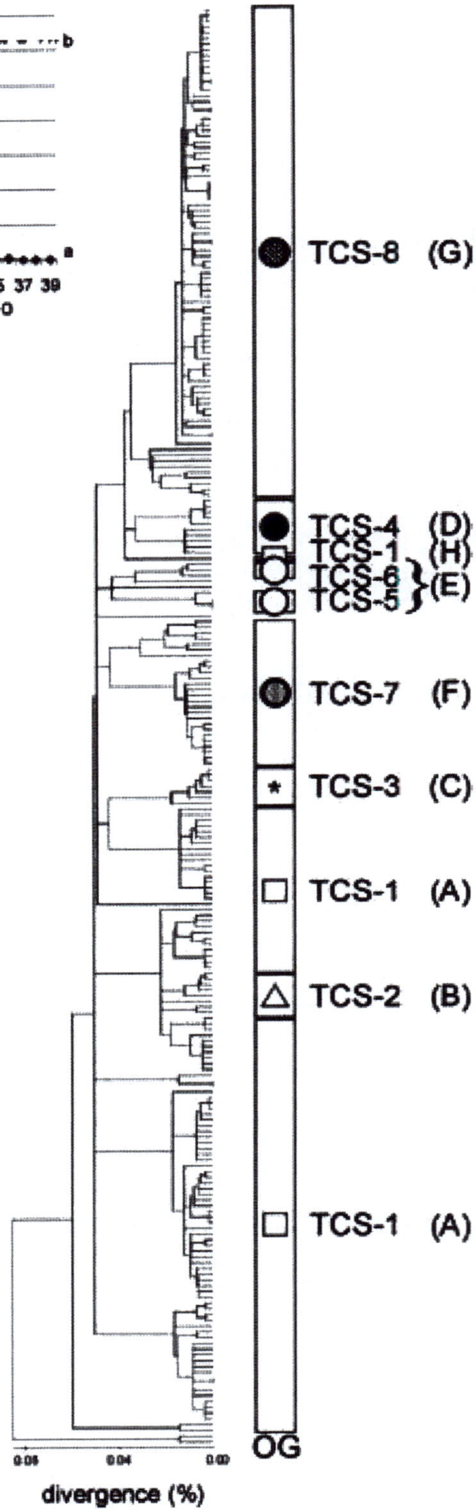
TCS-8 (G)
TCS-4 (D)
TCS-1 (H)
TCS-6
TCS-5
(E)
TCS-7 (F)
TCS-3 (C)
TCS-1 (A)
TCS-2 (B)
TCS-1 (A)
OG
divergence (%)

(a)

Figure 2. (a) Linearized tree based on TrN + Γ distances compiled with the computer program MEGA (Kumar et al., 2001). Clade symbols as in Figure 1b. (b) Line a: mismatch analysis based on number of base substitutions; Line b: cumulative representation of the number of sequence pairs with increasing K2P distances. The vertical steps in the curve indicate multiple sequence pairs with similar distance values.

segment of the control region in Baric et al. (2003). The two peaks of increased cladogenesis correspond to the 'primary radiation' and the 'tertiary radiation' of Baric et al. (2003). Their 'secondary radiation' corresponding to 3.51% (SD, 0.89%) K2P distance is not clearly evident in both the linearized tree and the mismatch distribution analysis, as the diversification events reconstructed in the time period between the primary and the tertiary radiation are not concentrated around one particular divergence estimate. Moreover, in contrast to the previous hypothesis, the subdivision of lineage 1-A into sub-lineages A1, A2, A3 and A4 seems to be more ancient than previously assumed and may in fact go back to the primary radiation, instead of the secondary radiation as inferred by Baric et al. (2003).

Phylogeographic patterns

TCS- lineages 1-A and 2-B

Both the phylogenetic and the phylogeographic substructure of TCS-lineage 1-A are highly complex. This lineage was subdivided into four sub-lineages (now designated 1-A1 to 1-A4) by Baric et al. (2003), as evident from the phylogram in Figure 3a. This substructure is also reflected in the TCS-network (Fig. 3b and c). Lineage TCS-1-A2 (Fig. 3c) represents the central haplotype cluster of the 1-A clade and covers a wide distribution from the southwestern edge of the northern basin (Kabimba, Kavala Island, Bilila Island, Kipampa, Kalemie), the western shore of the central basin (Kyeso), the northeastern and southeastern edge of the central basin (Bulu Point South, Bay Mahale Park, and Cape Kibwesa), and several localities at the eastern and southern shore of the southern basin (Mkagansi, Ikola, Mkombe, Wapembe, Kala, Kasanga, Kalambo, Mtondwe Island, Mbita Island, Mpulungu, Kasakalawe, Katoto, Katukula).

Despite its wide range of distribution, the TCS-1-A2 cluster shows considerable geographic structure, with all haplotypes found south of Kala resolved in their own clade (denoted by an arrow in Fig. 3b). A second clade consists of the haplotypes from the central basin and the northern part of the southern basin, but also holds haplotypes found in Wapembe, the location immediately north of Kala. Three sub-clusters derive from the central haplotype cluster in TCS-lineage 1-A (Fig. 3c). The first subcluster (TCS-1-A1) branches off one haplotype found at Mkombe Island (H174). The distribution of haplotypes assigned to this cluster is restricted to six populations (Kiti Point, Mkuyu, Kibwe, Segunga, Kabwe, Halembe) covering a stretch of about 70 km at the southeastern shore of the northern basin, bordered by the Malagarazi River in the North and by a stretch of sandy shore in the south towards the next rocky habitats at Bulu Point.

The distribution range of individuals assigned to the second subcluster TCS-1-A3 covers populations at the eastern shore line of the northern basin, ranging from Bulombora northwards, via Makasa and Ngombe to Nyanza Lac. The Malagarazi River separates the members of clade 1-A3 from lineage 1-A1 to the south, and a stretch of sand coast isolates them from lineage 2-B north of Nyanza Lac. In addition, a single isolated occurrence of a TCS-1-A3 haplotype was recorded at Bemba in the northwestern edge of the northern basin.

All specimens assigned to the third subcluster TCS-1-A4 were caught at the eastern and southern shore of the southern basin from Cape Mpimbwe southwards to Katoto (Cape Mpimbwe, Kalanswi, Manda, Nkondwe Island, Nvuna Island, Namansi, Wapembe North, Mtondwe Island, Mbita Island, Mpulungu, Kasakalawie and Katoto). One quite divergent haplotype from Wapembe North (H241) previously assigned to a separate lineage (H) by Baric et al. (2003) is now joined to this sub-cluster, albeit on a long branch of six mutation steps.

TCS-lineage 2-B (Fig. 3a) represents individuals sampled at the eastern and western shores of the northern basin from Minago to Rutunga. This lineage is separated from clade 1-A3 by a stretch of sandy coast north of Nyanza Lac to Minago. Members of TCS-2-B also occur at the opposite

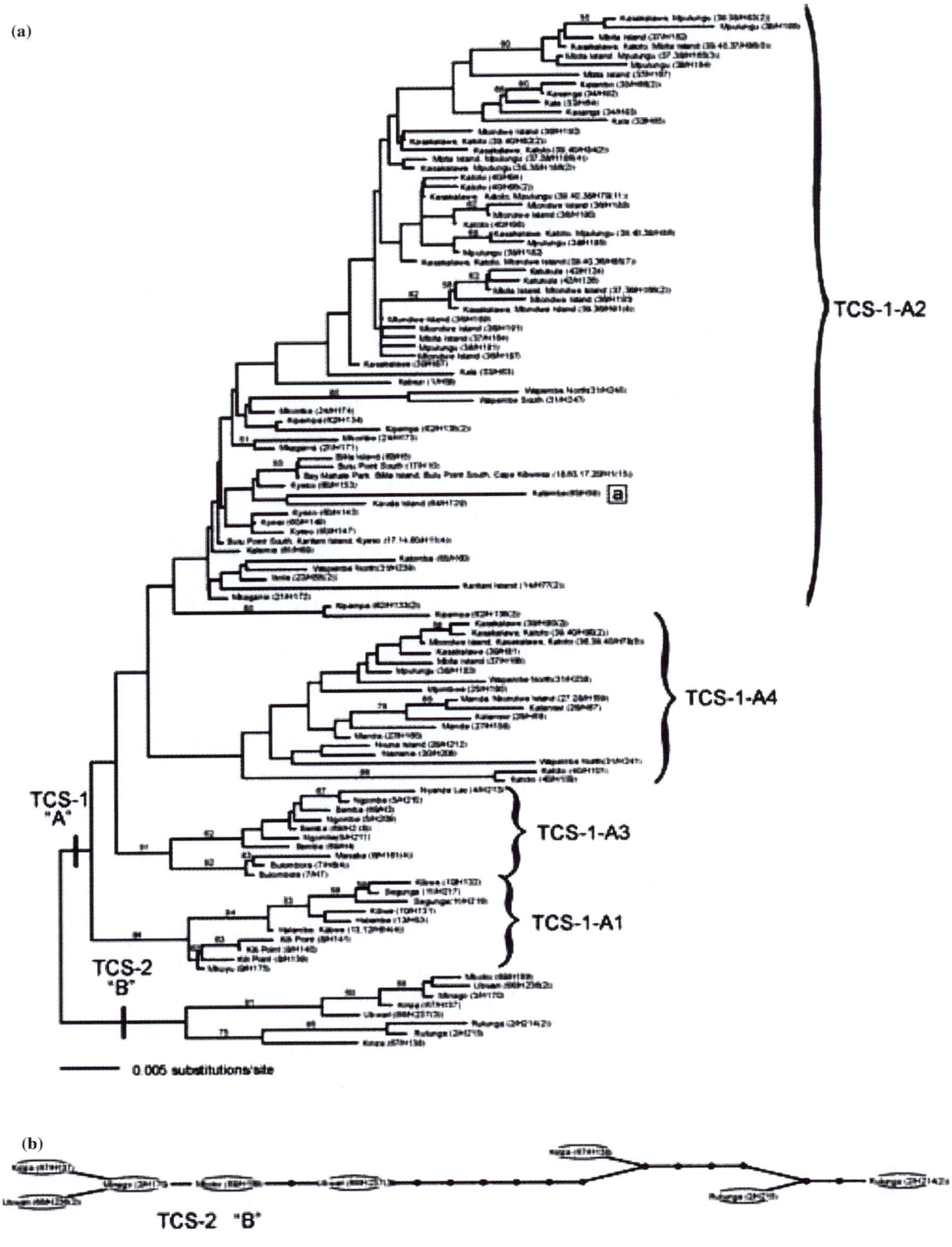

Figure 3 (a and b)

(c)

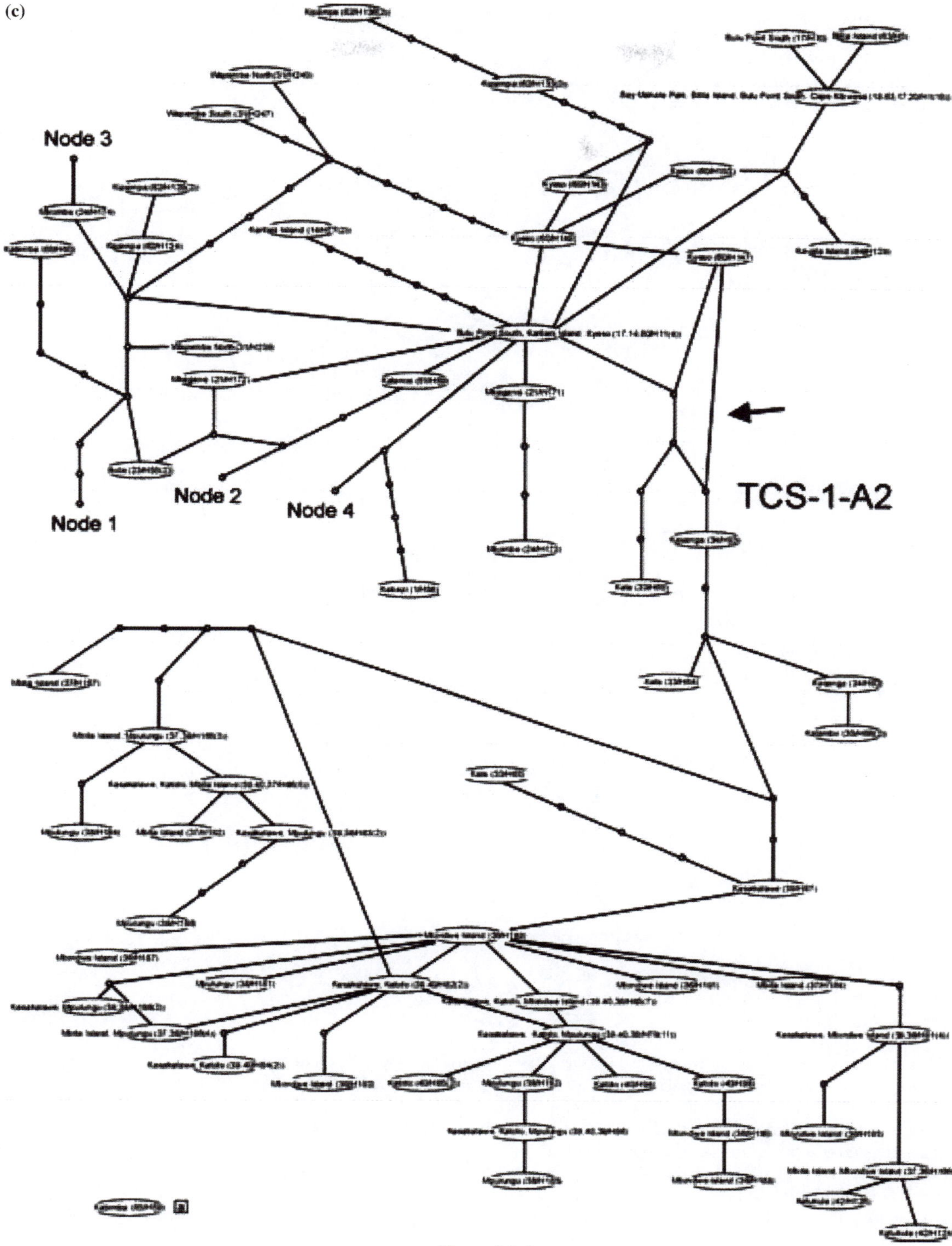

Figure 3 (c)

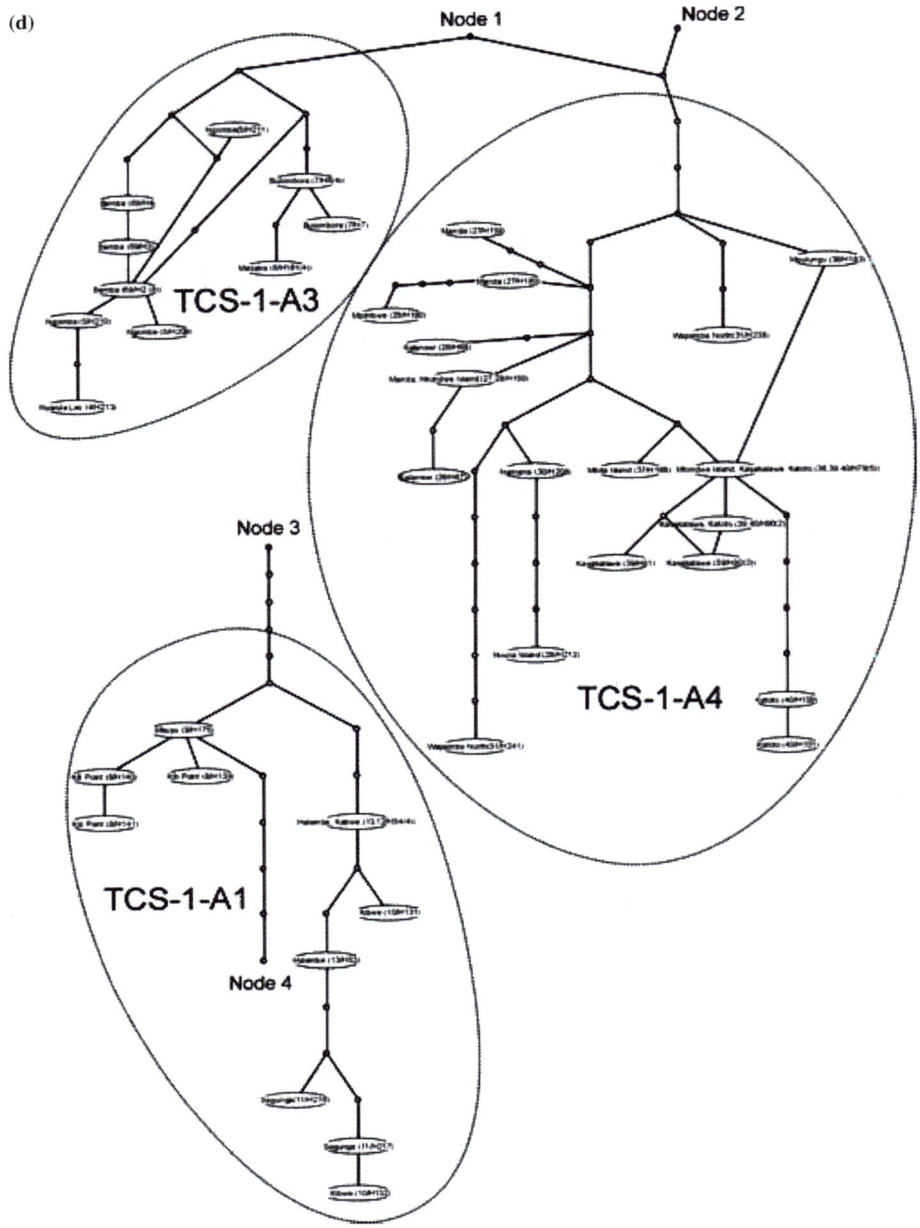

Figure 3. Gene genealogies of all individuals of *Tropheus* assigned to TCS-clades 1-A and 2-B. (a) Neighbor-joining phylogram. The model of sequence evolution applied was HKY + I + Γ (base frequencies, A, 0.33, C, 0.24, G, 0.13, T, 0.30; invariable sites, 0.55; Γ, 1.00). Haplotypes are specified by locality name, locality number, and haplotype number. For haplotypes found in more than one individual the number of identical individuals is denoted in parentheses next to the haplotype number. Bootstrap values equal or higher than 50 are given above the branches. The lower case letter 'a' indicates a haplotype, which was resolved as an independent clade by TCS. (b) Haplotype network of clade TCS-2-B. Black circles indicate missing intermediates. (c) Haplotype network of subclade TCS-1-A2. Nodes 1–4 connect the network to the remaining subclades of TCS-1-A shown in (d) with an overlap of the intermediate haplotypes among the partitioned networks. The arrow points to a phylogeographic separation between central and southern haplotypes. (d) Haplotype networks of subclades TCS-1-A1, TCS-1-A3 and TCS-1-A4.

shore at the Ubwari penninsula and northwards at Kiriza and Mboko.

TCS- lineages 3, 5–7

TCS-lineage 3-C corresponds to lineage C in Baric et al. (2003) and is exclusively found at Kyeso. Based on morphology, the individuals of this population were tentatively assigned to *T. annectens* (Baric et al., 2003). At this locality, the TCS-3-C 'annectens' individuals were found in sympatry with morphologically distinct individuals, which were assigned to TCS-1-A (clade 1-A2, see Fig. 4b).

TCS-lineage 5-E (part of lineage E in Baric et al., 2003) shows a clearly defined distribution range at the eastern shore of the central basin (Fig. 4c). It comprises haplotypes from the entire Kungwe Mountain Range (Siyeswe, Cape Kibwesa and other samples from the aquarium trade without exact localization summarized as 'Kungwe Mountain Range'). Morphologically, the individuals of this lineage are all assigned to *T. moorii* 'Kirschfleck'.

TCS-lineage 6-E (part of lineage E in Baric et al., 2003) comprises individuals from the southwestern edge of the northern basin at Kavala Island named *T. sp.* 'Kongole', and individuals known as *T. polli* from the Kungwe Mountain shores at the eastern coast of the central basin (Fig. 4c). The distribution of *T. polli* ranges from the northernmost edge of the Kungwe mountains to south of Mkagansi where it always lives in sympatry with at least one other *Tropheus*. The individuals from Kavala Island (*T. sp.* 'Kongole') phenotypically resemble *T. polli*.

The distribution of TCS-lineage 7-F is highly complex and can now be assessed in more detail due to the inclusion of new population samples. The lineage comprises populations from the Kungwe Mountain shores at the central basin (Kungwe Mountain Range, Siyeswe, Cape Kibwesa, Kalya), from two isolated localities at the eastern shore of the southern basin (Namansi and Wapembe South), and from the western shore of the southern basin (Lupota, Mvua, Moliro, Chipimbi, Chimba, Chisiki, Kachese, Sumbu, Chilanga and the Inangu Penninsula). The geographic distribution of the clade members is also reflected in the structure of the TCS-7-F haplotype network (Fig. 4d). Both peripheral clusters comprise individuals from the southern stretch of the Congolese coast and the Cameron Bay area (Mvua, Chimba, Kachese, Sumbu, and Chilanga in one cluster and individuals from Lupota, Moliro, Chipimbi, and Chisiki in the second cluster; denoted by stippled circles in Fig. 4). The central cluster comprises scattered populations from the eastern shoreline of the central and southern basin (Kungwe Mountain Range, Siyeswe, Cape Kibwesa, Kalya, Namansi, and Wapembe South), together with haplotypes from the western shore (Inangu Penninsula and Chilanga) – that branch off haplotypes from Namansi. The fish found at Kalya (comprised of TCS-lineages b, d, 7-F, 8-G), as well as the fish from Namansi (TCS-lineages 1-A, 7-F) each represent a new distinct phenotype. At other locations along the eastern shore of the central and southern basin, several populations contain haplotypes of lineage 7-F together with haplotypes of another lineage. Since these populations are phenotypically uniform (e.g. *T. moori Kirschfleck* at Kungwe Mountain Range, Siyeswe, and Cape Kibwesa), the presence of 7-F haplotypes suggests introgression after secondary contact of the divergent lineages. In contrast, multiple haplotype lineages at Wapembe South possibly correspond to sympatric populations of divergent morphologies, without indications of past introgression.

TCS-lineages 4 and 8

The individuals assigned to TCS-lineage 4-D show a clearly defined distribution range and have their center of diversity at the western shoreline between Moba and Kapampa (Fig. 5b). Members of this lineage were able to 'cross the lake' along the ridge between the central and the southern basin and colonized a stretch of about 500 m shoreline at Cape Kibwesa (Konings 1990), where it lives in sympatry with *T. polli* (TCS-lineage 6-E) and *T. moorii* 'Kirschfleck' (TCS-lineages 5-E and 7-F).

TCS-lineage 8-G has its center of diversity at the southwestern shore of Lake Tanganyika south of the Lufubu estuary (see Fig. 5a for the NJ-tree, and Fig. 5c and d for the TCS-network). It comprises individuals sampled at Lufubu (south of the estuary), Cape Chaitika, Nakaku, Tongwa, Katukula, Funda and Katoto. In addition, isolated occurrences of individuals of this lineage were recorded further north in Mvua (southern basin,

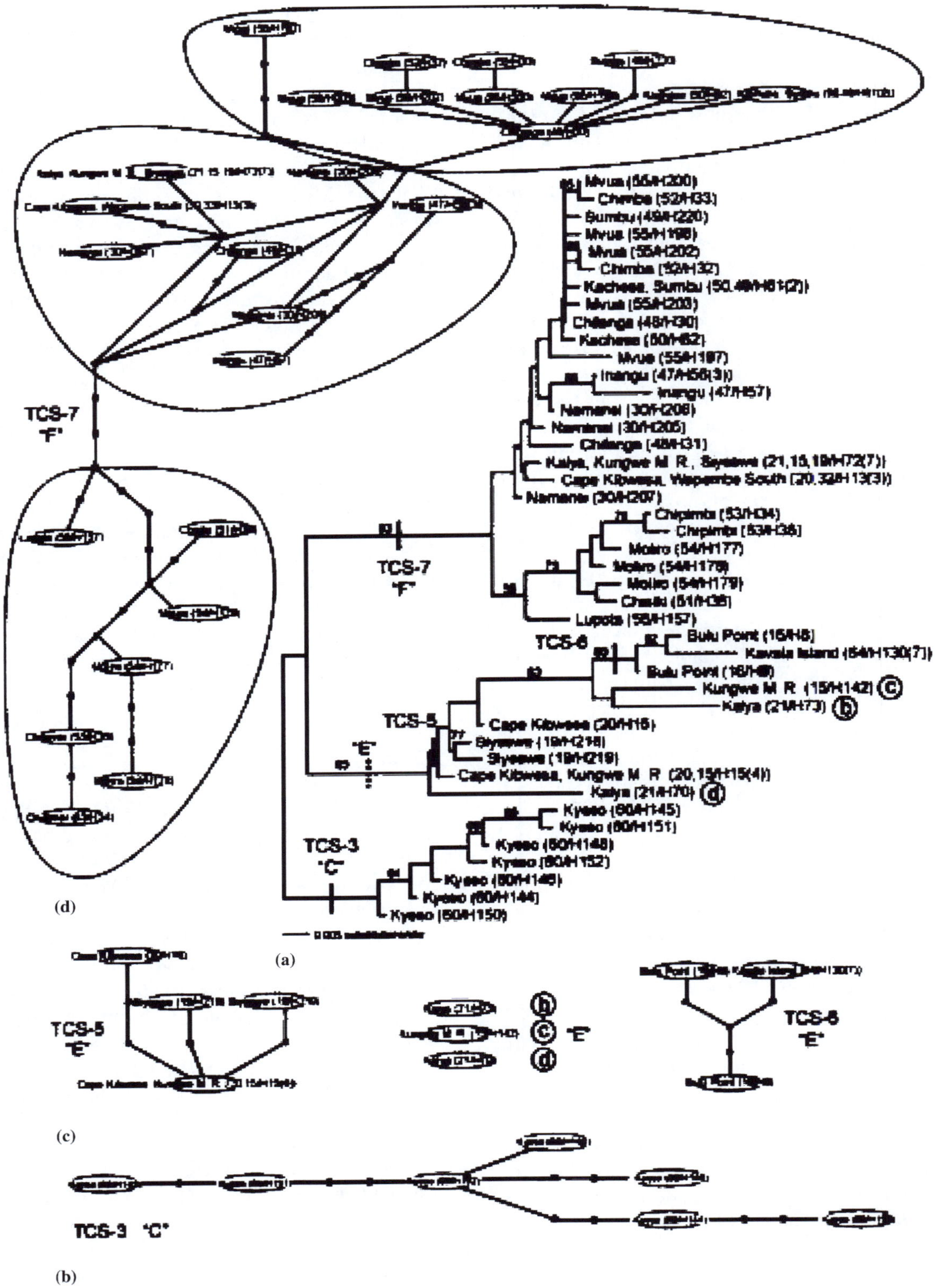
TCS-7
"F"
Mvua (55/H200)
Chimba (52/H33)
Sumbu (49/H220)
Mvua (55/H198)
Mvua (55/H202)
Chimba (52/H32)
Kachese, Sumbu (50,49/H81(2))
Mvua (55/H203)
Chilanga (48/H30)
Kachese (50/H82)
Mvua (55/H197)
Inangu (47/H56(3))
Inangu (47/H57)
Namansi (30/H206)
Namansi (30/H205)
Chilanga (48/H31)
Kalya, Kungwe M. R., Siyaswa (21,15,19/H72(7))
Cape Kibwesa, Wapembe South (20,32/H13(3))
Namansi (30/H207)
Chipimbi (53/H34)
Chipimbi (53/H35)
Moliro (54/H177)
Moliro (54/H178)
Moliro (54/H179)
Chisiki (51/H36)
Lupota (58/H157)
TCS-6
Bulu Point (16/H8)
Kavala Island (64/H130(7))
Bulu Point (16/H9)
Kungwe M. R. (15/H142)
Kalya (21/H73)
TCS-5
"E"
Cape Kibwesa (20/H16)
Siyaswa (19/H218)
Siyaswa (19/H219)
Cape Kibwesa, Kungwe M. R. (20,15/H15(4))
Kalya (21/H70)
TCS-3
"C"
Kyeso (60/H145)
Kyeso (60/H151)
Kyeso (60/H148)
Kyeso (60/H152)
Kyeso (60/H146)
Kyeso (60/H144)
Kyeso (60/H150)
(d)
(a)
TCS-5
"E"
"E"
TCS-6
"E"
(c)
TCS-3 "C"
(b)

Figure 4. (a) Neighbor-joining phylogram comprising haplotypes of clades TCS-3-C, TCS-5-E, TCS-6-E, TCS-7-F, TCS-b, TCS-c and TCS-d. The model of sequence evolution applied was HKY + Γ (base frequencies, A, 0.34, C, 0.21, G, 0.13, T, 0.32; Γ, 0.14). Haplotypes are specified by locality name, locality number and haplotype number. For haplotypes found in more than one individual, the number of identical individuals is denoted in parentheses next to the haplotype number. Bootstrap values equal or higher than 50 are given above the branches. (b) Haplotype network of clade TCS-3-C. Black circles indicate missing intermediates. (c) Haplotype networks of the five TCS clades previously combined in lineage 'E' by Baric et al. (2003). (d) Haplotype network of clade TCS-7-F. See text for circled clades.

western shore north of Cameron Bay in D.R. Congo) and along the eastern shoreline at Kalya (an underwater rift 500 m offshore south of the Kungwe Mountain Range), and Wapembe South (southern basin, eastern shore in Tanzania). Interestingly, at two locations (Mvua and Kalya), and possibly also at Wapembe South, the populations are phenotypically uniform and introgressed with various other TCS-lineages: The samples at Kalya comprise four TCS-lineages (b, d, 7-F, 8-G), and the samples from Wapembe South display haplotypes of three TCS-lineages (1-A, 7-F, 8-G). The samples from Mvua represent a distinct clade within lineage 8-G and also comprise haplotypes of TCS-lineage 7-F.

Discussion

Evolutionary implications

The genus *Tropheus* is member of the endemic Tanganyikan tribe Tropheini (Poll, 1986), a group of predominantly rock-dwelling cichlids feeding on epilithic algae and invertebrates. The Tropheini are subdivided into five major lineages that arose almost contemporaneously (Sturmbauer et al., 2003). The genus *Tropheus* is paraphyletic, in that *Tropheus duboisi* clearly represents a distinct entity, which should be described as a new genus. All remaining members of *Tropheus* form a monophyletic assemblage that rapidly underwent further diversification. The origin and spread of the genus *Tropheus* was reconstructed from mtDNA phylogenies in three previous studies (Sturmbauer & Meyer, 1992; Sturmbauer et al., 1997; Baric et al., 2003). These suggested three major diversification events, termed 'primary, secondary, and tertiary radiation'. The primary radiation, during which the major lineages were formed by rapid colonization of all rocky habitats throughout the lake, is likely to have taken place about 700–945 Ka, promoted by the rise of the lake level between 1.1 Ma and 550 Ka. The pattern of distribution was then re-shaped by the secondary radiation (about 540–399 Ka), possibly triggered by a lowstand between 390 and 360 Ka. Another lowstand between 190 and 170 Ka might have triggered the tertiary radiation (about 270–200 Ka). The fit to the two lowstands assumed the fast clock rate (Sturmbauer et al., 2001). Sturmbauer et al. (2001) found further genetic signatures of a much more recent climate change (about 18–12 Ka) in cichlids of Lakes Tanganyika, Malawi, and Victoria, resulting from a dramatic drop of the lake level in Lakes Tanganyika and Malawi, and from the desiccation of Lake Victoria (Johnson et al., 1996). Two of the three major cladogenesis events – the primary and the tertiary radiation – were corroborated by our study in the linearized tree and the mismatch distribution analysis (Fig. 2). According to the linearized tree and to the mismatch distribution, no simultaneous cladogenesis events that would fit to the 'secondary radiation' are evident, albeit various lineages underwent diversification between the primary and teriary radiation. Moreover, the linearized tree (Fig. 2a) suggests that the subdivision of lineage 1-A into clades 1-A1, 1-A2, 1-A3 and 1-A4 should be assigned to the primary radiation.

The present distribution of lineages is highly complex and shows a great degree of overlap in particular regions of the lake, especially at border regions of the three lake basins. It is evident from our data that some lineages were able to disperse across the lake along the ridges between the basins at time periods of extremely low lake level, which are likely to have split the lake into at least three sublakes. Such dispersal events across the lake were suggested to have happened during the secondary and tertiary radiation, and most probably also in the very recent past when Lake Tanganyika was again very low during the last glacial maximum 18–12 Ka (Lezzar et al., 1996; Cohen et al, 1997; Sturmbauer et al., 2001; Scholz et al., 2003).

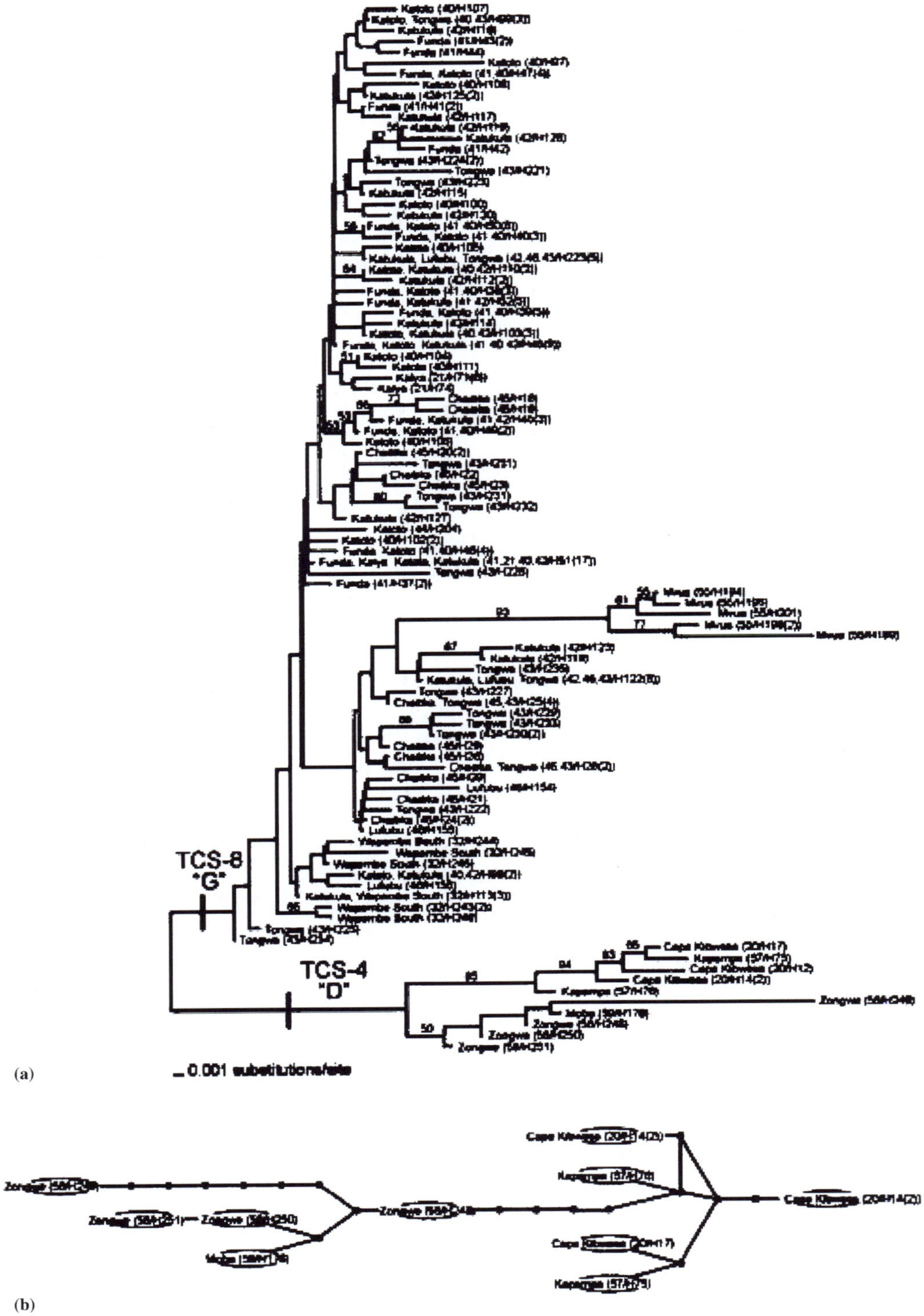

(a)

(b)

Figure 5 (a and b)

(c)

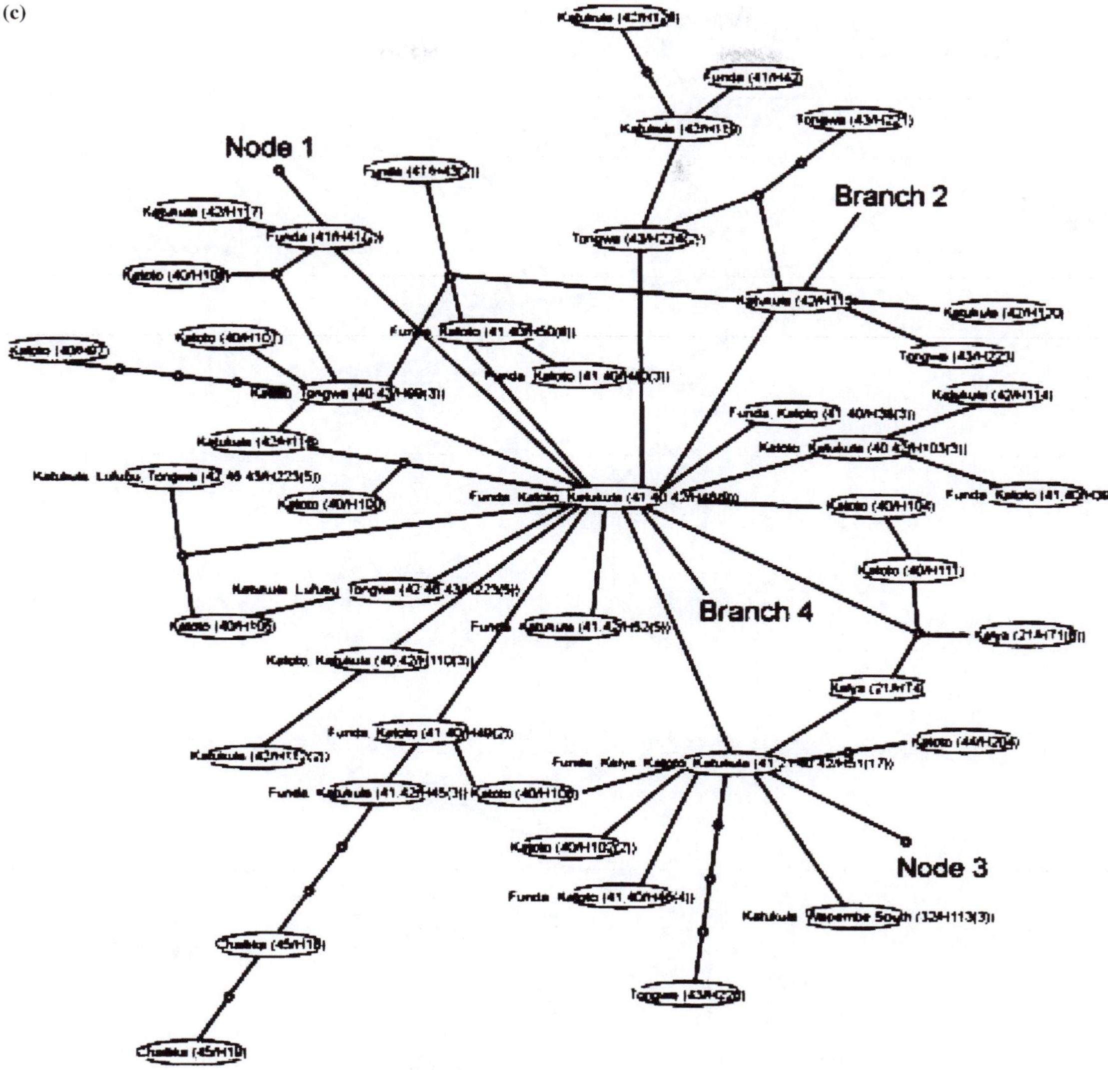

Figure 5 (c)

Using the median-joining network and TCS-based lineage definition of this study, the scenario for the colonization of Lake Tanganyika by *Tropheus* suggested by Baric et al. (2003) can be revised as follows. Given that lineage 1-A forms the most central haplotype cluster in the median-joining network, from which all other major mtDNA lineages are derived, it is likely that lineage 1-A represents the most ancestral cluster, marking the origin of the genus. The split of lineage 1-A into the four sublineages 1-A1 to 1-A4 most probably took place during the primary radiation, at about the same time as the diversification of the remaining lineages. The subclades of lineage 1-A may thus have arisen in consequence of primary colonization events.

Northern basin

Our new data allow for a more detailed hypothesis for the dispersal and distribution of the major mtDNA lineages (Fig. 1a). The northern basin is inhabited by four lineages (1-A, 2-B, 6-E, a). Of

(d)

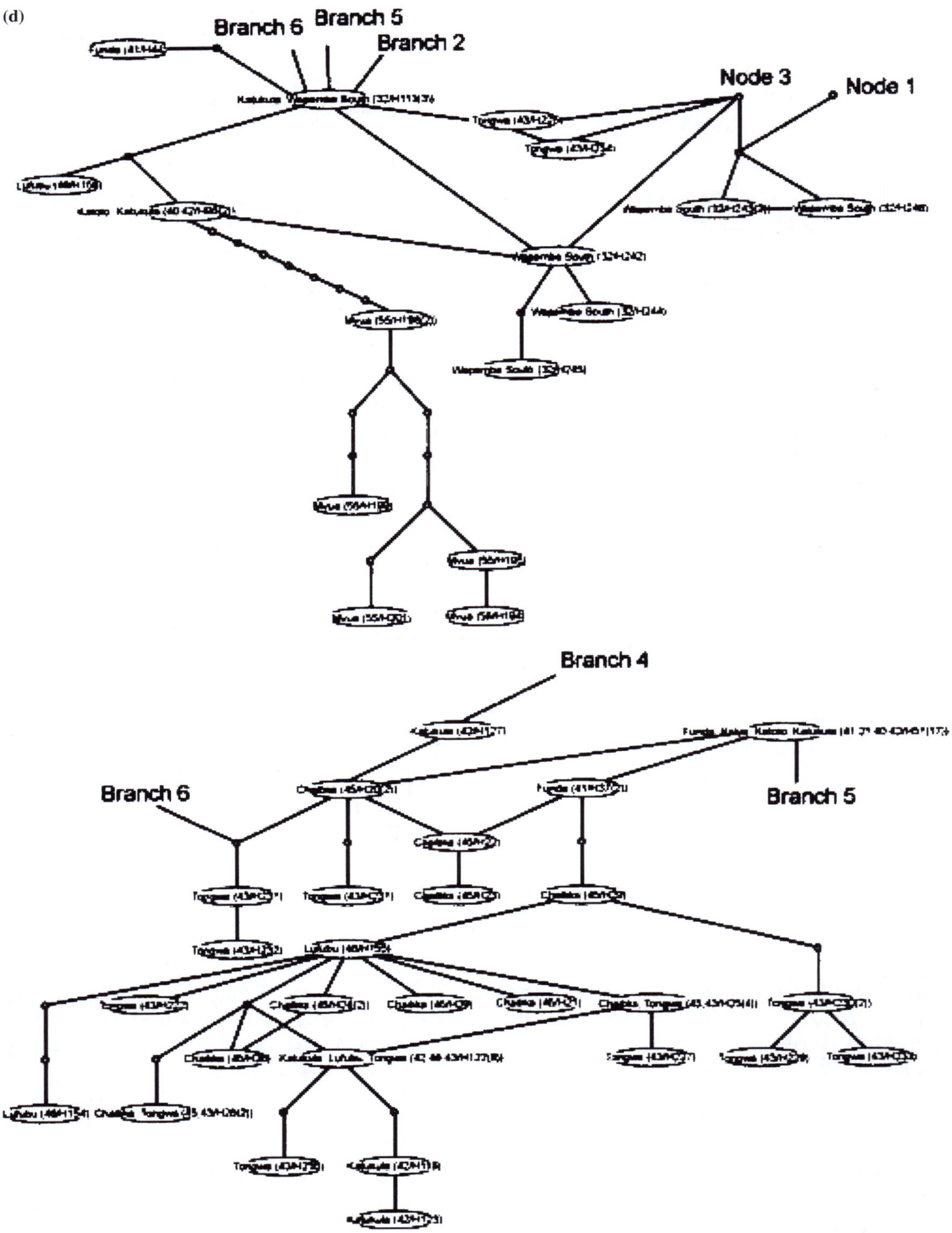
Branch 6
Branch 5
Branch 2
Node 3
Node 1
Branch 4
Branch 6
Branch 5

Figure 5 (d)

Figure 5. (a) Neighbor-joining phylogram comprising haplotypes of clades TCS-4-D and TCS-8-G. The model of sequence evolution applied was HKY + I + Γ (base frequencies, A, 0.36, C, 0.23, G, 0.12, T, 0.29; invariable sites, 0.56; Γ, 0.78). Haplotypes are specified by locality name, locality number and haplotype number. For haplotypes found in more than one individual, the number of identical individuals is denoted in parentheses next to the haplotype number. Bootstrap values equal or higher than 50 are given above the branches. (b) Haplotype network of clade TCS-4-D. Black circles indicate missing intermediates. (c) First part of the network of clade TCS-8-G. Nodes 1 and 3 correspond to missing intermediate haplotypes and overlap with part two of the TCS-8-G network, whereas branches 2 and 4 connect to sampled haplotypes in parts two and three (see (d)). (d) Parts two and three of the clade TCS-8-G. Node 1 and branches 2 and 4 overlap with the first part of the network shown in (c). The upper and the lower parts of the network in this figure (d) are connected by branches 5 and 6.

these, part of lineage 1 (clades 1-A1 and 1-A3), lineage a, and lineage 2-B inhabit distinct shore regions mainly at the eastern coast and do not share their distribution with another lineage, so that their colonization can be dated to the primary radiation.

Another clade of lineage 1-A (1-A2) is found at the southwestern edge of the northern basin (Kabimba, Kavala Island, Kipampa, Kalemie), suggesting that members of this clade were the primary colonizers of the southwestern shore of the northern basin. At present time clade 1-A2 has the most widespread area of distribution. It was suggested that it first colonized (south-) western shores of the northern basin from Kabimba to Kalemie, and later (during the secondary radiation) expanded its range of distribution to several other shore regions (see below).

At Kavala Island, two distinct *Tropheus* morphs belonging to two different haplotype lineages, are found sympatrically (TCS-1-A2 and TCS-6-E). The 6-E individuals have been described as *T.* sp. 'Kongole' and closely resemble *T. polli* of the opposite shore which also belong to lineage 6-E. We suggest that this island finding of members of lineage 6-E, which have their center of diversity at the Kungwe Mountain Range, results from a recent colonization event (18–12 Ka) by individuals crossing the lake in the opposite direction than the members of clade 1-A2 during a period of low lake level.

Central basin

The lineages inhabiting the central basin show a highly complex distribution with more than one species of *Tropheus* in sympatry in most localities, suggesting a great impact of lake level fluctuations on their present distribution (Fig. 1a). Along the western shoreline two lineages are found (1-A2 and 3-C). The members of the two mtDNA lineages are morphologically distinct and occur sympatrically at one of the sampling locations (Kyeso). The individuals grouped in lineage 3-C were tentatively assigned to *T. annectens* (characterized by 4–5 anal spines; see Baric et al., 2003). The individuals assigned to lineage 1-A closely resemble their genetic allies further north at the southern edge of the northern basin and near Bulu Point at the eastern shore, which are all greenish, have six anal spines, and a yellow pectoral fin.

Lineages 5-E, 6-E, b, c, and d have their center of diversity at the Kungwe Mountain shores on the eastern coast of the central basin, so that it seems highly likely that individuals carrying haplotypes of these mtDNA lineages are descendants of the original colonizers of the central eastern shore. The Kungwe Mountain Range represents the most diverse coastline in terms of sympatric *Tropheus* morphs and species. Its northernmost stretch is inhabited by *T. duboisi* (to be revised and placed in a separate genus), by a greenish *Tropheus* with yellow pectoral fins, by *T. polli*, and very few individuals (not analyzed genetically so far) resembling *T. moorii* 'Kirschfleck' with two yellow blotches instead of the orange ones observed in the 'Kirschfleck' morphs from further south. All greenish *Tropheus* with yellow pectoral fins are assigned to lineage 1-A (clade 1-A2), and haplotypes are shared across the opposite shores of the central basin with fish from Bilila Island and Kyeso (Sturmbauer et al., 2001). In terms of the taxonomic assignment, the individuals carrying haplotypes of lineage 6-E along the Kungwe Mountain Range are classified as *T. polli*. On the opposite shore at Kavala Island, individuals of the same lineage were tentatively classified as *T.* sp. 'Kongole', and closely resemble *T. polli*.

Further south after a sandy section of the eastern shore at Luagala Point (samples starting at

Siyeswe), *T.* 'Kirschfleck' replaces the greenish *Tropheus* and occurs in sympatry with *T. polli* (and *T. duboisi*). *Tropheus duboisi* only colonized the northernmost region around Luagala Point, while *T. polli* and *T.* 'Kirschfleck' inhabit the entire shoreline down to Cape Kibwesa. Along a stretch of about 500 m at Cape Kibwesa, a third *Tropheus* type (lineage 4-D) with a whitish blotch on the body sides is found sympatrically with *T. polli* and *T.* 'Kirschfleck'. Some of the individuals of *T.* 'Kirschfleck' were previously assigned to the same mtDNA lineage as *T. polli* (E, now to the closely related TCS-lineage c), and others to lineage 7-F, thus pointing to a secondary admixis event among native and invading individuals, most probably between the ancestors of lineage 5-E, 6-E, c, and invading *T.* 'Kirschfleck' ancestors of lineage 7-F (see Sturmbauer et al., 1997). The whitish blotched *Tropheus* 'Kibwesa' belong to lineage 4-D, which has its center of diversity at the opposite shore from Moba to Kapampa, so that it is likely that the ancestors of *T.* 'Kibwesa' dispersed along the ridge between the southern and the central basin during a recent period of low lake level (18–12 Ka).

Perhaps the most striking new finding concerns an isolated population at Kalya which is situated east of Cape Kibwesa at an underwater ridge, colonized by *T. polli* (not yet sampled from this location) and a dark reddish-brown *Tropheus* that has not been recorded anywhere else. This recently discovered small *Tropheus* population contains a striking mixture of mtDNA haplotypes (b, d, 7-F, 8-G) posing important phylogeographic implications. The haplotypes assigned to lineage b and d are quite divergent and cluster in two clades close to TCS-lineage 5-E and 6-E. The Kalya-haplotypes of lineage 7-F cluster with individuals from the Kungwe Mountain Range, Siyeswe, Cape Kibwesa, and from Wapembe South. The Kalya haplotypes assigned to lineage 8-G are identical or very closely related to those of geographically very distant individuals from the southernmost tip of the lake (Katoto, Funda, and Katukula) and thus represent a northern outpost of this lineage, which has its present center of diversity at the southwestern shore south of the Lufubu estuary at the Chaitika Mountain Range. Considering that individuals belonging to lineage 8-G have another isolated occurrence at the southeastern shore of the southern basin (Wapembe South), one must hypothesize a very recent long-distance dispersal, given sequence identity among individuals from Kalya in the north and the southern localities Katoto, Funda, and Katukula. As the Kalya population inhabits a remote underwater reef, human-mediated transfer seems unlikely.

Southern basin

The southern basin is inhabited by seven mtDNA lineages (clades 1-A2 and 1-A4 of lineage 1-A, lineages 4-D, 7-F, 8-G, see Fig. 1a). The distribution of lineage 4-D is restricted to the northwestern section of the southern basin from Moba to Kapampa, so that one can hypothesize this coastline as its primary distribution.

Clade 1-A4 is only found at the eastern and southern shores of the southern basin from Cape Mpimbwe to Katoto (Fig. 3c) and – contrasting the hypothesis of Baric et al. (2003) – may thus constitute the primary colonizer of this shore section in the course of the 'primary radiation'. As noted above, clade 1-A2 has the most widespread area of distribution in the present time (Fig. 3c). It was suggested that it first colonized (south-) western shores of the northern basin from Kabimba to Kalemie, and later (during the secondary radiation) expanded its range of distribution to several other shore regions, particularly in the western shores of the central basin (Kyeso) and by crossing the lake at the ridge between the central and southern basin to the eastern shore of the southern basin (Cape Kibwesa, Mkangansi, Ikola, Manda, and Wapembe North). Furthermore, clade 1-A2 comprises a cluster of haplotypes, which derives from sequences found along the central basin at Bulu Point South, Karilani Island, and Kyeso (denoted by an arrow in Fig. 3b) and is exclusively distributed in the southeastern and southern shores of the southern basin from Kala to Katukula. Clade 1-A2 may thus have moved secondarily into an area previously colonized by clade 1-A4.

Lineage 7-F has its center of diversity at the western shore of the southern basin from Lupota to Moliro, the Cameron Bay area, and the Inangu Penninsula. The Lufubu River estuary seems to represent a strict barrier against lineage 8-G, which is predominantly distibuted south of the estuary at the Chaitika Mountain Range. In addition, isolated occurrences of lineage 7-F were found at Wapembe South at the southeastern shore, and along the eastern central basin in Kalya and at the Kungwe Mountain Range (Siyeswe, Cape Kibwesa, see above).

Lineage 8-G has its present center of diversity at the southern shores of the lake, south of the Lufubu River. Further north at the western shore across the Lufubu-barrier, haplotypes of this lineage were only found at Mvua. These haplotypes form a divergent clade of lineage 8-G, pointing to a single dispersal event in the more distant past. The phenotypically uniform Mvua population thus comprises haplotypes of two major mtDNA lineages pointing to admixis between lineage 7-F and 8-G. Haplotypes of lineage 8-G were also found at Wapembe South, together with individuals belonging to lineage 1-A and 7-F. Whether individuals at Wapembe South assigned to different major mtDNA-lineages comprise a single population or represent sympatric species, must be clarified by morphological analyses in the future.

Inference of the colonization scenario of the southern basin

According to the pattern of distribution described above, we suggest the following revision of the colonization scenario of Baric et al. (2003) for the southern basin (Fig. 1a). Lineage 4-D reflects the primary colonizers of the shore section from Moba to Kapampa. The primary distribution of clade 1-A4 of lineage 1 is the eastern shore of southern basin from Cape Mpimbwe to Mpulungu. The primary distribution of lineage 7-F is at the western shores of the southern basin from Lupota to the Inangu Penninsula, according to the haplotype network (Fig. 4), while that of lineage 8-G originally ranged from Katoto to the Lufubu estuary south of the Inangu-Penninsula (Fig. 5b). This primary distribution was modified by a series of migration waves concerning predominantly the eastern and northeastern shores of the southern basin and also affecting the eastern shores of the central basin. The eastern shore section of the central basin is likely to have been repeatedly colonized during periods of low lake level by individuals of lineages 7-F and 8-G from the opposite shore, possibly along the underwater canyon of the Lufubu River, east of Cameron Bay. Remnants of these colonization events are the lineage 7-F-haplotypes at Kungwe, Siyeswe, Kalya, Namansi, and Wapembe South, and the lineage 8-G-haplotypes at Kalya and Wapembe South. Given that individuals from Cape Kibwesa and Wapembe South, as well as from Kalya, Funda, Katukula, and Katoto, as well as from Wapembe South and Katukula, share mtDNA haplotypes, these dispersal events are likely to be very recent, possibly triggered by the most recent lowstand of Lake Tanganyika 18–12 Ka (Lezzar et al., 1996; Cohen et al., 1997; Scholz et al., 2003; see also Sturmbauer et al., 2001).

Moreover, according to the structure of the haplotype network of lineage 1-A (Fig. 3b and 3c) haplotypes of clade A2 of lineage 1-A invaded from the central western shoreline across the ridge between the central and southern basin, to colonize the eastern shores from Cape Kibwesa to Mkombe, and to mix with the previously indigenous populations of clade 1-A4 south of Cape Mpimbwe. Since both divergent and identical or very closely related haplotypes were found among individuals of clade 1-A2 in distant locations such as Bilila Island, Kyeso, and Cape Kibwesa, one must assume more than one crossing of the ridge between the central and the southern basin. Cape Kasakalawe at the southern tip of the lake seems to separate lineage 1-A from lineage 8-G. The population at Katoto, east of Kasakalawe, originally comprised haplotypes of lineage 8-G but was introgressed by individuals carrying haplotypes of lineage 1-A (clades 1-A2, 1-A4). Populations north of Katoto consist almost exclusively of 8-G haplotypes (few lineage 1-A2-haplotypes were found at Katukula).

One island distribution of lineage 8-G is recorded from Mvua. About 50% of this population carry lineage 7-F-haplotypes and the rest those of lineage 8-G. Since all haplotypes of lineage 8-G from Mvua form a divergent clade within lineage 8-G, one must assume a single colonization event northwards across the Lufubu-Canyon in the more distant past.

When the pattern of distribution found in *Tropheus* is compared to that of species of the Eretmodini, another tribe of cichlid fishes endemic to Lake Tanganyika, (Verheyen et al., 1996; Rüber et al., 1998; Rüber et al., 2001), several parallels emerge. Clearly, the original distribution of lineages is basin-specific in both groups. Also, the original distribution was considerably modified by lake level fluctuations and both groups 'crossed the lake' at periods of low lake level. Moreover, not all lineages were equally prone to dispersal. A detailed comparison of these groups will be highly fruitful, and will be carried out in a separate work in the near future. Clearly, a major future task will be the analysis of nuclear data such as microsatellites or AFLP, to trace the complex scenario of replacement, admixis, introgression, and hybridization in this array of rock-dwelling species. Detailed knowledge of the phylogeographic structure can provide novel opportunities to investigate the role of character displacement and ecological segregation, and the relative importance of pre- and postmating isolation, for speciation in cichlid species flocks.

Acknowledgments

We wish to thank H. Phiri and the team at the Department of Fisheries at Mpulungu, Ministry of Agriculture and Cooperatives, Republic of Zambia, for their assistance during fieldwork. Special thanks to S. Baric for providing unpublished DNA sequences. We are further indebted to M. Taborsky, P. Henninger, T. Veall, and P. Schupke for providing specimens. We are also greatful to Lukas Rüber and one anonymous reviewer who made several valuable comments on the manuscript. This study was supported by the Austrian Science Foundation (grant P15239). N. D. and S. K. were also supported by the University of Graz. N. D. was further supported by a DOC-FFORTE-fellowship (Women in research and technology), and S. K. by a DOC-fellowship, both from the Austrian Academy of Sciences.

References

Bandelt, H. J., P. Forster & A. Röhl, 1999. Median-joining networks for inferring intraspecific phylogenies. Molecular Biology and Evolution 16: 37–48.

Baric, S., W. Salzburger & C. Sturmbauer, 2003. Phylogeography and evolution of the Tanganyikan cichlid genus *Tropheus* based upon mitochondrial DNA sequences. Journal of Molecular Evolution 56: 54–68.

Brichard, P., 1978. Un cas d'isolement de substrats rocheux au milieu fonds de sable dans le nord du lac Tanganyika. Revue Zoologique Africaine 92: 518–524.

Clement, M., D. Posada & K. A. Crandall, 2000. TCS: a computer program to estimate gene genalogies. Molecular Ecology 9: 1657–1659.

Cohen, A. S., K. E. Lezzar, J. J. Tiercelin & M. Soreghan, 1997. New palaeogeographic and lake-level reconstructions of Lake Tanganyika: implications for tectonic climatic and biological evolution in a rift lake. Basin Research 9: 107–132.

Fryer, G. & T. D. Iles, 1972. The Cichlid Fishes of the Great Lakes of Afrika: Their Biology and Evolution, TFH Publications, Neptune City, 641 pp.

Greenwood, P. H., 1984. African cichlid fishes and evolutionary theories. In Echelle, A. A. & I. Kornfield (eds), Evolution of Fish Species Flocks. University of Maine at Orono Press, Orono: 141–154.

Hasegawa, M., H. Kishino & T. Yano, 1985. Dating of the human-ape splitting by a molecular clock of mitochondrial DNA. Journal of Molecular Evolution 22: 160–174.

Johnson T. C., C. A. Scholz, M. R. Talbot, K. Kelts, R. D. Rickets, G. Ngobi, K. Beuni, I. Ssemmanda & J. W. McGill, 1996. Late pleistocene desiccation of Lake Victoria and rapid evolution of cichlid fishes. Science 273: 1091–1093.

Kawanabe, H., 1981. Territorial behaviour of *Tropheus moorii* (Osteichthyes: Cichlidae) with a preliminary consideration of territorial forms in animals. African Study Monographs 1: 101–106.

Kocher, T. D., W. K. Thomas, A. Meyer A, S. V. Edwards, S. Pääbo, F. X. Villablanca & A. C. Wilson, 1989. Dynamics of mitochondrial DNA evolution: amplification and sequencing with conserved primers. Proceedings of the National Academy of Sciences USA 86: 6196–6200.

Konings, A., 1990. The genus *Tropheus* from Lake Tanganyika, Tropical Fish Hobbyist 39: 66–89.

Konings, A., 1998. Tanganjika-Cichliden in ihrem natürlichen Lebensraum. Cichlid Press, El Paso, 272 pp.

Kosswig, C., 1947. Selective mating as a factor for speciation in cichlid fish of East African lakes. Nature 159: 604.

Kumar, S., K. Tamura, I. B. Jakobsen & M. Nei, 2001. MEGA2: Molecular evolutionary genetics analysis software. Bioinformatics 17: 1244–1245.

Lezzar K. E., J. J. Tiercelin, M. De Batist, A. S. Cohen, T. Bandora, P. Van Rensbergen, C. Le Turdu, W. Mifundu & J. Klerkx, 1996. New seismic stratigraphy and Late Tertiary history of the North Tanganyika Basin, East African Rift system, deduced from multichannel and high-resolution reflections seismic data and piston core evidence. Basin Research 8: 1–28.

Mayr, E., 1984. Evolution of fish species flocks: a comentary. In Echelle, A. A. & I. Kornfield (eds), Evolution of Fish Species Flocks. University of Maine at Orono Press, Orono: 3–11.

Meyer, A., 1993. Phylogenetic relationships and evolutionary processes in East African cichlid fishes. Trends in Ecology and Evolution 8: 279–284.

Meyer, A., J. M. Morrissey & M. Schartl, 1994. Recurrent origin of sexually selected trait in *Xiphophorus* fishes inferred from a molecular phylogeny. Nature 368: 539–541.

Poll, M., 1986. Classification des cichlidae du lac Tanganika, Tribus, genres et aspeces. Acad. Roy. De Belg. Memories de la classe des sciences. Collection in –8°–2° series. T XLV-Fascicule 2: 1–163

Posada, D. & K. A. Crandall, 1998. Modeltest: testing the model of DNA substitution. Bioinformatics 14: 817–818.

Rüber, L., A. Meyer, C. Sturmbauer & E. Verheyen, 1998. Lake level fluctuations and speciation in rock-dwelling cichlid fish in Lake Tanganyika, East Africa. In Grant, P. (ed), Evolution on Islands. Oxford University Press, Oxford: 225–240.

Rüber, L., A. Meyer, C. Sturmbauer & E. Verheyen, 2001. Population structure in two sympatric species of the Lake Tanganyika cichlid tribe Eretmodini: evidence for introgression. Molecular Ecology 10: 1207–1225.

Saitou, N. & M. Nei, 1987. The neighbor-joining method: a new method for reconstructing phylogenetic trees. Molecular Biology and Evolution 4: 406–425.

Schneider, S., D. Roessli & L. Excoffier, 2000. Arlequin 2.000: a software for population genetics data analysis. Genetics and Biometry Laboratory, University of Geneva, Switzerland.

Scholz, C. A., J. W. King, G. S. Ellis, P. K. Swart, J. C. Stager & S. M. Colman, 2003. Paleolimnology of Lake Tanganyika, East Africa, over the past 100 k yr. Journal of Paleolimnology 30: 139–150.

Schupke, P., 2003. Cichlids of Lake Tanganyika, Part 1: The species of the genus *Tropheus*, Aqualog, Rodgau, Germany: 190 pp.

Snoeks, J., L. Rüber & E. Verheyen, 1994. The Tanganyika problem: Taxonomy and distribution of its ichtyofauna. In Martens, K., B. Goddeeris & G. Coulter (eds), Speciation in Ancient Lakes. Advances in Limnology 44: 357–374.

Sturmbauer, C., 1998. Explosive speciation in cichlid fishes of the African Great Lakes: a dynamic model of adaptive radiation. Journal of Fish Biology 53 (Supplement A): 18–36.

Sturmbauer, C. & R. Dallinger, 1995. Diurnal variation of spacing and foraging behaviour in *Tropheus moorii* (Cichlidae) in Lake Tanganyika, Eastern Africa. Netherlands Journal of Zoology 45: 386–401.

Sturmbauer, C. & A. Meyer, 1992. Genetic divergence, speciation, and morphological stasis in a lineage of African cichlid fishes. Nature 358: 578–581.

Sturmbauer, C., E. Verheyen, L. Rüber & A. Meyer, 1997. Phylogeographic patterns in populations of cichlid fishes from rocky habitats in Lake Tanganyika. In Kocher, T. D. & C. Stepien (eds), Molecular Phylogeny of Fishes. Academic Press, New York: 97–111.

Sturmbauer, C., S. Baric, W. Salzburger, L. Rüber & E. Verheyen, 2001. Lake level fluctuations synchronize genetic divergence of cichlid fishes in African lakes. Molecular Biology and Evolution 18: 144–154.

turmbauer, C., U. Hainz, S. Baric, E. Verheyen & W. Salzburger, 2003. Evolution of the tribe Tropheini from Lake Tanganyika: synchronized explosive speciation producing multiple evolutionary parallelism. Hydrobiologia 500: 51–64.

Swofford, D. L., 2000. *PAUP**: Phylogenetic Analysis Using Prasimony (and other methods), beta version 4.0. Sinauer, Sunderland.

Takezaki N., A. Rzhetsky & M. Nei, 1995. Phylogenetic test of the molecular clock and linearized trees. Molecular Biology and Evolution 12: 823–833.

Tamura K. & M. Nei, 1993. Estimation of the number of nucleotide substitutions in the control region of mitochondrial DNA in humans and chimpanzees. Molecular Biology and Evolution 10: 512–526.

Templeton, A. R., K. A. Crandall & C. F. Sing, 1992. A cladistic analysis of phenotypic associations with haplotypes inferred from restriction endonuclease mapping and DNA sequence data. III. Cladogram estimation. Genetics 132: 619–633.

Verheyen, E., L. Rüber, J. Snoeks & A. Meyer, 1996. Mitochondrial phylogeny of rock-dwelling cichlid fishes reveals evolutionary influence of historical lake level fluctuations of Lake Tanganyika, Africa. Philosophical transactions of the Royal Society London B 351: 797–805.

Yanagisawa, Y., & M. Nishida, 1991. The social and mating system of the maternal mouthbrooder *Tropheus moorii* (Cichlidae) in Lake Tanganyika. Japanese Journal of Ichthyology 38: 271–282.

Appendix

Table A1. Characterization of the individuals studied, with information concerning sampling locality, GenBank accession numbers, and lineages and haplotypes to which individuals were assigned

Locality	Locality no.	Latitude (S)	Longitude (E)	Sample identification	Acession no.	Lineage	Haplotype
?				T. DUBOISI	Z12086	OG	
Ngombe	5			608T.duboisi-Ngombe	AJ489622	OG	
Bay Mahale Park	18	06°14′	29°43′	860	AY660812	1/"A"	H1
Bemba	69	03°37′	29°09′	111-BEMBA1	Z12099	1/"A"	H2
				99	Z12091	1/"A"	H2
				100	Z12100	1/"A"	H2
				127	Z12098	1/"A"	H2
				155	Z12097	1/"A"	H2
				158	Z12095	1/"A"	H2
				160	Z12093	1/"A"	H2
				168	Z12092	1/"A"	H2
				157-BEMBA2	Z12096	1/"A"	H3
				159-BEMBA3	Z12094	1/"A"	H4
Bilila Island	63	05°40′	29°24′	555	AJ295921	1/"A"	H1
				556	AJ295921	1/"A"	H5
Bulombora	7	05°02′	29°46′	Tr10	AY660826	1/"A"	H6
				Tr23	AY660827	1/"A"	H7
				Tr24	AY660828	1/"A"	H6
				Tr41	AY660844	1/"A"	H6
				Tr42	AY660845	1/"A"	H6
Bulu Point	16	06°01′	29°45′	813	AJ489698	6/"E"	H8
				815	AJ489699	6/"E"	H9
Bulu Point South	17	06°09′	29°40′	802	AY660806	1/"A"	H1
				806	AJ489682	1/"A"	H1
				807	AY660807	1/"A"	H1
				808	AJ489683	1/"A"	H1
				809	AY660808	1/"A"	H1
				810	AY660809	1/"A"	H1
				811	AY660810	1/"A"	H1
				812	AY660811	1/"A"	H1
				YELLFM1_CE	Z75706	1/"A"	H10
				YELLM15_CE	Z75707	1/"A"	H11
				YELLM16_CE	Z75708	1/"A"	H1
				YELLF14_CE	AY660815	1/"A"	H1
				YELLF15_CE	AY660816	1/"A"	H1
				YELLF23_CE	AY660817	1/"A"	H1
Cape Kibwesa	20	06°30′	29°56′	115-T.SPEC2Kib	Z12055	4/"D"	H12
				331	Z75697	7/"F"	H13
				333	Z75698	7/"F"	H13
				335-T.SPEC1Kib	Z75696	4/"D"	H14
				635	AJ489739	5/"E"	H15

Continued on p. 357

Table A1. (Continued)

Locality	Locality no.	Latitude (S)	Longitude (E)	Sample identification	Acession no.	Lineage	Haplotype
				636	AJ489697	5/″E″	H16
				637	AJ489738	5/″E″	H15
				638	AY660813	1/″A″	H1
				973	AY660814	4/″D″	H14
				974	AJ489696	4/″D″	H17
Chaitika	45	08°34′	30°47′	94-CHAITIK1	Z12079	8/″G″	H18
				95-CHAITIK2	Z12083	8/″G″	H19
				377	AJ295914	8/″G″	H20
				378	AJ489653	8/″G″	H21
				379	AJ489654	8/″G″	H22
				380	AJ489655	8/″G″	H23
				381	AJ489714	8/″G″	H24
				384	AY660818	8/″G″	H20
				385	AJ492122	8/″G″	H25
				387	AJ492121	8/″G″	H25
				390	AJ489656	8/″G″	H26
				391	AJ489657	8/″G″	H27
				392	AJ489713	8/″G″	H24
				393	AJ489658	8/″G″	H28
				394	AJ489659	8/″G″	H29
Chilanga	48	08°35′	30°31′	892	AJ489647	7/″F″	H30
				894	AJ489648	7/″F″	H31
Chimba	52	08°19′	30°32′	957	AJ489645	7/″F″	H32
				958	AJ489646	7/″F″	H33
Chipimbi	53	08°17′	30°34′	955	AJ489643	7/″F″	H34
				956	AJ489644	7/″F″	H35
Chisiki	51	08°20′	30°31′	959	AJ489705	7/″F″	H36
Funda	41	08°46′	30°59′	1048	AJ492072	8/″G″	H37
				1049	AJ492074	8/″G″	H38
				1050	AJ492077	8/″G″	H39
				1051	AJ492082	8/″G″	H40
				1052	AJ492075	8/″G″	H38
				1053	AJ492085	8/″G″	H41
				1054	AJ492087	8/″G″	H42
				1055	AJ492088	8/″G″	H43
				1056	AJ492089	8/″G″	H43
				1057	AJ492090	8/″G″	H44
				1058	AJ492078	8/″G″	H39
				1059	AJ492091	8/″G″	H45
				1060	AJ492062	8/″G″	H46
				1061	AJ492094	8/″G″	H47
				1062	AJ492086	8/″G″	H41
				1063	AJ492048	8/″G″	H48
				1064	AJ492039	8/″G″	H49

Continued on p. 358

Table A1. (Continued)

Locality	Locality no.	Latitude (S)	Longitude (E)	Sample identification	Acession no.	Lineage	Haplotype
				1065	AJ492027	8/"G"	H50
				1066	AJ492063	8/"G"	H46
				1067	AJ492079	8/"G"	H39
				1068	AJ492035	8/"G"	H51
				1069	AJ492136	8/"G"	H52
				1070	AJ492092	8/"G"	H45
				1071	AJ492028	8/"G"	H50
				1072	AJ492137	8/"G"	H52
				1074	AJ492095	8/"G"	H47
				1075	AJ492036	8/"G"	H51
				1076	AJ492083	8/"G"	H40
				1077	AJ492073	8/"G"	H37
				1078	AJ492096	8/"G"	H47
Halembe	13	05°44′	29°55′	Tr7	AY660771	1/"A"	H53
				Tr12	AY660775	1/"A"	H54
				Tr13	AY660776	1/"A"	H54
Ikola	23	06°41′	30°22′	87-IKOLA	Z12089	1/"A"	H55
				89	Z12078	1/"A"	H55
Inangu	47	08°29′	30°41′	721	AJ489632	7/"F"	H56
				722	AJ489707	7/"F"	H56
				723	AJ489708	7/"F"	H56
				724	AJ489633	7/"F"	H57
Kabezi	1	03°30′	29°20′	T74-KABEZI	Z75694	1/"A"	H58
Kabimba	65	05°33′	29°20′	T31-KABIMBA	AY660762	a/"A"	H59
				105-Canary Cheek	Z12077	1/"A"	H60
Kabwe	12	05°42′	29°54′	853	AJ468728	1/"A"	H54
				854	AJ489729	1/"A"	H54
Kachese	50	08°29′	30°28′	829	AJ489709	7/"F"	H61
				830	AJ489636	7/"F"	H62
Kala	33	08°08′	30°58′	78-KALA1	Z12076	1/"A"	H63
				83-KALA2	Z12075	1/"A"	H64
				93-KALA3	Z12074	1/"A"	H65
Kalambo	35	08°36′	31°11′	84-KALAMBO	Z12073	1/"A"	H66
				86	Z12072	1/"A"	H66
Kalanswi	26	07°18′	30°35′	867	AJ489700	1/"A"	H67
				868	AJ489701	1/"A"	H68
Kalemie	61	05°55′	29°11′	T48-KALEMIE	Z75695	1/"A"	H69
Kalya	21	06°28′	30°00′	Tr16	AY660779	d/"E"	H70
				Tr17	AY660780	8/"G"	H71
				2062	AY660787	8/"G"	H51
				2063	AY660788	8/"G"	H51
				2064	AY660789	8/"G"	H51
				2065	AY660790	8/"G"	H51
				2066	AY660791	7/"F"	H72
				2067	AY660792	8/"G"	H71

Continued on p. 359

Table A1. (Continued)

Locality	Locality no.	Latitude (S)	Longitude (E)	Sample identification	Acession no.	Lineage	Haplotype
				2071	AY660793	8/"G"	H71
				2072	AY660794	8/"G"	H71
				2073	AY660795	7/"F"	H72
				2074	AY660796	8/"G"	H51
				2075	AY660797	8/"G"	H71
				2076	AY660798	8/"G"	H51
				2077	AY660799	7/"F"	H72
				2078	AY660800	7/"F"	H72
				2079	AY660829	b/"E"	H73
				2080	AY660801	7/"F"	H72
				2081	AY660802	8/"G"	H71
				2082	AY660803	8/"G"	H74
				2083	AY660804	8/"G"	H51
				2084	AY660805	8/"G"	H51
Kapampa	57	07°30′	30°12′	Tr8	AY660772	4/"D"	H75
				Tr9	AY660773	4/"D"	H76
Karilani Island	14	05°58′	29°47′	Tr4	AY660768	1/"A"	H11
				Tr5	AY660769	1/"A"	H77
				Tr6	AY660770	1/"A"	H11
				Tr35	AY660839	1/"A"	H77
Kasakalawe	39	08°47′	31°04′	683	AJ492018	1/"A"	H78
				684	AJ491975	1/"A"	H79
				685	AJ491976	1/"A"	H79
				686	AJ492012	1/"A"	H80
				687	AJ492010	1/"A"	H81
				688	AJ491977	1/"A"	H79
				689	AJ491984	1/"A"	H82
				690	AY660819	1/"A"	H83
				691	AJ491986	1/"A"	H84
				692	AJ491988	1/"A"	H85
				693	AJ491978	1/"A"	H79
				694	AJ491989	1/"A"	H85
				695	AY660820	1/"A"	H86
				696	AJ491979	1/"A"	H79
				697	AJ491996	1/"A"	H87
				698	AJ491980	1/"A"	H79
				699	AJ491997	1/"A"	H87
				700	AJ491998	1/"A"	H87
				701	AJ489719	1/"A"	H88
				707	AJ492011	1/"A"	H80
				708	AJ491981	1/"A"	H79
				709	AJ492001	1/"A"	H89
				710	AJ491999	1/"A"	H87
				711	AJ492994	1/"A"	H86
				712	AJ492013	1/"A"	H90

Continued on p. 360

Table A1. (Continued)

Locality	Locality no.	Latitude (S)	Longitude (E)	Sample identification	Acession no.	Lineage	Haplotype
				713	AJ489723	1/"A"	H91
				714	AJ491990	1/"A"	H85
				715	AJ492000	1/"A"	H87
				716	AJ492002	1/"A"	H89
				717	AJ491991	1/"A"	H85
Kasanga	34	08°26′	31°08′	961	AJ489686	1/"A"	H92
				962	AJ489687	1/"A"	H93
Katoto	40	08°48′	31°01′	702	AJ492021	8/"G"	H50
				703	AJ492029	8/"G"	H51
				704	AJ492038	8/"G"	H49
				705	AJ492076	8/"G"	H38
				706	AJ492022	8/"G"	H50
				939	AJ492004	1/"A"	H94
				940	AJ491985	1/"A"	H82
				941	AJ492015	1/"A"	H78
				942	AJ491982	1/"A"	H79
				943	AJ491992	1/"A"	H85
				944	AJ492016	1/"A"	H78
				945	AJ491987	1/"A"	H84
				946	AJ491983	1/"A"	H79
				947	AJ491995	1/"A"	H86
				948	AJ492030	8/"G"	H51
				998	AJ492005	1/"A"	H95
				999	AJ492040	8/"G"	H48
				1000	AJ492017	1/"A"	H78
				1001	AJ295903	8/"G"	H96
				1002	AJ492003	1/"A"	H89
				1003	AJ491993	1/"A"	H85
				1004	AJ492080	8/"G"	H39
				1005	AJ492014	1/"A"	H90
				1006	AJ492049	8/"G"	H97
				1007	AJ492007	1/"A"	H98
				1008	AJ492066	8/"G"	H99
				1009	AJ492050	8/"G"	H100
				1010	AJ492006	1/"A"	H95
				1011	AJ492081	8/"G"	H39
				1016	AJ492023	8/"G"	H50
				1017	AJ492019	1/"A"	H101
				1018	AJ492051	8/"G"	H102
				1019	AJ492031	8/"G"	H51
				1020	AJ492053	8/"G"	H103
				1021	AJ492056	8/"G"	H104
				1022	AJ492041	8/"G"	H48
				1023	AJ492042	8/"G"	H48
				1024	AJ492057	8/"G"	H105

Continued on p. 361

Table A1. (Continued)

Locality	Locality no.	Latitude (S)	Longitude (E)	Sample identification	Acession no.	Lineage	Haplotype
				1025	AJ492043	8/"G"	H48
				1026	AJ492032	8/"G"	H51
				1027	AJ492058	8/"G"	H106
				1028	AJ492059	8/"G"	H107
				1029	AJ492060	8/"G"	H46
				1030	AJ492044	8/"G"	H48
				1031	AJ492033	8/"G"	H51
				1032	AJ492064	8/"G"	H108
				1033	AJ492020	1/"A"	H109
				1034	AJ492065	8/"G"	H99
				1035	AJ492068	8/"G"	H110
				1036	AJ492052	8/"G"	H102
				1037	AJ492034	8/"G"	H51
				1038	AJ492071	8/"G"	H111
				1039	AJ492024	8/"G"	H50
				1040	AJ491061	8/"G"	H46
				1041	AJ492025	8/"G"	H50
				1042	AJ492084	8/"G"	H40
				1043	AJ492069	8/"G"	H110
				1044	AJ492054	8/"G"	H103
				1045	AJ492045	8/"G"	H48
				1046	AJ492026	8/"G"	H50
				1047	AJ492097	8/"G"	H47
Katukula	42	08°43′	30°57′	818	AJ492125	8/"G"	H112
				819	AJ295912	8/"G"	H113
				820	AJ492070	8/"G"	H110
				821	AJ295913	8/"G"	H113
				822	AJ295902	8/"G"	H96
				823	AJ492093	8/"G"	H45
				824	AJ492129	8/"G"	H114
				825	AJ492130	8/"G"	H115
				826	AJ492131	8/"G"	H116
				1079	AJ492126	8/"G"	H112
				1080	AJ492139	8/"G"	H117
				1081	AJ492140	8/"G"	H118
				1082	AJ492037	8/"G"	H51
				1083	AJ492141	8/"G"	H119
				1084	AJ492132	8/"G"	H120
				1085	AJ492133	8/"G"	H52
				1086	AJ492046	8/"G"	H48
				1087	AJ492134	8/"G"	H52
				1088	AJ492114	8/"G"	H121
				1089	AJ492106	8/"G"	H122
				1090	AJ492138	8/"G"	H123
				1091	AJ492135	8/"G"	H52

Continued on p. 362

Table A1. (Continued)

Locality	Locality no.	Latitude (S)	Longitude (E)	Sample identification	Acession no.	Lineage	Haplotype
				1092	AJ492008	1/″A″	H124
				1093	AJ492142	8/″G″	H125
				1094	AJ492009	1/″A″	H126
				1095	AJ492047	8/″G″	H48
				1096	AJ492144	8/″G″	H127
				1097	AJ492055	8/″G″	H103
				1098	AJ492143	8/″G″	H125
				1099	AJ492145	8/″G″	H128
Kavala Island	64	05°38′	29°24′	63-KAVALLA	Z12071	1/″A″	H129
				Tr29	AY660835	6/″E″	H130
				Tr30	AY660836	6/″E″	H130
				2085	AY660830	6/″E″	H130
				2086	AY660831	6/″E″	H130
				2087	AY660832	6/″E″	H130
				2088	AY660833	6/″E″	H130
				2089	AY660834	6/″E″	H130
Kibwe	10	05°23′	29°46′	595	AJ489678	1/″A″	H131
				596	AJ489679	1/″A″	H132
Kipampa	62	05°42′	29°20′	930	AJ489732	1/″A″	H133
				931	AJ489695	1/″A″	H134
				932	AJ489733	1/″A″	H133
				933	AJ489734	1/″A″	H135
				934	AJ489737	1/″A″	H136
				935	AJ489736	1/″A″	H136
				936	AJ489735	1/″A″	H135
Kiriza	67	04°03′	29°13′	T40-KIRIZA1	Z75700	2/″B″	H137
				161-KIRIZA2	Z12070	2/″B″	H138
Kiti Point	8	05°17′	29°47′	856	AJ489676	1/″A″	H139
				857	AJ489677	1/″A″	H140
				952	AJ489675	1/″A″	H141
'Kungwe Mountain Range'			15	51-T.POLL1	Z12053	c/″E″	H142
				120	Z12087	5/″E″	H15
				146-KUNGWE1	Z12090	5/″E″	H15
				KUNGWE2	AY660764	7/″F″	H72
Kyeso	60	06°30′	29°29′	557	AJ295915	1/″A″	H143
				558	AJ489623	3/″C″	H144
				559	AJ489624	3/″C″	H145
				560	AJ489625	3/″C″	H146
				561	AJ295916	1/″A″	H147
				562	AJ295917	1/″A″	H11
				563	AJ489626	3/″C″	H148
				564	AJ295918	1/″A″	H149
				565	AJ489627	3/″C″	H150

Continued on p. 363

Table A1. (Continued)

Locality	Locality no.	Latitude (S)	Longitude (E)	Sample identification	Acession no.	Lineage	Haplotype
				566	AJ489628	3/"C"	H151
				567	AJ489629	3/"C"	H152
				568	AJ295919	1/"A"	H153
Lufubu	46	08°32′	30°44′	Tr18	AY660781	8/"G"	H154
				Tr19	AY660782	8/"G"	H155
				Tr20	AY660783	8/"G"	H122
				Tr21	AY660784	8/"G"	H156
				Tr22	AY660785	8/"G"	H121
Lupota	56	07°59′	30°26′	55-LUPOTA	Z12068	7/"F"	H157
Manda	27	07°24′	30°34′	987	AJ489702	1/"A"	H158
				988	AJ489703	1/"A"	H159
				989	AJ489704	1/"A"	H160
Masaka	6	05°02′	29°46′	993	AJ489742	1/"A"	H161
				994	AJ489741	1/"A"	H161
				995	AJ489740	1/"A"	H161
				996	AJ489743	1/"A"	H161
Mbita Island	37	08°46′	31°06′	788	AJ489670	1/"A"	H162
				789	AJ489717	1/"A"	H163
				791	AJ489671	1/"A"	H164
				792	AY660821	1/"A"	H165
				794	AY660822	1/"A"	H165
				795	AJ489724	1/"A"	H166
				796	AY663791	1/"A"	H89
				798	AJ489672	1/"A"	H89
				799	AJ489673	1/"A"	H167
				800	AJ489674	1/"A"	H168
Mboko	68	03°55′	29°05′	119-MBOKO	AY660763	2/"B"	H169
Minago	3	04°00′	29°25′	T40-MINAGO	Z75700	2/"B"	H170
Mkagansi	22	06°38′	30°19′	990	AJ489693	1/"A"	H171
				991	AJ489694	1/"A"	H172
Mkombe	24	06°58′	30°34′	953	AJ489684	1/"A"	H173
				954	AJ489685	1/"A"	H174
Mkuyu	9	05°28′	29°46′	597	AJ489726	1/"A"	H175
				598	AJ489727	1/"A"	H175
Moba	59	07°03′	29°47′	122-MOBA	Z75701	4/"D"	H176
Moliro	54	08°12′	30°34′	123-MOLIRO	Z12067	7/"F"	H177
				827	AJ489634	7/"F"	H178
				828	AJ489635	7/"F"	H179
Mpimbwe	25	07°08′	30°30′	148-MPIMBWE	Z12065	1/"A"	H180
Mpulungu	38	08°46′	31°06′	130-MPULUNG1	Z12066	1/"A"	H181
				131	Z12064	1/"A"	H181
				132-MPULUNG2	Z12063	1/"A"	H182
				133-MPULUNG3	Z12062	1/"A"	H83
				134-MPULUNG4	Z12061	1/"A"	H183
				135	Z12060	1/"A"	H79

Continued on p. 364

Table A1. (Continued)

Locality	Locality no.	Latitude (S)	Longitude (E)	Sample identification	Acession no.	Lineage	Haplotype
				137-MPULUNG6	Z12059	1/"A"	H79
				138-MPULUNG7	Z12058	1/"A"	H86
				139-MPULUNG8	Z12057	1/"A"	H163
				140-MPULUNG9	Z12056	1/"A"	H165
				496	AJ489660	1/"A"	H184
				497	AJ489661	1/"A"	H185
				498	AJ489715	1/"A"	H163
				499	AJ489718	1/"A"	H88
				500	AJ489662	1/"A"	H186
				501	AJ489716	1/"A"	H163
Mtondwe Island	36	08°42′	31°07′	502	AJ489721	1/"A"	H91
				503	AJ489663	1/"A"	H187
				504	AJ489664	1/"A"	H188
				510	AJ489665	1/"A"	H189
				511	AJ489666	1/"A"	H190
				512	AJ489722	1/"A"	H91
				513	AJ489667	1/"A"	H191
				514	AJ489720	1/"A"	H91
				515	AJ489668	1/"A"	H192
				516	AJ489669	1/"A"	H193
				517	AY660823	1/"A"	H85
				518	AY660824	1/"A"	H78
				520	AJ489725	1/"A"	H166
Mvua	55	08°05′	30°32′	639	AJ489649	8/"G"	H194
				640	AJ489650	8/"G"	H195
				671	AJ489711	8/"G"	H196
				672	AJ489638	7/"F"	H197
				673	AJ489639	7/"F"	H198
				674	AJ489696	8/"G"	H199
				676	AJ489640	7/"F"	H200
				677	AJ489652	8/"G"	H201
				678	AJ489712	8/"G"	H196
				679	AJ489641	7/"F"	H202
				680	AJ489642	7/"F"	H203
Nakaku	44	08°40′	30°54′	967	AJ489706	8/"G"	H204
Namansi	30	07°32′	30°36′	Tr1	AY660765	7/"F"	H205
				Tr2	AY660766	7/"F"	H206
				Tr3	AY660767	7/"F"	H207
				Tr31	AY660837	1/"A"	H208
Ngombe	5	04°40′	29°37′	605	AJ295922	1/"A"	H209
				606	AJ295923	1/"A"	H210
				607	AJ295924	1/"A"	H211
Nkondwe Island	28	07°23′	30°33′	865	AJ489689	1/"A"	H159

Continued on p. 365

Table A1. (Continued)

Locality	Locality no.	Latitude (S)	Longitude (E)	Sample identification	Acession no.	Lineage	Haplotype
Nvuna Island	29	07°26′	30°33′	864	AJ489688	1/"A"	H212
Nyanza Lac	4	04°20′	29°35′	104-NYANZA	Z12054	1/"A"	H213
Rutunga	2	03°40′	29°19′	56-RUTUNG1	Z12050	2/"B"	H214
				82	Z12049	2/"B"	H214
				121-RUTUNG2	Z12051	2/"B"	H215
Segunga	11	05°35′	29°51′	855	AJ489680	1/"A"	H216
				858	AJ489681	1/"A"	H217
Siyeswe	19	06°21′	29°48′	Tr11	AY660774	5/"E"	H218
				Tr25	AY660786	5/"E"	H219
				Tr32	AY660838	7/"F"	H72
Sumbu	49	08°31′	30°29′	681	AJ489710	7/"F"	H61
				682	AJ489637	7/"F"	H220
Tongwa	43	08°40′	30°53′	641	AJ492098	8/"G"	H221
				642	AJ492099	8/"G"	H222
				643	AJ492100	8/"G"	H122
				644	AJ492101	8/"G"	H122
				645	AJ492107	8/"G"	H223
				646	AJ492102	8/"G"	H122
				647	AJ492108	8/"G"	H224
				648	AJ492109	8/"G"	H224
				649	AJ295904	8/"G"	H225
				650	AJ492110	8/"G"	H226
				651	AJ492103	8/"G"	H122
				652	AJ492111	8/"G"	H121
				653	AJ492115	8/"G"	H227
				654	AJ492116	8/"G"	H228
				655	AJ492117	8/"G"	H229
				656	AJ492104	8/"G"	H122
				657	AJ492118	8/"G"	H230
				658	AJ492120	8/"G"	H231
				659	AJ492121	8/"G"	H25
				660	AJ492123	8/"G"	H232
				661	AJ492124	8/"G"	H233
				662	AJ492122	8/"G"	H25
				663	AJ492105	8/"G"	H122
				664	AJ492067	8/"G"	H99
				665	AJ295905	8/"G"	H234
				666	AJ492127	8/"G"	H28
				667	AJ492128	8/"G"	H235
				668	AJ492112	8/"G"	H121
				669	AJ492113	8/"G"	H121
				670	AJ492119	8/"G"	H230
Ubwari	66	04°11′	29°14′	T57-UBWARI	Z75702	2/"B"	H236
				Tr37	AY660840	2/"B"	H237
				Tr38	AY660841	2/"B"	H236

Continued on p. 366

Table A1. (Continued)

Locality	Locality no.	Latitude (S)	Longitude (E)	Sample identification	Acession no.	Lineage	Haplotype
				Tr39	AY660842	2/"B"	H237
				Tr40	AY660843	2/"B"	H237
Wapembe North	31	07°54′	30°49′	129-WAPEMB1	Z12048	1/"A"	H238
				169-WAPEMB2	Z75703	1/"A"	H239
				171-WAPEMB3	Z75704	1/"A"	H240
				128-WAPEMB4	Z75705	1/"H"	H241
Wapembe South	32	08°00′	30°53′	395	AJ489730	8/"G"	H242
				396	AJ489690	8/"G"	H243
				397	AJ489691	8/"G"	H244
				398	AJ489731	8/"G"	H242
				401	AJ295909	8/"G"	H245
				402	AY660825	8/"G"	H242
				403	AJ489692	8/"G"	H243
				404	AJ295910	8/"G"	H246
				405	AJ295911	8/"G"	H113
				Tr14	AY660777	1/"A"	H247
				Tr15	AY660778	7/"F"	H13
Zongwe	58	07°18′	30°08′	T66-ZONGWE1	Z75709	4/"D"	H248
				116-ZONGWE2	Z12047	4/"D"	H249
				831	AJ489630	4/"D"	H250
				832	AJ489631	4/"D"	H251

Hydrobiologia (2005) 542:367–378
H. Segers & K. Martens (eds), Aquatic Biodiversity II
DOI 10.1007/s10750-004-9461-0

Plankton richness in a eutrophic reservoir (Barra Bonita Reservoir, SP, Brazil)

Takako Matsumura-Tundisi* & José Galizia Tundisi
International Institute of Ecology, Rua Bento Carlos, n 750, Centro, CEP 13560-660, Sao Carlos, SP, Brazil
*(*Author for correspondence: E-mail: tundisi@zaz.co.br)*

Key words: eutrophic reservoir, horizontal heterogeneity, tributaries, plankton richness, cyanophycean blooms, nutrients concentration

Abstract

Species richness of plankton was studied in a eutrophic reservoir (Barra Bonita Reservoir) of the Middle Tietê River, São Paulo State, Brazil, during the period 1985–1986. This reservoir is formed by two rivers: the Tietê and the Piracicaba (the main tributaries), of which the Tiête is more eutrophic, having conductivity and nutrient concentration values twice those of the Piracicaba. In addition, the reservoir is fed by 114 smaller tributaries. Monthly sampling was carried out at three stations representing different environmental conditions: St1 on the Tietê, S2 on the Piracicaba, and St3 at the confluence of the two rivers. For the phytoplankton community, the Piracicaba River (St2) proved the richest site, with a listed 72 species with abundance of Chlorophyta, while St1 and St3 registered 59 and 50 species, respectively, with abundance of Cyanophyta. For the zooplankton community a great difference was not registered in species number at the three stations but the species composition and dominant species of rotifers and copepods were quite different. The occurrence of *Mesocyclops meridianus* and *Metacyclops mendocinus*, specific for St2; *Mesocyclops ogunnus* and *Notodiaptomus iheringi*, specific for St1, and that of these two species plus *Mesocyclops meridianus* at St3 shows that the conditions combining at this station were favorable to *Mesocyclops meridianus* but not to *Metacyclops mendocinus*. Both for phytoplankton and zooplankton, high values found of species richness were compared to species richness of natural lakes, e.g., Dom Helvecio, a monomictic stable lake in eastern Brazil, and another fifteen lakes in the same region.

In conclusion, this work shows that environmental gradients are strong selective factors that enhance plankton richness in eutrophic reservoirs exhibiting environmental instability. This fact could explain the presence of a high number of plankton species associated with a high number of individuals in Barra Bonita Reservoir, supporting the effects of the intermediate disturbance hypothesis.

Introduction

Species richness of plankton in a reservoir is related to its trophic state, spatial heterogeneity in vertical and horizontal structure (thermal, physical, and chemical),and frequency and degree of mixing and stratification of the water column.

Reservoirs are complex systems (Tundisi et al., 1998) with spatial gradients, produced by tributaries that influence local physical, chemical, and biological characteristics of water masses. A reservoir, according to Kimmel & Groeger (1984) presents three very distinct zones: the riverine zone, in the upper reservoir, that is subject to the influence of either the tributaries or river of origin;

the transition zone, downstream from the reservoir, that functions as an intermediate river-lake ecosystem; and the lacustrine zone, located further downstream. These are dynamic zones with interfaces that change periodically, depending on horizontal and vertical forces, and are influenced by wind and the effects of discharge at the dam site (by water spill devices, hydroelectric use, or water supply intakes).

Despite several descriptions, assumptions, theories, and hypotheses regarding spatial heterogeneity in reservoirs (Straskraba et al., 1993; Tundisi & Straskraba, 1999) and demonstrations such as those of Armengol et al. (1999), there is still a lack of data on the effects of these environmental gradients in plankton behavior and distribution affecting species richness in reservoirs.

Barra Bonita is the first of a series of six large reservoirs located in the middle and lower Tietê River, São Paulo State, Brazil. Several studies have been conducted in the last 25 years in order to understand its dynamic features and to provide a basis for its management (Barbosa et al., 1999; Matsumura-Tundisi et al., 1990, 2002; Tundisi & Matsumura-Tundisi, 1990; Tundisi et al., 1998, 2000). These studies show that the system is subject to strong inputs of climatological and hydrological factors that promote mixing processes affecting biogeochemical cycles and primary production. Spatial heterogeneity, resulting from the two main tributaries, the Piracicaba and Tietê rivers, discharging into the main body of the reservoir, has also been demonstrated. Barra Bonita is a eutrophic reservoir due to discharge of non-treated wastewater and agricultural fertilizers. Besides the two main tributaries, this reservoir has many small tributaries draining a vast watershed with diverse hydrogeochemistry and different inputs of chemical substances impacting them (Tundisi & Matsumura-Tundisi, 1990).

In the research reported in this paper, limnetic plankton richness was analyzed at three different limnological stations, each one representing a specific condition of the reservoir. These sites showed a diverse environmental conditions in relation to nutrient concentrations, conductivity, chlorophyll, and suspended matter. In searching for interrelationships between spatial heterogeneity and plankton species richness, the authors attempted to link limnetic zooplankton richness in the reservoir to spatial gradients and heterogeneity.

Material and methods

Barra Bonita Reservoir

This reservoir was built in 1963 by the damming of the Tietê and Piracicaba rivers with the aim of establishing a hydropower plant and, later on, a navigation system in the Tietê River. Located between coordinates 23° 31′ 49″ S and 48° 31′ 14″ W, its altitude is 430 m. Climatic characteristics are those of subtropical regions, with a dry period from May to October and a wet period from November to April. The reservoir is inserted into a hydrographic basin having 32 320 km^2 and whose drainage type is open (Tietê/Paraná River Basin). Reservoir area measures 310 km^2; total volume is 3.2 km^3; and dam length is 480 m. The total output of the hydroelectric power plant is 140 MW. During summer (wet season) the flushing rate is approximately 1500 m^3/s; in winter (dry season) it is approximately 200 m^3/s. The maximum depth is about 25 m with an average depth of 10 m. The retention time in summer is 37 days; in winter it is 137 days. The main uses are: hydroelectricity, navigation, recreation, tourism, fisheries, and aquaculture (Tundisi & Matsumura-Tundisi, 1990).

Sampling strategy, data collection, and period of sampling

For the purpose of this research three sampling stations were established: St 1 in the Tietê River, St2 in the Piracicaba River, and St3 at their confluence. Figure 1 shows the location of the reservoir in Brazil and São Paulo State, the satellite image of the reservoir with the Piracicaba and Tietê, location of the sampling stations, and the hydrographic basin with its many tributaries.

Monthly sampling was carried out from July 1985 to June 1986. At each sampling station, a complete vertical temperature profile was obtained, using a Toho Dentan thermistor. For other variables measured, i.e., dissolved oxygen, pH, electrical conductivity, and nutrient concentrations, water samples were collected at the surface

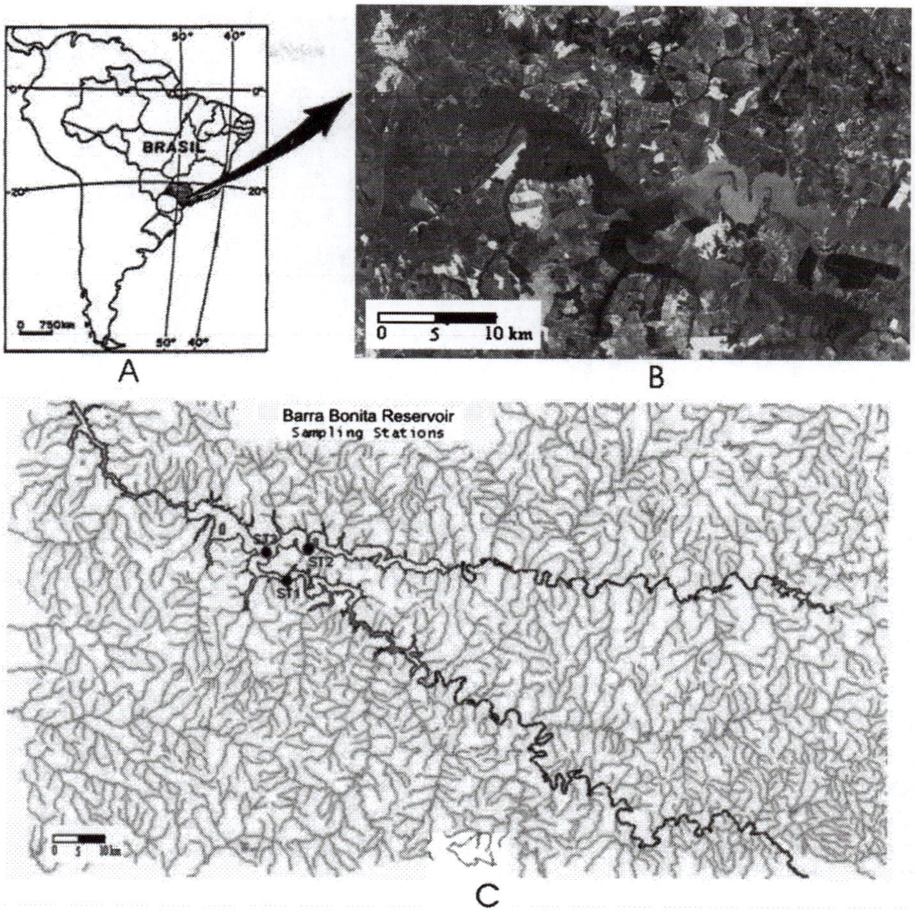

Figure 1. Map showing the location of Barra Bonita Reservoir in São Paulo State, Brazil (A); satellite image of reservoir (B); reservoir with main tributaries (Piracicaba and Tietê River) with the location of sampling stations (St1, St2 and St3) and a rich net work of small tributaries (C).

and bottom layers using a Van Dorn sampler. Chemical analyses were performed in the laboratory using methods described in Mackereth et al. (1978) and Golterman et al. (1978).

Plankton sampling

Chlorophyll data were obtained from surface water by sampling and filtering 1 l of water through a GF filter; chlorophyll was extracted using acetone 90%. The phytoplankton samples, collected from 250 ml of surface water, were fixed with lugol solution, and analyzed by the Uthermol technique using an inverted microscope. The zooplankton material was obtained by vertical hauls in the water column with a standard plankton net of 64 μm mesh size. The volume of filtered water was calculated using the formula: $V = \pi r^2 d$ where r is the radius of the net opening and d the distance covered by the net within the water column (m). The number of organisms was expressed by m^3.

After withdrawing a 1 ml subsample, small and abundant organisms such as rotifers and copepod nauplii were counted in Sedgewick Rafter cells by optical microscope with 200× of magnification. The microcrustaceans such as cladocerans and copepods were counted by stereo microscope with 80× magnification, withdrawing subsamples of variable volumes until reaching a minimum of 500 organisms for each sample.

Results

Spatial heterogeneity of the limnological conditions of the reservoir

The physical and chemical variables measured at the three stations (St1, St2, and St3) showed that St2, which is influenced by the Piracicaba River, differed widely from the other stations which were affected by the Tietê River. At St3, located at the confluence of both rivers, all data showed greater influence of the Tietê (St1) than was demonstrated for the Piracicaba (St2).

The temperature showed similar values for all three stations, with average minimum values of 17 °C in the dry period (spring/winter) and average maximum values of 28.5 °C in the wet period (autumn/summer). Figure 2 shows the temperature evolution during 1 year (July 1985–June 1986) at each site. In most periods of the year, the water column temperature was homogeneous, showing no thermal stratification. Even in November 1985 at St3, where the greatest difference between surface and bottom temperatures was registered (surface: 29.2 °C; bottom: 20.8 °C), there were no stable stratified layers.

Dissolved oxygen

Low oxygen concentration was registered in the bottom layer of St1, St2,and St3, mainly in the wet period of the year (October–January) (Fig. 3). At St1 in December 1985, high oxygen concentrations at the surface, such as 12.61 mg/l, were registered; these were probably due to the photosynthetic activity of phytoplankton, which showed high chlorophyll *a* concentration.

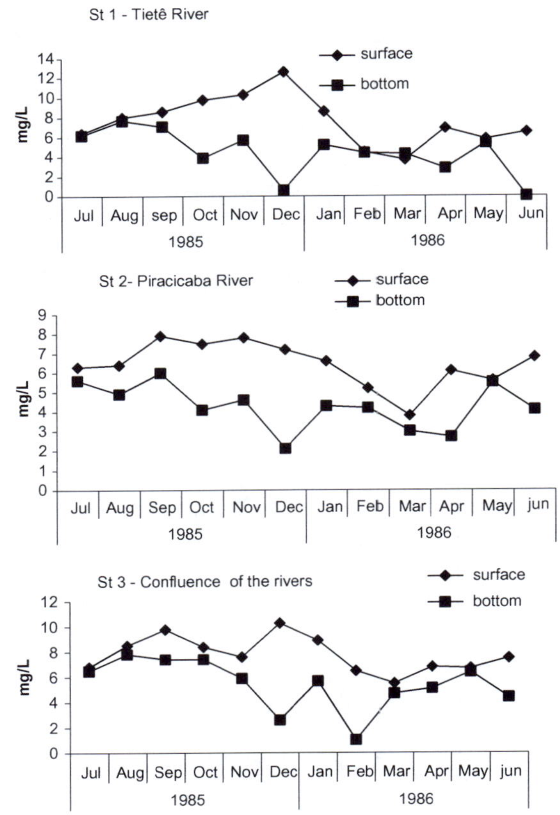

Figure 3. Variation of dissolved oxygen concentration at the surface and at the bottom of St1, St2 and, St3 of Barra Bonita Reservoir.

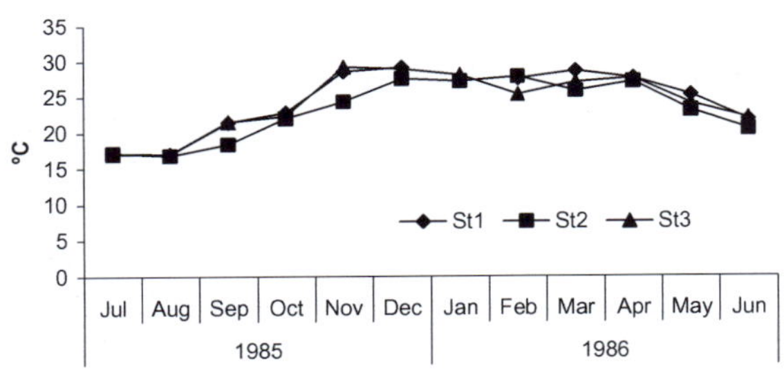

Figure 2. Variation of temperature (average value at the water column) at St1, St2 and, St3 of Barra Bonita Reservoir during the period of 1985–1986.

pH and conductivity

High pH values registered from October to January (wet season) for St1 and St3 (Fig. 4) followed the same distribution pattern presented by oxygen. The average pH values for St1 were 8.1, St2 – 7.1, St3 – 7.8. Conductivity was also high from October to January (Fig. 5).

Monthly variation of nutrients, phosphate, nitrate, and ammonium measured at the three stations are presented in Figures 6–8, respectively. The average values presented at the three stations for these nutrients were higher for the Tietê River for phosphate (25 μg/l) and for nitrate (1404 μg/l) than those for the Piracicaba River, where the average value of phosphate concentration was 11.8 μg/l, whereas for nitrate it was 473.5 μg/l. The concentration of ammonium was higher in the Piracicaba River (51.3 μg/l) than in the Tietê

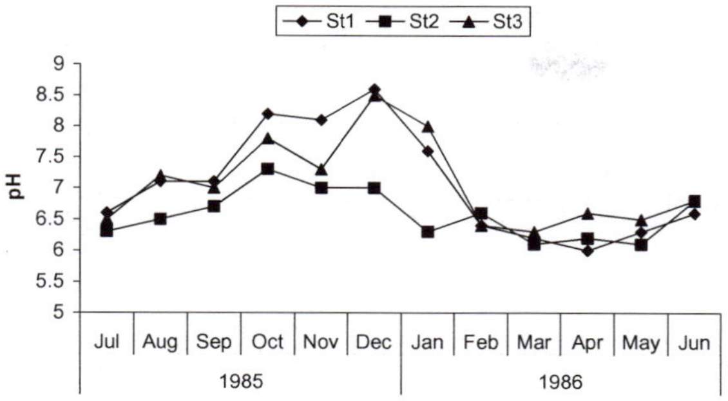

Figure 4. Monthly variation of pH at the St1 St2, and, St3 of Barra Bonita Reservoir.during the period of 1985–1986.

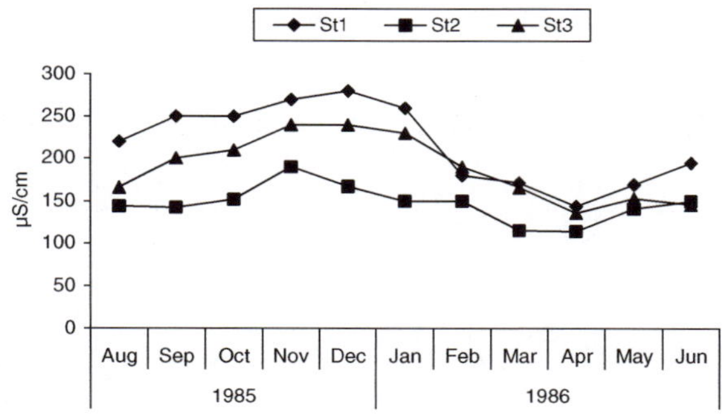

Figure 5. Monthly variation of conductivity at St1, St2, and, St3 of Barra Bonita Reservoir during the period of 1985–1986.

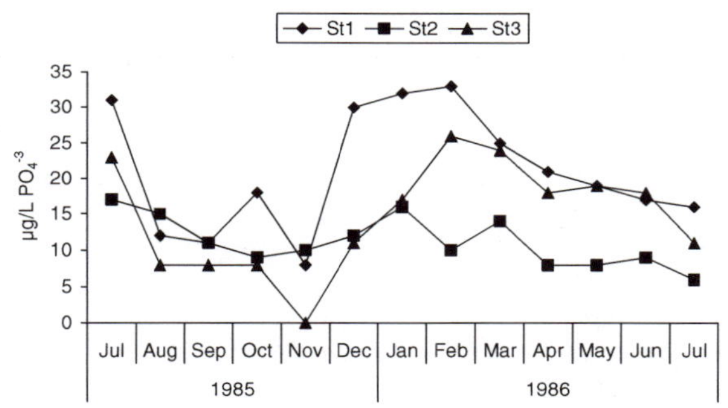

Figure 6. Monthly variation of dissolved phosphorous at St1, St2, and, St3 of Barra Bonita Reservoir during the period of 1985–1986.

(27.5 μg/l). Station 3 presented intermediate values, however it was much more influenced by the Tietê than by the Piracicaba, showing limnological characteristics of the former river (Table 1).

Chlorophyll a and phytoplankton composition

Figure 7 refers to the temporal variation of chlorophyll a at the three stations. At St1, the highest

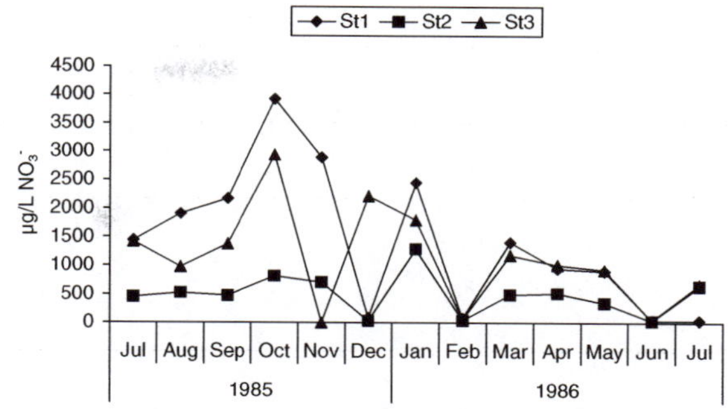

Figure 7. Monthly variation of nitrate at St1, St2 and, St3 of Barra Bonita Reservoir during the period of 1985–1986.

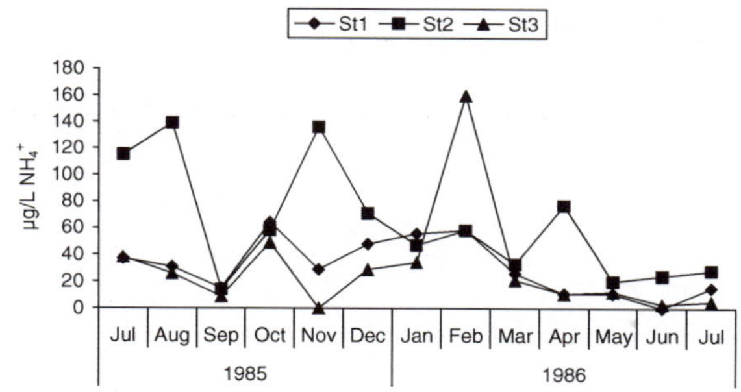

Figure 8. Monthly variation of ammonium at St1, St2, and, St3 of Barra Bonita Reservoir during the period of 1985–1986.

value (41.7 μg/l) was reached in December 1985 owing to the abundance of Cyanophyta, represented mainly by *Aanabaena spiroides* that constituted 99.4% of total phytoplankton. The first of the other two peaks occurred in October 1985 with 20 μg/l due to Chrysophyta (71.8%), represented mainly by *Aulacoseira granulata* and *Aulacoseira italica*; the other peak was reached in April 1986 with 25 μg/l of chlorophyll because of the contribution of Cyanophyta (39.6%), Chlorophyta (43.5%), and Euglenophyta (11.8%) (Table 2).

Station 2 showed a more homogeneous distribution of chlorophyll concentration during the year, with slight increases in September 1985, October 1985, and April 1986 (no more than 20 μg/l). In September, Chrysophyta contributed with 95.0% of total phytoplankton, represented mainly by *Aulacoseira italica curvata*, while in October, Chlorophyta contributed with 56.0% represented by *Staurastrum* and *Coelastrum,* and Pyrrhophyta contributed with 26.0% represented by *Peridinium*.

At the confluence of the Tietê and Piracicaba rivers, the bulk of chloropyll at St3 occurred

Table 1. Average values of physical, chemical and biological variables measured during the period of 1 year at the three stations of Barra Bonita Reservoir

	St1	St2	St3
Variables	(Tietê River)	(Piracicaba River)	(Confluence)
Temperature at surface (°C)	24.2	23.1	24.1
Oxygen at surface (mg/l)	7.6	6.5	7.8
Conductivity (μS/cm)	213.5	143.9	183.9
pH	6.8	6.4	6.8
PO_4^{-3} (μg/l)	25.0	11.8	16.1
NO_3^- (μg/l)	1404.0	473.5	1182.0
NH_4^+ (μg/l)	27.5	51.3	26.2
Chlorophyll *a* (μg/l)	14.7	11.7	9.3
No species phytoplankton	59	72	50
No species zooplankton	49	42	40

Table 2. Relative abundance of the phytoplankton taxa at St1, St2 and St3 during the periods of high concentrations of chlorophyll

Phytoplankton taxa	Months			
St1 (Tietê River)	Aug/85	Oct/85	Dec/85	Apr/86
Chrysophyta	90.2%	71.8%	0.4%	5.0%
Cyanophyta	8.3%	35.6%	99.5%	39.6%
Chlorophyta	1.5%	2.6%	0.1%	43.5%
Euglenophyta	–	–	–	11.8%
St 2 (Piracicaba River)	Sep/85	Oct/85	Apr/86	
Chrysophyta	95.0%	3.9%	30.7%	
Cyanophyta	2.1%	11.8%	16.3	
Chlorophyta	1.5%	56.2%	36.3	
Euglenophyta	–	2.0%	16.0%	
Pyrrhophyta	–	26.0%	–	
St3 (Confluence)	Dec/85	Jan/86	Feb/86	
Chrysophyta	51.7%	31.2%	59.2%	
Cyanophyta	44.9%	66.3%	38.7%	
Chlorophyta	2.9%	1.6%	1.8%	
Euglenophyta	–	–	–	
Pyrrhophyta	–	–	–	

between December 1985 and February 1986 (20–25 μg/l) due to the contribution of Cyanophyta (49.9%), represented by *Microcystis* and *Anabaena*, and also the Chrysophyta contribution (47.3%), represented by *Aulacoseira granulata*. The temporal variation of chlorophyll concentration for all three stations was related to phosphate variation but not with nitrate, indicating that phosphorous is a limiting factor for phytoplankton growth, as demonstrated in many papers (Fig. 9).

High phytoplankton diversity was observed for the Piracicaba (St2) with 72 species, the Tietê (St1) with 59 species, and St3 with 50 species (see Table 1).

Zooplankton composition

The zooplankton community of the Tietê River (St1) and that of the Piracicaba (St2) greatly differed both in abundance and composition, as seen

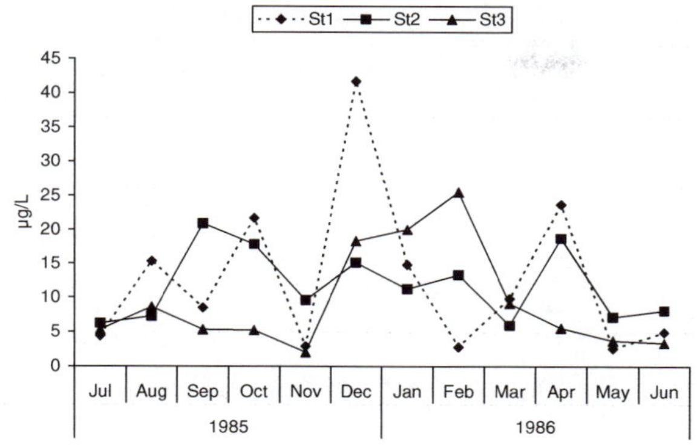

Figure 9. Monthly variation of surface chlorophyll *a* at St1, St2 and St3 of Barra Bonita Reservoir during the period of 1985–1986.

in Tables 3 and 4. The Tietê is richer than the Piracicaba, contributing with 32 species of Rotifera and 17 species of microcrustaceans. The Piracicaba contributed with 25 species of Rotifera and 17 species of microcrustaceans. But at the confluence of both rivers (St3), the number of rotifer species declined to 21 species. However, the microcrustaceans showed a slight increase (18 species).

Another significant difference observed in the three stations refers to the composition of copepods. The Tietê River (St1) was characterized by the presence of two species of Calanoida (*Notodiaptomus iheringi* and *Notodiaptomus conifer*) and four species of Cyclopoida (*Thermocyclops minutus, Thermocyclops decipiens, Mesocyclops ogunnus*, and *Mesocyclops longisetus*). At this site the proportion of Calanoida and Cyclopoida occurrence was similar (44% and 56%, respectively) (Fig. 10). At St2 (Piracicaba River) only Cyclopoida species occurred (*T. minutus, T. decipiens, Mesocyclops meridianus*, and *Metacyclops mendocinus*). At St3, all the species present in St1 and St2 occurred, increasing the richness of Copepoda species. Another special characteristic of this site was the dominance of Calanoida copepods (72%) over Cyclopoida (28%).

As for the cladocerans, all stations were characterized by dominance of *Diaphanosoma birgei, Ceriodaphnia cornuta + C. silvestrii, Bosmina longirostris*, and *Daphnia gessneri. Other species*, such as *Diaphanosoma brevireme, Daphnia ambigua, Moina minuta, Moina micrura, Bosmina hagmanni, and Bosmina tubicen* were represented.

In relation to the rotifer populations whose assemblage and temporal variation at these sites were studied (Matsumura-Tundisi et al., 1990), the Tietê River (St1) presented more richness of species than was found for the Piracicaba River (St2) (see Table 3). Eleven species were specific to the Tietê River (St1): *Brachionus angularis, Brachionus caudatus, Euchlanis dilatata, Euchlanis* sp., *Lecane cornuta, Lecane dorysa, Synchaeta patina, Trichocerca bicristata, Trichocerca elongata, Trichocerca mus*, and *Trichocerca pusilla*. For St2, only three species were specific: *Brachionus* sp., *Gastropus* sp., and *Synchaeta pectinata*. At the confluence of the two rivers (St3), the number of species fell, owing to the elimination of specific species.

Discussion

Plankton composition, species richness, and abundance of organisms in natural lakes are dependent on various factors such as lake origin, trophic state, colonization processes, and presence or absence of toxic substances or pollutants. Besides these factors, in man-made lakes, among them reservoirs, the contribution of tributaries can be a factor of considerable weight. In natural lakes, a patent difference exists in plankton richness between small and large lakes, as demonstrated by Patalas, 1975, with higher richness found in very large lakes. Both in temperate and tropical lakes the mean numbers per lake of rotifers, cladoceran, and copepods species have been recorded as 7, 4, and 2 for Lake Lanao (the Philippines) by Lewis, 1979. In addition, for comparing plankton richness of lakes, it is important to separate limnetic and littoral plankton because samplings at both spatial and temporal scales could affect the composite number of species. In relation to trophic states, according to Maitland (1978) oligotrophic lakes support a high number of phytoplankton and zooplankton species but represented only by a small number of individuals, while eutrophic lakes support small numbers of plankton species, but large numbers of individuals for each species.

The example given in this paper on Barra Bonita Reservoir could be considered as part of a more general picture on spatial heterogeneity in reservoirs and their influence on plankton richness

Table 3. Composition and abundance of Rotifera species at the three localities (St1, St2 and St3) of Barra Bonita Reservoir

	Tietê River (St1)	Piracaba River (St2)	Confluence (St3)
Ascomorpha ovalis	+	+	+
Asplanchna sieboldi	+	+	
Brachionus angularis	+		
Brachionus calyciflorus	+ + +	+	+ + + +
Brachionus caudatus	+		
Brachionus falcatus	+	+	+ + +
Brachionus sp.		+	
Conochiloides coenobasis	+ + +	+	+
Conochilus unicornis	+ + + +	+ + +	+ + + +
Euchlanis dilatata	+		+
Euchlanis sp.	+		
Filinia terminalis	+	+	+
Filinia sp.	+	+	
Gastropus sp.		+	
Hexarthra mira	+	+	+ + +
Keratella americana	+ + +	+	+ + +
Keratella cochlearis	+ + + +	+ + +	+ + +
Keratella lenzi	+	+	+
Keratella tropica	+ + +	+ + +	+ + +
Lecane cornuta	+		
Lecane dorysa	+		
Lecane sp.	+	+	+
Lepadella patella			+
Platyias patulus	+	+	+
Polyarthra vulgaris	+ + +	+ + + +	+ + +
Pompholyx complanata		+	+
Ptygura libera	+	+	+
Synchaeta pectinata		+	+
Synchaeta stylata	+	+ + +	+
Testudinella patina	+		
Trichocerca bicristata	+		
Trichocerca capucina	+	+ + +	+
Trichocerca elongata	+		
Trichocerca mus	+		
Trichocerca pusilla	+		
Trichocerca similis	+	+	+
Trichocerca sp.	+	+	
No. of species	32	25	21

+ – <1000 ind/m^3; + + + – 1000–10 000 ind/m^3; + + + + – >10 000 ind/m^3.

and diversity. Reservoirs are generally more complex than natural lakes, because of their interaction with the watershed and the influx of tributaries (Straskraba, 1997; Straskraba and Tundisi, 1999). Moreover, these reservoirs in southern Brazil are relatively shallow and, being polymictic, are subject to permanent mixing (Tundisi et al., 2004).

In a comparative study carried out in Barra Bonita Reservoir and Lake Dom Helvécio, a nat-

Table 4. Composition and abundance of microcrustacean species at the three localities (St1, St2 and St3) of Barra Bonita Reservoir

	Tietê River (St1)	Piracicaba River (St2)	Confluence (St3)
Cladocera			
Bosmina hagmanni	+	+ +	+ +
Bosmina tubicen		+ +	
Bosmina longirostris	+ + +	+ + + +	+ + +
Bosminopsis deitersi	+ +	+ + +	+ +
Ceriodaphnia cornuta	+ + +	+ + +	+ + + +
Ceriodaphnia silvestrii	+ + +	+ + +	+ + + +
Daphnia ambigua	+	+ +	+
Daphnia gessneri	+ + +	+ + +	+ +
Diaphanosoma birgei	+ + + +	+ + + +	+ + +
Diaphanosoma brevireme	+	+ +	+
Moina minuta	+ +	+ +	+ +
Moina micrura	+ +	+ +	+ +
Copepoda			
Notodiaptomus conifer	+ +		+ +
Notodiaptomus iheringi	+ + +	+	+ + + +
Thermocyclops minutus	+ +	+ + +	+ +
Thermocyclops decipiens	+ +	+ + + +	+ +
Mesocyclops ogunnus	+ +		+ + +
Mesocyclops longisetus	+		+
Mesocyclops meridianus		+ + +	+
Metacyclops mendocinus		+ + +	+
No. of species	17	17	19

+ – <100 ind/m^3; + + – 100–1000 ind/m^3; + + + – 1000–10 000 ind/m^3; + + + + – >10 000 ind/m^3.

ural warm monomictic lake in eastern Brazil, Tundisi and Matsumura-Tundisi (1994) showed that Barra Bonita has a greater number of limnetic zooplankton species (37 species: 20 Rotifera, 8 Cladocera, and 9 Copepoda) than does Lake Dom Helvécio (16 species: 6 Rotifera, 5 Cladocera, and 5 Copepoda). Despite the strong vertical gradients in water temperature and in the other physical and chemical variables, Dom Helvécio shows a less diverse community, probably because of its stability (8 months of thermal and chemical stratification). In another fifteen lakes of the same region, Tundisi et al. (1997) registered low plankton richness and, principally, low species numbers as well as a few abundance of rotifers in relation to microcrustaceans. In only a few lakes of this region a large abundance of rotifers (80–90% of total zooplankton) occurred. Barra Bonita, being a polymictic reservoir, is subject to permanent vertical mixing, and its 114 tributaries, both small (2–3 m^3/s discharge) and large (10–15 m^3/s), probably account for its higher diversity and species richness.

Changes in the plankton composition and diversity are observed frequently in reservoirs as eutrophication proceeds and planktonic species are substituted by other species adapted to the new conditions. Such changes in the last 20 years in the zooplankton in Barra Bonita Reservoir were demonstrated in a recent paper by Matsumura-Tundisi & Tundisi (2003). Despite these changes, species richness remained high. But it must also be taken into account that the sediment can contain a strong diversity component produced by resting zooplankton eggs. Thus, species richness at each station may also be influenced by resting egg development during periods of favorable water quality for each species. Development of and recruitment from resting eggs has been pointed out by Matsumura-Tundisi & Tundisi (2003) as a cause of shifts in species composition but not in

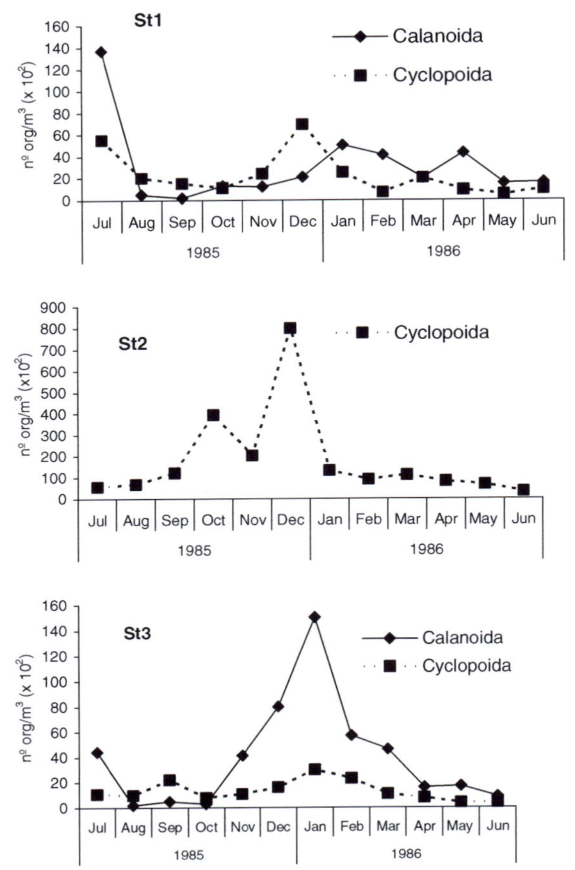

Figure 10. Abundance of Calanoida and Cyclopoida copepods at the three stations of Barra Bonita Reservoir.

species richness, as has been verified in a shallow reservoir (Lobo/Broa Reservoir) by Rietzler et al. (2002).

Barra Bonita Reservoir, besides being a dynamic system, has strong gradients in the limnological conditions and longitudinal processes along its two main tributaries: the Tietê and Piracicaba rivers. Spatially and temporally changing states promote gradients of conditions that are reflected in phytoplankton and zooplankton composition. In addition, physical and chemical changes interfere strongly with the dynamics of planktonic communities.

In this paper, it was shown that a very distinct pattern of planktonic community composition followed horizontal gradients. In reservoirs, these gradients are relatively common, as has already been shown extensively for other artificial aquatic ecosystems (Armengol et al., 1999; Straskraba, 1997; Straskraba & Tundisi, 1999).

The differing physical and chemical conditions in Barra Bonita Reservoir at each of the sampling stations is due to differential patterns of river circulation and flow. The gradients in this reservoir are typically controlled by mesoscale structures of 1.0–20.0 km (Tundisi et al., unpublished results). The existence of patches of phyto- and zooplankton, each with differing composition and properties, reflects an input of external energy consistent with the contemporaneous disequilibrium hypothesis set forth by Margalef (1967) and Reynolds (1997). According to this hypothesis the 'pelagic environment resembles a mosaic of microhabitats' (Legendre & Demers, 1984) where different species have distinct requirements resulting from environmental condition gradients. Margalef (1991) considers that the input of external energy breaks the regularity of the systems, leading them to decompose into horizontally heterogeneous units. The distribution of each 'cell' is related to the external energy input. In the Piracicaba and Tietê rivers, this external energy depends on the climatological and hydrological cycle, and on the input of nutrient and toxic substances promoting horizontal gradients. The many tributaries at Barra Bonita Reservoir promote a strong horizontal heterogeneity, increasing the horizontal discontinuities. Each tributary discharging into the reservoir produces a frontier of water masses which increases discontinuities and, therefore, high diversity of planktonic organisms. Thus, horizontal mixing and vertical instability augment diversity, which therefore boosts planktonic community responses over short periods and relatively small spatial scales.

The dominance of rotifers at all stations shows a pattern characteristic of reservoirs as has also been pointed out by Armengol et al. (1999). The rotifer dominance at Barra Bonita Reservoir has already been demonstrated by Matsumura-Tundisi et al. (1990) and is a consequence of an unstable and dynamic environment favoring the growth of *r*-strategists such as rotifers. As for rotifer diversity, it was higher for the Tietê River (St1) than for the Piracicaba (St2). However, for other groups such as cladocerans and copepods, there was no difference in number of limnectic species at the three stations. Station 2 is also characterized by the absence of Calanoida copepods and the presence of large-size species of Cyclopoida, e.g., *Metacyclops mendocinus* and *Mesocyclops meridi-*

anus. On the other hand, St1 (the Tietê) was characterized by the presence of *Notodiaptomus iheringi*, the smallest Calanoida species, and *Mesocyclops ogunnus*, smaller than any species of Cyclopoida present in the Piracicaba River (St2). The Tietê River is more eutrophic than the Piracicaba and the occurrence of Cyanophyta is more frequent and abundant. In this kind of environment, the zooplankton structure is composed of small-size organisms such as rotifers and small species of cladocerans and copepods that present a short-life cycle with energy spent on reproduction rather than growth. As shown by Rietzler et al. (2002), Notodiaptomus *iheringi* is an indicator of eutrophic systems, being a substitute of *Argyrodiaptomus furcatus*, a large species occurring in environments without Cyanophyta growth. *Notodiaptomus iheringi* feeds upon bacteria and free-floating small particles of detritus and phytoplankton $< 20\ \mu m$ size. Matsumura-Tundisi et al. (2002) related the zooplankton diversity in two localities of Billings Reservoir, a very eutrophic reservoir in the City of São Paulo, with periodic occurrence of Cyanophyta blooms. In the locality with frequent occurrence of Cyanophyta blooms the zooplankton structure was composed of small-size organisms and showed low diversity. In the other locality, without Cyanophyta blooms but presenting several species of Chlorophyta, zooplankton was composed by large species and showed high species richness. Thus, the zooplankton composition depends largely on phytoplankton composition.

In conclusion it seems clear from these results that, superimposed upon the eutrophication processes, horizontal gradients and vertical instability are strong factors enhancing plankton richness at Barra Bonita Reservoir, which can be an example of a more general pattern in other reservoirs showing the same characteristics. The high richness of planktonic species shown by this reservoir was probably due to its strong horizontal gradients which are promoted by the many tributaries flowing into the reservoir. Thus, horizontal mixing probably favors exploitation of new ecological niches by phytoplankton and zooplankton. In addition, vertical instability and horizontal variability may promote diversity, therefore enhancing responses of the planktonic community over short periods and at relatively small spatial scales. As stated by Tundisi & Matsumura-Tundisi (1994), changing temporal patterns of turbulence and intermediate disturbance factors are probably the forcing functions that provide Barra Bonita Reservoir with high richness of species in the pelagic environment.

Acknowledgements

This research is part of the Biota/FAPESP – Biodiversity Virtual Institute Program (www.biotasp.org.br). Process no 98/05091-2. The authors express thanks to FAPESP for financial support.

References

Armengol, J., J. C. Garcia, M. Comerma, M. Romero, J. Dolz, M. Roura, B. H. Han, A. Vidal & K. Simek, 1999. Longitudinal processes in canyon-type reservoirs: the case of Sal (N.E. Spain), In Tundisi, J. G. & M. Straskraba (eds), Theoretical Reservoir Ecology and its Applications, Braz. Acad. Science/Int. Inst. Ecology/Backhuys Publ., 585 pp.

Barbosa, F. A. R., J. Padsak, E. L .G. Espindola, G. Borics & O. Rocha, 1999. The cascading reservoir continuum concept (CRCC) and its application to the River Tietê Basin, São Paulo State, Brazil. In Theoretical Reservoir Ecology and its Applications, pp. 425–437.

Golterman, H. L., R. S. Clymo & M. A. M. Ohnstad, 1978. Methods for physical & chemical analysis of fresh waters. 2nd edn. Blackwell Scientific Publications, 213 pp.

Legendre, L. & S. Demers, 1984. Towards dynamic biological oceanography and limnology. Canadian Journal Fishery Aquatic Science, 41: 2–19.

Kimmel, B. L. & A. W. Groeger, 1984. Factors controlling primary production in lakes and reservoirs. A perspective. Lake and Reservoir Management. Proceedings of the Third Annual Conference, October 18–20. Knoxville, Tennessee, U.S. EPA, Washington, DC: 277–281.

Lewis, Jr. W. M., 1979. Zooplankton community analysis: studies on a tropical system. Springer-Verlag, New York, Inc: 163 pp.

Mackereth, F. J. H., J. Heron & J. F. Talling, 1978. Water analysis: some revised methods for limnologists. Freshwater Biological Association. Scientific Publication no. 36, 120 pp.

Maitland, P. S., 1978. Biology of fresh waters. London.

Margalef, R., 1991. Teoria de los Sistemas Ecológicos. Universitat de Barcelona Publ. 290 pp.

Margalef, R., 1967. Some concepts relative to the organization of plankton. Oceanography Marine Biology Annual Review 5: 257–289.

Matsumura-Tundisi, T., S. N. Leitão, L. S. Aguena & J. Miyahara, 1990. Eutrofização da represa de Barra Bonita: estrutura e organização da comunidade de Rotifera. Revista Brasileira de Biologia 50(4): 923–935.

Matsumura-Tundisi, T., J. G. Tundisi & O. Rocha, 2002. Zooplankton diversity in eutrophic systems and its relation to the occurrence of cyanophycean blooms. Verhandlungen Internationale Vereinigung Limnologie, 28: 671–674.

Matsumura-Tundisi & J. G. Tundisi, 2003. Calanoida (Copepoda) species composition changes in the reservoirs of São Paulo State (Brazil) in the last twenty years. Hydrobiologia 504: 215–222.

Patalas, K., 1957. Species composition of limnetic zooplankton communities. Limnology and Oceanography 2: 222–232.

Reynolds, S. C., 1997. Vegetation processes in the pelagic: a model for ecosystem theory. In Kinne, O. (ed.), Excellence in Ecology. Ecology Institute, D-21385 Oldendorf/Luhe, Germany, 371 pp.

Rietzler, A. C., T. Matsumura-Tundisi & J. G. Tundisi, 2002. Life cycle, feeding and adaptive strategy implications of the co-occurrence of *Argyrodiaptomus furcatus* and *Notodiaptomus iheringi* in Lobo-Broa Reservoir (SP, Brazil). Brazilian Journal of Biology 62(1): 93–105.

Straskraba, M., 1997. Limnological differences between reservoirs and lakes: management consequences. Proceedings: 7th International Conference on Lakes, Conservation and Management, San Martin de las Andes, Argentina.

Straskraba, M., J. G. Tundisi & A. Duncan (eds), 1993. Comparative Reservoir Limnology and Water Quality Management. Kluwer Academic Publishers, 291 pp.

Straskraba, M. & J. G. Tundisi, 1999. Reservoir ecosystem functioning: theory and application. In Tundisi, J. G. & M. Straskraba (eds), Theoretical Reservoir Ecology and its Applications, Brazilian Academy of Science/International Institute of Ecology/Backhuys Publishers, 585 pp.

Tundisi, J. G. & T. Matsumura-Tundisi, 1990. Limnology and eutrophication of Barra Bonita reservoir, S. Paulo State, Southern Brazil. Archiv für Hydrobiologie Ergebnisse der Limnologie 33: 661–676.

Tundisi, J. G. & T. Matsumura-Tundisi, 1994. Plankton diversity in a warm monomictic lake (Dom Helvecio, MG) and a polymictic reservoir (Barra Bonita, SP): a comparative analysis of the intermediate disturbance hypothesis. Anais da Academia brasileira de Ciências, 66: (Suppl. 1).

Tundisi, J. G., T. Matsumura-Tundisi, H. Fukuhara, O. Mitamura, S. M. Guillén, R. Henry, O. Rocha, M. C. Calijuri, M. S. R. Ibáñez, E. L. G. Espindola & S. Govoni, 1997. Limnology of fifteen lakes. In Tundisi, J. G. & Y. Saijo, (eds), Limnological Studies on the Rio Doce Valley Lakes, MG, Brazil. Brazialian Academy of Sciences/University of São Paulo, 513 pp.

Tundisi, J. G., O. Rocha, T. Matsumura-Tundisi & B. Braga, 1998. Reservoir management in South America. Water Resources Development 14(2): 141–155.

Tundisi, J. G. & M. Straskraba (eds), 1999. Theoretical Reservoir Ecology and its Applications. Brazilian Academy of Sciences/International Institute of Ecology/Buckhuys Publishers, 585 pp.

Tundisi, J. G., T. Matsumura-Tundisi, O. Rocha & E. L. G. Espindola, 2000. Limnology and integrated management of reservoirs in South America: recent developments and new perspectives. In Workshop on Dam Development and Environment. São Paulo, World Bank- University of São Paulo: 17–30.

Tundisi, J. G., T. Matsumura-Tundisi, J. D. Arantes Junior, J. E. M. Tundisi, N. F. Manzini & R. Ducrot, 2004. The response of Carlos Botelho (Lobo, Broa) Reservoir to the passage of cold fronts as reflected by physical, chemical and biological variables. Brazilian Journal of Biology 64(1): 177–186.

Hydrobiologia (2005) 542:379–390
H. Segers & K. Martens (eds), Aquatic Biodiversity II
DOI 10.1007/s10750-004-4941-9

Primary Research Paper

The effect of turbidity state and microhabitat on macroinvertebrate assemblages: a pilot study of six shallow lakes

Frank Van de Meutter*, Robby Stoks & Luc De Meester
Laboratory of Aquatic Ecology, Katholieke Universiteit Leuven, Ch. de Bériotstraat 32, B-3000 Leuven, Belgium
*(*Author for correspondence: Tel.: +32-16-32-38-57, Fax: +32-16-32-45-75,*
E-mail: frank.vandemeutter@bio.kuleuven.ac.be)

Key words: alternative stable states, macroinvertebrates, *Phragmites australis*, *Potamogeton pectinatus*, turbidity

Abstract

Shallow lakes can occur in two alternative stable states, a clear-water state and a turbid state. This is associated with separate assemblages of fish, zooplankton and plants. Little is known about whether macroinvertebrate assemblages differ across both stable states. This study investigated this in a connected set of three turbid and three clear-water shallow lakes. To overcome confounding effects of differences in spatial structure of macrophytes in turbid and clear-water lakes, we sampled three microhabitats that occurred in both alternative stable states: open water, sago pondweed (*Potamogeton pectinatus*) and reed (*Phragmites australis*). Univariate analyses indicated no differences in the number of organisms, taxon richness or diversity between turbid and clear-water lakes. Multivariate analysis, however, showed significant differences in the macroinvertebrate community structure of both stable states. Nine taxa explained a significant amount of the variation between both lake types, of which seven preferred the clear-water lakes. The number of organisms and the taxon richness were higher in reed than in the other microhabitats, but diversity and evenness did not differ among the microhabitats. Multivariate analyses could separate all three microhabitats. Eight taxa, mainly detritus feeders and collector–gatherers, explained most of the variation in the data and preferred the reed microhabitat. The effects of stable state (6.8% explained variance) and microhabitat (13.1% explained variance) on the macroinvertebrate assemblages were largely independent from each other (1.5% shared variance). Although macroinvertebrates are not implemented in the initial theory of stable states, our results show clearly different assemblages across both stable states.

Introduction

The structure and functioning of ecosystems can vary considerably across environmental gradients (e.g. hydroperiod: Wellborn et al., 1996; Stoks & McPeek, 2003; oxygen: Diaz & Rosenberg, 1995; nutrients: Jeppesen et al., 2003; Smith, 2003; ammonia: Krupa, 2003). The change in the structure and functioning of ecosystems, however, often does not parallel the change over the environmental gradient. Nonlinear behavior of ecosystems to a linear gradient in environmental factors, resulting in multiple ecosystem states, may be common in nature (Scheffer et al., 2001; Dent et al., 2002; McClanahan et al., 2002; Foley et al., 2003) and applies particularly well to shallow freshwater ecosystems (Scheffer et al., 1993; Hosper, 1998).

Shallow lakes can occur in two alternative stable states, depending on the nutrient load of the lakes: a clear-water state with abundant submersed vegetation at low nutrient loads and a turbid water state with high algal biomass at a high nutrient load (Scheffer et al., 1993; Scheffer, 1998). These alternative states have been found to be distinct attractors, each stabilized by multiple feedback mechanisms (Scheffer et al., 1993). The plankton and fish community of shallow lakes play a major role in the creation and stability of these alternative equilibria, and both exhibit a typical and well-studied community structure associated with each of the stable ecosystem states (Jeppesen et al., 2000; Cottenie et al., 2001). The epiphytic macroinvertebrate community was largely ignored in the initial model of Scheffer et al. (1993), but recent evidence suggests an important role for epiphytic macroinvertebrate periphyton grazers in maintaining the high macrophyte biomass of clear water lakes (Jones & Sayer, 2003). However, little is known on how the structure and dynamics of the epiphytic macroinvertebrate community vary with the alternative equilibria in shallow lakes.

Differences in food sources, predators and seasonal dynamics support a division of lentic macroinvertebrates in benthic and epiphytic macroinvertebrates (Diehl & Kornijów, 1998). The taxonomic diversity and density of epiphytic macroinvertebrates are related to (1) the density and growth form of the macrophytes they live on (Cyr & Downing, 1988; Lillie & Budd, 1992; Cattaneo et al., 1998; Taniguchi et al., 2003), (2) the abundance and functional diversity of the fish community (Osenberg & Mittelbach, 1989; Diehl, 1995) and their interaction with macrophytes (Crowder & Cooper, 1982; Diehl & Kornijów, 1998) and to a lesser extent (3) water turbidity (Timms & Boulton, 2001; Chessman, 2003). All these factors are directly linked with the theory of alternative stable states (Scheffer et al., 1993) and suggest that the epiphytic macroinvertebrate community may differ considerably between the alternative equilibria in shallow lakes.

This study investigates the epiphytic macroinvertebrate communities of three shallow lakes in the turbid state and three shallow lakes in the clear state. To minimize confounding variation from spatially related variables such as water quality, chemistry and dispersal, six neighboring and highly connected lakes were selected (Cottenie et al., 2001). This study focused on differences caused by turbidity state, without taking into account the effects of differences in macrophyte density and growth form between the alternative states that have been studied extensively before. Therefore, three microhabitat types were sampled that occurred in all lakes irrespective of its turbidity state: the open water, sago pondweed (*Potamogeton pectinatus*), and reed (*Phragmites australis*).

Material and methods

Study area

The study area is the nature reserve 'De Maten', a protected heath land of about 300 ha situated in the northeastern part of Belgium. The reserve comprises 34 shallow lakes, connected to each other by a complex network of rivulets and overflows (Cottenie & De Meester, 2001; Michels et al., 2001). The shallow lakes were created in the 15th century by peat-digging and the construction of dikes in the valleys of two parallel small streams. For centuries, the lakes were used for extensive fish farming, but this stopped in 1991.

Despite the high degree of connectivity, the shallow lakes display striking differences in turbidity, macrophyte cover and zooplankton community structure (Cottenie et al., 2001; Cottenie & De Meester, 2003). Based on a hierarchical clustering analysis of zooplankton data from all lakes in 'De Maten', Cottenie & De Meester (2001) classified the lakes into three different groups. This subdivision in lake groups based on zooplankton data was associated with striking differences in other biotic and abiotic factors such as turbidity, percentage submerged vegetation and fish density. According to the theory of alternative equilibria in shallow lakes (Scheffer et al., 1993), lakes with turbid water, a high fish density and sparse vegetation were classified as turbid water lakes; lakes with clear water, low fish densities and lush vegetation were classified as clear water lakes. A third group of lakes with intermediate characteristics exists in this system, but this will not be further discussed (Cottenie et al., 2001).

For this study, a cluster of six neighboring lakes was selected, three lakes (lakes 10, 11 and 12) of the turbid water group, and three lakes (lakes 13, 14 and 15) of the clear water group (Fig. 1). New environmental data collected during the

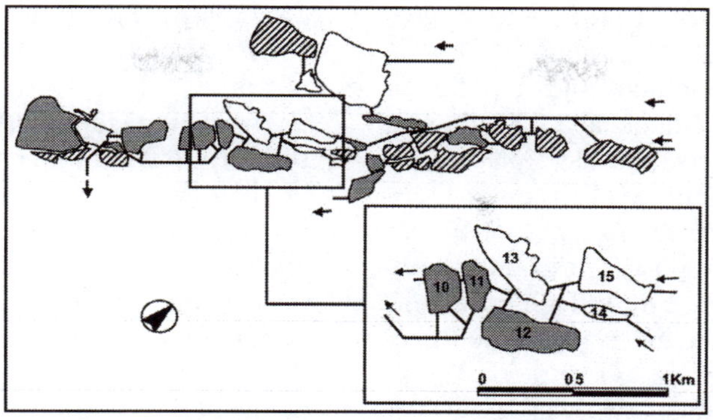

Figure 1. Map of the lake complex in the study area 'De Maten', with an enlarged detail of the six selected shallow lakes. Lakes are classified based on zooplankton community structure (Cottenie and De Meester 2001) as turbid water lakes (gray), intermediate lakes (shaded) or clear water lakes (white). Solid lines connecting lakes represent rivulets and overflows. Arrows indicate the directions of water flow.

sampling year 2001 confirmed the turbidity state of each lake (see Table 1).

Sampling

Samples were taken from 23 to 25 July 2001 in three different microhabitat types that occurred in both lake types: the open water zone, sago pondweed and reed. Each habitat type was sampled at three different localities within each lake. Reed was absent from lake 10 and sago pondweed could not be sampled in lakes 11 and 13. To take quantitative samples of macroinvertebrates, we used a large plastic sampling pipe (length 92 cm, diameter bottom 58 cm, diameter top 69.5 cm) that was open at both ends and that was placed vertically in the water column. To standardize the volume of water sampled, each sample was taken at a water depth of 80 cm. We selected sample localities with a standard vegetation density per pipe volume of approximately 7–10 g per pipe volume dry weight of sago pondweed, and a density of 15–20 stems of reed. The content of the sampling pipe was sampled with a net (mesh size 500 μm) and fixed on 4% formaldehyde.

Sample processing

In the laboratory, all invertebrates were sorted in white trays, identified and counted. Invertebrates were identified to the following taxonomic levels: Acari, Aranea, Asellidae, Chaoboridae,

Table 1. Turbidity state-related environmental variables for the six lakes measured on 23 July 2001

Lake	Turbidity state	Sneller depth (cm)	Chlorphyll a (μg l^{-1})	Submersed vegetation
10	Turbid	11	339.16	10
11	Turbid	11	172.36	0
12	Turbid	13	230.74	5
13	Clear	47	18.07	20
14	Clear	38	12.51	30
15	Clear	56	15.29	20

Sneller depth is the deepest point under water (in cm) at which a secchi disk, lowered in a gray PVC tube (8 cm diameter) filled with lake water, can be seen. 'Submersed vegetation' is the percentage cover of submersed macrophytes along a transect of 50 m parallel to the lake shore.

Chironomidae, Coleoptera, Coleoptera larvae, Culicidae, Diptera (comprising mainly Ceratopogonidae, Dixidae, Limoniidae and Tipulidae), Hirudinea, Lepidoptera, Oligochaeta, *Sialis* and Trichoptera. Ephemeroptera and Hemiptera were identified to species level and Mollusca and Odonata to genus level. Female *Sigara falleni* and *S. striata* water bugs were pooled into one group (*Sigara* females) since they are difficult to distinguish morphologically (Nieser, 1982). When needed, a stereomicroscope (Olympus, SZX-ILLK200) at 20× magnification was used for identification. The identification level for the different taxa is essentially a balance between the extra information obtained and the time that is needed for identification. Identification followed the key of De Pauw & Vannevel (1993) for most of the taxa and Gysels (1991) for Ephemeroptera, Nieser (1982) for Hemiptera and Heidemann & Seidenbush, (1993) and Norling & Sahlén (1997) for Odonata. For each sample, the following four characteristics were calculated selected: the number of organisms, the number of taxa (taxon richness) and the diversity and evenness measures of Shannon–Wiener (Shannon & Weaver, 1963).

Statistical analysis

We started with univariate analyses. Separate nested analyses of variance (Nested ANOVA) were used to identify differences between the different turbidity states and microhabitats for the four studied characteristics of macroinvertebrate assemblages. Turbidity state and microhabitat were included as fixed effects and lake nested in turbidity state was set as a random effect. The number of organisms and the diversity and evenness indices of Shannon–Wiener were log-transformed prior to analysis to meet the assumptions of ANOVA. Significant effects of the ANOVA were further explored with *post-hoc* Tukey HSD tests (Sokal & Rohlf, 1995).

For multivariate analysis, the abundances (ind./m^2) of all taxa were square-root transformed to minimize the effect of high densities (ter Braak & Smilauer, 2002). Detrended correspondence analysis (DCA) was used to explore patterns and compositional gradients in the macroinvertebrate assemblages of all samples (Hill & Gauch, 1980). To partition the variance in the macroinvertebrate assemblages into a turbidity state component and a microhabitat component, we used the procedure of Borcard et al. (1992) using Redundancy analysis (RDA). First, the RDA was performed with both turbidity state and microhabitat, as environmental predictors. This quantifies total variance explained by both variables. Second, separate partial RDA analyses were run with only turbidity state or microhabitat as the key variable, and the other variable as a covariable. This quantifies the conditional effects, i.e. the amount of variance explained by the variable that is not yet explained by the other variable, of both variables. To explicitly explore the variation in the macroinvertebrate assemblages associated with the a priori defined groups, two separate discriminant analyses (DA) were applied, one based on turbidity state and one based on microhabitat. A stepwise forward selection procedure was used to identify the taxa that best supported the subdivision by the environmental variables of the ordination model. Marginal rather than conditional effects of the taxa were calculated, to have a measure of the total amount of explained variance of each taxon.

The significance of the subdivision in clusters in the DA, of the variance explained by the selected taxa in the forward analysis and of the explained variance by the environmental predictors in the RDA, were determined by Monte Carlo randomization tests (9999 permutations). DCA, RDA and DA were performed in CANOCO 4.5 (ter Braak & Smilauer, 2002). Analysis of variance was performed using Statistica 6.0 (Statsoft 2001).

Results

Univariate analysis

On average 76 (SE: 11) organisms were collected in a sample. Samples of lakes with a different turbidity state did not support a different number of organisms ($F_{1,29} = 1.13$, $p = 0.346$) or different number of taxa ($F_{1,29} = 0.61$, $p = 0.479$). The mean number of organisms differed among microhabitats ($F_{2,29} = 13.15$, $p < 0.001$), and was significantly higher in reed than in the open water and sago pondweed (Fig. 2A). Taxon richness followed the same pattern and was also higher for reed samples than for open water and pondweed

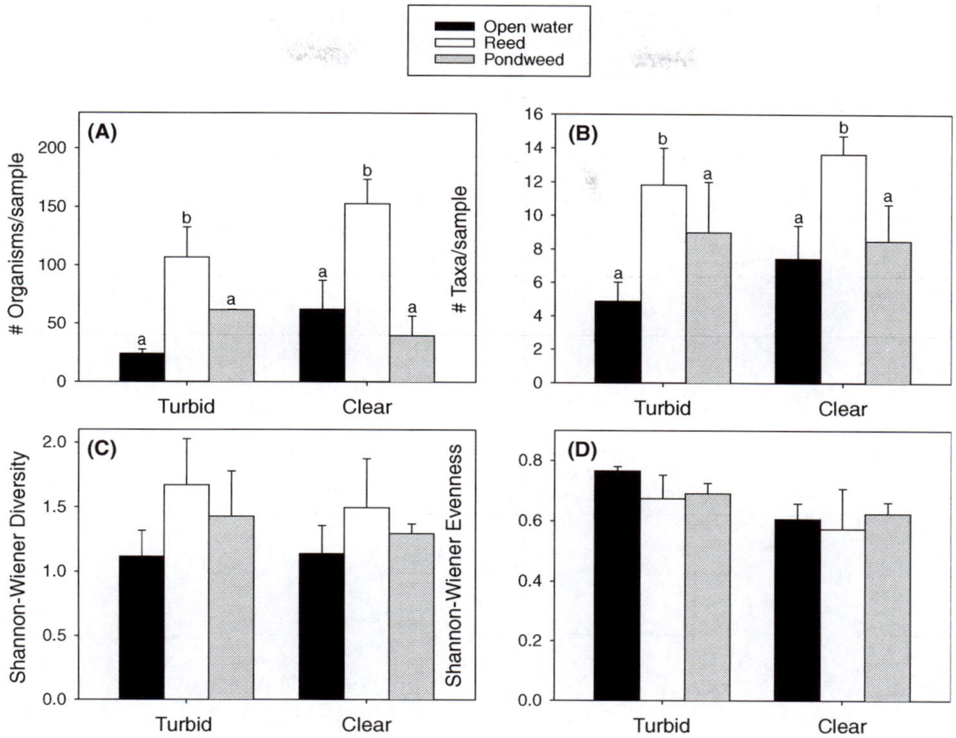

Figure 2. Characteristics of the macroinvertebrate assemblages in function of turbidity state and microhabitat. The following parameters are shown: (A) the number of organisms per sample, (B) taxon richness, (C) Shannon–Wiener diversity and (D) Shannon–Wiener evenness. When ANOVA indicated significant differences between the groups, each group differing in the *post-hoc* tests was given a different letter (a or b). Bars represent means (based on lake means) +1 standard error.

samples ($F_{2,29} = 14.37$, $p < 0.001$; Fig. 2B; see also Appendix 1 for and exhaustive list of taxon occurrence in the samples).

Lake turbidity and microhabitat state did not affect Shannon–Wiener diversity or evenness (all $p > 0.18$; Fig. 2C and D). Shannon–Wiener diversity displayed variable patterns over microhabitats depending on lake origin (lake (turbidity state) × microhabitat: $F_{5,29} = 5.20$, $p = 0.002$; Fig. 2C). Shannon evenness differed among lakes as a result of very low evenness values in lake 13 (lake(turbidity state): $F_{4,29} = 3.78$, $p = 0.011$; Tukey HSD lake 13 versus all other lakes all $p < 0.05$, other $p > 0.50$).

Multivariate analysis

The DCA ordination extracted two axes of maximum variation in the macroinvertebrate assemblages of all samples (Fig. 3). The first axis, explaining most of the variation (15.3%, eigenvalue 0.269) within the data, was only poorly correlated with the environmental variables of turbidity state and microhabitat. The second axis explained another 9.1% (eigenvalue 0.160). Along this axis, the DCA indicated a fairly distinct pattern within the macroinvertebrate samples according to turbidity state and between reed and the other microhabitats (Fig. 3), suggesting that both variables shape patterns in the macroinvertebrate assemblages.

RDA showed that Turbidity state and microhabitat together explained 21.4% of all variance in the data. Variation partitioning estimated the unique variance explained by turbidity state as 6.8% ($p = 0.006$) and by microhabitat as 13.1% of all variance ($p = 0.001$). Another 1.5% of the variance was shared between both variables.

DA on the macroinvertebrate samples with turbidity state as *a priori* grouping factor identified one well supported axis, along which both turbidity states separate ($p = 0.002$, Fig. 4). Nine taxa

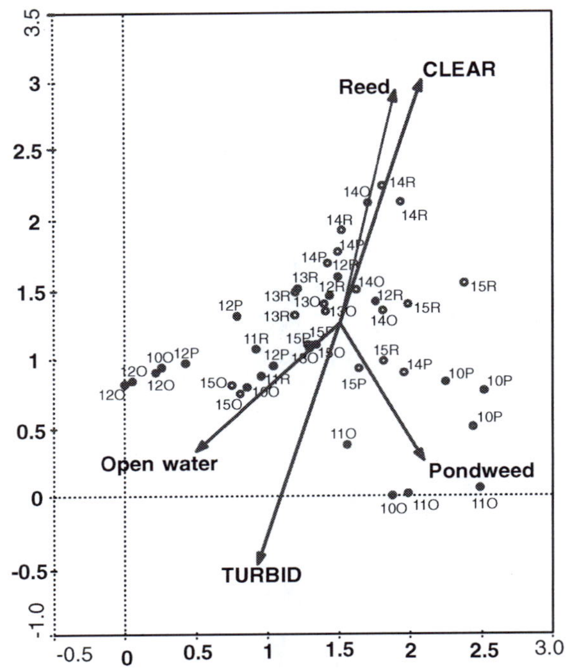

Figure 3. Biplot of the DCA-ordination of all samples. Superimposed on the biplot are the turbidity states and the microhabitats. Each circle represents a single sample; its code consists of pond number (see Table 1), followed by microhabitat (O: open water, R: reed, P: pondweed). Filled circles represent samples from lakes in the turbid state, open circles represent samples from lakes in the clear-water state.

explained a significant portion of the variance along this axis (Table 2). Apart from the water bug *Micronecta minutissima* and Oligochaeta, all of these taxa were more numerous in lakes of the clear water state.

The DA-ordination between the different microhabitats resulted in a subdivision of the reed group from the open water and pondweed group along axis 1 and the distinction of the open water (and reed) group from the pondweed group along axis 2 (Fig. 5). Both subdivisions were highly supported (axis 1: $p = 0.001$, axis 2: $p < 0.001$). Eight taxa explained a significant amount of variation between the three microhabitats (Table 3). All eight taxa had reed as the preferred microhabitat. Three of these taxa were also relatively abundant in other microhabitats than reed: females of the water bug *Sigara* in open water and the mayfly *Cloeon dipterum* and Chironomidae in pondweed.

Discussion

This study demonstrates that turbidity state and microhabitat generated structure in the macroinvertebrate assemblages of the studied shallow lakes. The effects of turbidity state and microhabitat were largely independent, as shown by their additive effects in the univariate analysis, and their low percentage of shared variation in the RDA. Assemblages of clear and turbid lakes were numerically and structurally similar, but were characterized by other taxa. Assemblages of different microhabitats were also structurally similar,

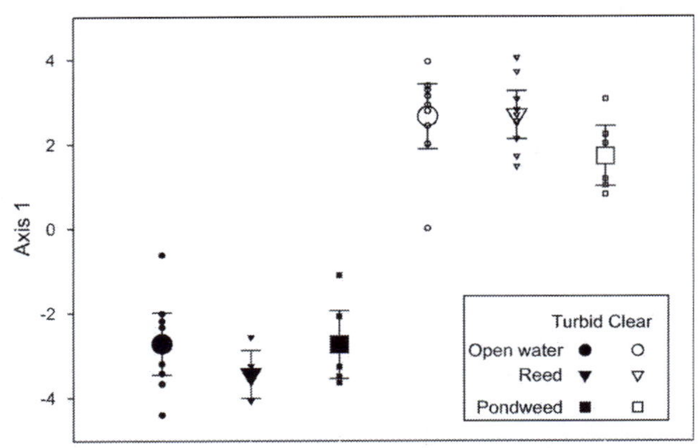

Figure 4. Scatterplot of the correlations with axis 1 of the DA based on turbidity state, for the different combinations of turbidity state and microhabitat. Small symbols represent individual samples. Large symbols represent the means (based on lake means) and are given ±2 standard errors.

Table 2. Results of the forward selection procedure on the DA with turbidity state as the a priori grouping factor

Taxon	Marginal explained variance	*p*	Correlation with axis 1
Chaoborus	0.23	0.002	0.476
Diptera	0.17	0.006	0.415
Caenis robusta	0.16	0.005	0.400
Chironomidae	0.13	0.012	0.361
Trichoptera	0.12	0.018	0.352
Gyraulus	0.12	0.012	0.347
Micronecta minutissima	0.11	0.036	−0.336
Oligochaeta	0.11	0.031	−0.330
Cloeon dipterum	0.1	0.042	0.301

Taxa that explain a significant portion of the variance between turbidity states are presented. *p*-Values were obtained with Monte Carlo randomization tests using 9999 permutations and test whether a taxon explains more than 0% of the variation between turbidity states. Positive and negative correlations of a taxon with DA axis 1 indicate a higher abundance in ponds of the clear and turbid water state, respectively (see Fig. 4).

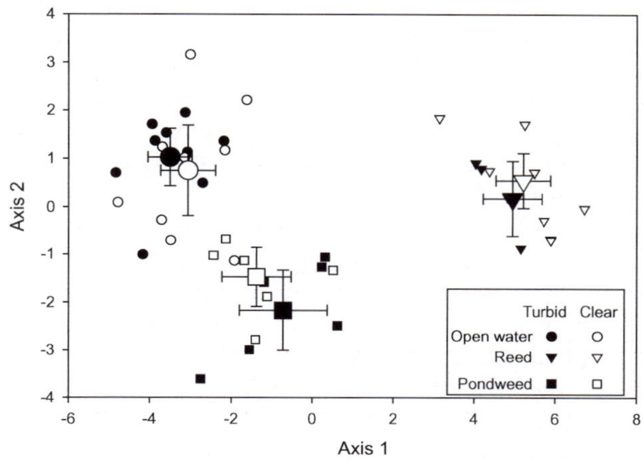

Figure 5. Scatterplot of the correlations of all samples with axis 1 and axis 2 of the DA-ordination based on microhabitat for each combination of turbidity state and microhabitat. Small symbols represent individual samples. Large symbols represent the means (based on lake means) and are given ±2 standard errors.

but differed numerically and in the presence of key taxa.

Although RDA suggested that microhabitat was a more important structuring force in macroinvertebrate assemblages than turbidity state, the effect of the latter may have been underestimated. First, the studied shallow lakes are highly connected by overflows and are situated closely to each other (Fig. 1; Michels et al., 2001). Preliminary results indicate high dispersal rates of macroinvertebrates between these lakes, both through the connecting overflows and through air (Van de Meutter, unpublished results). High dispersal rates could lead to source–sink dynamics (MacArthur & Wilson, 1967) or a 'rescue effect' (Brown & Kodric-Brown, 1977) between lakes of a different turbidity state. This may prevent some taxa from going extinct in their suboptimal habitat, making the distinction among lakes of a different turbidity state in the study system less pronounced than in more isolated habitats. Second, this study selected for similar microhabitats in different lakes types, not taking into account the fact that they differ in relative abundance according to lake type. Clear-water lakes usually have a considerably higher submerged macrophyte cover

Table 3. Results of the forward selection procedure on the DA with microhabitat as the *a priori* grouping factor

Taxon	Marginal explained variance	*p*	Correlation with	
			Axis 1	Axis 2
Caenis horaria	0.36	0.001	0.593	−0.083
Diptera	0.28	0.001	0.524	−0.065
Caenis robusta	0.27	0.002	0.517	−0.046
Chironomidae	0.25	0.003	0.452	−0.224
Ischnura	0.20	0.008	0.450	−0.041
Cloeon dipterum	0.18	0.012	0.316	−0.275
Sigara females	0.16	0.030	0.289	0.269
Coleoptera larvae	0.14	0.032	0.207	−0.086

Taxa that explain a significant portion of the variance between different microhabitats are presented. *p*-Values were obtained with Monte Carlo randomization tests using 9999 permutations. Positive correlations with axis 1 indicate higher abundances in reed; negative correlations with axis 2 indicate higher abundances in pondweed (see Fig. 5).

than turbid water lakes (Table 1; Scheffer et al., 1993). This should result in larger epiphytic macroinvertebrate communities with a higher number of taxa in clear-water than in turbid lakes (van den Berg et al., 1997; Piezcynska et al., 1998) as a result of species–area relationships (MacArthur & Wilson, 1967). In conclusion, these results do indicate that with a given volume of macrophyte-infested water, the two lake types did not differ in macroinvertebrate community estimates.

Despite the fact that macroinvertebrates may interact with the establishment and maintenance of alternative equilibria in shallow lakes (Jones & Sayer, 2003), studies that focus on macroinvertebrate community structure in lakes with a different turbidity state are still rare. Several studies have shown an effect of lake trophic status, which can be considered a rough proxy for turbidity state (Scheffer et al., 1993), on benthic macroinvertebrate communities (Kornijów, 1988; Brodersen et al., 1998; Brodersen et al., 2001; Tolonen et al., 2001). Similar studies on epiphytic macroinvertebrates are rare (but see van den Berg et al., 1997; Pieczynska et al., 1998). van den Berg et al. (1997) reported a change in macroinvertebrate community composition and density along with a turbidity-induced change in the dominating macrophyte species. They, however, did not disentangle the effect of changes in plant density and growth form, from that of turbidity state. The study by Pieczynska et al. (1998) found that macroinvertebrate assemblages and the number of macroinvertebrates retrieved from four species of submerged macrophytes, associated more closely with lake trophic state than with plant species.

Despite the distinct subdivision by DA of the macroinvertebrate assemblages between turbid and clear-water lakes, we could not demonstrate differences in the number of organisms, taxon richness or diversity between turbid and clear-water lakes. This suggests that the shift to the other turbidity state results in the substitution of individual taxa rather than in the net gain or loss of taxa per unit of volume. However, it should be noted that the use of a more detailed level of identification might reveal some community differentiation that remained hidden within the used higher-level taxonomic groups. Among the taxa that preferred the clear water state, *Chaoborus* exhibited the most clear-cut pattern. This contrasts with the studies that reported higher numbers of *Chaoborus* in turbid-water lakes than in clear-water lakes (Cuker, 1993; Jeppesen et al., 2003). The latter observation was explained by the lower predation pressure of visually hunting fish on *Chaoborus* due to reduced visibility with increasing turbidity (Gregory & Levings, 1998; Sweka & Hartman, 2003). Because *Chaoborus* species differ strikingly in their habitat preference (Berendonk et al., 2003), species differences among studies may explain the different pattern observed. The water bug *M. minutissima* and Oligochaeta were characteristic for the turbid water lakes. *M. minutissima* has been found to prefer lakes with elevated levels of particulate organic matter (Schmedtje & Colling, 1996). Oligochaeta are often found in elevated numbers in sediments that are rich in organic matter (Moroz, 1994).

The structuring role of microhabitats for the macroinvertebrate assemblages was mainly due to numerical and taxon differences between reed and the two other microhabitats. Reed housed the richest macroinvertebrate assemblage and was distinctly separated from the other habitats in the DA. Reed beds accumulate high amounts of detritus and macrophyte litter that may support a high number of detritus feeders and collector–gatherers (Varga, 2001; Warren et al., 2001). This was also reflected in the taxa that were found to prefer the reed microhabitat: the ephemeropterans *C. dipterum*, *C. horaria* and *C. robusta* apply a collector–gatherer feeding strategy (Elliott et al., 1988), and most of the Chironomidae genera and Diptera families (Ceratopogonidae, Dixidae, Limoniidae, Tipulidae) are detritus feeders (Schmedtje & Colling, 1996). Pondweed and open water had a comparable number of organisms, taxon richness and diversity, and differed mainly in the composition of their macroinvertebrate assemblages. This is surprising since the sampling method was biased towards epiphytic macroinvertebrates, which should be largely lacking in the open water. This unexpected similarity between pondweed and open water samples is likely due to small stands of *Chara* sp. (clear-water lakes) and filamentous algae (turbid lakes) that were unintentionally included in the sampling. Another explanation could be that the sampling procedure may have stirred up the upper sediment layer in open water samples, which is unprotected by vegetation or plant roots, to a higher extent than in samples in the vegetated microhabitats. However, the number and abundance of benthic taxa (Oligochaeta, Chironomidae) in open water samples did not differ from those in other samples, which invalidates the latter hypothesis.

The finding that macroinvertebrate assemblages differentiate between both stable turbidity states adds to the growing list of studies showing that assemblages of other organisms, like zooplankton and fish, clearly differentiate between both lake types (e.g. Cottenie & De Meester, 2001; Jeppesen et al., 2000). The major difference being that, according to the widely accepted model of Scheffer et al. (1993), the latter organisms are widely recognized to play a key role in maintaining the two stable states, while macroinvertebrates are not assumed to play such a role. Interestingly, Jones & Sayer (2003) provided data that support a model where fish are the prime determinants of community structure in shallow lakes, through a cascading effect of predation on grazing invertebrates influencing the biomass of periphyton and hence, plants. Therefore, the reported differentiation in macroinvertebrate assemblages may not just be a by-product of biotic and abiotic variables associated with the two stable states as suggested by the model of Scheffer et al. (1993), but may actually be part of the cascading mechanisms underlying the stable states. Another recent study, using mesocosms with three invertebrate species, indicated that predator–prey interactions among invertebrates, irrespective of fish, macrophytes and zooplankton, may also generate alternative stable community states in low and high productivity lakes (Chase, 2003). Clearly, our understanding of the key mechanisms structuring macroinvertebrate assemblages in shallow lakes is rapidly evolving. In our opinion, two types of studies are needed to allow further advances. There is a need for more field studies to evaluate the generality of discrete macroinvertebrate assemblages in turbid and clear-water shallow lakes, as reported here. Ideally, such studies should span several years to assess the stability of these assemblages. Next, mechanistic studies should clarify whether such assemblages are by-products or key components of stable equilibrium states of shallow lakes and whether or not their stability is purely maintained by intrinsic interactions.

Acknowledgements

FVDM is research assistant and RS postdoctoral fellow of the Fund for Scientific Research – Flanders (FWO Belgium). Naomi Baens is thanked for helping with the sampling, sorting and identification of the macroinvertebrates. This study was financially supported by project G.358.01 of the Fund for Scientific Research, Flanders. This study also benefited from EU project BIOMAN (EVK2-CT-1999-00046).

References

Berendonk, T. U., T. G. Barraclough & J. C. Barraclough, 2003. Phylogenetics of pond and lake lifestyles in *Chaoborus* midge larvae. Evolution 57: 2173–2178.

Borcard, D., P. Legendre & P. Drapeau, 1992. Partialling out the spatial component of ecological variation. Ecology 73: 1045–1055.

Brodersen, K. F., B. V. Odgaard, O. Vestergaard & N. J. Anderson, 2001. Chironomid stratigraphy in the shallow and eutrophic lake Sobygaard, Denmark: chironomid-macrophyte co-occurrence. Freshwater Biology 46: 253–267.

Brodersen, K. P., P. C. Dall & C. Lindegaard, 1998. The fauna in the upper stony littoral of Danish lakes: macroinvertebrates as trophic indicators. Freshwater Biology 39: 577–592.

Brown, J. H. & A. Kodric-Brown, 1977. Turnover rates in insular biogeography: effect of immigration on extinction. Ecology 58: 445–449.

Cattaneo, A., G. Galanti, S. Gentinetta & S. Romo, 1998. Epiphytic algae and macroinvertebrates on submerged and floating-leaved macrophytes in an Italian lake. Freshwater Biology 39: 725–740.

Chase, J. M., 2003. Experimental evidence for alternative stable equilibria in a benthic pond food web. Ecology Letters 6: 733–741.

Chessman, B. C., 2003. New sensitivity grades for Australian river macroinvertebrates. Marine and Freshwater Research 54: 95–103.

Cottenie, K., N. Nuytten, E. Michels & L. De Meester, 2001. Zooplankton community structure and environmental conditions in a set of interconnected ponds. Hydrobiologia 442: 339–350.

Cottenie, K. & L. De Meester, 2003. Connectivity and cladoceran species richness in a metacommunity of shallow lakes. Freshwater Biology 48: 823–832.

Crowder, L. B. & W. E. Cooper, 1982. Habitat structural complexity and the interaction between Bluegills and their prey. Ecology 63: 1802–1813.

Cuker, B. E., 1993. Suspended clays alter trophic interactions in the plankton. Ecology 74: 944–953.

Cyr, H. & J. A. Downing, 1988. Empirical relationships of phytomacrofaunal abundance to plant biomass and macrophyte bed characteristics. Canadian Journal of Fisheries and Aquatic Sciences 45: 976–984.

De Pauw, N. & R. Vannevel, 1993. Macroinvertebraten en waterkwaliteit. Stichting Leefmilieu, Antwerpen.

Dent, C. L., G. S. Cumming & S. R. Carpenter, 2002. Multiple states in river and lake ecosystems. Philosophical Transactions of the Royal Society of London Series B – Biological Sciences 357: 635–645.

Diaz, R. J. & R. Rosenberg, 1995. Marine benthic hypoxia: a review of its ecological effects and the behavioural responses of benthic macrofauna. Oceanography and Marine Biology 33: 245–303.

Diehl, S., 1995. Direct and indirect effects of omnivory in a littoral lake community. Ecology 76: 1727–1740.

Diehl, S. & R. Kornijów, 1998. Influence of submerged macrophytes on trophic interactions among fish and macroinvertebrates. In Jeppesen, E., Ma. Sondergaard & K. Christofferson (eds), The structuring role of submerged macrophytes in lakes. Springer-Verlag, New York: 24–45.

Elliot, J. M., U. H. Humpesch & T. T. Macan, 1988. Larvae of the British ephemeroptera: a key with ecological notes. Scientific Publications of the Freshwater Biological Association No. 49, Ambleside.

Foley, J. A., M. T. Coe, M. Scheffer & G. L. Wang, 2003. Regime shifts in the sahara and sahel: interactions between ecological and climatic systems in Northern Africa. Ecosystems 6: 524–539.

Gregory, R. S. & C. D. Levings, 1998. Turbidity reduces predation on migrating juvenile Pacific salmon. Transactions of the American Fisheries Society 127: 275–285.

Gysels, H., 1991. Haftenlarventabel: determineersleutel voor de larvale eendagsvliegen (Insecta Ephemeroptera) van het Benelux-gebied. Jeugdbondsuitgeverij, Utrecht.

Heidemann, H. & R. Seidenbusch, 1993. Die Libellenlarven Deutschlands und Frankreichs: Handbuch für Exuviensammler. Verlag Erna Bauer, Keltern.

Hill, M. O. & H. G. Gauch, 1980. Detrended correspondence-analysis – an improved ordination technique, Vegetatio 42: 47–58.

Hosper, S. H., 1998. Stable states, buffers and switches: an ecosystem approach to the restoration and management of shallow lakes in the Netherlands. Water Science and Technology 37: 151–164.

Jeppesen, E., J. P. Jensen, C. Jensen, B. Faafeng, D. O. Hessen, M. Sondergaard, T. Lauridsen, P. Brettum & K. Christoffersen, 2003. The impact of nutrient state and lake depth on top-down control in the pelagic zone of lakes: a study of 466 lakes from the temperate zone to the arctic. Ecosystems 6: 313–325.

Jeppesen, E., J. P. Jensen, M. Sondergaard, T. Lauridsen & F. Landkildehus, 2000. Trophic structure, species richness and biodiversity in Danish lakes: changes along a Phosphorus gradient. Freshwater Biology 45: 201–218.

Jones, J. I. & C. D. Sayer, 2003. Does the fish-invertebrate-periphyton cascade precipitate plant loss in shallow lakes? Ecology 84: 2155–2167.

Kornijów, R., 1988. Distribution of zoobenthos in littoral of two lakes differing in trophy. Polisch Archives of Hydrobiology 35: 185–195.

Krupa, S. V., 2003. Effects of atmospheric ammonia (NH_3) on terrestrial vegetation: a review. Environmental Pollution 124: 179–221.

Lillie, R. A. & J. Budd, 1992. Habitat architecture of *Myriophyllum spicatum* L. as an index to habitat quality for fish and macroinvertebrates. Journal of Freshwater Ecology 7: 113–125.

MacArthur, R. H. & Wilson E. O. 1967. The theory of island biogeography. Princeton University Press, Princeton, NJ, USA.

McClanahan, T., N. Polunin & T. Done, 2002. Ecological states and the resilience of coral reefs. Conservation Ecology 6: 18.

Michels, E., K. Cottenie, L. Neys, K. De Gelas, P. Coppin & L. De Meester, 2001. Geographical and genetic distances among zooplankton populations in a set of interconnected

ponds: a plea for using GIS modelling of the effective geographical distance. Molecular Ecology 10: 1929–1938.
Moroz, T. G., 1994. Aquatic oligochaeta of the dnieper-bug estuary system. Hydrobiologia 278: 133–138.
Nieser, N., 1982. De Nederlandse water- en oppervlaktewantsen (Heteroptera: Nepomorpha en Gerromorpha). Wetenschappelijke Mededelingen van de K.N.N.V. 155: 1–103.
Norling, U. & G. Sahlén, 1997. Odonata, dragonflies and damselflies. In Nilsson A. (ed.), The Aquatic Insects of North Europe 2. Apollo Books, Stenstrup; 13–65.
Osenberg, C. W. & G. G. Mittelbach, 1989. Effects of body size on the predator prey interaction between Pumpkinseed sunfish and gastropods. Ecological Monographs 59: 405–432.
Pieczynska, E., A. Kolodziejczyk & J. I. Rybak, 1998. The responses of littoral invertebrates to eutrophication-linked changes in plant communities. Hydrobiologia 391: 9–21.
Scheffer, M., 1998. Ecology of shallow lakes. Chapmann and Hall, London.
Scheffer, M., S. Carpenter, J. A. Foley, C. Folke & B. Walker, 2001. Catastrophic shifts in ecosystems. Nature 413: 591–596.
Scheffer, M., S. H. Hosper, M. L. Meijer, B. Moss & E. Jeppesen, 1993. Alternative equilibria in shallow lakes. Trends in Ecology and Evolution 8: 275–279.
Schmedtje, U. & M. Colling, 1996. Ökologische Typisierung der aquatischen Makrofauna. Informationsberichte des Bayer. Landesamtes für Wasserwirtschaft 4/96, München.
Shannon, C. E. & W. Weaver, 1963. The mathematical theory of communication. University of Illinois, Urbana.
Smith, V. H., 2003. Eutrophication of freshwater and coastal marine ecosystems – a global problem. Environmental Science and Pollution Research 10: 126–139.
Sokal, R. R. & F. J. Rolff, 1995. Biometry. Freeman, New York.
Statsoft. 2001. STATISTICA. In. Statsoft 6.
Stoks, R. & M. A. McPeek, 2003. Predators and life histories shape *Lestes* damselfly assemblages along a freshwater habitat gradient. Ecology 84: 1576–1587.
Sweka, J. A. & K. J. Hartman, 2003. Reduction of reactive distance and foraging success in Smallmouth Bass, *Micropterus dolomieu*, exposed to elevated turbidity levels. Environmental Biology of Fishes 67: 341–347.
Taniguchi, H., S. Nakano & M. Tokeshi, 2003. Influences of habitat complexity on the diversity and abundance of epiphytic invertebrates on plants. Freshwater Biology 48: 718–728.
ter Braak, C. J. F. & P. N. Smilauer, 2002. CANOCO reference manual and CanoDraw for Windows user's guide: software for canonical community ordination (version 4.5). Ithaca, New York.
Timms, B. V. & A. J. Boulton, 2001. Typology of arid-zone floodplain wetlands of the Paroo river (inland Australia) and the influence of water regime, turbidity, and salinity on their aquatic invertebrate assemblages. Archiv für Hydrobiologie 153: 1–27.
Tolonen, K. T., H. Hamalainen, I. J. Holopainen & J. Karjalainen, 2001. Influences of habitat type and environmental variables on littoral macroinvertebrate communities in a large lake system. Archiv für Hydrobiologie 152: 39–67.
van den Berg, M. S., H. Coops, R. Noordhuis, J. van Schie & J. Simons, 1997. Macroinvertebrate communities in relation to submerged vegetation in two *Chara*-dominated lakes. Hydrobiologia 342/343: 143–150.
Varga, I., 2001. Macroinvertebrates in reed litter. International Review of Hydrobiology 86: 573–583.
Warren, R. S., P. E. Fell, J. L. Grimsby, E. L. Buck, G. C. Rilling & R. A. Fertik, 2001. Rates, patterns, and impacts of *Phragmites australis* expansion and effects of experimental *Phragmites* control on vegetation, macroinvertebrates, and fish within tidelands of the lower Connecticut River. Estuaries 24: 90–107.
Wellborn, G. A., D. K. Skelly & E. E. Werner, 1996. Mechanisms creating community structure across a freshwater habitat gradient. Annual Review of Ecology and Systematics 27: 337–363.

Appendix 1. Frequency of occurrence of all taxonomic groups for every combination of microhabitat and pond number. Frequency ranges from 0 (present in none of the samples) to 3 (present in all of the samples). "*" indicates that only two samples were taken

Lake Type	Turbid							Clear							
Microhabitat	Open			Reed		Pondweed		Open			Reed			Pondweed	
Lake Nr.	10	11	12	11*	12	10	12	13	14	15	13	14	15	14	15
Oligochaeta	2	1	3	2	3	1	3	2	0	2	3	0	2	0	3
Hirudinea	2	0	1	2	0	0	1	0	0	0	0	1	1	0	0
Mollusca															
Gyraulus sp.	0	0	0	0	0	0	0	2	0	0	0	1	3	0	1
Lymnea sp.	0	0	1	0	0	1	0	1	1	0	1	1	2	1	1
Physa sp.	0	1	0	2	0	3	0	1	0	0	0	0	3	0	2
Sphaerium sp.	1	1	0	0	0	1	0	0	0	0	0	0	0	0	0
Acari	2	1	0	1	1	2	1	1	2	1	1	0	1	0	1
Aranea	0	0	0	0	1	0	0	0	0	0	0	1	0	0	0

Appendix 1. (Continued)

Lake Type	Turbid							Clear							
Microhabitat	Open			Reed		Pondweed		Open			Reed			Pondweed	
Lake Nr.	10	11	12	11*	12	10	12	13	14	15	13	14	15	14	15
Ephemeroptera															
Caenis horaria	1	0	0	2	3	3	2	3	3	0	3	3	3	3	3
Caen is robusta	0	0	0	0	2	1	1	1	2	0	3	3	3	1	2
Cloeon dipterum	0	0	0	2	1	0	0	3	0	0	3	3	2	1	1
Odonata															
Erythromma sp.	0	0	0	0	1	2	2	2	1	0	1	0	0	1	0
Ischnura sp.	0	0	0	2	3	3	0	2	3	0	2	3	1	2	0
Libellula sp.	0	0	0	0	0	1	0	0	1	0	1	2	0	0	0
Sympetrum sp.	0	0	0	0	0	0	0	1	0	0	2	1	0	0	0
Hemiptera															
Nepa cinerea	0	0	0	0	0	0	0	0	0	0	0	1	1	0	0
Micronecta meridionalis	1	0	0	2	0	0	1	0	0	0	0	0	0	0	0
Cymatia coleoptrata	2	1	0	2	1	1	1	0	0	0	0	1	1	0	1
Sigara falleni	1	1	0	1	0	0	0	1	0	0	1	0	1	0	1
Sigara striata	1	3	0	2	1	3	0	1	0	0	0	1	3	0	2
Sigara juveniles	2	2	0	0	3	3	0	3	2	3	3	1	3	1	3
Sigara females	1	3	0	2	1	3	1	0	0	0	1	1	3	1	2
Ilyocoris cimicoides	0	0	0	0	0	0	0	0	1	0	0	1	2	0	0
Diptera															
Chaoboridae	0	0	0	0	0	0	0	1	2	1	0	2	2	2	1
Chironomidae	3	3	3	2	3	2	3	3	3	3	3	3	3	3	3
Culicidae	0	0	0	0	0	0	0	0	0	0	0	1	1	0	0
Other Diptera families	0	0	0	2	3	1	0	1	2	1	3	2	2	2	3
Megaloptera															
Sialis sp.	0	0	0	0	0	0	1	0	0	0	1	0	0	0	1
Trichoptera	0	0	0	0	1	0	1	1	1	0	1	2	2	1	1
Coleoptera															
Adults	0	0	0	1	0	1	0	0	0	0	1	1	1	0	0
Juveniles	0	0	0	0	0	0	0	0	0	0	0	0	1	0	0

FSC
www.fsc.org
MIX
Papier | Fördert gute Waldnutzung
FSC® C083411

Kolibri 360 GmbH
Ferdinand-Jühlke-Straße 7
99095 Erfurt, Deutschland
produktsicherheit@kolibri360.de